Biosphere Origin and Evolution

Nikolay Dobretsov · Nikolay Kolchanov ·
Alexey Rozanov · Georgy Zavarzin
Editors

Biosphere Origin and Evolution

Editors

Nikolay Dobretsov
Inst. Geology, Mineralogy and
Geophysics, Institute of Geology and
Mineralogy, 3 Pr. Koptyuga Novosibirsk
630090, Novosibrisk, Russia

Nikolay Kolchanov
Institute of Cytology and Genetics
17 Lavrenteva Prospekt Novosibirsk
630090, Novosibirsk, Russia

Alexey Rozanov
Paleontological Institute of the Russian
Academy of Sciences, 123 Profsoyuznaya
Moskva 117647, Moscow, Russia

Georgy Zavarzin
Inst. Microbiology, Winogradsky
Institute of Microbiology of the Russian
Academy of Sciences, 7/2 Prospekt 60 Let
Oktyabrya Moskva 117312, Moscow,
Russia

Library of Congress Control Number: 2007935079

ISBN-13: 978-0-387-68655-4   e-ISBN-13: 978-0-387-68656-1

Printed on acid-free paper.

© 2008 Springer Science+Business Media, LLC

All rights reserved. This work may not be translated or copied in whole or in part without the written permission of the publisher (Springer Science+Business Media, LLC., 233 Spring Street, New York, NY 10013, USA), except for brief excerpts in connection with reviews or scholarly analysis. Use in connection with any form of information storage and retrieval, electronic adaptation, computer software, or by similar or dissimilar methodology now known or hereafter developed is forbidden. The use in this publication of trade names, trademarks, service marks, and similar terms, even if they are not identified as such, is not to be taken as an expression of opinion as to whether or not they are subject to proprietary rights.

9 8 7 6 5 4 3 2 1

springer.com

Preface

Modern natural science shows that the infancy of life on Earth experienced prebiotic evolution and included the emergence of primitive self-reproducing biologic forms and their systems. The subsequent coevolution of inorganic environment and biologic systems resulted in global propagation of life over the Earth and its enormous diversification. Diverse living organisms colonized the land, water, and atmosphere, as well as upper layers of the lithosphere, thereby forming the biosphere.

Formerly, it was thought that abiogenic synthesis of prebiotic matter occurred in the Earth's atmosphere on land surface. However, the presence of life signs in rocks more than 3.5 Gyr old suggests that the chemical matter evolution stage with synthesis of organic compounds from simple molecules is more likely to have occurred in the preglobal circumstellar disk together with the RNA world and the emergence of life itself. This notion removes the restriction imposed by the Earth's age and environmental conditions on the young Earth. It is favored by detection of intricate organic compounds on meteorites, comets, and (according to spectral data) in gas–dust nebulae. The detection of life traces in meteorite bodies of the same age as the Earth and the Solar System indicates that life in the latter can be older than the Earth as documented by geologic records.

The key features of living systems allowing their existence in time and space are self-reproduction and ability to evolve. The inseparability of these features was expressed by N.V. Timofeev-Ressovsky in the term "convariant reduplication." This means that the existence of a living system under constantly changing conditions demands that it should remember its own changes and stably reproduce them. At the same time, living systems should possess certain inertia, or rigidity, to prevent their dissolution in the environment. Two fundamentally different mechanisms perform stable reproduction in the nature. One of them involves control over the environment, and the result of this control is that formation of the system and its existence should occur in a certain way (top-down design). The other mechanism is execution of a program governing the development (bottom-up design). The former mechanism is of special importance for large and very large systems, which can influence the environmental conditions significantly and survive abrupt but short-term environmental fluctuations owing to their inertia. This mechanism is exemplified by geologic and geographic objects: intrusive bodies, ore bodies, and the ocean. In the living nature, this mechanism, greatly modified though, is implemented in biogeocenoses. The most prominent manifestation of this mechanism is observed in bacterial mats, which are of huge size in comparison with the size of their bacterial elements. However, the increase in size, supporting the stability of such systems, comes into conflict with their ability to change rapidly because of their inertia and the limited velocity of signal propagation. N. Wiener mentioned: "The community extends only so far as there extends an effectual transmission of information." As a result of growth in size, such a system should necessarily lose its lability. This fact determines its conservativeness and supports Lyell's uniformitarianism. As shown by G.A. Zavarzin, in living systems, the extreme conservativeness is demonstrated by prokaryotic communities, which have not changed significantly since the Archean.

Formation of a universal program coding for the mechanism of system development allows small systems to follow the way of stable reproduction. It is worth mentioning that the inorganic nature has analogs of self-reproducing systems (SR systems). A growing crystal is not living; however, it can time and again reproduce the shape of its growing surface. The crystal lattice is memory, which defines the structure of the subsequent layer of a growing crystal. Simultaneously, it performs some repair functions. An analog of mutation in this example is a change in the crystal lattice angle, recorded in the crystal structure. At the same time, the parameters of a growing crystal depend on the physical and chemical properties of the environment. Since classical crystals represent least free-energy systems, the limits of their variability can be calculated and possible shapes can be predicted. Once crystallization is under way, selection in the solution is in fact a competition among such shapes: the most stable shapes win and advance; no new shapes emerge.

Minerals that have crystal structures form part of live organisms as their skeletal, dermal and storage structures. Crystals have attracted the attention of experts in the origin of life for long, because their highly ordered structures could provide the base for abiogenic biopolymer synthesis, thereby giving an impetus to life emergence. However, live organisms represent systems that are in no equilibrium, and that is why interest is increasingly growing towards minerals like mixed layered mica or montmorillonite and organic crystalloids. They too are arranged into highly ordered structures capable of reproduction; however, unlike classical crystals, they represent systems that are in no equilibrium.

In contrast to crystals and crystalloids, which are defined as pure programs, the genome and phenotype are clearly recognized in living organisms. The genome is the medium for the program controlling the development of functional structures: organelles, organs, tissues, etc. The role of the phenotype is the reception of environmental signals. The environment action cannot alter the genetic program in a directional way, but it can ensure realization of a certain variant of its execution by changing gene expression regulation with a signal acting on gene network receptors. The segregation of functions between the genome and phenotype not only allowed flexible response of living organisms to environmental changes but also changed the evolution mode itself by distinguishing ontogeny (individual directional, or programmed, execution of a preformed program) and phylogeny (random process of alteration of this program by mutations, inherited in the succession of generations and tested by selection).

The evolution of a species is recorded in its genome, but it can be realized only via the phenotype. This is the phenotype that interacts with the environment and is ultimately tested by selection. The development of the phenotype most appropriate with regard to species survivability demands a certain stable environment. Starting from the beginning of biologic evolution, the divergence of living organisms was the cause of exponential sophistication of cooperative and competitive interactions between individual genetic factors within genomes and between living organisms. More and more econiches arose. In turn, colonization of new econiches in many taxa, although not in all, was accompanied by sophistication of organisms. The progressive sophistication occurred in different directions at different rates in different taxa. Random duplications of genes or gene elements provided a mechanism for this sophistication. In this way, genomes arose whose information capacity was sufficient for storing information necessary for reproduction and operation of complex life forms. Hereditary changes in the duplicated regions were accumulated under the action of natural selection and caused changes in the methods of genetic information storage, transmission, and realization, as well as changes in the related functions and differentiation lines of cells and organs. This process formed mechanisms for reliable reproduction of genetic information and fine and flexible regulation of the phenotypic manifestation of this information. The modular type of genome organization arose and developed. Random successive duplications and recombinations of modules supported an abrupt increase in the evolution rate and biodiversity.

Although the diversity of communities and taxonomic diversity are closely related, they were long studied separately by two different branches of science.

Construction of a generalized mathematical model of evolution became possible owing to the evolutionary synthesis theory. This theory was contributed by Morgan, Meller, Fisher, Wright, Chetverikov, Mayr, J. Huxley, Ohno et al. by the combination of Darwinism and classical population genetics. It absorbed Kimura's idea of the role of neutral mutations and notions by Goldschmidt, Shmalgauzen, and Waddington about the evolutionary significance of correlative linkages in ontogeny. The latter direction is now fruitfully developed as a part of the "evo-devo" concept.

The year 2006 saw the 80-year anniversary of the first publication of one of V. Vernadsky's most prominent book "The Biosphere," in which he presented the most complete version of his doctrine on "living matter" as an entity that includes all the organisms on our planet. The features of that "living matter" were treated as the features of a well-arranged integrated system, the biosphere, which is where the Earth's chemical elements travel the most and which is where energy exchange between the Earth and the Sun and the Cosmos is controlled from. What the Earth is as a celestial body, Vernadsky concluded, was shaped by life, which, upon emerging as a community of biocenoses and establishing a complex network of trophodynamic relationships, set the vectors of its own evolution and, in the long term, the vectors of the Earth's evolution. Thus, life provides for its own continuity in geographic space (the "film of life," the "living environment," *die Zebensphäre* after A. Humboldt; *die Biosphäre* after E. Suess) and time, keeping a global turnover of the chemical elements, biogens, going by enzymes catalysts (this point was first made by S. Vinogradsky). If it were not for the biogens, the Earth's surface would have been buried under "death's leftovers," that is, unrotten mortal remains. Modernly speaking, V. Vernadsky performed a synthesis of the ideas that had been proposed by his predecessors A. Humboldt, E. Suess, V. Dokuchaev (and certainly others) and came up with a research program that allows life to be treated as a phenomenon inseparable from its geological environment.

The great interest in biosphere studies presently expressed throughout the world is determined by not only trends in science but also the anxiety of the society about the consequences of the anthropogenic impact on global processes, including climate changes, biodiversity reduction, and general pollution. The fact that the society begins to perceive the depth of the problem is evidenced by the UN conferences on stable development in Rio de Janeiro in 1992 and in Johannesburg a decade later. The fundamental difference of civilization from natural processes, which determined the equilibrium between speciation and species extinction on Earth, is the tremendous growth of population. This factor, along with some others, will result in extinction of 40–60% of plant species in the nearest 20–30 years. This trend can and must be changed to save the whole biogeocenosis. This task demands comprehensive prediction of biogeocenosis development. Here, information on the remote past of our globe, reconstructed by geologists and paleontologists, comes to the aid. However, even possessing the precious knowledge of actual evolution, these sciences add little to understanding fine evolution mechanisms at the body, cell, and below-cell levels.

Recent numerous interdisciplinary studies of various taxa performed by methods of molecular phylogeny, in silico genomics, proteomics, cytogenetics, etc., and comparison of their results with data obtained by conventional methods of evolutionary morphology, paleontology, and branches of ecology revealed fundamental differences between evolution rates and modes at different biologic organization levels: genes, genomes, karyotypes, organisms, populations, and biogeocenoses. Thus, actual evolution cannot be reduced to evolution at any of the listed levels; it is an interference pattern, whose complexity increases with the number of interacting modules and hierarchical levels in the living systems and with the complexity of their relationships. Therefore, natural science demands a new synthesis.

This book covers notions by scientists of various branches on the evolutionary relationship between the biosphere and the geosphere, evolution features at various levels of living matter organization, and problems of prebiotic evolution and life origin. The data were collected in the course of the RAS program "Biosphere origin and evolution" (subprogram II) in 2003–2006. The objectives of this subprogram were (1) generalization of data related to problems of biosphere origin and evolution accumulated by geneticists, molecular biologists, zoologists, botanists, paleontologists, microbiologists, geologists, chemists, and archaeologists; (2) search for new interdisciplinary approaches to biosphere origin and evolution; (3) development of a "lingua franca" understandable by experts in various fields, which would allow apprehension of results concerning the topic obtained in allied sciences. The same objectives were pursued by sponsors of the conference of the same name held in Novosibirsk in the summer of 2005. A considerable portion of proceedings of this conference is present in this book.

Of course, it is impossible to compose a book that would consider all essential problems related to such a complicated topic as biosphere origin and evolution. Our aim was to clear the barriers of specialism in life sciences by providing an opportunity for intercommunication among experts in various fields. Another task of no lesser significance was the opportunity to discuss the efficiency of different methods applied to same problems, to find points of contact and differences in application of different approaches and/or result interpretation. Thus, the goal of this book can be expressed as preparation of a base for construction of an integrated notion of the biosphere, modes and mechanisms of its development, symptoms of global environmental crises, and, in a sense, the position of man in the biosphere. If this book is of assistance for advance in this field, the conference sponsors will account their mission completed.

The editors are grateful to V.V. Suslov, V.N. Snytnikov, N.Yu. Sournina, and O.P. Stoyanovskaya, whose constant effort made the publication possible.

Contents

Part I
Problems of Biosphere Evolution and Origin of Life

On Important Stages of Geosphere and Biosphere Evolution 3
Dobretsov N.L., Kolchanov N.A., Suslov V.V.

Microbial Biosphere 25
Zavarzin G.A.

Part II
Prebiological Stages of Evolution and RNA World on the Earth and in the Space

Astrocatalysis Hypothesis for Origin of Life Problem 45
Snytnikov V.N.

Comets, Carbonaceous Meteorites, and the Origin
of the Biosphere. 55
Hoover R.B.

Hierarchical Scale-Free Representation of Biological Realm—
Its Origin and Evolution. 69
Zhuravlev Yu.N., Avetisov V.A.

The Prebiotic Phase of the Origin of Life as Seen
by a Physical Chemist. 89
Parmon V.N.

Prebiotic Carbohydrates and Their Derivates 103
*Pestunova O.P., Simonov A.N., Snytnikov V.N.,
Parmon V.N.*

Theoretical and Computer Modeling of Evolution of Autocatalytic Systems in a Flow Reactor 119
Bartsev S.I., Mezhevikin V.V.

RNA World: First Steps Towards Functional Molecules 131
Lutay A.V, Zenkova M.A., Vlassov V.V.

***Trans* Hammerhead Ribozyme: Ligation vs. Cleavage** 143
Vorobjeva M.A., Privalova A.S., Venyaminova A.G., Vlassov V.V.

Paradoxical Bistate Status of a Prebiotic Microsystem: Universal Predecessor of Life 157
Kompanichenko V.N.

Part III
Archaen–Proterozoic Ecosystems: Their Interaction and Contemporary Analogous

The Ancient Anoxic Biosphere Was Not As We Know It 169
Fallick A.E., Melezhik V.A., Simonson B.M.

Evolutionary Aspects of Geochemical Activity of Microbial Mats in Lakes and Hydrotherms of Baikal Rift Zone 189
Namsaraev B.B., Gorlenko V.M., Namsaraev Z.B, Barkhutova D.D., Kozyreva L.P., Dagurova O.P., Tatarinov A.V.

On the concept for the Organization of the Modern Biosphere in the Terrestrial Subsurface 203
Oborin A.A., Rubinstein L.M., Khmurchik V.T.

Biomineralization and Evolution. Coevolution of Mineral and Biological Worlds 211
Barskov I.S.

Visualization of the Silicon Biomineralization In Cyanobacteria, Sponges and Diatoms 219
Likhoshway Ye.V., Sorokovikova E.G., Belykh O.I., Kaluzhnaya O.L.V., Belikov S.I., Bedoshvili Ye.D., Kaluzhnaya O.K.V., Masyukova Ju.A., Sherbakova T.A.

Transformational Changes in Argillaceous Minerals due to Cyanobacteria 231
Alekseeva T.V., Gerasimenko L.M., Sapova E.V., Alekseev A.O.

Part IV
Coevolution of Geological and Biological Events in Phanerozoe

Ecological Revolution Through Ordovician Biosphere (495 to 435 Ma ages): Start of the Coherent Life Evolution 245
Kanygin A.V.

Part V
Ecosystems and Molecular Genetic Factors of Organism Evolution

Evolution by Gene Duplications: from the Origin of the Genetic Code to the Human Genome 257
Rodin S.N., Rodin A.S.

Evolution of the Translation Termination System in Eukaryotes 277
Zhouravleva G.A., Tarasov O.V., Schepachev V.V., Moskalenko S.E., Abramson N.I., Inge-Vechtomov S.G..

The Hedgehog Signaling Cascade System: Evolution and Functional Dynamics 289
Gunbin K.V., Afonnikov D.A., Omelyanchuk L.V. and Kolchanov N.A

Approaches to the Resolution of Contradictions Between Phylogenetic Systems Based on Paleontological and Molecular Data 303
Rautian G.S., Rautian A.S., Kalandadze N.N.

Chromosomes and Speciation 315
Borodin P.M.

Biotic Turnover in Superorganism Systems: Several Principles of Establishment and Sustenance (Theoretical Analysis, Debatable Issues)... 327
Gubanov V.G. and Degermendzhy A.G.

Chromosomes and Continents 349
Kiknadze I.I., Gunderina L.I., Butler M.G., Wuelker W.F., Martin J.

Part VI
Biosphere and Human Being

Genetic Landscape of the Central Asia and Volga-Ural Region 373
Khusnutdinova E.K., Bermisheva M.A., Kutuev I.A., Yunusbayev B.B., Villems R.

Problems of Reconstruction of Paleoenvironment and Conditions of the Habitability of the Ancient Man by the Example of Northwestern Altai..................................... 383
Agadjanian A.K.

The Settling of the Ancient Man by the Example of North-Western Altai 395
Derevianko A.P., Shunkov M.V.

Evolutionary History of Wheats — the Main Cereal of Mankind 407
Goncharov N.P., Golovnina K.A., Kilian B., Glushkov S., Blinov A., Shumny V.K.

Subject Index 421

Contributors

N.I. Abramson
Institute of Zoology,
Russian Academy of Sciences,
199034, St. Petersburg,
Russia,
natalia_abramson@yahoo.com

D.A. Afonnikov
Institute of Cytology
and Genetics SB RAS,
Lavrentyev Ave. 10,
Novosibirsk, 630090,
Russia

A.K. Agadjanian
Paleontological Institute,
Russian Academy of Sciences,
aagadj@paleo.ru

A.O. Alekseev
Institute Physical,
Chemical and Biological Problems
of Soil Science RAS,
Pushchino

T.V. Alekseeva
Institute Physical,
Chemical and Biological Problems
of Soil Science RAS,
Pushchino,
alekseeva@issp.serpukhov.su

V.A. Avetisov
The Semenov Institute of Chemical
Physics,
Russian Academy of Sciences,
Moscow, Russia

D.D. Barkhutova
Institute of General
and Experimental Biology,
Siberian Branch of Russian
Academy of Sciences and Buryat
State University

I.S. Barskov
Paleontological Institute of the
Russian Academy of Science,
barskov@paleo.ru

S.I. Bartsev
Institute of Biophysics SB RAS,
Theoretical Biophysics Department,
bartsev@yandex.ru

Ye.D. Bedoshvili
S.I. Belikov

O.I. Belykh

M.A. Bermisheva
Institute of Biochemistry
and Genetics,
Department of Genomics

A. Blinov
Institute of Cytology and Genetics,
Siberian Branch of the Russian
Academy of Sciences,
10 Lavrentyev Avenue,
Novosibirsk, Russia

P.M. Borodin
Institute of Cytology and Genetics,
Siberian Department of Russian
Academy of Science,
10 Lavrentiev Avenue,
Novosibirsk 630090,
Russia
borodin@bionet.nsc.ru

M.G. Butler
University of North Dakota,
Fargo, ND

O.P. Dagurova
Institute of General
and Experimental Biology,
Siberian Branch of Russian
Academy of Sciences

A.G. Degermendzhy
Institute of Biophysics SB RAS,
Krasnoyarsk, Russia,
ibp@ibp.ru

A.P. Derevianko
Institute of Archaeology
and Ethnography,
Siberian Branch,
Russian Academy of Sciences,
derev@archaeology.nsc.ru

N.L. Dobretsov
Institute of Geology
Mineralogy,
SB, RAS,
Novosibirsk, Russia
arkair@uiggm.nsc.ru

A.E. Fallick
Scottish Universities Environmental
Research Centre,
Glasgow G75 0QF,
Scotland,
T.Fallick@suerc.gla.ac.uk

L.M. Gerasimenko
Institute of Microbiology RAS,
Moscow,
l_gerasimenko@mail.ru

S. Glushkov
Institute of Cytology and Genetics,
Siberian Branch of the Russian
Academy of Sciences,
10 Lavrentyev Avenue,
Novosibirsk, Russia

K.A. Golovnina
Institute of Cytology and Genetics,
Siberian Branch of the Russian
Academy of Sciences,
10 Lavrentyev Avenue,
Novosibirsk, Russia

N.P. Goncharov
Institute of Cytology and Genetics,
Siberian Branch of the Russian
Academy of Sciences,
10 Lavrentyev Avenue,
Novosibirsk, Russia
gonch@bionet.nsc.ru

V.M. Gorlenko
Winogradsky Institute
of Microbiology,
Russian Academy of Sciences,
vgorlenko@mail.ru

V.G. Gubanov
Institute of Biophysics SB RAS,
Krasnoyarsk, Russia
guban@ibp.ru

Contributors

K.V. Gunbin
Institute of Cytology and Genetics
SB RAS,
Lavrentyev Ave. 10,
Novosibirsk, 630090,
Russia,
genkvg@bionet.nsc.ru

L.I. Gunderina
Institute of Cytology and Genetics,
SB, RAS,
Novosibirsk, Russia
gund@bionet.nsc.ru

R.B. Hoover
Astrobiology Laboratory,
NASA/NSSTC,
Huntsville, AL
Richard.Hoover@NASA.GOV

S.G. Inge-Vechtomov
Department of Genetics
and Breeding,
St. Petersburg State University and St. Petersburg Branch,
Vavilov Institute of General Genetics,
Russian Academy of Sciences,
199034, St. Petersburg,
Russia
inge@mail333.ru

N.N. Kalandadze
Paleontological Institute,
Russian Academy of Sciences,

O.V. Kaluzhnaya

A.V. Kanygin
Trofimuk Institute of Petroleum
Geology and Geophysics,
Novosibirsk, Russia
fkanyginav@ipgg.nsc.ru

V.T. Khmurchik
Institute of Ecology and Genetics
of Microorganisms of Russian
Academy of Sciences,
khmurchik@iegm.ru

E.K. Khusnutdinova
Institute of Biochemistry
and Genetics,
Department of Genomics,
ekkh@anrb.ru

I.I. Kiknadze
Institute of Cytology and Genetics,
SB, RAS,
Novosibirsk, Russia
kiknadze@bionet.nsc.ru

B. Kilian
Max-Planck-Institut für
Züchtungsforschung,
Department of Plant Breeding
and Genetics,
Germany

N.A. Kolchanov
Institute of Cytology and Genetics,
SB RAS,
Lavrentyev Ave. 10,
Novosibirsk, 630090,
Russia
kol@bionet.nsc.ru

V.N. Kompanichenko
Institute for Complex Analysis
at Birobidzhan,
Russia and University of California
at Santa Cruz,
Department of Chemistry and
Biochemistry,
USA
kompanv@yandex.ru,
vladk@soe.ucsc.edu

L.P. Kozyreva
Institute of General
and Experimental
Biology, Siberian Branch
of Russian Academy
of Sciences

I.A. Kutuev
Institute of Biochemistry
and Genetics,
Department of Genomics

Ye.V. Likhoshway
Limnological Institute
SB RAS, Irkutsk,
Russia,
Ulan-Batorslaya st, 5, 664033

A.V. Lutay
Institute of Chemical Biology
and Fundamental Medicine
SB RAS, Novosibirsk,
Russia
lutay_av@niboch.nsc.ru

J. Martin
University of Melbourne,
Victoria, Australia

Ju. A. Masyukova

V.A. Melezhik
Geological Survey of Norway,
N-7491 Trondheim,
Norway
victor.melezhik@ngu.no

V.V. Mezhevikin
Institute of Biophysics SB RAS,
Bacterial Bioluminescence
Department,
vlad_me@akadem.ru

S.E. Moskalenko

B.B. Namsaraev
Institute of General
and Experimental Biology,
Siberian Branch of Russian
Academy of Sciences and Buryat
State University,
bair_n@mail.ru

Z.B. Namsaraev
Winogradsky Institute
of Microbiology,
Russian Academy of Sciences

A.A. Oborin
Institute of Ecology and Genetics
of Microorganisms of Russian
Academy of Sciences

L.V. Omelyanchuk
Institute of Cytology and Genetics SB
RAS,
Lavrentyev Ave. 10
Novosibirsk, 630090, Russia

V.N. Parmon
Boreskov Institute of Catalysis,
Novosibirsk, Russia
Parmon@catalysis.ru

O.P. Pestunova
Department of Nontraditional
Catalytic Processes,
Boreskov Institute of Catalysis
and Department of Natural Sciences,
Novosibirsk State University
oxanap@catalysis.ru

A.S. Privalova
Institute of Chemical Biology and
Fundamental Medicine,
Novosibirsk, Russia
privalova.anna@gmail.com

Contributors

A.S. Rautian
Paleontological Institute,
Russian Academy of Sciences

G.S. Rautian
Paleontological Institute,
Russian Academy of Sciences
gsrautrian@mtu-net.ru

A.S. Rodin
School of Public Health,
University of Texas,
Houston, TX
andrei.s.rodin@uth.tmc.edu

S.N. Rodin
Beckman Research Institute of the
City of Hope,
Duarte CA
srodin@coh.org

L.M. Rubinstein
Institute of Ecology and Genetics
of Microorganisms of Russian
Academy of Sciences

V.V. Schepachev
Department of Genetics and
Breeding,
St. Petersburg State University,
199034, St. Petersburg
Russia

E.V. Sapova
Institute of Microbiology RAS,
Moscow

T.A. Sherbakova

V.K. Shumny
Institute of Cytology and Genetics,
Siberian Branch of the Russian
Academy of Sciences,
10 Lavrentyev Avenue,
Novosibirsk, Russia

M.V. Shunkov
Institute of Archaeology
and Ethnography,
Siberian Branch,
Russian Academy of Sciences,
shunkov@archaeology.nsc.ru

A.N. Simonov
Department of Nontraditional
Catalytic Processes,
Boreskov Institute of Catalysis
and Department of Natural Sciences,
Novosibirsk State University

B.M. Simonson
Oberlin College,
Bruce.Simonson@oberlin.edu

V.N. Snytnikov
Department of Nontraditional
Catalytic Processes,
Boreskov Institute of Catalysis
and Department of Natural Sciences,
Novosibirsk State University

E.G. Sorokovikova

V.V. Suslov
Institute of Cytology and Genetics
SB, RAS,
Novosibirsk, Russia
valya@bionet.nsc.ru

O.V. Tarasov
Department of Genetics
and Breeding,
St. Petersburg State University,
199034, St. Petersburg,
Russia

A.V. Tatarinov
Geological Institute,
Siberian Branch of Russian
Academy of Sciences,
gin@bsc.buryatia.ru

A.G. Venyaminova
Institute of Chemical Biology
and Fundamental Medicine,
Novosibirsk, Russia
ven@niboch.nsc.ru

R. Villems
Estonian Biocentre,
Department of Evolutionary Biology,
rvillems@ebc.ee

V.V. Vlassov
Institute of Chemical Biology
and Fundamental Medicine,
Novosibirsk, Russia
valentin.vlassov@niboch.nsc.ru

M.A. Vorobjeva
Institute of Chemical Biology
and Fundamental Medicine,
Novosibirsk, Russia
kuzn@niboch.nsc.ru

W.F. Wuelker
University of Freiburg,
Freiburg, Germany

B.B. Yunusbayev
Institute of Biochemistry
and Genetics,
Department of Genomics

G.A. Zavarzin
Winogradsky Institute
of Microbiology,
Russian Academy of Sciences,
Moscow, Russia
zavarzin@inmi.host.ru

M.A. Zenkova
Institute of Chemical Biology
and Fundamental Medicine SB RAS,
Novosibirsk, Russia

G.A. Zhouravleva
Department of Genetics
and Breeding,
St. Petersburg State University,
199034, St. Petersburg,
Russia
zhouravleva@rambler.ru

Yu.N. Zhuravlev
Institute of Biology and Soil Science,
Russian Academy of Sciences,
Far Eastern Branch,
Vladivostok, Russia
zhuravlev@ibss.dvo.ru

Part I
Problems of Biosphere Evolution and Origin of Life

On Important Stages of Geosphere and Biosphere Evolution

N. L. Dobretsov, N. A. Kolchanov, and V. V. Suslov

Abstract The necessary conditions for the existence of protein–nucleic acid life are the presence of liquid water, some protection against high-amplitude temperature jumps and cosmic factors (these may be the atmosphere and or a thick layer of water or same rocks) and the accessibility of biogenes, which are macroelements and microelements. Two geosphere-related canalizing vectors of biosphere evolution can be discerned. One is associated with an irreversible cooling and oxygenation of the planet and the associated complex pattern of interplaying endogenous cycles, which affect climates as well as the amount and composition of the biogenes in the "liquid water zone." Change of the convection mode in the mantle between 3 and 2 Byr ago had the most important implications for the biosphere: the formation of plate tectonics (a deep ocean and continents), enrichment of the chemical composition of the effusive material and the "plume dropper," which changes the oceanic-to-continental area ratio and the mantle-to-island-arc volcanism intensity ratio every 30 Myr. The World Ocean operates as a homeostatic system: it tempers climates, distributes biogene concentrations evenly over the globe and provides the hydrosphere with direct biogene supply from the mantle, which is how the second vector of biosphere evolution is set. Life is a homeostatic system too—not due to a tremendously high buffer's capacity, but due to high rates of chemical reactions and a special program (the genome), which warrants autonomy from the environment. Reduction in methane concentrations and increase in atmospheric O_2 in the course of the Earth's geological evolution caused the extinction of chemotrophic ecosystems. Autotrophic photosynthesis provided the biosphere with a source of energy that was not associated with the geosphere and helped the biosphere for the first time to gain independence (autonomization) from the geosphere. As a result, the biosphere develops a solid film of life spread out over the continents, pelagic and abyssal zones, and the geosphere

---

N. L. Dobretsov
Institute of Geology and Mineralogy, SB, RAS, Novosibirsk, Russia
e-mail: arkair@uiggm.nsc.ru

supplemented its geochemical cycles with biogeochemical ones which are comparable, if not by the mass of the matter involved, by annual balance.

The necessary condition for the existence of DNA/RNA/protein-based life is the presence of liquid water, an atmosphere and the accessibility of biogenes: macroelements (O, C, H, N, Ca, P, S, K, Mg, as well as Si and Al) and microelements (Fe, Ni, Mn, W, Mo, V, Zn, Cu, Co, Se, Cr) in the form of soluble substances. It was not before these conditions were established in the course of the Earth's evolution that the biosphere could start or, if it is of cosmic origin, resume its evolution. Due to gravitational separation, the primary material began to arrange itself into a crust enriched in light elements and a core, into which heavy elements had been migrating. The process of separation of the metal core into a stand-alone entity played an important role in the Earth's temperature dynamics: it is responsible for the meltdown of the mantle and crust at the Earth's earliest, moon-like stage (4.6–4 Byr ago). The heat accumulated during that process accounts for $\sim$35% of the Earth's current total, a major portion of which dissipates and is lost into space, and a minor portion of which is accumulated by the biota and is in part preserved in dead fossil organic matter (in particular, caustobiolites, including hydrocarbons, are nothing else than the preserved portion of the Earth's thermal energy). The heat provided by the solidification of the Earth's growing inner core composed of a solid iron–nickel alloy with some diamond admixtures accounts for additional 15%, the growth of the outer core accounts for additional 10–15% (this is due to separation of Fe and Ni from the mantle) and radioactive decay accounts for the rest (Trubitsin, Rykov, 2001). The inner core grows due to the material coming from the outer liquid metal core. The outer liquid core supports the magnetic field, the vanguard protection network of the biosphere, and plays an important role in heat transfer in the Earth's interior. Over 4.5 billion years, the average mantle temperature dropped from 3000 to 2100 °C, and the heat flow reduced. The curve q(t) (Fig.1A) allows the integral heat losses to be estimated as

$$Q = \int_0^{4.6} S_0 q(t) dt$$

Given the hot Earth model and assuming that the Earth's area, S_0, has been subject to little variation, we obtain an estimate for the heat lost over the first 150 million years: 9 % of the total heat lost over the Earth's history (6 % per 100 million years). Over 650 million years that followed and were associated with an intensive separation of the core and intensive one-layered mantle convection, the heat loss amounted to 28 % (or 4.3 % per 100 million years). Over 1.1 billion years that followed and were associated with the separation of a liquid core from a solid core and one-layered mantle convection, the heat loss amounted to 26 % (2.5 % per 100 million years). Over the period between 2.6 and 1.2 billion years associated with the transition to two-layered mantle convection and a reduction in the rate of core solidification, the heat loss amounted to 17% (the heat loss rate reduced to 1.3 % per 100 million years). Finally, over the past 1.2 billion years, which are associated with two-layered mantle convection and a slow-paced core solidification with periodic faster-paced laps, the heat loss amounted to 11% (0.9% per 100 million years) (Dobretsov, Kirdyashkin, 1998). Thus, irreversible trends in the Earth's evolution are its cooling, which proceeds with periodic variations on the background of total slowdown (Tajika, Matsui, 1992, Dobretsov, Kovalenko, 1995), and change of the ratio between the mobile and bound oxygen in rocks and the atmosphere,[1] which have resulted in rock oxidation and atmosphere oxygenation (Dobretsov and Chumakov, 2001). As a result of the cooling, the moon-like stage of the Earth's history gave way to the nuclear one. As long ago as 4.3–4.2 Byr, the Earth had a thin crust, sufficiently cool (no hotter than 100 °C) for the formation of the hydrosphere. This time is deduced from findings of corroded zircon grains (de Laeter and Trendall, 2002). The first traces of life, probably, prokaryotic, are recorded in 3.8–3.7 Byr old rocks of earthly origin (Schidlowski, 1988). Hence, at least since that time, two conjugated systems existed: the biosphere and the geosphere, and geosphere evolution determines the direction of irreversible evolution (Fig. 1).

There are two aspects to the concept of evolution: (1) the process of de novo formation of an archetype[2] (biologically speaking, phylogenesis); and (2) the process of the canalized (pre-programmed) individual development of an existing archetype (biologically speaking, ontogenesis). Discussion of the possible relevance of ontogenesis and phylogenesis to geology was started by V.I. Vernadsky, E.S. Fedorov and Grigoryev D.P., but reasoning has never been perfected into any scientific concept (Grigoryev, 1956; Rundkvist, 1968; Rundkvist et al., 1971; Izokh, 1978), except for those occasional events in which the concepts of ontogenesis and phylogenesis have been applied to analyze the genesis of mineral and ore associations. It was proposed to apply the concept of the phylogenesis of minerals (ore bodies, parageneses, mineral species and others) to the geological processes that span over time and space

---

[1] It should be noted that before photosynthesis, rock oxidation was determined mainly by hydrogen dissipation, directly depending on the Earth's temperature. Thermochemical degradation of mobile hydrogen-containing compounds is accompanied by dissipation of hydrogen and binding of oxygen to metals (in particular, to iron, to generate magnetite crystals). Photosynthesis is also degradation of hydrogen-containing compounds (hydrogen sulfide in the anoxic bacterial photosynthesis and water in oxygenic photosynthesis by cyanobacteria and plants). Therefore, since the beginning of photosynthesis, metals have been oxidized by biogenic oxygen as well.

[2] The archetype is assumed to be a set of traits and characters that make a particular group of members, or individuals, that share them stand alone as a species among all the others groups (Grigoryev, 1956; Liubischev, 1982).

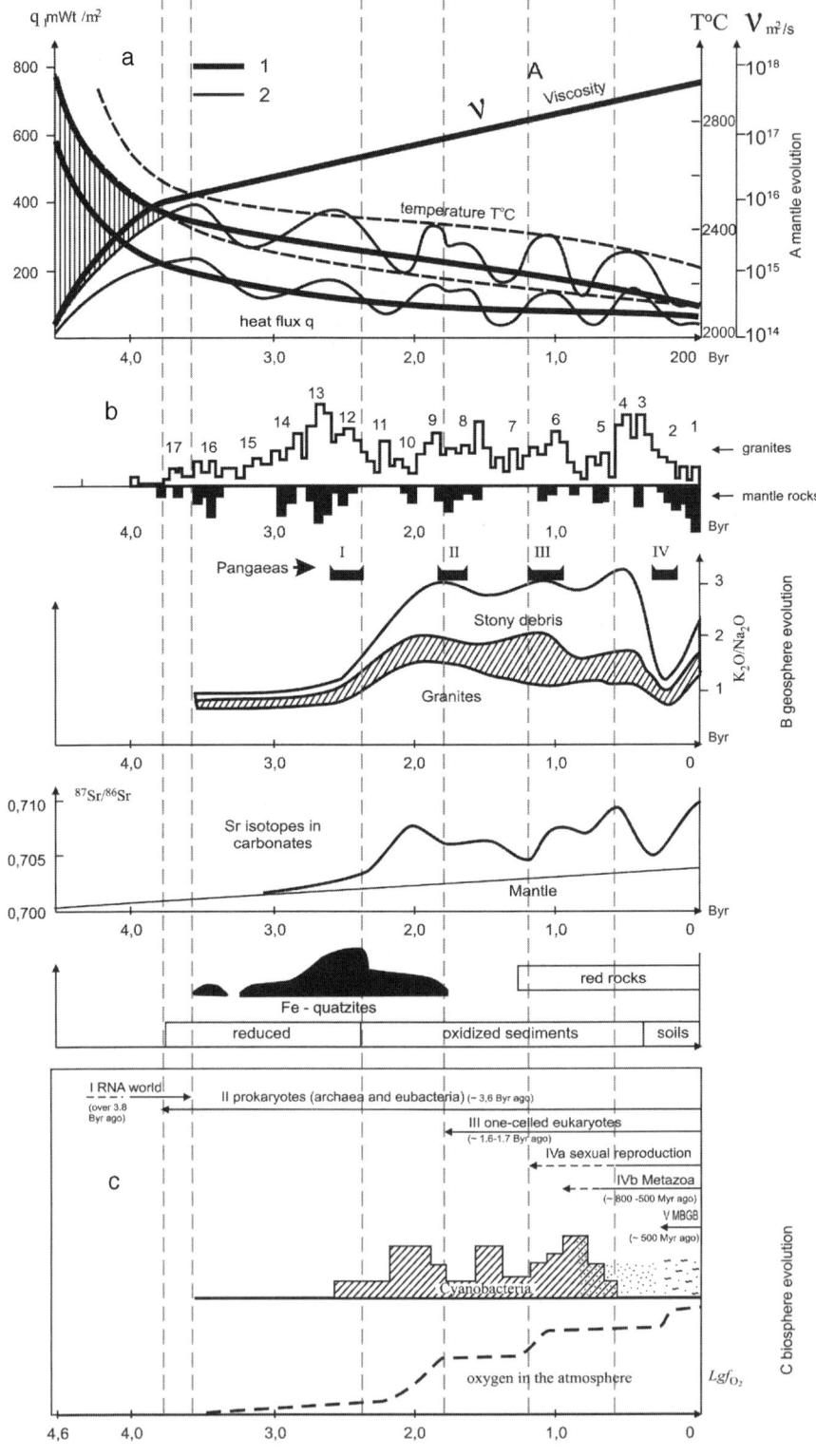

intervals considerably (by a factor of in excess of dozens) exceeding both the age of any particular ore body and all the room it has ever required (Rundkvist et al., 1971). These processes shape environments so that the development of particular ore bodies can only go the way it does. Here the canalization is obviously very much similar to that in biology; however, the mechanisms underlying it are quite different.[3] In biology, the canalization of ontogenesis is largely performed by a program made in the form of a special structure, the genome (Kolchanov et al., 2003). This mechanism of canalization "from inside" rather than "from outside" allowed the biological forms to embark upon a course of development independent of the rule of the environment (Shmalgauzen, 1968) and eventually to form an independent vector of biosphere evolution.

By saying "a code," we mean any type of monomer context that carries, within a polymer, information, the significance of which for a particular function is set not directly, but by matching rules. Is there anything like the genome in geological bodies? The lattice not only has a program to carry but a function to perform. The first step toward the emergence of the code in the course of evolution was perhaps associated with the formation of the feedback between two functional structures like these. Feedback sets matching rules. For instance, the structure of the

---

Fig. 1 The evolution of the mantle (**a**), the geosphere (**b**) and the biosphere (**c**). (**a**) The calculated variations in the mean temperature, heat flux and viscosity of the mantle after Tajika and Matsui (1992) (*bold line*), and variations in temperature and heat flux after Dobretsov and Kovalenko (1995) (*fine lines*), reflecting the processes of Earth's total cooling (no matter what initial state) and convection slowdown as viscosity grows. Periodical first-order variations for temperature and heat flux are comparable with max and min for granites (see 1b). (**b**) The most important indices of geosphere evolution: *upper row*—histograms for the granite age distribution (*gray bars*) and mantle rock age distribution (*blue bars*) in the Earth's crust; digits—the most important endogenous cycles reflected by the max for granites; *middle row*—K_2O/Na_2O ratio in granites, compared to the emergence of the supercontinents Pangaea I, Pangaea II, Pangaea III, Pangaea IV; *lower row*—variation of Sr isotope concentrations in carbonate sediments (Condie, 1989, refined after Semikhatov, 1993) on the background of the typical signals of oxidation and continental emergence: the crust-wide distribution of ferruginous quartzites, red rocks, reduced and oxidized sediments and soils (Zavarzin, 2003a). (**c**) Biosphere evolution: *upper row*—biosphere evolution milestones; *middle row*—the age-related distribution of cyanobacteria (Zavarzin, 2001); *lower row*—atmospheric oxygen evolution (Rozanov, 2006)

---

[3] Different as canalization is in biology and geology, this process uncovers common developmental features, such as the geogenetic law (evolutionary parallelism at all levels), which is similar to the biogenetic law (ontogenesis is a reduction of phylogenesis), and von Baer's law of corresponding stages (Izokh, 1978; Rundkvist, 1968). At the same time, because of a high level of environmentally independent development of biological forms, no analogy to the corollary to the geogenetic law can be drawn (ontogenesis sets pattern for future pylogenesis). What there is is only a less strict version of the law of homologous series (Vavilov, 1967) and Cope's rule of a less specialized ancestor (Shmalgauzen, 1968).

minerals formed by aluminosilicate clays is a multiplicity of stacked layers which are stabilized (as DNA and RNA) by stacking interactions. The layers could only grow laterally, where they were washed by the nutrient solution. In a flow system composed of many microchannels, on whose walls aluminosilicates grow, the most long-lived ("fit") are the stacks of layers that do not clog the microchannels and are not taken away with the solution (Cairns-Smith, 2005). The "phenotype" of that system, sensing the signal from the environment (nutrient solution saturation), is the ion exchange properties; the "genotype" of the system is the putting together of the layers maintained by stacking interactions. Matching rules exist between these two characteristics; therefore, it is possible that the earliest code was not linear, but conformational and it still is present in modern organisms like prions and the stacking interactions of DNA and RNA.

A linear polymer can afford many more conformational rearrangements. The discovery of the spontaneous enzyme-free recombination of RNA oligonucleotides (Chetverin, 1999) provides an insight into how the RNA world could emerge from short oligonucleotides abiogenically synthesized on montmorillonite. SELEX experiments demonstrated that an RNA molecule ensembles with enzyme activities sufficient to provide for the process of self-reproduction of an RNA matrix entirely: from nucleotide synthesis (Unrau and Bartel, 1998) to RNA polynucleotide synthesis on an RNA matrix (Johnston et al., 2001) can be obtained by selection from among a pool of random RNA polymers. As montmorillonite dries and wets, natural selex in RNA colonies, which apparently were the earliest co-evolving cell-free ensembles, is a possibility[4] (Chetverin, 1999) (Fig. 2). The evolution of such ensembles was accelerated because they could share RNA molecules through the air even at long distances (Chetverina and Chetverin, 1993).

The working structures of ribozymes are loops, linked by many complementary pairs of nucleotides. These loops are conservative, because for them to undergo rearrangements, more than one mutation should occur (Aleshin and Petrov, 2003). By contrast, for a protein enzyme to change functions, one or two mutations replacing one or two amino acid radicals in the active center are enough[5] (Ivanisenko et al., 2005). Therefore, refusal of performing functions by

---

[4] Without going into panspermy (see Hoover "Comets, Carbonaceous Meteorites, and the Origin of the Biosphere" in this book), we hold that it is oligonucleotides that can (for example, frozen into ice of whatever kind (Chyba and McDonald, 1995)) survive traveling in the outer space. Upon entering a favorable environment, these oligonucleotides start reproducing life.

[5] For the information on the origin and evolution of the triplet genetic code, which is beyond the scope of this paper, the reader is referred to the article by Zhouravlev et al. "Evolution of the Translation Termination System in Eukaryotes" in this book. The membrane, which isolated the cell from the environment, may have emerged in the RNA world. Experiments revealed RNA molecules that bind to phospholipid layers by arranging them into vesicles and modulating the permeability of such vesicles, which is the required condition for their stability (see Lutay et al. "RNA World: First Steps Towards Functional Molecules" in this book).

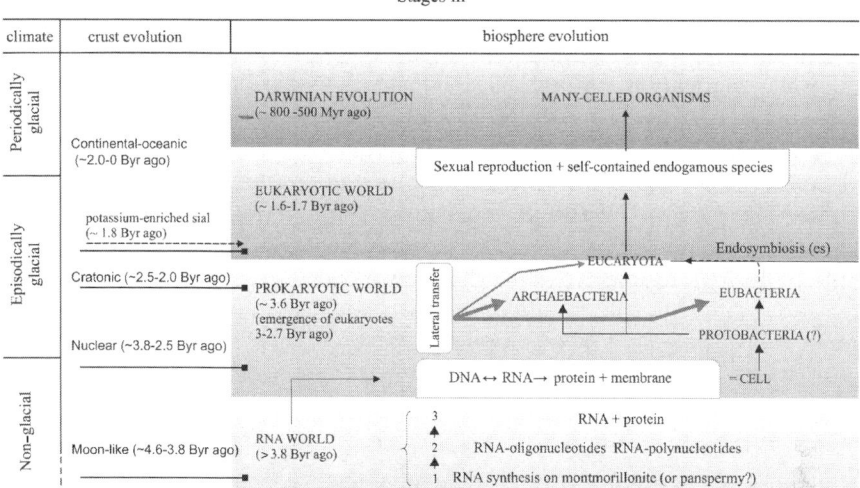

Fig. 2 Scenario of the basic stages of climate evolution, earth's crust evolution and biosphere evolution

the systems that use the triplet code allows such systems to change functions rather easily, which makes these systems extremely flexible from the evolutionary point of view. The protein–nucleic life is more responsive to environmental change than the RNA world. It is advantageous, from the evolutionary point of view, that the enzymes be proteins and that the conserved processes associated with genome functioning be run by RNA structures (rRNA and regulatory RNA).

Montmorillonite and other aluminosilicate derivatives are able to adsorb the ions of metal biogenes (see above) and the most primitive organic matter (amino acids, peptides, sugars) and organic molecules, clay minerals are capable of arranging them into complex ordered molecular ensembles (are they the ancestors to the active centers of enzymes?). Being the product of the weathering of magmatic rocks, clay minerals may be enriched with phosphor and sulfur (Ferris, 2005; Hazen, 2005). Thus, the key chemicals of life (nucleic acids, amino acids, biogenes metals, biogenes non-metals, water) are spatially united in montmorillonite from weathering crust, which facilitates their co-evolution and suggests a "clay-silicon cradle" for life. J.D. Bernall proposed the word "equilibrosphere" for the predecessor to the biosphere, that is, a sphere in which pre-biological evolution might be under way (Bernal, 1967); also, he made the point that this is a spatial region in which, for some physical and chemical reasons, liquid, solid and/or gas phases may come in contact. This is the only kind of sphere in which matter exchange could start and go on.

So, how could the evolution of the geosphere set the vector of the evolution of life? Gravitational separation had elements separated: heavy elements would leave the liquid water zone for the core zone, light elements would largely be concentrated in cratonic crust. Access of biogenes to the hydrosphere is limited by the rate of continental rock weathering, the intensity of volcanism, which brings back part of the elements that have migrated to the mantle and core, and the water solubility of biogenic compounds, which depends on water temperature and pH. At present, biogene enrichment of the "life zone" is due to a global endogenous cycle associated with plate tectonics. The scale of oceanic volcanism associated with either sea floor spreading (the birth of new crust) at mid-oceanic ridges or the subduction of core plated in island arcs is about 10 times the scale of continental volcanism (Lisitsin, 1980, 2001). Importantly, the respective environments, to which these two kinds of volcanism are confined, are quite different. All the processes associated with oceanic volcanism are running in the medium of a natural electrolyte, marine water, at temperatures of up to 400 °C and pressures of 30,000–50,000 kPa on the ocean floor. Passing in through a network of cracks and getting heated up to 300–400 °C by hot rocks, marine water transforms into a high-temperature fluid, which leaches basalts of a large group of elements (including Fe, Mn, Zn, Cu), and so they become part of the solution. The total amount of water entering the World Ocean's hydrothermal system per unit time is ~5.7 thousand tons per second: geologically speaking, the entire World Ocean's water passes through the hydrotherms just instantly, over 3–8 Myr (Lisitsyn, 1993). Another pathway of the endogenous cycle is associated with the volcanic activity of the island arcs in the subduction zones, where the biogenes coming from continental crust occur either in volcanic products or in the mantle. Crust recycling takes from 60 to 600 Myr to complete (Dobretsov and Kirdyashkin, 1998). The endogenous cycle is supplemented with an exogenous cycle associated with the transfer of gases and dispersed effusive material from the lithosphere, through the atmosphere, to the Earth's crust and the hydrosphere. Both cycles contribute comparably; a low capacity of the exogenous cycle is compensated for by a rapid turnover. Both cycles can operate only at a certain regime of convection cells in the upper and lower mantles. At the current figure for heat flow, the lower mantle cells take ~400–570 Myr to complete the cycle, which is comparable with the Wilson cycles ("from Pangaea to Pangaea"). The upper mantle cells take 30–60 Myr to complete the cycle, which is comparable with or divisible by the cycles of magnetic inversions (the so-called Stille and Bertram cycles)[6] and the duration of paleontological periods distinguished by typical faunas (Dobretsov and Chumakov, 2001; Dobretsov and Kovalenko, 1995).

It should be noted that the Raley and Prandtl numbers rather than the Reynolds number control thermochemical gravitational convection. For the Archaean, the Rayleigh number (Ra)[7] for the lower mantle is estimated to be

---

[6] Even assuming convection mantle-wide, the cycling time remains the same.

[7] $Ra = \beta g \Delta T l^3 / a\nu$, β is the coefficient of volumetric expansion, g is the acceleration of gravity, ΔT is the superadiabatic difference in the mantle, l is mantle thickness, a is thermal diffusivity and ν is dynamic viscosity.

$Ra = 10^8$–10^9. Studies of the conditions of emergence of turbulent thermal gravitational convection in a horizontal layer heated from below and cooled from above showed that the turbulent mode of free convection occurs at $Ra > 10^6$, not depending on the Prandtl number at $Pr >100$. Thus, in the Riphean and Hadean (before 3 Byr ago), the convection was fast-paced and constantly turbulent, and the modern plate tectonics could not exist. At present, Ra for the lower mantle varies from 5×10^5 to 3×10^6, and the flow in cells is non-steady or clearly turbulent (Dobretsov et al., 2001).

Intensive volcanism largely associated with the chemically depleted upper mantle was unable to provide biogene supply, but it contributed to the cratonization of the crust (Fig. 1). In a variety of areas of the Earth's surface, 4.0–3.9 Byr old continental type rocks have been found. Khain (2003) opines that by the beginning of the Late Archaean (3.0 Byr ago) strong, heat-resistant continental crust plates had formed in large numbers, but small in size. Those plates were separated by basins with floors composed of oceanic type crust. As early as 2.5 Byr ago, change of the convection mode in the mantle gave rise to the first supercontinent in the Pangaea series, and a new, continental–oceanic stage of the crust evolution began 2.0 Byr ago (Fig. 1b, 2). This stage was characterized by the modern type of plate tectonics with a deep ocean and continents, enrichment of the chemical composition of acidic effusive and granitic rocks and a "plume dropper" (see below). These events had profound consequences for the biosphere evolution.

The migrating biogenes can be held captive in the liquid water zone and jumps in their concentration can be avoided, if there is a more or less limited cycle of chemical reactions and a buffer's capacity high enough for those elements to circulate within. The World Ocean, this still operating powerful thermostat and stabilizer of the chemical composition of the atmosphere and the upper part of the Earth's crust, was the system that could keep things going. The invariability of the chemical content of the ocean is supported by the cooperation of this hydrothermal chemical reactor and the supply of carbonates and other compounds of weathering products supplied by continental waters. Thus, the World Ocean is one of the most important exchange systems for the Earth's main spheres. The fact that the chemical content of the oceans has remained invariable at least over the past 1 Byr (see below) indicates that this global exchange system is ancient; this is also yet another piece of evidence, which supports the hypothesis that the ancient World Ocean was deep and defies the hypothesis that ancient oceans were shallow. The invariance of the amount and the chemical content of the World Ocean water suggests that the World Ocean's depth has been mostly invariant, with cyclic and evolutionary fluctuations not exceeding 30% (Dobretsov et al., 2001; Khain, 2003).

As modern plate tectonics came to the scene, not only was a hydrothermal oceanic reactor built, but also, more importantly, the chemical content of the effusive rocks changed, which is well indicated by the change in K_2O/Na_2O ratio (Fig. 1b). It was about 3 Byr ago that the rocks in the tholeiite and calc-alkali series, low in K and Na, were replaced by magmatic associations in the subalkali and alkali series. In the paleontological record, that was at about the same time when the growth of the biodiversity of the earliest prokaryotic systems occurred (3–2.5 Byr ago) and the eukaryotes emerged (3–2.7 Byr ago); molecular phylogeny also puts the emergence of eukaryotes at 2.7 Byr ago and the emergence of cyanobacteria at 2.5 Byr ago (paleontologists opine that cyanobacteria had emerged earlier; however, 2.5 Byr ago is when the role of those organisms in the biosphere became important) (Fig. 1c, 2). Mantle enrichment processes could be under way due to the recycling of crustal material; however, this does not explain the emergence of potassium-rich sial ∼1.8–1.2 Byr ago, when the crustal rocks were rapidly enriched in K, and Na became part of the solution, the marine water, the chemical content of which has since remained invariable. The emergence of potassium-rich sial is a mystery. What can be said at the moment is that a large-scale commitment of the lower mantle, enriched in many elements, to the endogenous cycle was a factor. Increased mantle viscosity should have initiated a rearrangement in the global mantle convection. Since this system is extremely slow responsive, the rearrangement should have occurred in a saltation-like manner, that is, convection slowdowns would alternate with releases of the heat accumulated in the core, which rapidly (on a geological scale) gave birth to a series of plumes (it is possible that convection, for some short period of time, went mantle-wide and then again specific to separate mantle layers). That could co-occur with an active formation of magmatic rocks after a more or less calm period 2.5-2 Byr ago (Fig. 1a, b).

The global change in the K/Na ratio affected the biosphere. The very existence of a closed lipid membrane-enabled cell would have been impossible, if it had not been for the molecular pumps: the cell would have been torn apart by the osmotic pressure (Polevoy, 1985). The key role in the photochemical system of the plants is played by the proton pump, which is the only feature they have perfected: the K/Na ratio in the modern plants is comparable with that in the geosphere. The heterotrophs had to adapt themselves to the environment (potassium prevails over sodium even compared to the geospheric ratio (Natochin, 2005)); however, the evolution of the K/Na pump promoted a fast-paced evolution of the nervous system (by the molecular clock, animals and plants diverged ∼1500 Myr ago, animals and fungi diverged ∼1208 Myr ago (Hedges and Kumar, 2003, 2004)).

The alteration of the convection mode in the mantle could also result in the formation of a "plume dropper" in the Phanerozoic (Fig. 3). Although Phanerozoic eons are associated with the periods of existence of sustainable faunal communities, the timeline of how these communities replaced one another is worth looking at: the Cambrian, the Silurian, the Triassic, the Paleogene, the Neogene each lasted 30 Myr; the Ordovician, the Carboniferous, the Jurassic each lasted ∼30·2 Myr; the Devonian, the Permian each lasted ∼30+30/2 Myr. The timeline associated with 30 Myr long periods, correlates well with the periods of magnetic field inversions and periodic increase in plume volcanism,

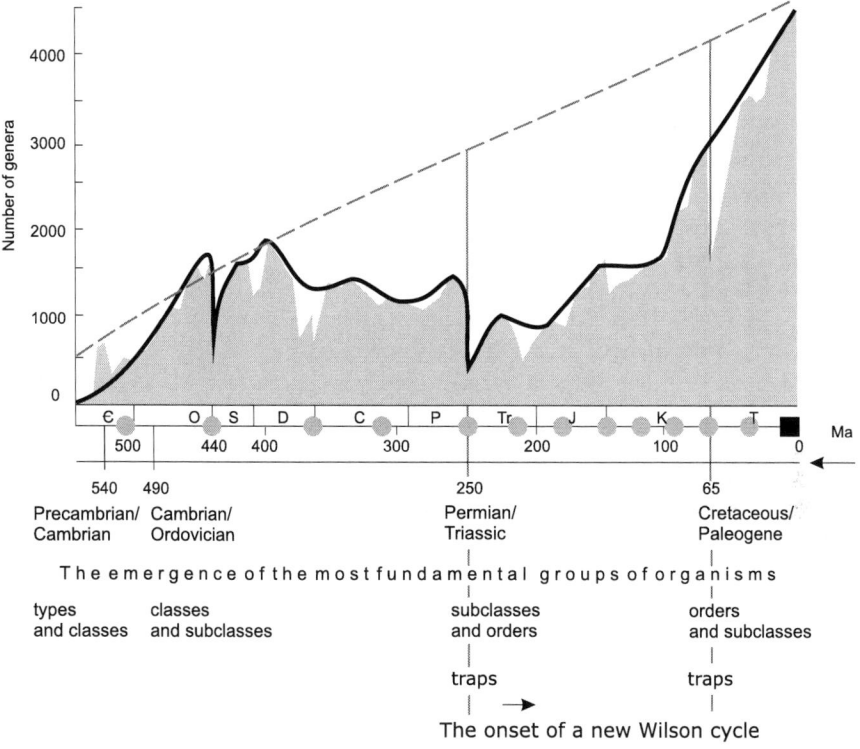

Fig. 3 Biodiversity in motion according to J.J. Sepkoski (Sepkoski, 1994; Sepkoski, 1996). Moments of high taxons fixation in paleontology according to S.V. Rozhvoz (Rozhnov, 2005). Circles stand for plume droppers

commonly followed by increasing island-arc volcanism. As shown by calculations, the periodic pattern of volcanism can be explained by the suggestion that the rate of heat transfer in the outer liquid core is not compensated by convection in the lower mantle (the Rayleigh number for the outer liquid core in the Phanerozoic is $Ra = 10^{18}$, and in the mantle, $5 \times 10^5 – 3 \times 10^6$). The overheat of the outer core enhances the instability of its flows. Mantle vortex funnels arise and roam about the core–mantle boundary of the liquid core under the influence of the Coriolis force, like sunspots (Kirdyashkin and Dobretsov, 2001). Thermochemical plums are separated in the mantle funnels, where a chemical dope reducing the melting temperature of the mantle matter is supplied, and generate hotspots similar to the Hawaii one, with effusion of liquid lavas enriched in biogenic elements. The chemical dope reducing the mantle matter melting temperature can be formed, for example, by the reaction of hydrogen and/or methane with iron-containing minerals of the lower mantle, such as perovskite and magnesiowustite. Gas release increases with the overheat of the outer core. As a series of such plumes is separated from the core–mantle boundary, the excess heat comes to the surface, causing an outbreak of plume volcanism, reviviscence of rifts and rearrangement of Asthenospheric flows. This, in turn, affects the continent drift. The result is an outbreak of island-arc volcanism with specific explosive eruptions, whose gases and ash can affect atmosphere transparence and circulation.

In particular, the beginning of the cycle, which co-occurred with violent effusions of lava in the Siberian Traps, could initiate the break-up of Pangaea, thus pushing biocoenotic rearrangements toward the Permian/Triassic extinction, while a high-level activity of mantle–plume volcanism in the Mesozoic contributed to a prolongation of the Mesozoic non-glacial interval (Fig. 3).

So, there are two geosphere-related canalizing vectors of biosphere evolution. One is associated with the irreversible cooling of the Earth and the resulting endogenous cycles of various origin and periodicities, which changes the amount and the "item list" of accessible biogenes in the "liquid water zone." Another is associated with the formation of the World Ocean, which, with its tremendously high buffer capacity, distributes biogene concentrations evenly over the globe and provides the hydrosphere with direct biogene supply from the mantle. As a result, the World Ocean set up the conditions for the "film of life" to conquest the globe. Before the World Ocean, the biosphere apparently had been huddling around "hotspots of biogenes" (hydrotherms, upwellings, valley trains, etc.).

In summary, it should be noted that the geochemical postulates advanced by Vernadsky (1987) on the expansion of the biosphere and its evolution toward an intensive closed migration of the atoms of biogenes and a broadening of diversity of such atoms result from the fact of the Earth's evolution. Gravitational separation and endogenous cycling changes the accessibility of the chemical elements, and the cooling and oxidation of the Earth changes the accessibility of the dissolved compounds these elements form part of. A hypothetical succession of metals coming into the evolution of enzymes (Fedonkin, 2003) is confirmed by geological and, to a lesser degree, microbiological data. In particular, elevated concentrations of Mo and V are found in ancient black shales. If the hypothesis about the evaporite origin of the Archaean gneisses with scapolite and scheelite $CaWO_4$ is true (Dobretsov and Chumakov, 2001), then the ancient biosphere could very probably be rich in W.

Life also represents a relatively limited number of chemical reactions. This cycle maintains the stability of concentrations (homeostasis) of the elements passing through it, but here the events unfold in a phase-separated system with the aid of enzymes: organic or metalorganic catalysts, providing the intensity of cycling ten times as much as geochemical process do. The genome grants life the "convariant reduplication" ability: mutation to enzymes can be fixed by selection, allowing the carriers to keep track of the changes in the geochemical environment. Noteworthy, the "evolutionary flexibility-to-conservatism" ratio is more optimal in protein–nucleic acid life than in the RNA world. A low mass of the elements involved in the turnover is compensated for by such a high rate of cycling that geological process may never attain. Carbonates contain $0.6 \times \cdot 10^{23}$ g of carbon, while the biosphere only $0.5 \times \cdot 10^{18}$ g, and that is why the carbon cycle is first of all the carbonate cycle (Dobretsov and Kovalenko, 1995; Tajika and Matsui, 1992). Nevertheless, the high rates of biogeochemical cycling make its C_{org} annual balance comparable with that in the carbonate cycle, although their respective absolute values are 10,000-fold different. A high rate of biogeochemical cycles was the first mechanism in helping the biosphere to gain autonomy from the geosphere. That mechanism was best used by bacteria. Various estimates put the microbial biomass at 50–90% of the Earth's total (Vinogradov, 2004; Zavarzin, 2003a). This figure includes not only terrestrial bacteria, symbiotic eukaryotic bacteria, soil bacteria, nanoplankton and other bacteria inhabiting the ocean, but also prokaryotes living in such extreme biotopes as hydrotherms, salt and soda lakes and even archaea in nuclear reactors. The prokaryotes control the main biogenic cycles and this, in combination with a tremendously large total prokaryotic biomass, allows the microbes to shape the environment "from outside" like the geological bodies do (see Zavarzin "Microbial Biosphere" in this book). This abrogates the need for further adaptive evolution in prokaryotes. The genome began to play a regulatory and a repairing function, supporting adaptive dynamics (species balance) in the prokaryotic communities and restoration of these communities wherever they were adversely affected (by volcanism, glaciations and later eating-up by eukaryotes). Molecular phylogeny has made an interesting point: all the known energy-producing enzyme systems appeared very early, at the dawn of evolution, from a common ancestor (Castresana and Moreira, 1999), which seems to be a miracle. It is more likely that there is no such thing as a common ancestral organism—the "common ancestor" is a biogeocoenosis (Zavarzin, 2001; see Zavarzin "Microbial Biosphere" in this book), and the elements developing in various parts of its enzyme systems were later dispensed among taxa by horizontal transfer, so popular with the prokaryotes (Shestakov, 2005). That is, the main enzyme systems may have evolved by recombination. This scenario is also supported by that surprising fact that among the modern prokaryotes the highest levels of diversity and the best balance of biogeochemical cycles (and, therefore, stability) are features of mixed archaean/eubacterial communities: methanogenic archaea interact with fermenting bacteria, hydrolytic clostridia interact with dissipatrophic spirochete, spirochete interact with sulfate-reducing proteobacteria (Zavarzin, 2001). Thus, the evolution of prokaryotes is quite different from the evolution as per Darwin (Fig. 2).

With the advent of photosynthesis, the second and most important mechanism of gaining autonomy from the geosphere, the biosphere gained access to a source of energy that was not associated with the geosphere, and for the first time set its own, biotic vector of evolution, which prepared the geochemical settings (in particular, the ozone layer) suitable for life to come to the continents and to continue expansion. There is an opinion that oxygenic photosynthesis could be older than 3.5–3.3 Byr, and some enzyme systems even more. However, any calculations using the molecular clock with only single key genes of the photosynthetic machinery dealt with make little sense. These genes could have existed long before photosynthesis and they performed other functions, for example, synthesis of protective pigments against sunrays or/and active oxygen of non-biogenic origin, by water photolysis. What is required is molecular-genetic calculations for many key genes for photosynthesis at a time, including consideration to their co-adaptive evolution.

The Archaean biosphere was probably anoxygenic and chemotrophic, with oxidized "spots" of oxygenic photosynthetics. The inversion of that biosphere with irreversible oxygenation of the atmosphere by cyanobacteria began ~2 Byr ago, after large amounts of Fe_3O_4 ore mass were buried (Fig. 1). Then the iron cycle in the biosphere was replaced by the sulfur cycle. These cycles are incompatible: sulfate reduction brings all free iron into sulfides (Zavarzin, 2003a). Before that, sulfur was likely to concentrate in the crust, for example, in Precambrian volcanite in the form of sulfides. So the immensely large iron "reservoir," which used to bind free oxygen beginning to accumulate in the Earth's atmosphere, vanished due to geosphere/biosphere interactions (Fig. 1).

The ecosystems that formed in a reducing atmosphere had to seek to escape from the poisonous effects of oxygen to refugia, first to the deep-water zones of the ocean, which was anoxic until the Ordovician, then to others, where they still persist. We do not know of any perfect analogies to such ecosystems in today's world. It is possible that in the ocean, the archaean or the methanotrophic biota was similar to what we can see these days around the hydrothermal vents also known as "black smokers." Now is the right time to say a few words about the vestimentifera, the most typical inhabitants of such biocoenoses. Most of its body is occupied by the trophosome, a special organ, whose large cells are literally stuffed with chemosynthetic bacteria, which consume methane and hydrogen sulfide supplied by the vestimentiferan blood system, whose hemoglobin equally binds oxygen and hydrogen sulfide: oxygen binds to heme, and hydrogen sulfide, to the protein-made part (Malakhov and Galkin, 1998). It is possible that the dual system of energy/matter exchange in the vestimentifera is associated with their ancient origin (Malakhov and Galkin, 1998; Maslennikov, 1999). Apparently, the conservatism of the vestimentiferan structure is somehow associated with the conservatism of their habitat (see also Kanygin "Ecological Revolution Through Ordovician Biosphere (495 to 435 Ma Ages): Start of the Coherent Life Evolution" in this book).

An important implication of oxygenation was the formation of a eukaryotic add-on in the biosphere, which apparently took place 1.7–1.9 Byr ago (Cavalier-Smith, 2002a; Knoll, 1994; Sergeyev et al., 1996) (Fig. 2). The oxygenation of the atmosphere affected the accessibility of many biogenes by turning them to oxides and silicates: on the global scale, the balance and rates of the geochemical cycles of biophilic elements should have undergone dramatic change. The biosphere productivity should have dropped around that time, as the changes in sedimentation patterns suggest (Fig. 1). The eukaryotes introduced consumers that were new to the ecosystems and thus considerably elongated food chains, which improved the restraint and efficiency of biogeochemical cycles (Zakrutkin, 1993).

The main mechanism of reproduction in prokaryotes is through fission. They possess some analogies to sexual reproduction: conjugation, transformation and transduction; but these are only triggered as backup processes in critical environmental situations, when all else fails (Prozorov, 2002). Therefore, the "sexual process" in the prokaryotes is only casually associated with reproduction, and, as far as the prokaryotes are concerned, there is no such thing as sexual reproduction in the sense it exists in the eukaryotes. By contrast, in most eukaryotes, reproduction is necessarily preceded by DNA fragment exchange in the form of a sexual process (Cavalier-Smith, 2002b) as the main mechanism for generating diversity. Non-sexual reproduction, which occurs in some eukaryotic groups, is only optional and serves either to rapidly increase the biomass (parthenogenesis in plant louses, vegetative propagation in plants) or to raise the probability of reproduction in special conditions, when chances of meeting a partner are low (parthenogenesis in lizards, apomixis in plants) (Ruvinsky, 1991; Vasilyev et al., 1983). Thus, the sexual process has since some time become not only the leading mechanism of reproduction in the eukaryotes but also one of the major factors of speciation, for it gave birth to self-contained endogamous species (Ruvinsky, 1991; Starobogatov, 1985). It is possible that many-celled eukaryotes, most of which are self-contained endogamous species, appeared due to that important role of sexual reproduction in speciation (Starobogatov, 1985). Thus, 1.7–1.9 Byr ago is the most important period of time in the development of biodiversity (Fig. 2). The genome size increased from 10^4–10^5 to 10^9–10^{10} bp following transition from prokaryotes to eukaryotes, and the gene number increased from 470 (*Mycoplasma genitalium*) to 30–40 thousand (many-celled eukaryotes). However, while the organizational complexity of the bacteria generally correlates with the genome size and gene number, the eukaryotic situation tells a different story, for there is no correlation between their organizational complexity, genome size or gene number. In the insects, which are noted for the highest level of biodiversity on the Earth, and in the amphibians, which have the lowest level of biodiversity among the vertebrates, genome size fluctuations never exceed 100-fold. At the same time, *Drosophila melanogaster* numbers 13,600 genes, while *Caenorhabditis elegans*, which is a more primitive organism, 19,000. Surprisingly, the man and the fish fugu have about the same number of genes, 30,000–40,000 (Carroll, 2001; Taft and Mattick, 2003). It looks like we face a fundamental biological law, which predetermined the evolution of the prokaryotes, whose progress from certain point on became dependent on the genome size, and the evolution of the eukaryotes, in which this dependence at certain point was abolished, perhaps, because of a complex organization and regulation of the genome.

In conclusion, the reader is invited to a general schematic for the evolution of the biosphere and critical biological innovations in the Precambrian–Phanerozoic. There are five most important stages to it: (1) Archaean, throughout which the prokaryotes dominated, and the environments were all reducing (probably, with some occasional spots of oxidation); (2) Proterozoic, throughout which one-celled eukaryotes emerged, and the environments were oxygen (about 2.1–2.0 Byr ago); (3) Vendian, throughout which many-celled Metazoa throve for the first time in shallow well-aerated littoral waters of cold seas. The Metazoa forms their own ecosystems with, for the time being, short trophic chains (Sokolov and Fedonkin, 1988); (4) Cambrian, which is noted for the emergence of the skeleton, which provided a rapid development of all the main phylogenetic stems of marine invertebrates, the Metazoa acquired a high morphogenetic potential through multiple gene duplications in the *Hox* cluster (~600 Myr ago, by the "molecular clock") (Balavoine et al., 2002; Peterson and Eernisse, 2001). As a result, the trophic chains of the eukaryotic ecosystems elongated because most of the vacant potential ecological niches had been occupied by new forms. A tight packaging of the niches limited the space of logical possibilities of evolution: thus was a coherent[8] biocoenotic environment for the first time formed in the eukaryotic world. From then on, taxa from the type up are rare (they are occasionally formed due to regressive evolution) (Rozhnov, 2005); however, the competition-based relationships that are established within the existing types activate diversification of classes, orders, etc. The morphological innovations in the Cambrian arranged for the "globalization" of the eukaryotic biosphere in the Ordovician; (5) the emergence of nearly all classes and main families of oxyphilic hydrobionts initiated colonization of the benthic and pelagic zones, withdrawal of anoxygenic communities and a growth of biodiversity. The Earth was shrouded in the film of life!

The formation of a coherent global biocoenotic environment planetwide enhanced biosphere resistance to external impacts and internal fluctuations[9] and also caused a strict canalization of evolutionary pathways which made for long-term evolutionary pathways in separate phyletic lineages, and considerably accelerated evolution caused a strict canalization of evolutionary pathways (Krasilov, 1986). Long-term evolutionary pathways in separate phyletic

---

[8] Coherent evolution is controlled by a developing stable structure of a high-competition ecological community. By contrast, non-coherent evolution is attributed to degrading, low-competition ecological systems (Krasilov, 1986).

[9] The similarity is contributed to both "from the inside," by a common origin (and, therefore, biochemistry) and "from the outside," by million years of co-evolution in a globalized biosphere. It should be remembered, however, that it is ecosystems that would recover, not the initial species composition. For instance, the extinction of the Pleistocenic herbivorous megafauna in Eurasian steppes due to glaciation comebacks and human hunting activities did not affect the steppen biome, because the function of the main herbivores was now performed by mid-size animals (saiga, Asiatic wild ass), small-size animals (vole) and very small-size animals (acridoids) (Lockwood et al., 1994; Zhegallo et al., 2001). Reduction in biodiversity implies that a lot of ecological niches have become vacant again (Krasilov, 1986; Zherikhin, 1986). Unless they are occupied by migrants from other ecosystems, diversification of aboriginal taxa into new species is warranted ("splendid isolation" in South America, marsupial faunas in Australia). Thus the biosphere defends itself either by assembling new ecosystems from the remains of the old ones (a quick-response mechanism) or by initiating diversification in "depressed" ecosystems (a slow-response mechanism).

lineages were established (one of such trends, encephalization, would lead to the emergence of intellect and *Homo sapiens*, its today's carrier). On the other hand, the improved stability of the global biosphere made its further evolution staged. The eukaryotic ecosystems, which had emerged from the "Cambrian skeletal revolution," obliterated the pre-existing Vendian biota (Sokolov and Fedonkin, 1988), but did not last too long and succumbed to the Cambrian/Ordovician biocoenotic crisis (~490 Myr ago). Nevertheless, there was no general reduction in taxonomical biodiversity: during globalization (the exploration of a new tremendously great biotope—marine pelagic and benthic zones), the rates of the formation of new taxa were higher than the rates of their extinction (Kanygin, 2001). By contrast, the biocoenotic crises of the "globalization epoch" (Ordovician/Silurian, Devonian/Carboniferous, Permian/Triassic, Triassic/Jurassic, Cretaceous/Paleogene) are noted for abrupt reductions in biodiversity and considerable simplifications of community structures (Kalandadze and Rautian, 1993; Rozhnov, 2005). However, if the Paleozoic crises from Ordovician to Permian had the taxonomic diversity curve oscillate around a level (the Paleozoic plateau), the post-Permian ones are noted for a growth of biodiversity. Thus, as the biosphere globalization began, the devastating effects of biocoenotic crises were first sharply enhanced and then slowly reduced, which may be interpreted as a growth of the total (within the entire biota) fitness of the organisms in the course of evolution. Paleontological data suggest that that this growth is observed indeed: in particular, the intervals between biocoenotic crises become longer (Sepkoski, 1994, 1996), an exponential growth in the total time of existence of genera is observed (Markov, 2002), which, although interrupted by crises, revealed, after the Permian (~250 Myr ago) a tendency toward reduction in devastating crises and a growth in biodiversity (Markov, 2001). Thus, the following can be stated: life, which was ab initio absolutely dependent on geospheric conditions, could, in the course of evolution, acquire the ability of relying on an external energy supply (photosynthesis) (1), which considerably enhanced the rates of biogene turnover (2) and build their biospheric depot (3) in the living matter and mort-masses of the biosphere. Finally, in the course of the "globalization" of the biosphere, a network-like back-up system of similar ecosystems was created, which made the biosphere less responsive to both external impacts and (after prolonged evolution) internal "age of puberty" (4). As a result, the dependence of the biosphere on the geosphere became considerably weaker, although the biosphere is still unable to return the biogenes that have migrated to the core and are buried in the rocks and kerogen.

When the man began to mine for minerals, he created the first combo (natural–man made) ecosystems (urbocoenoses), which are capable of immediately extracting buried organics from the nature, kerogen and/or various biospheric depots, which granted the urbocoenoses the ability to expand ceaselessly by giving them access to energy and matter sources, bypassing the main biogeochemical cycles. Although any urbocoenosis is geographically limited, the trade ties, communication routes and, later, information communication systems (from most ancient, that is, letters, to modern, that is electronics-based) allow it to gain control of dozens of times as vast territories (Braudel, 1990). In this respect, an urbocoenosis is similar to the communities of migrating animals, but there is a difference, too: potentially, it has no limitations in the form of either geographical barriers or the mass of the matter being transported or transportation rates. Technical progress is the only barrier-setting factor to each stage of human evolution. Throughout the New Era, the urbocoenoses are gradually sneaking away from the control of the biogeocoenoses and are trapped within and settle on the surface of a geographic sphere which is, on Zavarzin's suggestion, known as the cacosphere (from the Greek kakós— worthless, of a bad nature, destructive) (Zavarzin, 2003b). Naming it "the cacosphere," Zavarzin assumes all the man-made biogeocoenoses, whose typical features are low taxonomic diversity, loss or misfunction of regulatory associations and a low degree of closure (associated with both misfunction of biocoenotic associations and a large amounts of matter transported from the lithosphere in bypass of biogeochemical cycles). Since the degree of closure is low, the cacosphere is unable to support itself and is prone to self-contamination, hence the need for self-decontamination biospheric mechanisms (first of all, bacterial filters of soils, water bodies and woods) and/or dump sites (Zavarzin, 2003b). A low taxonomic and biocoenotic diversity and an exponential growth of the cacosphere are the factors that together make it somewhat questionable that the cacosphere will ever be able to evolve independently, always improving the total fitness, for so long as the biosphere has been evolving since the Cambrian: hundreds of millions of years. Over a similar span of time, the external resources that support the cacosphere (soils, woods, and accessible minerals) would be depleted, and the way the cacosphere has gone global over the past 80 years leaves no hope for exploration of new areas or resources: believing that the Cambrian/Ordovician situation (a crisis which did not end up with total extinction) would ensue again makes little sense.

Acknowledgment We acknowledge our gratitude to A.Yu. Rozanov and A.V. Kanygin for fruitful discussion and to A.V. Kharkevich of the IC&G for his design work. We are in debt to V. Filonenko for translating this manuscript from Russian into English. This work is supported by the Russian Foundation for Basic Research (No. 04-01-00458, 05-04-49068, 05-07-90274, 06-04-49556) and the Program of the Presidium of RAS "Biosphere Origin and Evolution."

References

Aleshin, V.V. and Petrov, N.B. (2003) Conditionally neutral characters. Priroda 12, 25–34 (in Russ.).
Balavoine, G., de Rosa, R. and Adoutte, A. (2002) Hox clusters and bilaterian phylogeny. Mol. Phylogenet. Evol. 24. 366–373.
Bernal, J.D. (1967) The Origin of Life. World, New York.

Braudel, F. (1990) La Méditerranée et le monde méditerranéen à l'époque de Philippe II, tome 2: Destins collectifs et mouvements d'ensemble. Armand Colin, Paris.

Cairns-Smith, A.G. (2005) Sketches for a mineral genetic material. Elements 1, 157–161.

Carroll, S.B. (2001) Chance and necessity; the evolution of morphological complexity and diversity. Nature 409, 1102–1109.

Castresana, J. and Moreira, D. (1999) Respiratory chains in the last common ancestor of living organisms. J. Mol. Evol. 49, 453–460.

Cavalier-Smith, T. (2002a) The phagotrophic origin of eucaryotes and phylogenetic classification of Protozoa. Int. J. Syst. Evol. Microbiol. 52, 297–354.

Cavalier-Smith, T. (2002b) Origins of the machinery of recombination and sex. Heredity 88, 125–141.

Chetverin, A.B. (1999) The puzzle of RNA recombination. FEBS Lett. 460, 1–5.

Chetverina, H.V. and Chetverin, A.B. (1993) Cloning of RNA molecules in vitro. NAR 21, 2349–2353.

Chyba, C.F. and McDonald, G.D. (1995) The origin of life in the solar system: current issues. Annu. Rev. Earth Planet. Sci. 23, 215–249.

Condie, K.C. (1989) Plate Tectonics and Crustal Evolution. Pergamon Press, Oxford.

Dobretsov, N.L. and Chumakov, N.N. (2001) Global periodical variations in litologspheric and biospheric evolution. In: N.L. Dobretsov and N.I. Kovalenko (Eds), Global Environmental Changes. SO RAN, filial GEO, Novosibirsk, pp. 11–26 (in Russ.).

Dobretsov, N.L. and Kirdyashkin, A.G. (1998) Assessment of global matter exchange between the Earth's layers: comparing geological and theoretical data. Geol. Geofiz. 39, 1269–1279 (in Russ.).

Dobretsov, N.L. and Kovalenko, N.I. (1995) Global environmental changes. Geol. Geofiz. 36, 7–30 (in Russ.).

Dobretsov, N.L., Kirdyashkin A.G. and Kirdyashkin, A.A. (2001) Depth Geodynamics. Geya, Novosibirsk (in Russ.).

Fedonkin, M.A. (2003). The origin of the Metazoa in the light of the Proterozoic fossil record. Paleontological Research 7, 9–41.

Ferris, J.P. (2005) Mineral catalysis and prebiotic synthesis: montmorillonite-catalyzed formation of RNA. Elements 1, 145–149.

Grigoryev, D.P. (1956) Further insights into mineralogical objects; minerals as per A.K. Boldyrev. Zap. Vses. Mineral. O-va. 85, 463–471 (in Russ.).

Hazen, R.M. (2005) Genesis: rocks, minerals and the geochemical origin of life. Elements 1, 135–137.

Hedges, S.B. and Kumar, S. (2003) Genomic clocks and evolutionary timescales. Trends Genet. 19, 200–206.

Hedges, S.B. and Kumar, S. (2004) Precision of molecular time estimates. Trends Genet. 20, 242–247.

Ivanisenko, V.A., Pintus, S.S., Grigorovich, D.A. and Kolchanov, N.A. (2005) PDBSite: a database of the 3D structure of protein functional sites. NAR 33, D183–D187.

Izokh, E.P. (1978) Assessment of the Ore-Bearing Capacity of Granitoid Formations with a View to Making Predictions. Nedra, Moscow (in Russ.).

Johnston, W.K., Unrau, P.J., Lawrence, M.S., Glasner, M.E. and Bartel, D.P. (2001) RNA-catalyzed RNA polimerization: accurate and general RNA-templated primer extension. Science 292, 1319–1325.

Kalandadze, N.N. and Rautian, A.S. (1993) Symptomatology of ecological crises. Stratigr. Geol. korrel. 1, 3–8 (in Russ.).

Kanygin, A.V. (2001) The Ordovician explosive divergence of the earth's organic realm: causes and effects of the biosphere evolution. Russ. Geol. Geophis. 42, 599–633.

Khain, V.E. (2003) Main Challenges in Modern Geology. Nauchnyy Mir, Moscow (in Russ.).

Kirdyashkin, A.G. and Dobretsov, N.L. (2001) The effects of the structure of convective flows and plume flows in the Earth's mantle on the periodicity of endogenous processes. In: N.L. Dobretsov and N.I. Kovalenko (Eds), Global Environmental Changes. SO RAN, filial GEO, Novosibirsk, pp. 27–41 (in Russ.).

Knoll, A.H. (1994) Neoproterozoic evolution and environmental change. In: S. Bengtson (Ed.), Early Life on Earth. Columbia Univ. Press, New York, pp. 439–449.

Kolchanov, N.A., Suslov, V.V. and Shumny, V.K. (2003) Molecular evolution of genetic systems. Paleontol. J. 37, 617–629.

Krasilov, V.A. (1986) Unsolved Problems of Evolution Theory. FERSAS SSSR, Vladivistok (in Russ.).

de Laeter, J.R. and Trendall, A.F. (2002) The oldest rocks: the Western Australian connection. J. R. S. West. Aust. 85, 153-160.

Lisitsyn, A.P. (1980) The history of oceanic volcanism. In: A.S. Monin and A.P. Lisitsyn (Eds), The Geological History of the Ocean. Nauka, Moscow, pp. 278–319 (in Russ.).

Lisitsyn, A.P. (1993) Hydrothermal systems of the World Ocean as a supplier of endogenous matter In: A.P. Lisitsyn (Ed.), Hydrothermal Systems and Sediment Formations of Mid-oceanic Ridges. Nauka, Moscow, pp. 147–247 (in Russ.).

Lisitsyn, A.P. (2001) The lithology of lithospheric plates. Geol. Geofiz. 42, 522–559 (in Russ.).

Liubischev, A.A. (1982) On the Form, Systematics and Evolution of Organisms. Nauka, Moscow (in Russ.).

Lockwood, J.A., Bomar, C.R., Williams, S.E., Dodd, J.L., Quan, M. and Li, H. (1993) Insect ecology on the Asian and North American steppes: striking differences and remarkable similarities. In: Li Bo (Ed.), Proceedings of the International Symposium on Grassland Resources. August, 1993. Agricultural Scientech Press, Beijing, pp. 513–527.

Malakhov, V.V. and Galkin, S.V. (1998) The Vestimentifera: Acoelic Invertebrates of the Deep. KMK, Moscow (in Russ.).

Markov, A.V. (2001) Dynamics of the marine faunal diversity in the Phanerozoic: a new approach. Paleontol. J. 35, 1–9.

Markov, A.V. (2002) Mechanisms responsible for the increase in the taxonomic diversity in the Phanerozoic Marine Biota. Paleontol. J. 36, 121–130.

Maslennikov, V.V. (1999) Sedimentogenesis, Halmyrolysis and the Ecology of Pyritiferous Paleohydrothermal Fields: a South Urals Case. Geotur, Miass (in Russ.).

Natochin, Yu.V. (2005.) The role of sodium ions as a stimulus for the evolution of cells and multicellular animals. Paleonotol. J. 39, 358–363.

Peterson, K.J. and Eernisse, D.J. (2001) Animal phylogeny and the ancestry of bilaterians: inferences from morphology and 18S rDNA gene sequences. Evol. Dev. 3, 170–205.

Polevoy, V.V. (1985) The living state of the cell. In: V.V. Polevoy and Yu.I. Maslov (Eds), The Evolution of Function in Plants. LGU, Leningrad, pp. 36–45 (in Russ.),

Prozorov, A.A. (2002) Altruism in the bacterial world? Uspekhi Sovremennoy Biol. 122, 403–413 (in Russ.).

Rozanov, A.Yu. (2006) Precambrian geobiology. Paleontol. J. 40, S434–S443.

Rozhnov, S.V. (2005) Morphological patterns in the formation and evolution of higher taxa of echinodermata. In: E.I. Vorobjeva and B.R. Striganova (Eds), Evolutionary Factors of the Formation of Animal Life Diversity. KMK Scientific Press, Moscow, pp. 156–170 (in Russ.).

Rundkvist, D.V. (1968) Issues in mineral research. Zap. Vses. Mineral. O-va. 97, 191–209 (in Russ.).

Rundkvist, D.V., Denisenko, V.K. and Pavlova, I.G. (1971) Greisen Deposits (Ontogenesis and Phylogenesis). Nedra, Moscow (in Russ.).

Ruvinsky, A.O. (1991) Sex, meiosis and progressive evolution. In: V.K. Shumny, N.A. Kolchanov and A.O. Ruvinsky (Eds), Problems of Genetics and Evolutional Theory. Nauka, Novosibirsk, pp. 214–228 (in Russ.).

Schidlowski, M. (1988) A 3,800 million-year old record of life from carbon in sedimentary rocks. Nature 333, 313–318.

Semikhatov, M.A. (1993) The most recent scales for general division of the Precambrian: a comparison. Stratigr. Geol. korrel. 1, 6–16 (in Russ.).

Sepkoski, J.J. (1994) Limits to randomness in paleobiologic models: the case of Phanerozoic species diversity. Acta Palaeontol. Polon. 38, 175–198.
Sepkoski, J.J. (1996.) Patterns of Phanerozoic extinction: a perspective from global data bases. In: O.H. Walliser (Ed.), Global Events and Event Stratigraphy. Springer, Berlin, pp. 35–51.
Sergeyev, V.N., Noll, E.H. and Zavarzin, G.A. (1996) The first three billion years of life: from prokaryotes to eukaryotes. Priroda 6, 54–67 (in Russ.).
Shestakov, S.V. (2005) Contribution of genomics to investigation of prokaryotic evolution. In: A. Yu Rozanov and V.N. Snytnikov (Eds) Proceedings of the International Workshop on Biosphere Origin and Evolution. IC SB RAS, Novosibirsk, pp. 24–25.
Shmalgauzen, I.I. (1968) Evolutionary Factors (Stabilizing Selection Theory). Nauka, Moscow (in Russ.).
Sokolov, B.S. and Fedonkin, M.A. (1988.) Modern Paleontology. Nedra, Moscow (in Russ.).
Starobogatov, Ya.I. (1985.) Aspects of Speciation. VINITI, Moscow (in Russ.).
Taft, R.J. and Mattick, J.S. (2003) Increasing biological complexity is positively correlated with the relative genome-wide expansion of non-protein-coding DNA sequences. Genome Biol. 5. P1. Epub.
Tajika, E. and Matsui, N. (1992) Evolution of terrestrial proto-CO_2-atmosphere coupled with thermal history of Earth. Earth Planet. Sci. Lett. 113, 251–266.
Trubitsin, V.P. and Rykov V.V. (2001) Numerical models of mantle convection's evolution: In: N.L. Dobretsov and N.I. Kovalenko (Eds), Global Environmental Changes. SO RAN, filial GEO, Novosibirsk, pp. 42–55 (in Russ.).
Unrau, P.J. and Bartel, D.P. (1998) RNA-catalyzed nucleotide synthesis. Nature 395, 260–263.
Vasilyev, V.P., Vasilyeva, E.D. and Osipov, A.G (1983). First evidence favoring the main hypothesis of net-like speciation. Dokl. AN SSSR 271, 1009–1012 (in Russ.).
Vavilov, N.I. (1967) The law of homologous series in the inheritance of variability. In: I.A.Rappoport (Ed.), Selection from the Works of N.I. Vavilov, Vol.1. Nauka, Leningrad, pp. 7–61 (in Russ.).
Vernadsky, V.I. (1987) The Chemical Composition of the Earth and its Surroundings. Nauka, Moscow (in Russ.).
Vinogradov, M.E. (2004) Biological Productivity of Oceanic Ecosystems. Nauka, Moscow (in Russ.).
Zakrutkin, V.E. (1993) On the scale of organic matter accumulation in the Precambrian and Phanerozoic. In: A.Yu. Rozanov (Ed.), Problemy Doantropogennoy Evoliutsii Biosfery. Nauka, Moscow, pp. 202–212 (in Russ.).
Zavarzin, G.A. (2001) Formation of biosphere. Vestnik RAS 71, 988–1001 (in Russ.).
Zavarzin, G.A. (2003a) Lectures on Natural Resource Microbiology. Nauka, Moscow (in Russ.).
Zavarzin, G.A. (2003b) The antipode of the noosphere. Vestnik RAS 73, 627–636 (in Russ.).
Zhegallo, V.I., Kalandadze, N.N., Kuznetosva, T.V. and Rautian, A.S. (2001) The fate of megafauna in the Late Anthropogene. In: The Mammoth and its Neighborhood: 200 Years of Research. Geos, Moscow, pp. 287–306 (in Russ.).
Zherikhin, V.V. (1986) Biocoenotic regulation of evolution. Paleontol. Zh. 1, 3–12 (in Russ.).

Microbial Biosphere

G. A. Zavarzin

Abstract Evolution of prokaryotic biosphere is considered from the system point of view. It starts with the appearance of first organisms, the ~3.5 Ga date makes a border between the observed and imagined biosphere. Prokaryotic community dominated from Archaean to Mesoproterozoic. Prokaryotes make a sustainable community due to the cooperative action of specialized forms. The main route for establishing community is made by trophic links. Structure of the trophic links in prokaryotic community making a trophic network is an invariant, with secondary adaptive deviations. Material balance is the ultimate requirement for a long living self-supporting system. The system of biogeospheric cycles is dictated by the constancy of biomass composition establishing quantitative ratio between C_{org}:N_{org}:P_{org}. Biospheric processes are driven by the C_{org}-cycle. Carbon assimilation is limited by the illuminated moist surface populated by producers, and this means that the C_{org}-production remains within an order of magnitude of 10^2 Gt/yr. Evolution of the prime producers makes a stem for evolution of the biospheric–geospheric system but cyanobacteria integrated as chloroplasts remain to be its driving force. Decomposition of organic compounds is performed by organotrophic destructors, anaerobic being less effective. Destructors determine the residual C_{org} accumulation. Recalcitrant C_{org} remaining in the sedimentary record is equilibrated by O_2 and other oxidized compounds such as Fe-oxides or sulfates. The geospheric and biotic interactions include both direct and biotically mediated processes; the most important is the weathering–sedimentation pathway. The prokaryotic community makes a sustainable frame into which all other more complex forms of life fit in. That makes the prokaryotic biosphere a permanent essence of the whole system. New actors might come in and substitute for functional components only when they fit to the existing system. Evolution of the large system is additive rather than substitutive. "All originated from cyano-bacterial community" – is the slogan.

G. A. Zavarzin
Winogradsky Institute of Microbiology, Russian Academy of Sciences, Moscow, Russia
e-mail: zavarzin@inmi.host.ru

1 Introduction

Coming-into-being of the biosphere begins with appearance of the first organisms. All preceding events do not fall into the concept of the biosphere. Habitability precedes habitation. The Earth entering the corridor of habitability depends from both astronomical and planetary factors (Franck and Zavarzin, 2004). This implies the primacy of geospheric conditions. Link between geospheric and biospheric processes is realized by the system of biogeochemical cycles propulsed by the coupling with the cycle of organic carbon. The system of biogeochemical cycles was realized by activity of prokaryotic microorganisms. Development of the biosphere via cycling mechanisms needs functional diversity of actors and cannot be reduced to the single common ancestor. Evolution of the biosphere cannot be substituted for by the evolution of biota.

The organic carbon cycle is initiated by the primary producers and could be quantified in grams of C_{org}. This is why the evolution of the biosphere depends primarily on the evolution of photoautotrophic organisms, which, in turn, depend on the availability of the Sun light, and to some extent on chemosynthetic microorganisms, which depend on endogenic redox reactions in the geosphere. Evolution of the biospheric–geospheric system as a whole can be described as biogeochemical succession, driven by the incomplete balance in the production–destruction cycle and coupled to direct and mediated cycles. Since everything new for its establishment should be installed into already existing, the subsequent evolution of biota ought to be mounted into the frames of the prokaryotic biosphere. In this sense the evolution is additive, not substitutive. Substitution of components occurs inside the functional niche within the sustainable system. Since new system is installed into existing, old should be retained. The prokaryotic biosphere is a permanent constituent of the biosphere

The cyanobacterial community represents an entity with almost closed cycles, destructors being responsible for regenerative recycling of biogenic elements. General principles of the functional organization of the biosphere can be studied on the autonomous microbial communities in extreme environment. That is why it is possible to reveal the functional structure of the prokaryotic biosphere by means of studying extant relict microbial communities in extreme habitats, devoid of non-prokaryotic producers and consumers.

Practically, this means that interactions in the primitive biosphere can be studied actualistically with cyano-bacterial mat (Swiss "Matte") as a test system, where "cyano-" refers to the main prime producers and "-bacterial" – to the prokaryotic destructors. This system can be described chemically by the trophic substrate–product interactions.

The prokaryotic biosphere is complete and sustainable, values confirmed by its being a permanent basis of biotic system throughout the geological history.

Post-prokaryotic evolution of the biosphere has as its prerequisite successional changes, caused by accumulation of the products due to the incomplete destruction, the leading is stoichiometric pair C_{org} of the kerogen and oxygen introduced into the atmosphere, and consumed mainly for the formation of sulfates and iron oxides.

Evolution of biota proceeded within the meta-sustainable biosphere due to the rising of organisms complexity with limitation of possibilities for their versatility. The primary photosynthetic unit—cyanobacterium—was integrated into new entities and reductively transformed into chloroplasts. Subsequent evolution of prime producers occurred in the frames of cyto-morphological complication aimed at creation of appropriate physical environment for photosynthetic units. Deformation of the entirely prokaryotic biosphere is manifested by appearance of phagotrophy—nutrition by particles—based on cytology of protists and than multicellular metazoans like conspicuously symbiotic parazoans as spongy or, probably, archaeociates. The next step, zootrophy, represents variations on the theme "bowel with gonads" based on intestinal microbiota in a tubular chemostate-like structure.

The trophic approach to the evolution of biota unifies the biological and geological concepts in terms of chemical units. It gives possibility for the system to approach the Earth's biosphere, which evolves from its pioneering state by the coupled interactions in the serial quasistable states. This approach needs understanding of the system of interlinked events within unity of time and space rather than their order. The evolution, understood as a sequential order, does not explain everything as evolutionists claim it. The philosophical reason for persistent mis-comprehension is concentration on the singular instead of multiple in the Abendlandische Phylosophie (Heidegger, 1984). The interaction of bio- and geospheric systems is better understood with landscape in mind as A. Humboldt suggested it in XIX and V.N. Sukachev in XX century (biogeocoenosis principle). The aim of this work is not a comprehensive review with extensive bibliography but an attempt to have a frame for the system approach based on balances. Quatatis indicate on the subjects where I have my own experience. Priorities in the study are dictated by masses of material involved in the processes and turnover rates for reservoirs.

For the geospheric–biospheric system, the functional traits of its components are of primary importance. As a consequence, the virtual world of ribosomal phylogeny remains out of scope as well as "RNA-world." The present-day "sunflower" topology of prokaryotic phylogeny makes it difficult to represent evolution in old-fashioned "order" (Woese, 2004). Functional morpho-physiological traits, which are needed for the system analysis, do not correlate with the phylogenetic position and representatives of various phylogenetic lineages might have similar functions. An empirical rule, as I found, is that trophically interact phylogenetically distantly related organisms, constructing a cooperative functional entity.

2 Limits for Actualistic Principle

To what extent the actualistic principle is applicable for the study of Early Proterozoic and Archaen biosphere? This is the question, which should be taken into consideration each time. The greatest limitation is the composition of the present day atmosphere and the ocean equilibrated with it. Both of them are products of the biosphere. The simplest approach is that the atmosphere was formed by endogenic degazation and transformed by bacteria and photochemical processes. The bacterial community is particularly specialized in transformation of gaseous compounds and its cooperative action suffices to the transformation of the chemical composition of the atmosphere except inert gases (Zavarzin, 1984). It can be stated as an empirical evidence that the atmosphere of contemporaneous type was formed by cooperative action of the prokaryotic biota already in existence much earlier 2 Ga. There was no significant accumulation of oxygen until 2.4 Ga as it is concluded from various geochemical indicators, which does not exclude acting sources of O_2, which was scavenging from the atmosphere. That is the approximate date when the actualistic principle could be applied with some degree of certainty. Reconstruction of the Archean environment is more problematic. Did the iron-dominated cycles characterize it? What was the form of sulfur?

The history of the biosphere is the history of organic carbon. There is firm evidence that sedimentary organic carbon has biotic origin. Biologically mediated carbon isotope fractionation persisted over 3.5 if not 3.8 Ga, early deviation of $^{13}C/^{12}C$ ratio in Isua is interpreted as a result of high-temperature metamorphism (Schidlowski, 2002). The history of biota remains uninterrupted from that time with the autotrophic assimilation of CO_2 as the main income into the biogeochemical system. Organic carbon in biomass makes a minor dynamic reservoir leading to recalcitrant forms from humus to kerogen and dispersed reduced carbon of sedimentary rocks. Kerogen was balanced by oxygen in the Earth's system.

The origin of the Archean atmosphere supposes two sources: a remnant of primordial atmosphere, as it is indicated by depletion of rare gases from the cosmic ratio, and continuous degazation, as volcanic exhalations indicate it. The source of water remains enigmatic, including supposed transport by comets. It is assumed that the prebiotic atmosphere was neither reductive, no oxidative with N_2, H_2O, and some CH_4. It should contain also products of photochemical reactions. Two main sources of photochemical transformation should be considered: photochemical reactions of water vapor with production of hydroxyls and photochemical reactions of methane with possible production of various C-compounds. Could they serve as a source for organotrophic bacteria? It should be noted that fears about the lethal UV are not in accordance with presumed high iron content in Achaean water since iron serves as a perfect shield for 220–270 nm UV (Phoenix et al., 2001).

The traditional view on the early atmosphere is based on CO_2- dominated atmosphere and wollastonit equilibrium $CaSiO_3 + CO_2 \leftrightarrow CaCO_3 + SiO_2$ as geologically most important CO_2-sink (e.g. Schwartzman, 1999 and literature there). Calcium and magnesium carbonate deposits in conjunction with weathering and remnant clay minerals formation provide an evidence for the main route of mineral deposits evolution in sedimentary record. Carbonates are obvious reservoirs of the atmospheric CO_2 sink produced in an essentially subaerial process. The other pathway is subaqueous carbonatization of basalts, which was characteristic for the ancient ocean hydrochemistry (for review see Westall, 2003). Salinity and sulfate content of the Archaean ocean might be most different from the recent waters (Melezhik et al., 2005). Carbonates are responsible for the neutral environment on the Earth's surface. The geological context for early life should be based on paleogeographic mosaic, large portions of protocontinental blocks were actually submerged.

Biota–geosphere feedback was expressed first by the change of the chemical composition of the atmosphere as the small dynamic reservoir with the substitution of O_2 for CO_2 and transition to oxidative state. Presently it is supposed that there was methane-rich atmosphere during 2.8–2.6 Ga and at the same period oxygenic photosynthesis evolved. Great oxidation occurred during 2.2–2.0 period as it is represented by jatulian deposits in Karelia (Melezhik et al., 1988). BIF formation ceased at 1.8 Ga after their presence since 3.5 Ga. About 1.9 Ga or even earlier eukaryotes appeared with coming to dominance around 1.2–1 Ga.

3 Relict Microbial Communities

A pragmatic assumption is that the prebiotic atmosphere was like gas streams from volcanoes. However, which kind of volcano? Gaseous emissions from volcanoes of border islands like Kuryls and Kamchatka coming to the surface are influenced by thermal transformation of overlaying sedimentary deposits with injection of meteoric water and air in gas stream. The composition of emissions is close to the thermodynamic equilibrium. Submarine vents undergo even stronger influence of convective exchange with the deep water of the modern sulfate-rich ocean. May be a better example of ancient microbial community is the deep subterranean habitat where water comes into equilibrium with porous rocks. The presence of the prokaryotic life in the hydrothermal environment is stated for 3.3–3.5 Ga Barberton sediments (Westall et al., 2001). Two types of extant microbial biogeocoenoses are related to endogenous exhalations.

Hydrogenotrophic chemosynthetic microbes represent the first type. These microbes use H_2 produced by reduction of water by superheated iron of igneous rocks with partial oxidation of iron. An immediate consequence is that these organisms include hyperthermophiles that develop over the zone of condensation of the vapor of hydrothermal fluids at approximately 100 °C. The trophic problem in H_2-rich environment is the access to an oxidant. There are two possible oxidants: CO_2 and S_0, which give trophic niche for methanogenic archaea by the reaction $4H_2 + CO_2 = CH_4 + 2H_2O$ and sulfur reducing sulfidogenic archaea: $H_2 + S_0 = H_2S$. A large number of hyperthemophiles of these types were described by group of C. Stteter and other students (for review see Bonch-Osmolovskaya et al., 2004). Sulfate-reducing *Archaeoglobus* could acquire this ability in the sulfate-rich environment of recent hydrotherms. Another possibility is fermentation of organic substances by *Desulfurococcus*-type organotrophs, which need elimination of excessive H_2 by reaction with S_0 ("S-dependent archaea") or simply by the gas stream. At present the source of organic compounds seems to be boiled biomass transported by convective stream in shallow-water hydrotherms. The product of incomplete reduction of CO_2–carbon monoxide—might be oxidized by water: $CO + H_2O = CO_2 + H_2$ by extensively studied now *Carboxydothermus*, which was the first representative of a novel group of anaerobic lithotrophic eubacteria (Svetlichny et al., 1991). Of course, H_2 escaping to the surface is oxidized by thermophiles with different growth temperature ranges (Miroshnichenko, 2004). In modern O_2- environment H_2 can be oxidized by extremely thermophilic "Knallgasbakterien" like *Hydrogenobacter, Calderobacterium, Aquifex* and it should be noted that, phylogenetically, they are among the most deep-branching eubacteria. The question is if "molecular clock" for thermophiles are adjusted to the same path of time as for other creatures? Below-the-ground biogeocoenosis may be considered as the oldest persisting relict, especially in porous space of cratons because of presumably changing composition of the oceanic water. The problem for scaling-up activity of methanogenic and sulfidogenic ancient microbes comes from the need to close biogeochemical cycle by oxidation of CH_4 with its strong greenhouse effect. Large-scale photochemical production of organic compounds from CH_4 in anoxic atmosphere washed out from the atmosphere could make a source of nutrition for anaerobic microbial community, a possibility, which as far as I know, was not yet explored. Excessive H_2S does not make such a problem for the atmosphere since it might be bound by iron, presumably present in the ancient ocean; however no geological evidences for such type of ancient sedimentary iron sulfide depositions are reported—iron remains as oxides. Studies of microbiota from deep drills to about 3 km until now gave the same types of microbes, which are known from volcanoic regions and thermal vents (Bonch-Osmolovskaya et al., 2004). If estimation of subterranean microbiota in 10^{30} organisms (Whitman et al., 1998) is correct, then we can assume that life did not develop numerically.

Oxidized iron, remaining from H_2O reduction, might be reduced to magnetite by thermophilic iron-reducing bacteria (FeRB) (Slobodkin et al., 1995) or to siderite. Magnetite is formed by thermophiles under high pCO_2, excess of $Fe(OH)_3$, and the absence of organic colloids, while siderite is formed under high pCO_2, limited iron hydroxide, the presence of utilizable organic compounds (Zavarzina, 2004). The iron cycle driven by geochemical reactions at high temperatures seems to be the key process for Archean–Early Proterozoic. Is it possible to consider the period before accumulation of sulfate in the ocean and sulfate reducers as a period of biospheric "iron cycle"? Iron cycle catalyzed by long known oxidative "iron bacteria" and reductive ferric reducers comprehensively studied by D. Lovley represents a possibility for microbial life in the Archean–Early Proterozoic. Anaerobic production of Fe(III) by phototrophs, which is known both for cyanobacteria and anoxygenic non-sulfur bacteria, could serve as an oxidant (Widdel, 1993). Banded Iron Formations (BIFs) make an existing evidence for the possibility of such hypothetical cycle. The iron cycle is compatible neither to sulfur cycle nor to the present type oxygen cycle. Phototrophic microbial actors for all three possibilities, that is Fe, S, O are in hands. Even patches of appropriate extant communities are found somewhere. To what extent these possibilities of anoxygenic prime producers are consistent to Archaen–Paleoproterozoic $\delta^{13}C \approx -25‰$ kerogen accumulation record?

Another type of hydrothermal environment on the Earth's surface leads us to the stem of biospheric evolution. Low temperature, approximately less than 65°C, hydrotherms characterized by CO_2 stream ("mophetes") are inhabited by moderately thermophilic cyanobacterial communities dominated by mat-forming cyanobacteria like *Mastigocladus laminosus* and *Phormidium laminosum,* the species epithet indicates the structure of community. Studied by Brock (1986) group in Yellowstone and by our group in Kamchatka and Kuryls (Zavarzin et al., 1989) indicate that thermophilic cyanobacterial communities seem to be trophically complete with closed cycles. In addition to cyanobacterial prime producers they include various thermophilic organotrophs, which comprise trophic chains leading to methanogens and sulfate reducers. In methanogenic trophic chain thermophilic acetogens and acetoclastic methanogens as *Metanosaeta* (formerly *Methanothrix*) *thermoacetophila* are present. However, thermophilic methanotrophs are remarkably absent here, biogenic oxidation of methane starts at lower temperature. The sulfur cycle is initiated by thermophilic sulfate reducing *Thermodesulfobacterium*, which represents a separate phylogenetic lineage, but functionally is equivalent to conventional desulfovibria. Oxidation of H_2S is mediated by various thionic bacteria with spectacular streamers of *Thermothrix* capable of oxidizing H_2S into S_0 aerobically and of reducing sulfur in anaerobic conditions. Excessive sulfur is oxidized outside the cyanobacterial community by extremely acidophilic aerobic microbial associations, which are the main producers of sulfuric acid and initiate geochemical events resulting in rock-leaching by sulfuric acid and acid iron-bearing streams ending with limonite deposition. Highly acidophilic microbial communities that include both thermophilic archaea and/or mesophilic eubacteria are most characteristic for the thermal fields marked by snow-white clays in place of weathered lava rocks. Aerobic sulfur oxidizers are mighty producers of sulfuric acid. However, it should be left outside the present discussion since these communities are oxygen dependent. In anoxic environment anaerobic production of sulfates is performed by phototrophic sulfur bacteria (see Gorlenko, here) and it is tempting to imagine "purple palaeoocean." Sulfur dependent phototrophs are absent at elevated temperature.

The thermophilic cyano-bacterial community that develops in CO_2 dominated mophetes represents an actualistic example of how gaseous exhalations were transformed into modern type atmosphere. The photosynthesizing cyanobacterial community, studied both in situ or in vitro, exposed under the light to volcanic exhalations or their imitates in the laboratory transforms composition of gases into air-like mixture with about 25% of O_2, which is the upper limit. Remarkably, methane behaves here as an inert gas—it is not oxidized by the community (Gerasimenko and Zavarzin, 1982). The question is scaling up of the process: cyanobacterial communities in hydrotherms on the day-surface represent local sites. Is it possible to extend the application of results to the ancient Earth's surface? Landscape of what kind and extension is needed for such extrapolation?

3.1 Formation of Landscape in Prokaryotic Biosphere

Let us define landscape as an entity of biota and geographic environment, large enough to consider it as uniform. Landscapes are classified hierarchically from large climatic and orographic zones to small elementary habitats. The Earth's surface never was as uniform as well-mixed atmosphere, even ocean with its bays and border seas can be considered as more uniform in comparison to the terrestrial environment. There was always a mosaic of environments during the history of Earth's surface. Generalist gross approach of homogenic environment on Earth during certain period of geological time is entirely unacceptable for microbiologist whose objects live in microhabitats, and some landscapes could serve as refugia even for "Snow-ball" events. The reason is that microbes are easily disseminated by wind in aerosols including non-habitable surfaces like Antarctic, which makes the statement doubtful that biosphere is everywhere where life forms are found. The biosphere is where microbes are actively involved in the geochemical processes. Microbes have a short life cycle and that makes adaptive dynamics with the change of species, not adaptation, the main mechanism for formation of microbiogeocoenoses. Landscapes, which could have global influence, should have large enough moist Sun-illuminated surface for accommodation of chlorophyll containing photoautotrophic prime producers. Since assimilation of CO_2 is proportional to the active chlorophyll and wet surface was approximately the same during the Earth's history it is realistic to accept the present annual production of 10^{18} grams C_{org} as approximately constant value within the order of magnitude for the biosphere of modern type.

Ocean of course represents the largest wet surface. At present it produces about a half of primary production. For the prokaryotic biosphere, cyanobacterial picoplankton is of primary importance. Quantitatively, it delivers a quarter of present annual global C_{org} production. Was ocean a cradle of life?

There are certain limitations in discussing the role of the Archaean ocean as a continuous body of water since our understanding of its chemistry and geography is most illusive. Of course it worked as a sink for deposits, its surface was in dynamic equilibrium with the atmosphere, and most probably its bottom represented the main route for volcanic degazation. However, the ocean is too large for being a cradle. Life could not be concentrated there. The distances, except for veils in stagnant zones, limit interactions between organisms. Another drawback of the microbial evolution in the ocean is the permanency of physical and chemical conditions: there is no need to change and adaptation facilities are oriented to restrictive evolution. Nevertheless the presumed role of ancient cyanobacterial picoplankton as the possible source of oxygen seems to be undisputable. It is the source large enough for oxidation of iron in Banded Iron Formations and it is able to produce enough C_{org} to reduce part of it into magnetite by anaerobic dissimilatory iron reducers. At least two known limitations for cyanobacterial picoplankton of nowadays were absent: limitation by iron and limitation by filtrating zooplankton which keeps the concentration level below 10^5 cells/ml, but cyanophagy with threshold about 10^4 could exist. The presence of this type of cyanobacteria is not yet proved by paleontology. It would be quite difficult to identify minute 0.5–0.8×0.7–1.6 μm^2 coccoid cells of *Prochlorococcus* morphotype as microfossils of cyanobacteria. In this case the time gap between BIF and first undisputable trichomic cyanobacteria (Sergeev, 2006; Sergeev et al., 2002) as O_2-producers might be illusive. Is it possible to consider nannoplankters in the limited by biogens contemporary "blue ocean" as remnants of an ancient community?

Amphibial landscapes look most promising as a cradle for microbiota. Exchange between the gaseous, water, and solid phases occurs here most easily. These landscapes with shallow water of marine or lacustrine origin usually represent concentration of life with strong aerobic–anaerobic interaction. They are densely populated by diverse microbiota. This facilitates trophic interaction between microorganisms. Such landscapes are unstable and promote diversification and adaptive changes both for subaqual and subaerial development. Shallow deposits of continental slope contains spectacular evidences of the ancient microbial life.

Terrestrial plains of humid climatic zones, which are now occupied by meadows and forests, seem to be an appropriate environment for cyanobacterial mats. Nowadays there are only patches of mats in ephemeral bodies of water, puddles on the ground. These sites are immediately occupied by oscillatorian cyanobacteria. Nostocacean dominates in a drier climate. Production of C_{org} by soil cyanobacteria is highly variable but considerable. This is an indication to the possibility of the ancient subaerial terrestrial life, which is not as yet supported by the data from paleontologists or geologists. The crucial point for these habitats is the formation of water proof bottom under the mat either by calmatation from allochtonous deposits or, that is much more promising, by microbially mediated transformation of lithogenic minerals into clays with smectites as the most suitable ground material.

It brings us to the most crucial problem of weathering in Proterozoic. Since it was stated by Retallack (1990) and confirmed by many other observations, Precambrian soil profiles are considered to be much the same as at present. In Russian usage since pioneering works of B.B. Polynov the term humified soil *sensu stricto* differs from mineral "weathered rock" while English usage is soil *sensu lato*. The mechanism of chemical weathering includes substitution for metals in initial minerals by protons retaining Si-Al backbone stabilized in clay minerals. Acid in question is mainly carbonic acid and the process represents the main sink for CO_2 with carbonates as the synthesized end product and clays in the rest. Weathering was reviewed with emphasis on kinetics at high temperature and it was stated that this process is too slow at mild temperatures (see Schwartzman, 1999). Additional kinetic factor is needed. Formation of clays occurs very rapidly on thermal fields not only in the streams of hydrothermal steam but mainly due to the intervention of acid-forming bacteria at mild temperatures (Karpov et al. 1984). Cations are transported in the humid climate zone to the sea by run-off or accumulated in the depositional basins of endorheic arid regions. In this case soda lakes are formed.

To my understanding, scavenging of CO_2 from the atmosphere is essentially subaerial process, while endogenic generation of CO_2 most probably occurs mainly via deep-sea subaqual degazation. If so, the ocean was netto-source of atmospheric CO_2, while moist terrestrial subaerial surface represented netto-sink. In this case soda lakes of endorheic regions and its biota are relevant to the global scavenging of atmospheric CO_2.

There are two main problems for weathering at mild temperatures. The first is the concentration of carbon, which takes place during autotrophic assimilation and is followed by the release of CO_2 locally in the sites of decomposition of dead biomass ("mortmass"). Organic acids might be produced as additional strong leaching agent. Decomposition of organic matter makes the main mechanism of leaching. Another problem is transformation of inert CO_2 into carbonic acid as a leaching species. The chemical process is drastically enhanced by enzyme carbonic anhydrase (CA): $CO_2 + H_2O + CA \leftrightarrow [CO_2-CA-H_2O] \leftrightarrow HCO_3^- + H^+ + CA$, which is universally present in living cells to transport CO_2 in and out of the cells. The only report on extracellular CA known to me is its location in glycocalix of halophilic *Microcoleus* (Kuprriianova et al., 2004).

The illuminated wet surface indicates shallow water bodies as preferable sites for cyanobacterial community. What does shallow mean? Limited by the photic zone? Being more restricted, let us define such an area as an amphibial landscape, where the solid and liquid phases are in approximate ratio 1:5 and an easy exchange with the atmosphere is available. Two types of amphibial landscapes come immediately into mind: marine or thalassic and terrestrial or athalassic. As examples of the thalassic type lagoons, tidal flats, sabkhas, marshes, German "Watte" of different kinds with spectacular microbial "Farbstreifbandsandwatt" can be mentioned. For athalassic types ombrotrophic bogs as autonomous, and swamps as subordinate type, lakes and dry lakes as accumulating sites, takyrs on peneplaine come into mind. All of them are quite diverse. Thalassic amphibial landscapes are particularly important for epicontinental seas that developed in thalassocratic epochs on the passive margin of continents. Epicontinental (epeiric) seas are the beloved sites for both paleontologists and sedimentologists. Geochemically, marine and terrestrial landscapes differ in domination of chlorine in marine environment ("halite-ocean") and bicarbonate in terrestrial waters ("soda continent"). Transport of aerosols, which is important for terrestrial environment, is usually forgotten. Alkaliphilic microbial community of soda lakes was studied in detail and its trophic system is now being described (Zavarzin, 1993 Zavarzin, 2006, Alkaliphilic microbial communities, 2007).

The landscape is a receptacle for both microbiota and products of microbial activity. The landscape is changed under the influence of microbiota. The visible landscape extends for microbiota by an invisible landscape of intersticional porous waters and mineral surfaces. Due to SEM studies it is clear now that all surfaces are covered by biofilms (see Fossil and Recent Biofilms, 2004). The problem is to establish to what extent biofilms are causative agents of particular geochemical transformations. The surface of minerals is covered by organo-mineral cutanae originating from the soil solution. Pedogenic minerals are transformed accordingly to the thermodynamic equilibrium with microbially modified soil solution, both by dissolution and by precipitation. Interaction of microbes with the solid surface, including direct or mediated catabolic processes, still remains an obscure field. Microbes, as iron-oxidizers, for example, at the first step produce disperse amorphous insoluble compounds, ferrihydrite in this case, which is stabilized during early diagenesis into magnetite or siderite in the order of dynamic of crystallization (Zavarzina, 2004). In the colloid environment electrostatic interactions dominate. Coagulation of mineral colloids is an important process not only for purification of potable water in water supply systems but also as a large-scale natural process, designated as A. Lysitcin's "marginal filter."

Texture of deposits is modified due to the colloid matrix formed by microbes, and biofilms lead to multilayered laminated structure as in stromatolites. The process can be simulated in the laboratory when cyanobacterial film develops on sediments. When new portion of sediment covers the surface, part of cyanobacteria (hormogonia) move to the surface. As an exotic example of such a process, diurnal layers of *Phormidium laminosum* in thermal spring in Uzon, Kamchatka with layers of colloid sulfur can be observed. In a large scale the same process takes place with carbonate precipitates in evaporative environment. For alkaliphilic communities of Central Asia and halophilic communities of marine-dependent Satonda lake, the role of glycocalix in carbonate lamination and deposition was argued by Arp (1999).

Formation of landscape on macrolevel is illustrated by the well-known formation of stromatolites. It includes the formation of reefs as biogerms in hydrodynamically active outer slope, the formation of lagoons and tidal flats leading to carbonate platforms as described by Grotzinger (1989). The outer line of biogerms protects the shallow plane from hydrodynamic impacts.

Cyanobacteria develop here in hard ground, protected by mineral crust from mechanical stresses as opposed to the soft ground (Sergeev, 2006). It is important to interpret this total event as an evaporative process: carbonate deposits are formed as evaporites on the shallow or dry surface due to degazation of CO_2 from bicarbonate solution and rising concentration of dissolved solids to the beginning of carbonates precipitation (more than 10-fold concentration over saturation index).

For microbiologist this means that a moderately halophilic community should dominate the habitat. The halophilic community is not an exotic assemblage of extremophiles but a component of the evaporative environment where chemical sedimentogenesis occurs. Diagenesis in the sites of chemical sedimentation involves a halophilic microbial community, moderately halophilic for the early stages of precipitation and extremely halophilic for formation of evaporites. This community differs both from marine and freshwater communities. Nowadays *Microcoleus chthonoplastes* mats dominate most of these haline environments. A new biofilm up to 2 mm thick develops on the layers of older and dead cells as a kind of multilayered cyano-bacterial "peat." Sites of chemical sedimentation and cementation might serve for burial of allochtonous microbiota, brought in both by water currents and by wind. These dispersed remnants differ from aborigenal biophilms formed by halophilic microorganisms. A large renewed reservoir for chemical precipitation obviously should be below the precipitation point.

Stromatolites correlate with dolomite and magnesite formation epochs. Their formation occurred within carbonate provinces, being restricted by the photic zone. It was supposed that the decline of stromatolites after 2 Ga on the Fennoscandian shield occurred due to transformation of the landscape with numerous shallow-water lakes saturated with Ca and Mg carbonates as terrestrial basins, devoid from sulfates, to a deep-water sea during "oceanization" of the region (Melezhik et al., 1997, 2001), but now formation of sulfates, either local or marine, is recognized. The high productivity, which is known for inland bodies of water with high mineral content, coupled to the partial suppression of decomposition due to the limitation in closing sulfur cycle could be the cause of carbon-rich shungite formation.

There is an old idea that carbonates are deposited by the photoautotrophic activity due to the rise of pH. In this case the ratio $C_{org}:C_{carb} \approx 1:1$ should be as for corals and *Halimeda* lime formation. However, in our experiments with alkaliphilic cyanobacteria *Microcoleus* it was found that living photosynthesizing bacteria do not precipitate carbonates even under chemically favorable conditions of pH. By contrast, the mineral deposits encrust rapidly the dead bacteria turning the sheaths into microfossils, "sarcophagus." This observation excludes photosynthesis as the driving force of lithification in the case under study but makes an emphasis on the role of glycocalix as the site for nucleation. This is in accordance with the observations in situ by Arp (1999). The reason for the difference between functions of glycocalixes in living and dead cyanobacteria might be the presence of CA in the former, since CA serves as bicarbonate ion producing mechanism (Kuprriianova et al., 2004), preventing carbonatization. On the other hand, production of CO_2 from bicarbonate stimulates formation of carbonate-ion needed for Ca precipitation. A possible role of CO_2 volatilization in warm water should be considered.

3.2 Trophic Structure of Cyano-bacterial Community

Cyano-bacterial communities are the most persistent components of the modern type biosphere. Persistence is guaranteed by almost closed cycles of biogenic elements. Light makes an ultimate energy source; for early stages of biosphere evolution the lower Sun luminosity and different spectral transparency of the atmosphere should be considered. Volatile compounds such as CO_2 are consumed from the atmosphere. Waste compounds such as O_2 and minerals formed in biotically mediated reactions—the list should be specially considered by geologists since it includes not only typomorphic minerals as Ca and Mg carbonates, iron and manganese oxides but also clays—make the driving force for the geochemical succession in sedimentary record. The autonomous community has approximately closed cycles of productive and destructive branches; the latter is sometimes called "regenerative cycle."

The trophic structure follows the Winogradsky's rule: each natural compound has its specific microbial consumer. If chemical compounds are listed in one column, then in another column corresponding consumers should be listed. Producers have constant composition of biomass with the atomic ratio close to $C_{org}:N_{org}:P_{org} \approx 106:16:1$ and approximately the same bulk composition of aminoacids, nucleotides, lipids, carbohydrates. The difference is mainly in storage compounds and structural components. The chemical diversity, which makes a ground for chemotaxonomy, deals mainly with minor compounds and is out of scope in our gross scale. Similarity in chemical composition leads to involvement of similar trophic groups of destructors, which have their trivial designations according to the substrates consumed and products formed. The system is complete. Philosophical generalization formulated by Winogradsky was that "the circle of Life" represents "one huge organism."

The trivial functional classification of bacteria, elaborated in its general lines at the end of XIX century by Winogradsky, Beijerinck and their followers, is what is needed for practical use in natural sciences. There are groups like photoautotrophs divided into conventional oxygenic and microbial anoxygenic, proteolytic and saccharolytic, aerobic and fermentering decomposers, anaerobic methanogens and sulfidogens. That makes their firm knowledge needed to each natural scientist. Detalization leads to the number of other functional groups, which make the basis of general microbiology. The functional trophic structure is much the same for different microbial communities in spite of diversity of their habitats. It gives a possibility to predict organisms from missing links in the graph of trophic network. The trophic structure may be treated as a systemic invariable throughout the history of the biosphere. Each of these functional groups contains a lot of representatives adjusted to special habitat. Bacteriologists know more than 2×10^4 molecular clones (ribotypes) from which less than 6000 species are validated. Diversity of known microorganisms started to expand extremely fast when molecular methods came into wide use with description of taxa by the methods of discrete mathematics. Cladistic tree graph nowadays does not represent any acceptable topology for the microbial diversity, which is presented now for prokaryotes by a set of multiple radiating phila (Woese, 2004). My old understanding is that the microbial diversity is best represented by combinatorial matrix (Zavarzin, 1974).

The functional trophic structure for different microbial communities is organized toward the universal lines. The general scheme of trophic network in prokaryotic community includes groups of prime producers, aerobic hydrolytics and dissipotrophs (organisms utilizing low molecular weight compounds dissipating from the places of origin), anaerobic microorganisms starting with particulated organic matter (POC) sinking through oxicline created by depletion of O_2 by aerobes. The anaerobic pathway is started by hydrolytics, which decompose insoluble cell wall constituents, fermentative dissipotrophs like spirochaete, syntrophic associations decomposing non-fermentable fatty acids in conjunction with hydrogen-scavenging lithotrophs, represented by sulfate reducing bacteria (SRB) or methanogens. The cycle is closed by aerobic gas-consuming methanotrophs and sulfide-oxidizing bacteria, which protect their anaerobic benefactress from oxygen. As a variant, the trophic loop is closed by anaerobic anoxygenic phototrophic bacteria, responsible for purple layers on the illuminated surfaces. Their development is most pronounced in mineral-rich water and even in brines. The trophic network is spatially organized in benthic cyano-bacterial mats with their regular structure of similar architecture. The upper layer up to 2 mm is occupied by cyanobacteria, followed by white sulfur oxidizers, below are purples, and in the bottom is black mud with sulfidogens. The structure is repetitive due to the burial. It was studied in hypersaline lagoons and many times reviewed (Microbial mats, 1989, 1994; Biostabilization of sediments, 1994; Fossil and recent biofilms, 2003). Mat is considered as the predecessor of stromatolites, which represent a large-scale geological evidence for domination of these type of cyano-bacterial communities during Proterozoic.

Of special interest is the same structure of mats in soda lakes, which represent an extreme case of the terrestrial athalassic environment (Gerasimenko and Orleanski, 2004; Arp, 1999). This biocoenosis has come to attention only recently and representatives of many functional groups have been isolated and described during last decade (for review: Alkaliphilic microbial communities, 2007). Most of the isolated alkaliphilic bacteria belong to new genera (Zavarzin et al., 1999; Zavarzin and Zhilina, 2000). Since soda lakes are the end basins they may be considered as some kind of natural lysimetrs for watershed and weathering thereof in continental environment. Architecture of cyano-bacterial mats follows the principle of construction of phototrophic community on solid surface which could be traced up to the moss cover as in *Sphagnum*-bog with 5 cm of green layer, 5–10 cm of decomposing "white moss" and peat below. Young offspring's grow on dead bodies of the parents. This style of architecture ceases with vascular plants cover, which "switch in" the evapotranspiration and principally changes the atmospheric hydrological cycle on the terrestrial surfaces.

More important is the fact that cyano-bacterial community is autonomous that leads to its persistence on the Earth. It is the most persistent biocoenosis on the Earth's surface. Chemical cycles of essential elements are closed within the community. The cyano-bacterial community is sufficient to catalyze all biogeochemical cycles (Zavarzin, 2003a,b). One exception should be taken in mind: the cyano-bacterial community works as a sink for phosphorus when it is responsible for the formation of phosphorite deposits. Anaerobic remobilization of phosphates seems to be not sufficient to close the phosphorus cycle (for overview see: Keasling et al., 2000). The cyano-bacterial community is sustainable as a complete system.

The role of biotically mediated reactions in the prokaryotic biosphere is still unclear. There are only examples of microbial participation in diagenetic processes, guesses about weathering but no full-scale picture, which needs more geological and geochemical knowledge on early C_{org}-cycle. The pendulum of interpretation of stromatolites as organo-sedimentary structures based on cyano-bacterial activity swing up to the interpretation of formation of laminae based mainly on inorganic evaporative precipitation followed by rapid colonization of the surface by biofilms responsible for the texture of the rock. Formation of dolostones and magnesite deposits, which correlate with high ^{13}C-content carbonates, are ascribed to playa evaporative environment of high bioproductivity. The most important fact is the pronounced negative correlation between C_{org}-poor carbonates depositions with Fe(III) as indicator of oxidative status and rich C_{org} deposits of fine sediments in anaerobic highly productive non-marine basins (Melezhik et al., 1988, 2001). Let us keep in mind: spatially separated from photosynthetic production anaerobic decomposition processes, which became to be responsible for netto-oxidative state of the entire geochemical system, favor misbalance of O_2/C_{org}. Paradoxically anaerobes created oxidative state due to the incomplete oxidation and accumulation of reduced carbon. What was the role of microbially mediated processes transforming the chemical environment?

The skeleton-forming ability came into action beginning from the appearance of Protists. With the development of radiolarians, diatoms, spongy Si became involved in the biological cycle and became a limiting element in the ocean; Sr is used by acantharians. But the main change is in Ca-recycle, which came totally under the biotic control in the skeleton forming $CaCO_3$ ability and biotically mediated dissolution during weathering in the biosphere.

As a result of completeness of biogeochemical cycles in the prokaryotic biosphere all subsequent evolution of biota is submerged into the essential trophic structure created by prokaryotes. Sustainable development retains what has been created before. Everything new can survive if it fits to the old, already existing. The evolution of biota is rather additive than substituitive at least in the functional aspect. That gives a general explanation why primitive, I should prefer to use word "pioneer," organisms remain and proliferate in spite of growing biodiversity of seemingly fittest more complex organisms. The line of partial substitution of prime producers from cyanobacteria by protists, by multicellular algae, by mosses, by vascular plants makes the backbone for the evolution of biota and to lesser extent to the geochemistry of the biosphere due to the full-scale C_{org} production in each moment. In feedbacks it gives new possibilities to microbes mainly in regenerative cycles. As a slogan we can state: "we all came from the cyanobacterial community."

References

Alkaliphilic Microbial Communities (2007) Trudy Winogradsky Institute of microbiology, vol. 14 (Russ.).
Arp, G. (1999) Calcification of non-marine cyanobacterial biofilms (USA, PR China, Indonesia, Germany)—Implications for the interpretaion of fossil microbialites. Dissertation Gottingen, 1999 (delivered by the courtesy of the author).
Arp, G., Reimer, A. and Reitner, J. (1999) Calcification in cyanobacterial biofilms of alkaline salt lakes. Eur. J. Phycol. 34, 393–403.
Bonch-Osmolovskaya, E.A., Miroshnichenko, M.L., Sokolova, T.G. and Slobodkin, A.I. (2004) Thermophilic microbial communities: new physiological groups, new habitats. Proceedings of Winogradsky Inst. Microbiol., Moscow, Nauka Publ., vol. 12, pp. 29–40 (Russ.).
Brock, T.D. (1986) Thermophiles. General, molecular, and applied microbiology. Wiley, New York, 316 p.
Ehrenreich, A. and Widdel, F. (1994) Phototrophic oxidation of ferrous minerals—a new aspect in the redox microbiology of iron pp. 393–402. Microbial mats: structure, development and environmental significance. In: L.J. and P. Caumette (Eds), Springer-Verlag, Berlin, NATO ASI Ser., v. G., vol. 35.
Fossil and Recent Biofilms. (2003) A natural history of life on Earth. In: W.E. Krumbein, D.M. Paterson and G.A. Zavarzin (Eds), Kluwer Ac. Press, Dordrecht, 482 pp.
Franck, S.A. and Zavarzin, G.A. (2003) What are the necessary conditions for origin of life and subsequent planetary life-support systems? In: H.J. Schellenhuber, P.J. Crutzen, W.C. Clark, M. Claussen and H. Held (Eds), Dahlem Konferenzien 91. Earth System Analysis for Sustainability. MIT Press, Cambridge Massachusetts, London UK, pp. 74–90.
Gerasimenko, L.M. and Orleansky, V.K. (2004) Actualistic paleontology of cyanobacteria. Proceedings of Winogradsky Inst. Microbio., Moscow, Nauka Publ. vol. 12, pp. 80–108, (Russ.).
Gerasimenko, L.M. and Zavarzin, G.A. (1982) Metabolism of H_2, CO_2, O_2, CH_4 in cyanobacterial communities. Microbiologia 51, 718–722 (Russ.).
Grotzinger, J.P. (1989), Facies and evolution of Precambrian carbonate depositional systems: Emergence of the modern platform archetype. In: P. Crevello, R. Sarg, J.F. Read and J.L. Wilson (Eds), Controls on Carbonate Platform and Basin Development: SEPM Special Publications, no. 44, pp. 79–104.
Heidegger, M. (1984) Was heißt Denken? Vorlesung Wintersemestr 1951/52. Philipp Reclam jun. Stuttgart, 80 p.

Karpov, G.A., Eroschev-Shakh, V.A. and Zavarzin, G.A. (1984) Role of biogenic factor in formation of environment for the zone of argillization in the contemporary hydrothermal systems and solfataric fields. Volcanol. Seismol. 2, 64–74.

Keasling, J.D., van Dien, S.J., Trelstaad, P., Renninger, N. and McMahon, K. (2000) Application of polyphosphate metabolism to environmental and biotechnological problems. Biokhimia 65, 394–404 (and other papers in this special issue).

Knoll, A.H. Life on a young planet. The first three Billion years of evolution on earth 2003 Princeton Univ. Press, 277p.

Kuprriianova, E.V., Markelova, A.G., Lebedeva, N.V., Gerasimenko, L.M., Zavarzin, G.A. and Pronina, N.A. (2004) Carbonic anhydrase of the alkaliphilic cyanobacterium *Microcoleus chthonoplastes*. Microbiology 73, 307–311.

Melezhik, V.A., Basalaev, A.A., Predovsky, A.A. et al. (1988) Carbanaceous deposits of early stages of the Earth's evolution (Geochemistry and conditions of accumulation on the Baltic shield) Leningrad, "Nauka," 200 p. (Russ.).

Melezhik, V.A., Fallick, A.E., Makarikhin, V.V. and Lyubtsov, V.V. (1997) Links between Paleoproterozoic paleogeography and rise and decline of stromatolites: Fennoscandian Shield. Precambrian Res., 82, 311–348.

Melezhik, V.A., Fallick, A.E., Medvedev, P.V. and Makarikhin, V.V. (2001) Paleoproterozoic magnesite: lithological and isotopic evidence for playa/sabkha environments. Sedimentology 48, 379–397.

Melezhik, V.A., Fallick, A.E., Rychanchick, D.V. and Kuznetzov, A.B. (2005) Paleoproterozoic evaporites in Fennoscandia: implications foe seawater sulphate, the raise of atmospheric oxygen and local amplification of the $\delta^{13}C$ excurse/Terra Nova 17, 141–148.

Microbial Mats (1989) Physiological ecology of benthic microbial communities. Y. Cohen and E. Rosenberg (Eds), ASM, Washington, DC, 494 p.

Microbial Mats (1994) Structure, development and environmental significance. In: L.J. Stal and P. Caumette (Eds), NATO ASI ser., v.G Ecological sciences. Springer-Verlag, Berlin, vol. 35, pp. 443–452, 463 p.

Miroshnichenko, M.L. (2004) Thermophilic microbial communities of deep-sea hydrotherms. Rev. Microbiol. 73, 5–18.

Phoenix, V.R., Konhauser, K.O., Adams, D.G. and Botrell, S.H. (2001) Role of biomineralzation as an ultraviolet shield: implications for Achaean life. Geology 29(9), 823–826.

Retallack, G.J. (1990) Soils of the Past: An Introduction to Paleopedology. Unwyn Hyman, Boston.

Schidlowski, M. (2002) Sedimentary carbon isotope archives as recorders of early life: implications for extraterrestrial scenarios. In: G. Palyi, C. Zucchi and L. Caglioti (Eds), Fundamentals of Life, Elsevier, pp. 308–329.

Schwartzman, D.W. (1999) Life, temperature, and the Earth: the self-organizing biosphere. Columbia University Press, New York, 241 p.

Sergeev, V.N. (2006) Precambrian Microfossils in cherts: their paleobiology, classification and biostratigraphic usefulness. Transactions of the Geological institute, vol. 567, Moscow GEOS, 280 p.

Sergeev, V.N., Gerasimenko, L.M. and Zavarzin, G.A. (2002) Proterozoic history and present state of cyanobacteria. Mikrobiologia 71(6), 725–740.

Slobodkin, A.I., Eroschew-Shakh, V.A., Kostrikina, N.A., Lavrushin, V. Yu., Dayniak, L.G. and Zavarzin, G.A. (1995) Formation of magnetite by thermophilic anaerobic microorganisms. Dokladi AN SSSR 345, 694–697 (Russ.).

Svetlichny, V.A., Sokolova, T.G., Gerchardt, M., Ringpfeil, M., Kostrikina, N.A. and Zavarzin, G.A. (1991) *Carboxydothermus hydrogenoformans* gen.nov., sp.nov. a new CO-utilizing thermophilic anaerobic bacterium from hydrothermal environments of Kunashir Island. System. Appl. Microbiol. 14, 254–260.

Westall, F. (2003) The geological context for the origin of life and the mineral signatures of fossil life. In: B. Bachier et al. (Eds), The Traces of Life and the Origin of Life. Springer-Verlag, .

Westall, F., de Witt, M.J., de Ronde, J. and Gerneke, D. (2001) Early Archean fossil bacteria and biofilms in hydrothermally-influenced sediments from the Barberton greenstone belt, South Africa. Precambrian Res. 106, 93–116.

Whitman, W.D., Coleman, D.C. and Wiebe, W.J. (1998) Prokaryotes: the unseen majority. PNAS 95(12), 6578–6583.

Widdel, F. (1993) In: L.C. Stal and P. Caumette (Eds), NATO ASI ser.G Ecological sciences. Springer-Verlag, Berlin, vol. 35, 463 p.

Woese, C.R. (2004) A new biology for a new century. Microbiol. Mol. Biol. Rev. 68(2), 173–186.

Zavarzina, D.G. (2004) Formation of magnetite and siderite by thermophilic Fe(III)-reducing bacteria. Paleontol. J. 38(6), 585–589.

Zavarzin, G.A. (1974) Phenotypic systematics of bacteria: the space of logic possibilities. Nauka, Moscow, 143 pp.

Zavarzin, G.A. (1984) Bacteria and the composition of the atmosphere. Nauka, Moscow, 192 p. (Russ.).

Zavarzin, G.A. (1993) Epicontinental soda lakes as supposed relict biotopes for the formation of terrestrial biota. Microbiologia 62(5), 789–800.

Zavarzin, G.A. (2003a) Diversity of cyanobacterial mats. In: W.E. Krumbein, D.M. Paterson and G.A.Zavarzin (Eds), Fossil and Recent Biofilms. A Natural History of Life on Earth. Kluver Academic Publishers, Dordrecht, pp. 141–150.

Zavarzin, G.A. (2003b) Coming-into-being of the system of biogeochemical cycles. Paleontological J. 6, 16–24.

Zavarzin, G.A. (2003c) Lectures in environmental microbiology. Nauka, Moscow, 348 p. (Russ.).

Zavarzin, G.A. (2006) Alkaliphilic microbial community as an analog of terrestrial biota in Proterozoic. In: Evolution of the Biosphere and Biodiversity. KMK-Press, Moscow, pp. 97–120 (Russ.).

Zavarzin, G.A. and Zhilina, T.N. (2000) Anaerobic chemotrophic alkaliphiles. In: J. Seckbach (Ed.) Journey to Diverse Microbial Worlds, Kluwer Academic Publishers. The Netherlands, pp. 191–208.

Zavarzin, G.A., Gerasimenko, L.M. and Zhilina, T.N. (1993) Cyanobacterial communities in hypersaline lagoons of lake Sivash. Microbiologia 62(6), 579–599.

Zavarzin, G.A, Karpov, G.A., Gorlenko, V.M., Golovacheva, R.S., Gerasimenko, L.M., Bonch-Osmolovskaja, E.A. and Orleanski, V.C. (1989) Calderic Microorganisms. Nauka, Moscow, 121p. (Russ.).

Zavarzin, G.A., Zhilina, T.N. and Kevbrin, V.V. (1999) The alkaliphilic microbial community and its functional diversity. Microbiologia 68(5), 503–521 (Engl.).

Part II
Prebiological Stages of Evolution and RNA World on the Earth and in the Space

Astrocatalysis Hypothesis for Origin of Life Problem

V. N. Snytnikov

Abstract Analysis of the available natural science data has allowed the astrocatalysis hypothesis to be formulated. The hypothesis indicates the pre-planetary circumstellar disk as the most probable time and place of the primary abiogenic synthesis of prebiotic organic substances from simple molecules along with the "RNA world" and the life origin. The sequence of self-organization stages that gave rise to the Earth biosphere is determined. Results of computational experiments with supercomputers are used to determine conditions of abiogenic organic compounds in the Earth's biosphere. In handling the problem of the origins of the Earth's biosphere, it is necessary to establish and study the role of abiogenic synthesis of prebiotic compounds in the sequence of key stages of the self-organization of matter. The sequence is the evolution of the surroundings from the point of Big Bang towards organic life and further to humans.

1 Evolution of the Surroundings and the Life Origin

A large body of data on the evolution of the universe has been accumulated in the modern natural science. The universe, along with its time, space and matter, emerged in the Big Bang 13.7 billion years ago. The inflation and annihilation stages determined the baryons' mass as equal to 4% of a total of gravitating matter in the universe. The baryons exist primarily as free H and He. Stars that emerged in the next stage of the universe's self-organization and evolution comprise, in modern estimations, a mass of the order of 10% of a total of the baryon mass. The stellar nucleosynthesis provided emergence of no more than 2% of chemical elements heavier than He, most of them being carbon, oxygen and nitrogen. Composing molecules by these elements favored the formation of next generation long-living stars in

---

V. N. Snytnikov
Boreskov Institute of Catalysis, Novosibirsk and Novosibirsk State University Russia

molecular clouds. These stars could form along with their protoplanet disks; the disk was of the order of 0.1 of the protostar in mass. The disk around the Sun, one of these stars that emerged approximately 4.56 billion years ago in the Milky Way galaxy, evolved into a planet system. The system's mass is about 10^{-3} of the Sun's mass, and the Earth's mass equals ca. 10^{-3} of the planet system's mass. The natural processes on the Earth gave rise to the organic biological life with the mass of about 10^{-10} of the planet's mass. Evolution of the Earth's biosphere took no less than 3.8–3.9 billion years. There are trophic chains in the biosphere, part of them being related to geochemical matter cycles on the Earth's surface. No more than 10% of base products leave from the stable trophic chains.

It is seen from the available data that the evolution of matter is based on the emergence of a new attribute carrier at each self-organization stage with the mass smaller than 10^{-1} of the initial system's mass. There is no reason to consider that the chemical and prebiological stages of the life and biosphere origins on the Earth are principally different, in terms of general self-organization laws, from speciation in biology or nucleosynthesis of heavy elements in astrophysics. Hence, we shall treat the problem of abiogenic synthesis of complex prebiotic compounds as a stage of the self-organization process.

The identified and above-mentioned stages of matter's evolution allow us to understand when and where the life originates in the universe and on the Earth. When? Not earlier than first stars emerged in the universe because there were no chemical life elements—carbon, oxygen, nitrogen—but only hydrogen and helium. Where? Life on Earth originated in the Solar system during the earliest, the very mysterious 600 million years after the initial point of the system formation. If otherwise, there would be some mechanism to transfer huge masses of biological compounds from planet to planet or from star to star (on the Earth, this would be no less than 2.5×10^{18} g of carbon comprised in the dry biological matter). Such a transfer seems very doubtful in the evolution of stars and processes in the interstellar medium.

While studying abiogenic synthesis of first organic substances on the Earth, the prime question to be answered is if this is a planetary (the Oparin–Holdein hypothesis) or cosmic (the Arrhenius or "panspermia" hypothesis) problem. It is shown (Snytnikov, 2006) that there may be no more than several self-organization steps in the chemical stage of the evolution from simple chemical compounds, such as CH_2O or NH_3, towards the "RNA world" and then to the biosphere. This conclusion, along with available biogeochemical data (Zavarzin, 2000), makes the Oparin hypothesis doubtful but leads to suppose that "the life origin is definitely displaced to the space" (Zavarzin, 2000). At the same time, the idea of "panspermia" of the eternal life existence and its transfer from star to star (Warmflash and Weiss, 2005) also disagrees with the above-mentioned natural science results. Among the probable solutions of the problem is an assumption that the abiogenic synthesis of primary organic matter for the Earth's biosphere occurred during pre-planetary evolution of the Solar system at the "astrocatalysis" stage (Snytnikov, 2006).

2 Astrophysics and Planetology

Let us consider the protostar–protoplanet stage of the Solar system formation that is dated back to 4.56 billion years using the chronological label based on the content of products of plutonium and aluminum decay. Original isotopes of these elements in the nature are generated by supernova explosion that stimulates the process of star formation in molecular clouds.

In modern concepts (Syunyaev, 1986), a protostar starts forming in a cold molecular cloud with the elemental composition characteristic of their occurrence in the space. Elements heavier than helium are the dust components. The formation process goes along with the development of the Jeans gravitational instability towards gravitational matter collapse into nuclei of future protostars. At the last stage of the evolution of gravitational instability, the star formation zone is broken down. The formed stars may escape the zone and disconnect from one another. *The stage of protostar formation* at the gravitational collapse takes up to hundred thousand years. During this time, dust falling on the equatorial plane along with the entrained gas gives rise to the formation of a protoplanet accretion disk turning round the nucleus. The matter comes through the gas–dust disk towards the nucleus to increase the mass. As soon as the protostar mass reaches ca. 0.1 of the Solar mass, thermonuclear reactions are ignited in the deep that are favored by the presence of carbon and other elements. Radiant energy is supplied from the protostar surface into the protoplanet disk and the parental molecular cloud to raise the temperature of the surrounding matter.

When the protostar mass becomes comparable to the Solar mass, the increasing radiation and stellar wind throw the surrounding matter away to make the star visible in the optical region. This is the finish of the *stage of star formation and existence of the protoplanet gas–dust disk*. The matter comparable in mass with the star may be transferred through the protoplanet disk. A star with Solar-like mass is formed during the time of the order of a million years; the estimated diameter of the protoplanet disk around a Solar-type star is 100–200 AU. In some astrophysical models (Makalkin, 2003), the gas temperature reaches 100–200 K at the Jupiter's orbit and 1000 K at the Earth's orbit at a pressure more than 10^{-4} bar. Protoplanets and primary stellar clustering bodies emerge in the disk under these conditions.

The stage of star formation and emergence of protoplanets with the cloud of primary bodies is followed by a comparatively long (of the order of

60 million years in the Solar system) *stage of planet formation* through collisions with the primary bodies. At this stage, the disk and internal planets lose the predominant gas components (hydrogen and helium), while the molecular cloud is destructed or the star escapes it. The final state at this stage is a young Solar system with the disk remnants such as planets with satellites, disk structures like rings of Saturn, Uranium and other remote planets.

The loss of hydrogen and helium in the Solar system, space bombardment, degassing and differentiation of Earth's matter, tidal effects and other physico-chemical processes led to the next stage of self-organization related to the *emergence of the Earth crust* during geologically non-documented 600 million years. The last documented 3.9 billion years are the stage of the Earth's geobiological evolution to form the *present biosphere*, as well as lithosphere, atmosphere and hydrosphere.

3 Astrocatalysis, "RNA World" and Life Origin

Let us discuss in more detail how protoplanets and stellar clustering bodies develop at the stage of star formation and existence of a protoplanet gas–dust disk.

In the molecular cloud, dust particles are 0.1 µm in characteristic diameter and have a multilayer structure (Levasseur-Regourd, 2004). Their internal nucleus of ca. 10 nm in size consists of refractory inorganic compounds of silicon, magnesium and iron (the elements following nitrogen, carbon and oxygen in abundance). The nucleus is covered by condensed organic compounds and hydrides of nitrogen, carbon, oxygen, in particular water, in the order corresponding to their volatility. While moving from the protoplanet disk periphery towards the protostar by spiral or more intricate trajectories, the dust particles successively lose the volatile components upon temperature rise to uncover the inorganic nucleus, the particles and heavy organic compounds accumulated in the accretion disk due to a very strong centrifugal effect with respect to hydrogen and helium as gas carriers. Earth's and meteorite-originated compounds of iron and silicon at the ratio approximately similar to their space abundance were used to demonstrate experimentally (Khasin and Snytnikov, 2005) a good catalytic activity of the nanoparticles in the Fischer–Tropsch synthesis. Note that the classical industrial catalyst for this reaction is iron supported on silica. However, these compositions are also active to synthesize ammonia from hydrogen and nitrogen and cyan hydride (the latter is in the presence of methane). Carbon monoxide, which follows hydrogen and helium in abundance in the space, reacts with hydrogen to give paraffin hydrocarbons, including high-boiling heavy compounds, olefins, alcohols, water:

$$(2n+1)H_2 + nCO \rightarrow C_nH_{2n+2} + nH_2O,$$

$$2nH_2 + nCO \rightarrow C_nH_{2n} + nH_2O,$$

$$2nH_2 + nCO \rightarrow C_nH_{2n+2}O + (n-1)H_2O.$$

As complex hydrocarbons are formed on the particle's surface, the dust particles start easily adhering to one another like plasticine clots. This is a kind of "snow ball" adhesion to result in an increase of the bodies in size up to 1–10 m. The bodies not smaller than these in size start moving around the protostar; they move at relative space speed higher than the sonic speeds in condensed matter and rarely collide with one another. The collisions under these conditions result in disintegration and destruction of the bodies.

The absence of the collision mechanism for bodies' enlargement to several hundred kilometers was the principal obstacle in the earlier scenarios of planet origin (Safronov, 1969). To handle this problem, we have suggested another putative mechanism based on the development of gravitational instability in a two-phase system of gas and solid bodies larger than 1 m in diameter (Snytnikov et al., 2002). The number of these bodies in the Solar system can be estimated using the planet masses corrected for the gas component quantity of giant gaseous planets. Let us estimate the characteristic Jeans length for gas: $\lambda_J^{-1} = 4\pi G m_g \rho_g / T_g$, $\lambda_J \approx 4 \times 10^{12}$ cm per 1 AU = 1.5×10^{13} cm. When gas alone is in the disk, the development of gravitational instability at this characteristic length is strongly hindered due to high temperatures T_g and oriented orbital speed of the gas orbiting at the protostar (ρ_g is the density of gas with average molecular mass m_g, G is the gravitation constant). Again, the instability cannot develop in clustering bodies in the absence of gas. The reason is high relative speeds $<V>$ of the bodies, which are close in quantity to their orbital speeds round the protostar. However, when bodies of moderate mass m_p move in a decelerating gas, the difference between their relative speeds $<V>$ decreases down to $m_p<V>^2 \approx T_g$ at the identical orbital speed. As the body's mass m_p increases, the effective Jeans length λ_{pg} can decrease considerably. Within the given speed range, the decrease occurs at the ratio $(m_g/m_p)^{1/2}$:

$$\lambda_{pg}^{-2} = 4\pi G m_g \rho_g / T_g + 8\pi G m_p N_p / <V>^2 \approx 8\pi G m_p N_p / T_g,$$

$$\lambda_{pg} / \lambda_J \approx (m_g \rho_g / m_p^2 N_p)^{1/2} \approx 10 (m_g/m_p)^{1/2}.$$

where N_p is the number of bodies in unit volume. When the Jeans length λ_{pg} equals some critical quantity, which is smaller or comparable to the circumstellar disk thickness, conditions emerge again for gravitational self-compression of the medium to favor the collective body clustering. The newly formed clusters are comparable to the protostar in size.

Computational experiments using supercomputers (Snytnikov, 2006) revealed the matter clustering which results from the development of instability. The formed clusters are comparable in size to the protostar. Namely the gravitational instability regions are the sites of nucleating large (many kilometers in size) primary bodies and protoplanets (Snytnikov et al., 1996).

The gravitation field causes an increase in the pressure of the main components of gas—hydrogen and helium—in the clusters during their formation. The pressure is sometimes higher than dozens of atmospheres depending on the distance from the protostar. The high pressures provide high absolute contents of oxygen, nitrogen, carbon compounds and non-volatile substances in the clusters. However, while hydrogen and helium are highly thermoconductive, their high pressure means moderate temperatures. Helium and hydrogen, when in the proportion larger than 20%, provide the energy supply to endothermic reactions and heat withdrawal from exothermic reactions. At the same time, the clusters contain a mobile solid phase of bodies with the catalytically active surface. Hence, there exist conditions in some clusters for operation of a chemical reactor similar to industrial catalysts with a fluidized catalyst bed (Boreskov, 2003). This is one of the most effective but complex operation reactors. The space reactor was close in pressure and reactant temperatures to lab-scale high-pressure catalytic reactors. These reactors are effectively used for abiogenic synthesis of organic compounds (ammonia, hydrogen cyanide, hydrocarbons, alcohols, aldehydes water, and other simple and complex compounds) in reductive media. Fragments of these compounds are detected now in comets. The process traces can be found in meteorites and asteroids, on planet's satellites and in the objects that did not undergo a geological evolution. While the times of chemical reactions of organic syntheses are usually no longer than hours or days under normal conditions, the chemical evolution rates (Parmon, 2002) at that stage were determined by the rates of reactant feeding to the reaction zone. Hence, they were very fast, of the order of several years. Presumably, the "RNA world" existed under those conditions.

This conclusion follows from analysis of conditions of emergences of primordial species (Spirin, 2005). For example, as Spirin states, the "RNA world" needs "drying-up pools" (water shortage but not excess), the presence of magnesium, wet clay surfaces, periodical intensive mixing of substances from different pools. These are conditions that occur in the space-fluidized catalyst reactor. There, the meter-sized bodies consist (in accordance to the space abundance of elements) of silica, magnesia and iron oxide. Water is formed on the porous body surfaces through the above-mentioned synthesis of primary carbon-containing compounds due to catalytic properties of these compositions. When the bodies escape the favorable synthesis regions, water and other volatile compounds are evolved but high-molecular compounds rest on the bodies. When the bodies re-enter the regions with the favorable conditions, water is formed again on the surface, penetrates deep into the porous bodies and intensifies the internal mass transfer. The merging, collision and intensive mixing of bodies at the stage of development of gravitational instability make the mechanism of blending of colonies of RNA and other compounds. "Systematic evolution through exponential enrichment" (Spirin, 2005) happens to the bodies. That the evolving matter in the space reactor is larger by many orders of magnitude than the Earth and the primordial bodies are formed initially from space dust nanoparticles are of principal importance. The total surface area, both internal and external, of the bodies built up by the nanoparticles is so large and diversified that, undoubtedly, it allows active centers with catalytic properties appropriate for any synthesis to be formed. The most effective organic compounds—autocatalysts—can be preferably and commonly used due to the intense mass exchange under conditions of this reactor. In other words, "the colonies with most active and most complimentary RNA molecules grow faster than others" (Spirin, 2005). The high pressure (dozens or more atmospheres) of hydrogen and helium as carriers of different substances is a distinctive feature of the space reactor. It seems like the primordial SELEX is preferable to study under these conditions. In addition, excess water emerged at later stages upon enlargement of the bodies and shrinkage of their internal surface areas. Probably excess water as an effective solvent destroyed the "RNA world" in the enlarged bodies. Only those RNA molecules and their colonies which managed to create membranes for not only abating but also employing the aqueous medium appeared suitable for further evolution. The world of primitive monocellulars, primeval species and the life was created.

The author is unaware of the conditions on the Earth or in the present Solar system which would resemble the pre-planetary conditions of the "RNA world" in the space reactor. However, primordial monocellular remnants may be kept in pores and fissures very deeply in the Earth's crust at high pressures and in non-oxygen media. Of some interest in this connection may be microorganisms that occur in porous rocks in dry regions at high temperatures or at temperatures below the melting point of ice. But the works to discover traces of microorganism colonies in meteorites are of most importance (Gerasimenko et al., 1999). If the results of these studies are confirmed, they will be an evidence of life origins at the pre-planetary stage in the Solar system. The further periods of time may then be considered as the struggle of living species for existence under conditions of disastrous changes in the external conditions, including planet formation, due to amplification of their organization and creation of an organisms community—biosphere—while the overwhelming portion of the synthesized primordial organic substances was lost.

Further evolution of matter in the cluster with the reactor for chemical synthesis is evident enough. As the cluster of organic compounds increased in mass, it could collapse into a matter clot confined by gravitation, Van der Waals or chemical forces, planetesimals or protoplanets. These processes, each with its specific features, could propagate progressively from the Sun's proximity to the periphery. In the Sun's proximity, the clusters and bodies moved to lose hydrogen, helium and light organic compounds under the action of solar wind and radiation. The rates of gas lost decreased in the regions of the Oort Cloud, remote cold planets and Jupiter. Then the protoplanets accumulated other bodies and participated in different processes to transform into planets and to enter the eon of geological or, if under suitable conditions, geobiological evolution. In the primary synthesis zone, most of the organic compounds and methane transferred to the Sun and drove away to the space. However, heavy and complex organic compounds were in huge concentrations and could survive further disasters and be the basis for originating and feeding to the biological community.

While density clusters—"life sources"—emerge at the Venus' orbit, the medium temperature defines the conditions of evaporation of water and organic compounds. At the Mars' orbits, the low gravitation field and comparatively low temperatures favor sublimation and loss of organic compounds and water to the surrounding space. Probably, the conditions at the Earth's orbits favor the losses of the reductive atmosphere (hydrogen and helium) but hold water and other gases. In further considerations, it is important to determine the probable "zone of life expansion" in the Solar system. In particular, does Mars fall into this zone? The result will be checked using automated space stations.

4 Conclusions

The conditions of the "space catalytic reactor" in a protoplanet disk were most appropriate for abiogenic synthesis of primordial organic compounds. That was a fluidized solid-phase reactor containing a reductive hydrogen–helium atmosphere at the pressure of less or more than 10 atm. The solid "granules" were of the order of 1–10 m. The solid phase consisted mainly of SiO_2–MgO–Fe. The solid phase exposed a vast catalytically active surface where water and other compounds were condensed at some distance from the protostar. At different time periods, the reactor was as large as hundreds to dozens times of the Sun diameter. Energy was supplied to the reactor zone by the protostar radiation to warm up the surface of the protoplanet disk. The zone of synthesis by exothermic reaction was cooled by hydrogen and helium. The existence time of the reactor for synthesis of chemical compounds was ca. 10 years. The chemical final state meant high-molecular organic compounds, H_2O and other element hydrides, probably, the "RNA world." The physical final state was emergence of bodies of many kilometers in size in the reductive atmosphere. The number of such "life sources" as the reactor for synthesis of chemical compounds and the "RNA world" depended on specific features of the development of gravitational instability in the two-phase medium of the disk. At the early stages of the gravitational instability, there might be several "sources" or more. Hence, synthesis of complex organic compounds directly initiated the formation of primary bodies. Protoplanets were the result of simultaneous confluence of numerous bodies upon development of collective gravitational instability. Catalytic processes gave rise to the planet formation.

Acknowledgment The studies were supported by Programs of the Presidium of the Russian Academy of Sciences N 8-12 "Origin and Evolution of Biosphere" and "Origin and Evolution of Stars and Galaxies," projects NSh-6526.2006.3 and RNT.2.1.1.1969 of the Ministry of Education and Science of the Russian Federation.

The author is grateful to academician V. Parmon and the corresponding member of RAS, A. Rosanov, for their assistance to him and the team of his colleagues from the research institute of the Siberian branch of the Russian Academy of Sciences.

References

Boreskov, G.K. (2003) Heterogeneous Catalysis. Nova Science Publishers, New York.

Gerasimenko, L.M., Hoover, R.B., Rozanov, A.Yu., Zhegallo, E.A. and Zhmur, S.I. (1999) Bacterial paleontology and studies of carbonaceous chondrites. Paleontol. J. 33(4), 439.

Khasin, A.A. and Snytnikov, V.N. (2005) Peculiarities of the product composition of the Fischer–Tropsch synthesis over the meteorite Tsarev material. Abstracts, International Workshop Biosphere Origin and Evolution, Novosibirsk, Russia, June 26–29, pp. 158–159.

Levasseur-Regourd, A.C. (2004) Cometary dust unveiled. Science 304(5678), pp. 1762–1763.

Makalkin, A.B. (2003) Problems of protoplanet disks evolution. In: M.Ya. Marova, Sovremennye problemy Mekhaniki I Fiziki Kosmosa: K 70-letiyu. Fizmatlit, Moscow, pp. 402–446.

Parmon, V.N. (2002). The origin of life: the prebiotic phase. Herald Russ. Acad. Sci. (Vestnik Rossiiskoi Akademii Nauk) 72(6), 592.

Safronov, V.S. (1969) Evolution of Protoplanetary Cloud and the Origin of the Earth and Planets. Nauka, Moscow, Russia, English translation, NASA TTF-667, 1972.

Snytnikov, V.N. (2006) Astrocatalysis as the first step of geobiological processes. In: A.Yu. Rozanov, Evoluzsiya biosfery I bioraznoobraziya: K 70-letiyu. KMK, Moscow, pp. 49–59.

Snytnikov, V.N., Vshivkov, V.A. and Parmon, V.N. (1996) Solar nebula as a global reactor for synthesis of prebiotic molecules. 11th International Conference on the Origin of Life, Orleans, France, July 5–12, p. 65.

Snytnikov, V.N., Dudnikova, G., Gleaves, J., et al. (2002) Space chemical reactor of protoplanetary disk. J. Adv. Space Res. 30(6), 1461–1467.

Spirin, A.S. (2005) Origin, possible forms of being, and size of the primeval organisms. Paleontol. J. 39(4), 364.

Syunyaev, R.A. (red.) (1986) Fizika kosmosa: malen'kaja e'nciklopedija. Moscow, Russia, 783 pp.

Warmflash, D. and Weiss, B. (2005, November) Did life come from another world? Scientific American.

Zavarzin, G.A. (2000). The non-Darwinian domain of evolution. Herald Russ. Acad. Sci. (Vestnik Rossiiskoi Akademii Nauk) 70(3), 252.

Comets, Carbonaceous Meteorites and the Origin of the Biosphere

R. B. Hoover

Abstract Evidence for indigenous microfossils in carbonaceous meteorites suggests that the paradigm of the endogenous origin of life on Earth should be reconsidered. It is now widely accepted that comets and carbonaceous meteorites played an important role in the delivery of water, organics and life-critical biogenic elements to the early Earth and facilitated the origin and evolution of the Earth's biosphere. However, the detection of embedded microfossils and mats in carbonaceous meteorites implies that comets and meteorites may have played a direct role in the delivery of intact microorganisms and that the biosphere may extend far into the cosmos. Recent space observations have found the nuclei of comets to have very low albedos ($\sim$0.03) and these jet-black surfaces can become very hot ($T \sim 400$ K) near perihelion. This chapter reviews recent observational data on comets and suggests that liquid water pools could exist in cavities and fissures between the internal ices and rocks and the exterior carbonaceous crust. The presence of light and liquid water near the surface of the nucleus enhances the possibility that comets could harbor prokaryotic extremophiles (e.g., cyanobacteria, sulfur bacteria and archaea) capable of growth over a wide range of temperatures. The hypothesis that comets are the parent bodies of the CI1 and the CM2 carbonaceous meteorites is advanced. Electron microscopy images will be presented showing forms interpreted as indigenous microfossils embedded in freshly fractured interior surfaces of the Orgueil (CI1) and Murchison (CM2) meteorites. The size range and morphological characteristics of these forms are consistent with known representatives of morphotypes of all five subsections (orders) of Phylum Cyanobacteria. Energy dispersive X-ray spectroscopy (EDS) elemental data show that the forms in the meteorites have anomalous C/N and C/S as compared with modern extremophiles and cyanobacteria. These images and spectral data indicate that the clearly biogenic and embedded remains cannot be interpreted as recent biological contaminants and therefore are indigenous microfossils in the meteorites.

R. B. Hoover
Astrobiology Laboratory, NASA/NSSTC, Huntsville, AL, USA
e-mail: Richard.Hoover@NASA.GOV

1 Introduction

The origin and extent of the biosphere and the temporal and spatial distribution of life represent the most profound and fundamental problems of the rapidly emerging multidisciplinary field of astrobiology. Astrobiologists are concerned with the origin and the distribution of life and seek answers to the questions:

Is life on Earth endogenous or exogenous?
Is life restricted to Earth or is life a Cosmic imperative?

The recent discovery that microbial extremophiles thrive in hot rocks deep within the Earth's crust (Frederickson and Onstott, 1996) and in hydrothermal fluids of deep-sea vents has established that the deep, hot biosphere first proposed by Thomas Gold does indeed exist (Gold, 1992). The detection of viable, cryopreserved microorganisms in deep ice cores from the Central Antarctic Ice Sheet (Abyzov, 1993; Abyzov et al., 2004), ancient permafrost (Gilichinsky et al., 1992) and in deep marine sediments has confirmed the existence and established the significance of the deep cold biosphere. It is now known that the vast deep, cold biosphere (cryosphere) contains the majority of the prokaryotic cells on Earth (Whitman et al., 1998). Cell counts indicate that the deep-sea sub-seafloor sediments harbor 1.3×10^{29} cells which represent more than 50% of all prokaryotic cells on Earth (Schippers et al., 2005). It has been shown that the living component of the Earth's biosphere is comprised primarily of prokaryotic chemoautotrophs and chemolithotrophs of the deep (hot and cold) biosphere rather than more obvious larger multicellular eukaryotes (plants and animals) that are more popularly recognized as "life." The composition and the spatial and temporal distribution of the biosphere of our planet is profoundly different than that which was widely accepted only a few decades ago as based on the pioneering work of Vernadsky (1926, 1998).

Chemical and molecular biomarkers indicate that life appeared very early (~3.8 Ga) on the primitive Earth (Schidlowski, 1988, 2001). The majority of the earliest definitively recognizable microfossils in ancient Earth rocks have sizes and detailed morphological features that are virtually indistinguishable from modern cyanobacteria and filamentous prokaryotes of benthic mats and hydrothermal vent communities (Rasmussen, 2000; Walsh and Lowe, 1985). Fossils from the 2.15 Ga Belcher Group, Canada (Hofmann, 1976), were sufficiently well preserved as to be recognized as members of an extant clade of endolithic cyanobacteria. These ancient colonial microfossils exhibited recognizable packaging of cells within a mucilage envelope and the surface of the colony was considered to have been darkened by pigmentation and the forms were interpreted as microfossils of cyanobacteria. The fossil forms were designated *Eoentophysallis* to reflect their similarity to living representatives of the genus *Entophysallis* (Knoll and Golubic, 1992). The colonial habit within a carbonaceous envelope is important for the recognition of the air biogenicity, since spherical forms are produced by many physical processes and clearly abiotic forms resembling cocci or diplococci are present in lunar dust (Storrie-Lombardi et al., 2006). Carotenoids, chlorophyll and other pigments are common in cyanobacteria where they play a crucial role in photosynthesis and protection of the internal cells from UV radiation (Castenholz and Garcia-Pichel, 2000; Hoiczyk and Hansel, 2000).

Filamentous cyanobacteria and sulfur bacteria have far more distinctive and unambiguously biological morphologies than the coccoidal forms (even when the coccoidal forms are in colonial assemblages). Many of the filamentous cyanobacteria have sheaths and exhibit polarized filaments that taper or have distinctive basal and apical termini. The precise sizes and shapes of the cells within the filaments are often of definitive diagnostic value when combined with other morphological features of the organism. Compared with other bacteria and archaea, filamentous cyanobacteria are often very large and exhibit complex, unambiguous and recognizable morphological characteristics that historically have formed the basis for their taxonomy. The filaments may be a simple linear chain of cells (trichome) that is often ensheathed within a mucilaginous polysaccharide envelope that is variously called the sheath, capsule, or glycocalyx depending upon the consistency of the mucilage or slime. Filamentous cyanobacteria often possess distinctive differentiated cellular structures for nitrogen fixation (basal or intercalary heterocysts) and reproduction (hormogonia, baeocytes, akinetes and spores). Sheaths that envelope trichomes can be thick, thin, laminated or unlaminated, uniseriate (with one trichome), multiseriate (containing two or more trichomes). The filaments can also exhibit different types of false or true branching. Cyanobacterial sheaths are extracellular fibrillar carbohydrates and complex polysaccharides. Taphonomic studies have shown that the sheaths of cyanobacteria are better preserved in the fossil record, in preference to their peptidoglycan-rich walls of the cyanobacterial cells (Bartley, 1996). Bacterial paleontology studies have established that fossil bacteria and cyanobacteria can be extremely well preserved in a wide variety of Proterozoic and Archaean rocks and carbonaceous meteorites (Hoover and Rozanov, 2005; Zhegallo et al., 2000).

The morphological characteristics of cyanobacteria have remained phenomenally stable since they first appeared on Earth over 3 billion years ago. Ruedemann (1918) was the first to observe "arrested evolution" when he recognized many hundred million year old fossils that were morphologically indistinguishable from their modern analogs. This phenomenon was termed "bradytely" (Simpson, 1944). Schopf (1987) introduced the term "hypobradytely" to "refer to the exceptionally low rate of evolutionary change exhibited by cyanobacterial taxa, morphospecies that show little or no evident change over many hundreds of millions of years and commonly over more than one or even two thousand million years" (Schopf, 1992). Knoll and Golubic (1992, p. 453) noted "Essentially all of the salient morphological features used in the taxonomic classification of living cyanobacteria can be observed in well-preserved microfossils." The similarity of size and detailed morphological characteristics are so great that names of fossil cyanobacteria are often formed by adding the prefixes (palaeo-, eo-) or the suffixes (-opsis, -ites) to the genus name of the modern morphological counterpart (e.g., *Palaeolyngbya, Palaeopleurocapsa, Eomicrocoleus; Oscillatoriopsis,* etc.). Over 40 genera of fossil cyanobacteria have been named using this convention (Schopf, 2000).

2 Comets, Meteorites, and the Origin and Distribution of Life

Comets and carbonaceous meteorites are some of the most interesting bodies of the solar system. Evidence continues to mount that the CI and CM meteorites are fragments of cometary nuclei with most of the volatiles removed (Ehrenfreund et al., 2001; Hoover, 2006a) and that these meteorites contain the mineralized remains of indigenous microfossils of cyanobacteria and prokaryotic mats (Hoover, 1997, 2006b; Zhmur et al., 1997). Therefore comets and carbonaceous meteorites may have played a crucial role in the origin and evolution of the biosphere and the distribution of life throughout the cosmos. It is generally accepted that the volatiles of the nuclei of comets are primordial ices that condensed on carbonaceous material and mineral dust grains in the proto-solar nebula (far away from the hot central region) during the formation of the solar system. The "dirty snowball" model was first advanced by Fred Whipple (1950) who also recognized the close relationship between comets and meteors (Whipple, 1951, 1963). Comets do not develop comae or tails until they come closer to the Sun than the orbit of Jupiter ($\sim$5 AU) where solar radiation heats the nucleus to a temperature in excess of 200 K. Beyond this "snow line" the volatiles are solidified and the coma and tail disappears and the comet looks like an asteroid. It is a widely accepted hypothesis that although comets are rich in water ice, their porosity is so great the water ice sublimes directly to water vapor without transition through the liquid phase. The widely accepted hypothesis that liquid water cannot exist on comets has led to the conclusion that they must be sterile, since metabolism and active growth of all known life forms appears to be predicated upon the presence of liquid water. However, the phase diagram shows that the transition of ice into liquid water will occur whenever the temperature exceeds 273K as long as the pressure exceeds 6.1 millibars. Spontaneous flaring and the eruption of geyser-like jets from comets Halley and Temple 1 indicate pressures in excess of 6.1 millibars (and hence liquid water) can occur in cometary nuclei.

Impacts of comets, meteorites and asteroids during the Hadean delivered water and ice to the Earth and Mars and may have played a significant role in the formation of early oceans of our planet (Delsemme, 1997). The deuterium to hydrogen ratio for water molecules of comets Halley, Hale–Bopp and Hyakutake was found to be of an order of magnitude higher than the D/H ratios for Saturn, Jupiter and the interstellar medium but the D/H ratios not only overlap the Earth's oceans (1.6×10^{-4}) but they are also consistent with the values found for the indigenous water of the carbonaceous meteorites (Delsemme, 1998; Eberhardt et al., 1987). Chyba and Sagan (1997) pointed out that the cometary delivery of bulk of the Earth's oceans is quantitatively consistent with the Earth deriving its inventory of carbon from comets as well. Comets arrive from the volatile-rich outer regions of the solar system, which led to the suggestion that comets delivered pre-biotic organic molecules along with water and carbon to the inner planets (Oró, 1961; Oró, Mills, and Lazcano, 1995).

The concept that comets are sterile continues to be widely held even though recent discoveries have shown that living microorganisms exist in ice (Abyzov, 1993; Abyzov et al., 2004; Pikuta et al., 2005) and that the surface of a comet nucleus becomes hot within the orbit of Mars (1.5 AU). Hoyle and Wickramasinghe (1980) argued that comets might periodically deliver intact and possibly yet viable microorganisms to Earth and other bodies of the solar system. Hoover et al. (1986) advanced the hypothesis that diatoms, bacteria and other microorganisms could live in ice and water beneath the organic-rich carbonaceous crusts of comets. Consequently comets could thereby have played a crucial role in the origin of life on Earth as well as in the distribution of life throughout the cosmos.

Spacecraft observations of comet P/Halley in March 1986 provided important data about the nature, behavior, surface temperature and chemical composition of the comet nucleus and coma. The Vega Dust Mass Spectrometer found the Halley nucleus material to have a composition similar to carbonaceous meteorites with water as the primary volatile (75–80%). The Vega 1 IR spectrometers found the dayside surface of the $16 \times 8 \times 8$ km nucleus of comet Halley at perihelion was astonishingly hot ($T \sim 400$ K) (Emerich et al., 1987). ESA's Giotto spacecraft determined the Halley nucleus was one of the darkest objects in the solar system with an albedo ~ 0.03, that is comparable to CI and CM carbonaceous meteorites and P- and D- class asteroids (Hiroi et al., 2004). An inert black crust may build up as comets lose ice, and dust and carbonaceous materials concentrate on the surface (Wallis and Wickramasinghe, 1991; Wickramasinghe and Hoyle, 1999). Giotto obtained spectacular images from 600 km showing the full contour of Halley's jet-black nucleus as two very bright geyser-like jets of water vapor, dust and ice particles were being ejected. Giotto measured the D/H ratio of the water of comet Halley at 5.5×10^{-4} which is similar to seawater (1.5×10^{-4}); carbonaceous meteorites (~ 1.4–6.2×10^{-4}) but very different from the D/H ratio of Jupiter (1.5–3.5×10^{-5}); diffuse interstellar medium (1.5×10^{-5}) and protosolar nebula (2.1×10^{-5}) (Eberhardt et al., 1987). Evidence for microfossils in CI and CM carbonaceous meteorites may be interpreted as supporting the hypothesis that comets were the parent bodies of these meteorites (Hoover, 2006a,b; Hoover et al., 2004, Hoover and Pikuta, 2004).

Sunshine et al. (2006) found exposed regions of water ice on the surface of comet Temple 1 at 1.5 AU (near the orbit of Mars). Their measurements indicated the temperature was slightly above 273 K (melting temperature of water ice) over a large area of the surface of the nucleus with the maximum of 330 K in the hottest regions exposed to direct sunlight. Infrared spectra of comet Temple 1 as observed by the *Deep Impact* spacecraft showed a tremendous increase in the 3.2 μm CH–X band when compared with the 3.2 μm H_2 band and the 4.25 μm CO_2 band indicating a strong surge of material with high organic content from the inner regions of the nucleus. These observations provide additional evidence in support of the hypothesis that liquid water can exist on cometary nuclei (Sheldon and Hoover, 2006). Hence, cyanobacterial and prokaryotic mats could form within regimes between a semipermeable carbonaceous crust and the interior ice where pressure and temperature parameters allow liquid water films or pools to exist.

3 Microfossils of Cyanobacteria in the Orgueil and Murchison Meteorites

High-resolution scanning electron microscopy imaging of meteorites and living and fossil cyanobacteria and microbial extremophiles has been carried out at the Astrobiology Laboratory of the NASA/Marshall Space Flight Center (MSFC) and the National Space Science and Technology Center (NSSTC). This research employed the ElectroScan environmental scanning electron microscope (ESEM) and the FEI Quanta 600 FEG and Hitachi S-4100 field emission scanning electron microscopes (FESEM) operating with secondary and backscattered electron detectors. Uncoated, freshly fractured, interior surfaces of meteorites, proterozoic rocks, Vostok ice, living and fossil cyanobacteria and other prokaryotic and eukaryotic extremophiles were studied. All these instruments have energy dispersive X-ray spectroscopy (EDS) systems capable of obtaining quantitative elemental composition (spot and 2D X-ray maps) for elements with atomic number above boron. Several thousand images and EDS spectra have been produced during the past decade

Mineralized and embedded remains of clearly biogenic, large, filamentous forms have been found in all of the CI1 (Orgueil, Alais, Ivuna); CI Ungrouped (Tagish Lake); CM2 (Murchison, Murray, Mighei, Nogoya) and some CO3 (Rainbow, DAG 749) and CV3 (Efremovka) meteorites that have been investigated. Similar forms were not detected in studies at NASA/MSFC of the CV3 (Allende); CK5 (Karoonda); L4 (Barratta; Nikolskoye); L/LL6 (Holbrook); Diogenites (Tatahouine); Nakhlites (Nakhla; Lafayette) or Nickel/Iron (Antarctic) meteorites studied. The CI and CM carbonaceous meteorites contain a large number of complex and embedded coccoidal and filamentous forms consistent in size, shape and morphology with known species of cyanobacteria, sulfur bacteria and other morphologically convergent prokaryotes. Many of the forms are clearly embedded in the meteorite rock matrix and have C/N and C/S ratios dramatically different from those measured in living and dead cyanobacteria, prokaryotic extremophiles and eukaryotes (fungi, diatoms, wood and moss). Consequently, many of these complex and recognizable filaments have been interpreted as indigenous microfossils of cyanobacteria in the meteorites rather than recent microbial contaminants (Hoover, 2006a, 1997; Hoover and Rozanov, 2005; Hoover et al., 2004; Zhmur et al., 1997). Representatives of all five orders of cyanobacteria recognized by Castenholz and Waterbury (1989) (corresponding to Sections I–V of Boone et al., 2001) have been found embedded in the carbonaceous meteorites and a few examples are presented herein.

3.1 Morphotypes of Cyanobacteria in Carbonaceous Meteorites

Coccoidal and nanometric scale spherical microstructures have been found in all the meteorites studied. Since simple spheres can be produced by many abiotic mechanisms most of the spherical forms found in the meteorites were not considered to be unambiguously biological in origin and therefore they were not interpreted as microfossils. Unicellular Chroococcalean cyanobacteria (e.g., *Synechococcus, Gloeothece, Gloeocapsa*) reproduce by binary fission. Even though coccoidal forms can be abiotic, the forms shown in Fig. 1 are encapsulated within a carbonaceous sheath and they are interpreted as being biogenic in origin (Gerasimenko et al., 1999). These forms exhibit the appropriate sizes and size ranges of morphotypes of cyanobacteria of the Order Chrococcales, such as are found encased within carbonaceous sheaths (Fig. 1a and b) or as colonial assemblages in carbon-rich mucilage envelopes (Fig. 1c). Other irregular or polygonal coccoidal forms have been found in pseudofilaments encapsulated within carbonaceous sheaths such as unicellular members of the Order Pleurocapsales. Representatives of this Order reproduce by multiple fission yielding "baeocytes" of diverse sizes such as are seen in Fig. 1d.

By far the most abundant and recognizable microfossils found in the carbonaceous meteorites are morphotypes of filamentous non-heterocystous cyanobacteria that divide in only one plane. These are the cyanobacteria that belong to the Order Oscillatoriales (Section III of the Bacteriological Code). Morphotypes interpreted as representative of species of the genera *Oscillatoria, Spirulina, Arthrospira, Microcoleus, Phormidium, Lyngbya* and *Plectonema* have all been encountered in the Orgueil meteorite. Microfossils of Oscillatorialean cyanobacteria are shown in Fig. 1e from Murchison and Fig. 1f from the Orgueil meteorite.

The most complex and highly distinctive are the Nostocalean cyanobacteria. Examples of these in the meteorites include the morphotype of filamentous *Nostoc* sp. with emergent hormogonia and hollow empty sheath (Fig. 1g) in the Murchison CM2 meteorite. Figure 1h shows the funnel-shaped apical regions of three almost completely embedded filaments beside a thin embedded benthic mat of several tapered filaments that are attached to the Orgueil meteorite rock matrix. These smooth bulbous structures are interpreted as basal heterocysts,

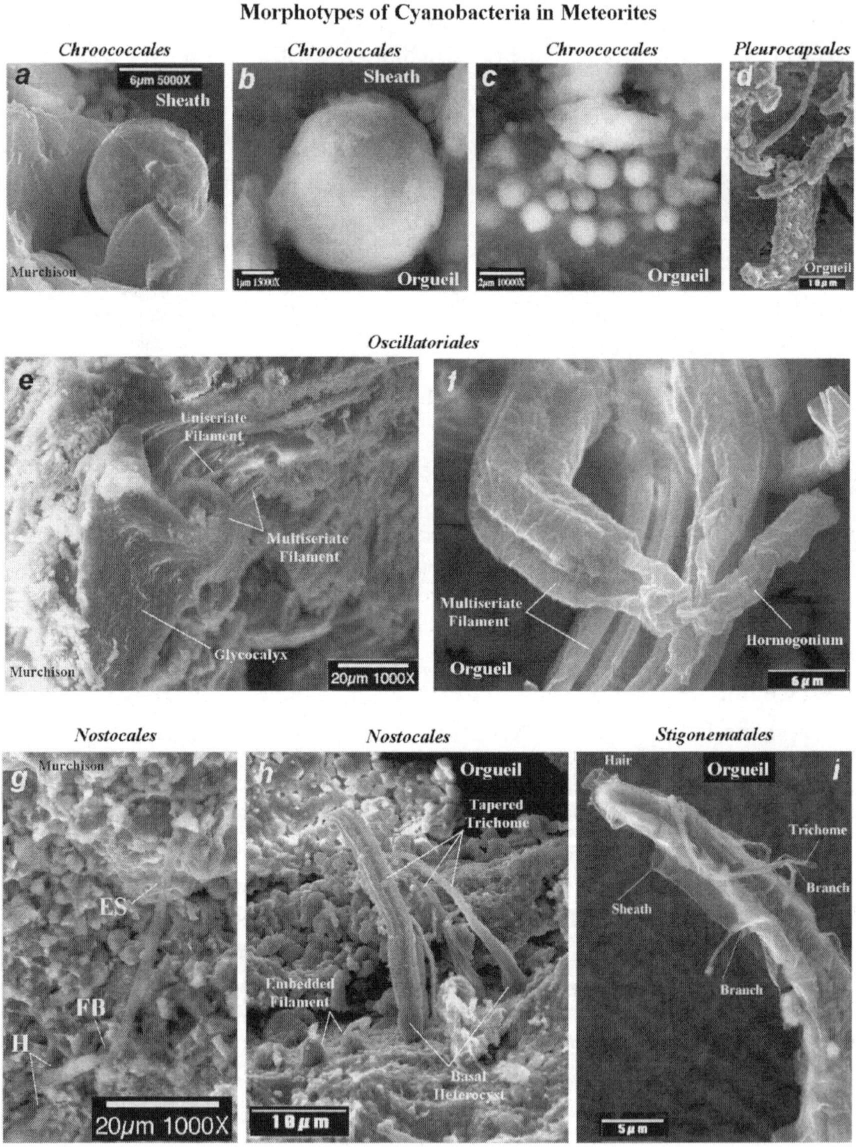

Morphotypes of Cyanobacteria in Meteorites

Fig. 1 Well-preserved mineralized microfossils of morphotypes of all five orders of *Cyanobacteriaceae* (Chroococcales (**a–c**); Pleurocapsales (**d**); Oscillatoriales (**e, f**); Nostocales (**g, h**); and Stigonematales (**i**)) embedded in freshly fractured interior surfaces of the Murchison CM2 (**a, e, g**) and the Orgueil CI1 (**b, c, d, f, h, i**) carbonaceous meteorites

which are well known in modern cyanobacteria of the family *Rivulariaceae*. These forms are consistent with size, morphology and ecological habit with modern morphotypes of *Calothrix* spp. Figure 1i shows a morphotype of a

Comets, Carbonaceous Meteorites and Biosphere Origin

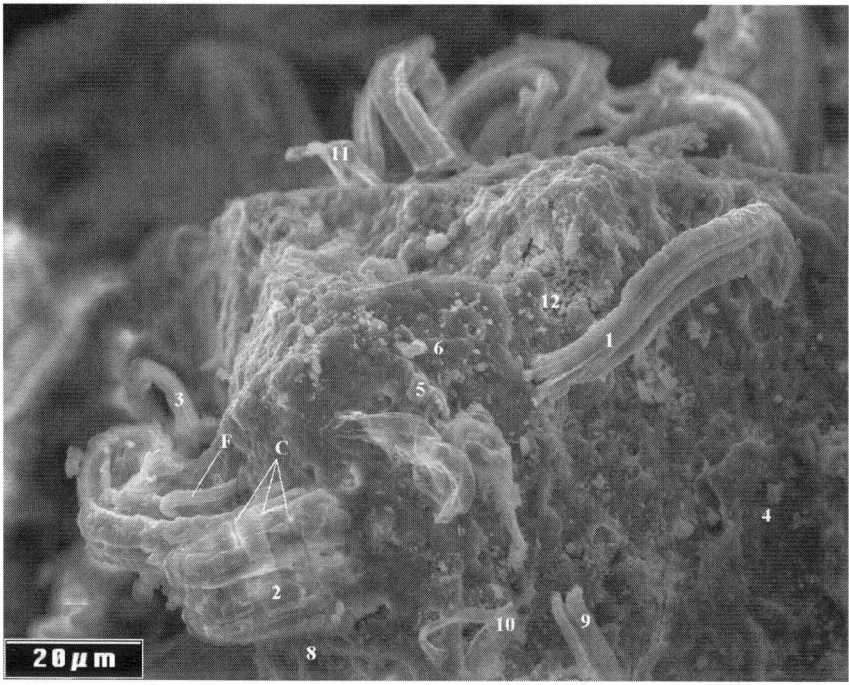

Fig. 2 Cyanobacterial filaments embedded in the Orgueil meteorite

complex trichomic filament of *Stigonemalalean* cyanobacteria with a terminal hair and both true and false branchings.

Figure 2 is 1000 X Hitachi FESEM image of a densely populated region of Orgueil with several different well-preserved embedded filaments and electron transparent sheaths that exhibit complex morphological features consistent with the known characteristics of several major groups of cyanobacteria. The numbers correspond to the filaments and sheaths for which the C/N and C.S ratios are given in Fig. 3. The values are typical of the filaments and sheaths measured in many samples of Orgueil from different museums. Irregular longitudinal striations of filaments 1 and 2 indicate multiseriate trichomes consistent with the Oscillatorialean genus *Microcoleus*. The smaller multiseriate filament 1 and the solitary filament F (2 μm diameter hook-shape with a narrowed terminus) are interpreted as representing morphotypes of the genus *Trichocoleus* Anagnostidis which has trichomes $\sim$ 0.5–2.5 μm diameter. Figure 3 provides a comparison of the C/N and C/S ratios of the numbered filaments and sheaths from the Orgueil meteorite fragment of Fig. 2 with these ratios as measured for a dried herbarium sample (*Bangia punctata*) collected in 1816 and several genera of living cyanobacteria grown in culture at the NSSTC Astrobiology Laboratory. Nitrogen was below the EDS detection limit ($< 0.5\%$ atomic) for many of the Orgueil filaments and the value shown in the plots are

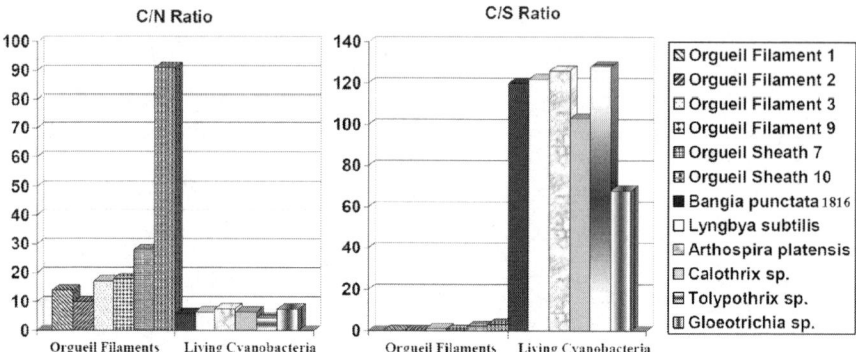

Fig. 3 Comparison of C/N and C/S ratios of the numbered Orgueil meteorite filaments and sheaths (from Fig. 2) with a filament in an 1816 herbarium sample and living cyanobacteria grown in culture at the NSSTC Astrobiology Laboratory

lower limit. The 1816 filament and living cyanobacteria had Nitrogen levels in the the range of 3–15% (atomic) but Sulfur levels often below the (<0.5 %) detection limit and the C/S ratios represent lower limits vor the living forms. Sulfur was always present (often at astonishingly high values) in the Orgueil filaments resulting in extremely low C/S ratios. The dramatically different C/N and C/S plots permit the Orgueil filaments to be readily distinguished from modern biological materials. Hence, recent contaminants can be ruled out and these filaments are interpreted as representing indigenous microfossils of cyanobacteria or morphologically convergent trichomic prokaryotes embedded in the carbonaceous meteorite matrix.

4 Conclusions

The discovery of indigenous microfossils of cyanobacteria in carbonaceous meteorites clearly indicates that the biosphere of Earth is open rather than closed. The possibility that biological matter has been transferred into space by the impact of large comets and asteroids onto shallow marine sediments must be considered as well as the possibility that terrestrial microbial extremophiles arrived on Earth during the Hadean bombardment from other regimes of the cosmos. The images provided in Figs. 1 and 2 are unambiguously biological in nature. These filaments, sheaths and mats are clearly recognizable as associated with morphotypes of highly distinctive and well-known polarized and differentiated benthic filamentous cyanobacteria. The Orgueil meteorite is immediately destroyed by contact with liquid water (Cloëz, 1864, Hoover, 2006b). It is impossible that these photosynthetic microorganisms (which require liquid water to grow and a water/substrate interface in order to form mats) could have constructed the observed mats and benthic ecological consortia within the dry black Orgueil rocks subsequent to the arrival of the Orgueil meteorite on Earth. Furthermore, there are dramatic differences between the C/N and C/S ratios of the Orgueil filaments and sheaths than those found in living cyanobacteria and old dead forms encountered in herbarium material. Therefore it is concluded that the indigenous microfossils of cyanobacteria found in the CI and CM carbonaceous meteorites provide strong evidence that the biosphere is open and is not restricted to the planet Earth.

Acknowledgment I express gratitude to Gregory Jerman and James Coston of NASA/MSFC for electron microscopy support and William Birch, Victoria Museum (Murchison); Paul Sipiera, Dupont Meteorite Collection (Orgueil, Ivuna); Martine Rossignol-Strick and Claude Perron, Musée Nationale d'Histoire Naturelle, Paris (Orgueil) for the meteorite samples used in this research. I also thank Elena V. Pikuta, NSSTC for cultures of extremophiles and Alexei Yu. Rozanov, N. C. Wickramasinghe, Max Wallis, Academician Georgi Zavarzin, Academician Erik Galimov, Rosemarie Rippka, Ludmilla Gerasimenko and the late Sir Fred Hoyle for many helpful discussions regarding cyanobacteria, bacterial paleontology, comets and carbonaceous meteorites.

References

Abyzov, S.S. (1993) Microorganisms in the Antarctic ice. In: E.I. Friedman (Ed.), Antarctic Microbiology. Wiley-Liss, New York, pp. 265–296.

Abyzov, S.S., Hoover, R.B., Imura, S., Mitskevich, I., Naganuma, T., Poglazova, M. and Ivanov, M.V. (2004) Use of different methods for discovery of ice-entrapped microorganisms in ancient layers of the Antarctic glacier. Adv. Space Res., Cospar 33, 1222–1230.

A'Hearn, M. F., Belton, M. J. S., Delamere, W. A., Kissel, J., Klaasen, K. P., McFadden, L. A., Meech, K. J., Melosh, H. J., Schultz, P. H., Sunshine, J. M., Thomas, P. C., Veverka, J., Yeomans, D. K., Baca, M. W., Busko, I., Crockett, C. J., Collins, S. M., Desnoyer, M., Eberhardy, C. A., Ernst, A. M., Farnham, T. L., Feaga, L., Groussin, O., Hampton, D., Ipatov, S. I., Li, J. -Y., Lindler, D., Lisse, C. M., Mastrodemos, N., Owen, Jr., N. M., Richardson, J. E., Wellnitz, D. D., and White, R. L., (2005). Deep Impact: Excavating Comet Tempel 1, Science 310 258–264.

Bartley, J.K. (1996) Actualistic taphonomy of cyanobacteria: implications for the Precambrian fossil record. Palaios 11, 571–586.

Boone, D. R., Castenholz, R. W., Garrity, G. M. Eds. (2001) *Bergey's Manual of Systematic Bacteriology*. Volume One. The *Archaea* and the Deeply Branching and Phototrophic *Bacteria*. Springer-Verlage, New York, N.Y., pp. 473–600.

Castenholz, R.W. and Garcia-Pichel, F. (2000) Cyanobacterial response to EV-radiation. In: B.A. Whitton and M. Potts (Eds), The Ecology of Cyanobacteria: Their Diversity in Time and Space. Kluwer Academic Publishers, Dordrecht, pp. 591–611.

Castenholz, R. and Waterbury, J. B. (1989) Introduction to Cyanobacteria. In: N. R. Krieg and J. G. Holt (eds.), Bergey's Manual of Systematic Bacteriology. Vol. 3 Williams and Wilkens, pp. 1710–1728.

Chyba, C.F. and Sagan, C. (1997) Comets as a source of prebiotic molecules for the early Earth. In: P.J. Thomas, C.F. Chyba and C.P. McKay (Eds), Comets and the Origin and Evolution of Life. Springer-Verlag, New York, pp. 147–168.

Cloëz, S. (1864) Note sur la composition chimique de la pierre météorique d Orgueil. Compt. Rend. Acad. Sci., Paris 58, 986–988.

Delsemme A.H. (1997) The origin of the atmosphere and of the oceans. In: P.J. Thomas, C.F. Chyba and C.P. McKay (Eds), Comets and the Origin and Evolution of Life. Springer-Verlag, New York, pp. 29–67.

Delsemme, A.H. (1998) The deuterium enrichment observed in recent comets is consistent with the cometary origin of seawater. Planet. Space Sci. 47, 25–131.

Eberhardt, P., Dolder, U., Schulte, W., Krankowsky, D., Lämmerzahl, P., Berthelier, J.J., Woweries, J., Stubbemann, U., Hodges, R.R., Hoffman, J.H. and Illiano, J.M. (1987) The D/H ratio in water from comet P/Halley. Astron. Astrophys. 187, 435–437.

Ehrenfreund, P., Glavin, D.P., Botta, O., Cooper, G., Bada, J. (2001) Extraterrestrial amino acids in Orgueil and Ivuna: tracing the parent body of CI type carbonaceous chondrites. Proc. Natl Acad. Sci. 98, 2138–2141.

Emerich, C., Lamarre, J.M., Moroz, V.I., Combes, M., Sanko, N.F., Nikolsky, Y.V., Rocard, F., Gispert, R., Coron, N., Bibring, J.P., Encrenaz, T. and Crovisier, J. (1987) Temperature and size of the nucleus of comet P/Halley deduced from IKS infrared Vega 1 measurements. Astron. Astrophys. 187, 839–842.

Frederickson, J.K. and Onstott, T.C. (1996) Microbes deep inside the Earth. Sci. Am. 275, 68–73.

Gerasimenko, L.M., Hoover, R.B., Rozanov, A.Yu., Zhegallo, E.A. and Zhmur, S.I. (1999) Bacterial paleontology and studies of carbonaceous chondrites. Paleontol. J. 33, 439–459.

Gilichinsky, D.A., Vorobyova, E.A., Erokhina, L.G., Fedorov-Davydov, D.G. and Chaikovskaya, N.R. (1992) Long-term preservation of microbial ecosystems in permafrost. Adv. Space Res. 12, 255–263.

Gold, T. (1992) The deep, hot biosphere. Proc. Natl Acad. Sci. 89, 6045–6049.

Hiroi, T., Pieters, C.M., Rutherford, M.J., Zolensky, M.E., Sasaki, S., Ueda, Y. and Miyamoto, M. (2004) What are the P-type asteroids made of? Lunar Planet Sci. XXXV, 1616.

Hofmann, H.J. (1976) Precambrian microflora, Belcher Islands, Canada: significance and systematics. J. Palaeontol. 50, 1040–1073.

Hoiczyk, E. and Hansel, A. (2000) Cyanobacterial cell walls: news from an unusual prokaryotic envelope. J. Bacteriol. 182, 1191–1199.

Hoover, R.B. (1997) Meteorites, microfossils and exobiology. SPIE 3111, 115–136.

Hoover, R.B., Hoyle, F., Wallis, M.K., and Wickramasinghe, N.C., (1986) "Can Diatoms Live on Cometary Ice?" in *Asteroids, Comets and Meteors II.*, Proceedings of Meeting at Astronomical Observatory of Uppsala University, June, 1985, (C. I. Lagerkvist, Ed.), pp. 359–352.

Hoover, R.B. (2006a) Comets, asteroids, meteorites, and the origin of the biosphere. SPIE 6309, 63090J 1–12.

Hoover, R.B. (2006b) Fossils of prokaryotic microorganisms in the Orgueil meteorite. In: R.B. Hoover, A.Yu. Rozanov and G.V. Levin (Eds), Instruments, Methods and Missions for Astrobiology, IX. SPIE, 6309, 02 1-17.

Hoover, R.B. and Pikuta, E.V. (2004) Microorganisms on comets, Europa, and the polar ice caps of Mars. SPIE 5163, 191–202.

Hoover, R.B. and Rozanov, A.Yu. (2005) Microfossils, biominerals and chemical biomarkers in meteorites. In: R.B. Hoover, R. Paepe and A.Yu. Rozanov (Eds), Perspectives in Astrobiology, Vol. 366 NATO Science Series: Life and Behavioural Sciences. IOS Press, The Netherlands, pp. 1–18.

Hoover, R.B., Hoyle, F., Wickramasinghe, N.C., Hoover, M.J. and Al-Mufti, S. (1986) Diatoms on Earth, comets, Europa and in interstellar space. Earth Moon Planets 35, 19–45.

Hoover, R.B., Pikuta, E.V., Wickramasinghe, N.C., Wallis, M.K. and Sheldon, R.B. (2004) Astrobiology of comets. SPIE 5555, 93–106.

Hoyle, F. and Wickramasinghe, N.C. (1980) Comets—a vehicle for panspermia. In:C. Ponnaperuma (Ed.), Comets and the Origin of Life. Reidel, Dordrecht, pp. 222–239.

Knoll, A. H. and Golubic, S. (1992) Proterozoic and living cyanobacteria. In: M. Schidlowski, S. Golubic and M.M. Kimberley (Eds), Early Organic Evolution: Implications for Mineral and Energy Resources. Springer, Berlin, pp. 450–462.

Oró, J. (1961). Comets and the formation of biochemical compounds on the primitive Earth. Nature 190, 389–390.

Oró J., Mills, T. and Lazcano, A. (1995). Comets and life in the Universe. Adv. Space Res. 15, 81–90.

Rasmussen, B. (2000) Filamentous microfossils in a 3,235-million-year-old volcanogenic massive sulphide deposit. Nature 405, 767–679.

Ruedemann, R. (1918) The paleontology of arrested evolution. N.Y. State Mus. Bull. 196, 107–134.

Schidlowski, M. (1988) A 3,800 million-year-old record of life from carbon in sedimentary rocks. Nature 333, 313–318.

Schidlowski, M. (2001) Carbon isotopes as biogeochemical recorders of life over 3.8 Ga of Earth history: evolution of a concept. Precambr. Res. 106, 117–134.

Schippers, A, Neretin, L.N., Kallmeyer, J., Ferdelman, T.G., Cragg, B.A., Parkes, R.J. and Jorgensen, B.B. (2005) Prokaryotic cells of the deep sub-seafloor biosphere identified as living bacteria. Nature 433, 861–864.

Schopf, J.W. (1987) "Hypobradytely": Comparison of rates of Precambrian and Phanerozoic Evolution, J. Vertebr. Paleontol. 7, Suppl. 3, 25.

Schopf, J.W. (1992). Tempo and Mode of Proterozoic Evolution in: *The Proterozoic Biosphere, A multidisciplinary study*, (Schopf, J.W. and Klein, C., eds.) Cambridge University Press, New York, p. 596.

Schopf, J.W. (2000). The fossil record: tracing the roots of the cyanobacterial lineage. In: B.A. Whitton and M. Potts (Eds), The Ecology of Cyanobacteria: Their Diversity in Time and Space. Kluwer Academic Publishers, Dordrecht, The Netherlands, pp. 13–35.

Sheldon, R.B. and Hoover, R.B. (2006) Evidence for liquid water on comets. SPIE 5906, 59060E 1–19.

Simpson, G.G. (1944) Tempo and Mode in Evolution. Columbia Univ. Press, New York, pp. 1–237.

Sunshine, J.M., A'Hearn, M.F., Groussin, O., Li, J.-Y., Belton, M.J.S., Delamere, W.A., Kissel, J., Klaasen, K.P., McFadden, L.A., Meech, K.J., Melosh, H.J., Schultz, P.H., Thomas, P.C., Veverka, J., Yeomans, D.K., Busko, I.C., Desnoyer, M., Farnham, T.L. Feaga, L.M., Hampton, D.L., Lindler, D.J., Lisse, C.M. and Wellnitz, D.D. (2006) Exposed water ice deposits on the surface of Comet 9P/Tempel 1. Science 311, 1453–1455.

Storrie-Lombardi, M.C., Hoover, R.B., Abbas, M., Jerman, G., Coston, J., and Fisk, M. (2006). Probabilistic classification of elemental abundance distributions in Nakhla and Apollo 17 lunar dust samples, SPIE 6309, 630906:1–12.

Vernadsky, V.I. (1926) *Biosfera*, Leningrad, Nauka, pp. 1–24.

Vernadsky, V.I. (1998) The Biosphere (Translation by D.B. Langmuir, Revised and Annotated by M.A.S. McMenamin). Copernicus, New York, pp. 1–192.

Wallis, M.K. and Wickramasinghe, N.C. (1991) Structural evolution of cometary surfaces. Space Sci. Rev. 56, 93–97.

Walsh, M.M. and Lowe, D.R. (1985) Filamentous microfossils from the 3,500-Myr-old Onverwacht Group, Barberton Mountain Land, South Africa. Nature 314, 530–532.

Whipple, F.L. (1950) A comet model. I. The acceleration of comet Enke. Astrophys. J. 111, 134–141.

Whipple, F.L. (1951) A comet model. II. Physical relations for comets and meteors. Astrophys. J. 113, 464–474.

Whipple, F.L. (1963) On the structure of the cometary nucleus. In: B.M. Middlehurst and G.P. Kuiper (Eds), The Moon, Meteorites, and Comets. Univ. of Chicago Press, Chicago, pp. 639–664.

Whitman, W.B., Coleman, D.C. and Wiebe, W.J. (1998) Prokaryotes: the unseen majority. Proc. Natl Acad. Sci. USA 95, 6578–6583.

Wickramasinghe, N.C. and Hoyle, F. (1999) Infrared radiation from comet Hale–Bopp. Astrophys. Space Sci. 268, 379–381.

Zhegallo, E.A., Rozanov, A.Yu., Ushatinskaya, G.T., Hoover, R.B., Gerasimenko, L.M. and Ragozina, A.L. (2000) Atlas of Microorganisms from Ancient Phosphorites of Khubsughul (Mongolia), NASA/TP 209901 (In English and Russian), pp. 1–167.

Zhmur, S.I., Rozanov, A.Yu. and Gorlenko, V.M. (1997). Lithified remnants of microorganisms in carbonaceous chondrites. Geochem. Int. 35(1), 58–60.

Hierarchical Scale-Free Representation of Biological Realm—Its Origin and Evolution

Y. N. Zhuravlev and V. A. Avetisov

Abstract In this work we develop the concept of biological referents to analyze the origin of complexity of biological systems. The concept, as we demonstrate, can be formalized by classes of the objects which constitute hierarchic scale-free patterning at different levels of biological complexity. By this reason, ultrametric relationships between these classes are assumed to be relevant to the referent representation. To explore this idea, we realize particular formalization of the referent concept including construction of objects and classes, construction of ultrametric space of classes, and description of the ultrametric space by a field of p-adic numbers. We discuss how a notation "evolution" can be introduced through ultrametric formalism. An example of ultrametric evolutionary equations is presented. Finally, we demonstrate that different aspects of the origin and evolution of the Biosphere (such as macroevolution and development) being verified in the frame of the referent concept acquire new contours and interpretations.

1 Introduction

Description of evolution of any system is based upon a "dynamic paradigm" presupposing formalization of such notions as "system," "state of system," and "control function" which specifies the "dynamical law" and governs dynamic (evolutionary) trajectories. At first sight, all these notions could be somehow defined with respect to the biological realm. However, it must be mentioned that physical description of systems' dynamics is based upon well-defined *primary elements* (e.g., elementary particles, atoms, molecules, etc.) which constitute a system and do not transform during the evolution of the system itself. At any instant, the system is completely represented by subsets of the given set

---

Y. N. Zhuravlev
Institute of Biology and Soil Science, Russian Academy of Sciences, Far Eastern Branch, Vladivostok, Russia
e-mail: zhuravlev@ibss.dvo.ru

of primary elements, and only by them. By contrast, the question concerning existence of something like a "primary element" of the biological realm remains to be open, moreover, it is not perfectly understood if it is possible to define "primary elements" of the living nature using the reduction principle as used by physical description of objects and processes in the "non-living" nature.

Nonetheless, let us assume for a moment that *primary elements* of the biological realm are somehow definable. In this case the biological evolution could be presented as temporal variation of subsets of primary elements and their interrelations presenting a current system state. Following the dynamic paradigm, to describe the evolution it is necessary to introduce also a control function, i.e. a *system potential*. In physical systems, the *potential* is, for example, the energy getting minimized. It could be determined on the basis of "standard" (fundamental) interrelations well-established for physical "primary elements." In the biological world, it is not yet possible to find such an equivalent. Nor it is possible to define, more or less satisfactory, *biological "fundamental interrelations"* and that *optimization principle* which governs the biological evolution. Consequently, an application of the dynamic paradigm for a description of the biological evolution in the form in which this paradigm exists in physics is complicated by the ambiguity of definition of all main notions necessary for that.

It has developed historically that the formation of evolutionary ideas in biology was tightly connected to the species concept (Vorontsov, 1999). Later on, the place of the species concept was taken up by the population concept (Mayr, 1984). But are these categories able to claim for the role of the primary biological elements?

During recent decades, many researchers have expressed dissatisfaction with current evolutionary views. Based on the species concept, the synthetic theory of evolution (STE) dominates this field (Dobzhansky, 1937; Mayr, 1984), while as many as 20 (and perhaps even more) species concepts have been suggested (see the review in Hey (2001) and the list of species concepts provided by Mayden (1997)). In reality, the STE is the most detailed modern theory of evolution that claims to embrace the entire biological realm; however, its depth and logic appear to be insufficient in terms of integrating the many different areas of evolutionary research. As documented in many previous publications (long lists of such papers can be found in the following: Gould, 2002; Mayr and Provine, 1980; Vorontsov, 1999; however, also see counterarguments in Ayala, 2000; McShea, 2004, and many others), some important aspects of evolution are not included in the evolutionary synthesis. Notably missing is any consideration of the patterns and processes of evolution among microorganisms, the problem of mass extinction, and the apparent rapidity of major radiations.

Scientific progress has led to the identification of further problems with the species concept. Recently, Sites and Marshall (2003), after comparing nine methods of species delimiting, concluded that speciation processes create "fuzzy" boundaries, and that all methods will occasionally fail or contradict each other. It has been stated that "the species problem is the long-standing failure of biologists to agree on how we should identify species and how we should define the word 'species' " (Hey, 2001, p. 326). A number of authors agree that the word "species" means only a "category" in the taxonomic hierarchy; i.e., this term only has any sense in a classificatory context. Similarly, Dupré (2002, p. 428) proposed "that we should reserve the term species for the basic units of classification." However, nobody has proposed anything new to operate in the dynamic world; accordingly, 4 years ago it was noted that: "The study of evolution now stands at a point comparable with that in the 1930s" (Carrol, 2002, p. 912).

Presumably, the situation can be explained by the existence of fundamental methodological biases in a number of general concepts that underlie evolutionary views. Nevertheless, it seems obvious that a single additional concept of species or paper criticizing the STE will not prove to be decisive in this regard. We need a new worldview and a new organization of our knowledge; perhaps even, as stated by Poli (2001, p. 281), the "modification of the metaphysics implicit in a large part of contemporary science and philosophy."

Thus, it is necessary to admit: (1) that the categories of species and population, being mostly frequently used in evolutionary constructions, do not suit for the role of the primary biological elements; (2) that primary biological elements are far from being so simple as the physical ones, and (3) that different "primary elements" correspond to different levels of biological systems' organization. With account for that, we have formulated a concept of biological referents, in accordance to which the sought "primary" (but not simple) elements of the biological realm are presented by *the referents*. Being organized in various configurations, these referents determine the structure of life and its parts, whereas their interactions with each other and with environments determine the processes that are generally interpreted as the evolution of life.

2 Identification of Referents

Many biologists and philosophers, including M. Conrad, S. Salthe, and H. Pattee (e.g., Eldredge and Salthe, 1984; Pattee, 1995; and the detailed review in Rocha, 2001) have appealed to the terms *reference* and *referent*, but no systematic analysis of the biological realm has been undertaken on this basis.

The referent concept accepted here is based on the definition from the American Heritage Dictionary that considers a referent as: **1.** Something that refers, especially a linguistic item in its capacity of referring to a meaning. **2.** Something referred to.

We will restrict this chapter to the consideration of physically tangible referents and attempt to show that, in contrast to the species concept, the referent concept is capable of answering the following questions: (1) "what does the word 'referent' mean," and (2) "how should we identify referents?"

Moreover, the term "referent" can be used in a description of the total diversity of the systems and structures that represent the existing biological realm; however, before reaching this goal, we must clarify many points.

As the dictionary definition is too broad to be used in the field of material referents of life structure, it is modified as follows:

A referent is the simplest element of a certain set of elements that can be considered as a basic element of the set.

Following this definition, the material world as a whole and any individual subdivision of this world can be considered as a referent. This definition remains too broad but it is suitable for the purpose of this chapter.

The clearest referents are organisms; however, even these referents require very careful consideration, as all biological referents are neither strictly discrete nor identical. Cells are referents when they are individual organisms; however, the cells that make up the tissue of a multicellular organism are only parts of this organism, yet, only the organism is a referent that belongs to the higher level. Generally, we cannot consider, for example, a muscle as a set of cells because neither the muscle nor its cells are discrete and individual entities. Genes can be considered as referents because of their discrete nature; however, to be consistent with the above definition, we must endow genes with the capability to form sets. This problem will be discussed further in the following subsection.

If we accept Koestler's (1967) holon concept, then holons of the same level can be considered as referents in a given holonic set; however, it should be made clear that while every holon set is constructed of referents, there are many referent sets that are not holonic in nature. The significant difference between holons and referents is that holonic interactions between sub-elements in a holonic set are assigned in advance, whereas the nature of referents in a certain set can be predefined only in the context of the set under consideration; it may differ in other contexts. Any referent can simultaneously be a member of two or more sets or any sequence, thus demonstrating the universal, all-penetrating capability of the referent concept. Moreover, this property of referents makes them capable of creating the biological realm in all its diversity of structures and interconnections. One can conclude that the concept of referents is more universal than the species concept because of its applicability to any level of biological complexity.

2.1 Defining Sets of Referents

According to the above definition of a referent, there are no referents that are common to all of the levels of biological complexity, whereas it is possible to distinguish a number of groups of referents or sets of referents that correspond to the levels of complexity and that define the number of such levels. The problem of the apportionment of the levels of biological complexity is far from trivial (Haken, 1988 (2005, p. 24)), and different researchers use different numbers of transitions or levels. Thus, Smith and Szathmáry (1995) indicated eight main transitions in evolution, while Michod and Herron (2006) indicated only six. As there is no criterion for the assignment of the levels of complexity, the number of levels is generally defined arbitrarily. As a general guideline, some investigators use the assumption that the levels of biological complexity must be separated in terms of their endogenous cycle rates by one or more orders of magnitude (Salthe, 2004). As the idea of endogenous cycle rates is rather vague, we decided to use an interval of a certain number of orders of magnitude to obtain the lowest number of levels of complexity.

As units of matter, referents can be characterized by number, size, and connections, thereby enabling us to define the sets of referents using any of these three characteristics. We start with a definition in terms of size, as it is known that any increase in differentiation is dependent upon size (Changizi et al., 2002). If we take a molecule as a unit of a certain size, it is possible to calculate that one simple gene contains about 10^{3-4} of such units (e.g., about 400 nucleotides are required for the coding of the 118 amino acids that constitute a molecule of the structural protein of the tobacco mosaic virus). The simplest cell contains approximately 3×10^2 genes (e.g., *Mycoplasma genitalium*, which can grow and divide under laboratory conditions (Hutchinson et al., 1999), probably possesses between 265 and 350 genes). However, as many as 3×10^4 genes were identified in the human genome (Claverie, 2001). Most vertebrate animal organisms consist of about 10^{10-13} cells, while an average European state consists of approximately 10^{7-8} human beings.

Thus, the size difference between referents of different levels represents a number of orders of magnitude; and a definition on the basis of size provides a trivial result: all of the referents can be subdivided into five groups: molecules, genes, unicellular organisms, multicellular organisms, and communities of organisms. However, as we will see later in the text, the contents of these groups are far from trivial.

A definition on the basis of numbers is more difficult, especially for molecules and genes. The number of genes is difficult to estimate, especially when genes are considered as referents of gene sets. This problem stems from our tradition of considering genes as units at a certain loci of DNA; we then transfer the same terminology to extra-chromosomal genetic elements. According to the definition of a referent provided above, we are unable to indicate the genes-referents and sets of genes-referents among "traditional" genes. We are unaware of examples of genes that currently exist independently, and not all researchers agree that such genes existed in the past. Therefore, we face a choice: either we consider genes as referents and admit to their autonomy to some degree, or we exclude the possibility of their autonomy and reject the reality of a gene level of referent complexity. The second alternative appears to be nonsense; however, the first option also requires caution.

One can assert that genes are discrete and thereby autonomous; numerous experiments in transgenesis provide the best evidence of gene autonomy.

However, this evidence is deficient: such discrete genes cannot form sets of genes. To invest the genes with the capability of set-forming, we must consider the genes as certain pre-organismal units. As a case in point of the (past) existence of such forms of life, we can consider recent viruses, plasmids, and similar entities, as they possess a certain autonomy even without being true organisms; however, this action creates a new domain of genetic units that is located between true organisms and a pre-biotic form of evolution. This domain is absent from the most widely known three-domain-based tree of life (Woese, 1998, 2002; Cavalier-Smith, 2004, and many others). Nevertheless, some of scientists with Forterre at the head (see Zimmer, 2006, p. 870) believe that "Maybe each of these viruses is a remnant of domain that has disappeared." This view is in good agreement with the fact that, after molecules, viruses are the most abundant non-cellular entities on Earth (Drake et al., 2001; Forterre, personal communication 2005; Wommack and Colwell, 2000; Zimmer, 2006). Despite a lack of metabolic apparatus, viruses nevertheless undergo evolution that in some respect is identical to that of some "true" organisms (Noda et al., 2006). The completion of the sequencing of the *Mimivirus* genome enables the proposition that this virus lineage could have emerged prior to the individualization of cellular organisms (Suhre et al., 2005). Given this understanding, the genes form the sets of genes and represent the level of gene complexity.

Providing "pre-organismal genes" with autonomy and segregating viruses, phages, viroids, and plasmids into a separate domain, we must consider that autonomous "pre-organismal genes" existed during the early period of biological evolution and at that time differed dramatically from recent non-organismal units that are the result of a billion years of evolution. We must also distinguish these autonomous units from true genes that are now fused to produce the joined genomes of organisms, whereby they lost most of their autonomy because they belong to higher level.

This idea appears strange in terms of viruses and plasmids, but a similar consideration appears self-evident when we consider molecules that existed at the time of the transition from chemical to biological evolution. As these molecules are also absent today, we are unable to state that "these molecules were precursors of these genes"; therefore, the problem of estimating the numbers of molecules–referents also exists.

Further defining on the basis of numbers will reveal the fact that the number of organisms is 2–4 (or much more) orders of magnitude less than the number of genes. A similar ratio holds true for organisms and societies, as the number of parts is always greater within a more inclusive whole. Although defining on the basis of numbers is not so evident as defining on the basis of sizes, it produces the same referent sets.

Assuming that these five meta-sets of referents correspond to at least five levels of biological complexity, we will attempt to construct the five-level model of the biological realm.

3 Referents and Hierarchy of Complexity

3.1 From a Linnaean Hierarchy to the Hierarchy of Complexity

The notion that the biological realm is organized hierarchically was quickly assimilated by biologists, as biological systems with a hierarchical architecture were assumed to evolve more rapidly, enable greater stability, and thus be favored by natural selection (Koestler, 1967; Pattee, 1970; Simon, 1962).

Numerous attempts were made to present an accurate picture of the hierarchy of the living world and to show the different types of such a hierarchy. The main results of these investigations were (1) an understanding of the hierarchical structure of many evolutionary systems, (2) the concepts of their "levels" of organization, and (3) the rough classification of hierarchies into structural (constitutive) and functional (associative and control) types (e.g., see Mayr, 1984; Pattee, 1995; Valentine and May, 1996). Thus, Eldredge and Salthe (1984) considered two types of parallel and partly overlapping hierarchies: one (genealogical) comprising genes, chromosomes, organisms, demes, species, and monophyletic taxa, and the other (economic or ecological) dealing with molecules, organelles, cells, tissues, organs, organ systems, organisms, populations, and certain ecological units.

A new wave of interest in hierarchical structures in biology has developed in recent decades. Recently, the very detail of the sequence, from a molecule to gene, nucleus, cell, and so on, right up to species, was constructed to capture the essence of the biological hierarchy (Korn, 2002). However, many additional (and non-universal) explanations were required to understand the novel origin implicit in such a hierarchy (Korn, 2005). At the same time, Gould (2002) proclaimed the need for a hierarchy in which the units are individuals, physically nested one within another. This type of hierarchy is partly related to the hierarchies of (McShea, 2001; Simon, 1962; Valentine and May, 1996); however, Gould rejected the idea of the dual nature of a biological hierarchy and proposed a single hierarchy of individuals that included genes, cells, multicellular organisms, demes, species, and higher monophyletic taxa (Gould, 2002). In his review of Gould's book, McShea (2004) emphasized difficulties with the identification of individuals and emphasized that many biological objects, such as ecological associations ranging from microbial mats to ecosystem-scale units, are missing in Gould's schema. McShea (2001) proposed the use of different hierarchical constructions in accordance with research programs. This kind of pluralism is somewhat similar to the pluralism of Ereshefsky (2001) in his interpretation of species concepts, but the pluralistic approach is the only way to conceal our inability to find an adequate solution to the problem. In any case, either "structural" or "economic," or even "pluralist" types of presentations of hierarchies, produced surprisingly limited additional knowledge of life complexity. As a result, the biological theory has no clear and valid criterion to use in identifying the levels of the hierarchy of life.

Simultaneously, different researchers began to call for new ideas for reconstructing the architecture of life complexity. Thus, it was stated that life systems and subsystems gravitate toward the level or stratum organization, and can be described in terms of the holon concept (Honma et al., 1998). A rather long discussion followed the proposal to segregate two components in the hierarchical structure of organisms: (1) a number of levels when entities are nested within entities; and (2) a degree of individuation of the top-level entity (for relevant references, see McShea, 2001; Santelices, 2004; Valentine, 2003; Wu and David, 2002).

A simple complexity pyramid composed of the various molecular components of the cell (genes, RNAs, proteins, and metabolites) was presented to introduce the idea of network organization in the cell hierarchy (Oltvai and Barabási, 2002). Notably, as declared in the title of the paper, "Life's complexity pyramid," the authors avoided a description of species and higher taxa levels in life's complexity. It is necessary to draw attention to the fact that a species is an important element of many hierarchical constructions, and this might lead to the failure of reconstructions. Despite the fact that the species concept operates successfully and possesses the power of prediction in many fields of evolutionary biology, it contains serious weaknesses, and one might anticipate the appearance of a new worldview that incorporates all of the achievements of the STE and numerous phenomena that are currently outside the realm of this concept (Vorontsov, 1999). To achieve this goal, it is necessary to revise our views of the structure of the biological realm in respect of the central position of the species (Fig. 1a). This revision, which we call for, does not involve the complete rejection of all of the species concept-based knowledge and achievements; instead, it is aimed at the reorganization of factual material to construct a new and less contradictory structure of the biological realm. The essence of such a reorganization is the use of the referent concept and referents as the basic

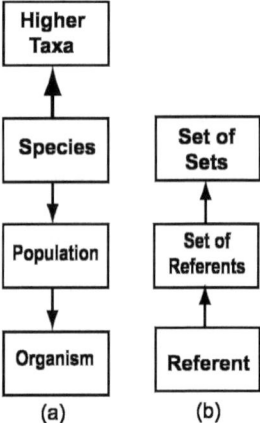

Fig. 1 Two logics of descriptions of biological realm: species-defined schema with the species in the central position (**a**) and referent-defined down-up schema (**b**)

elements of more complex sets that comprise all of the divisions and subdivisions of the biological realm (Fig. 1b)

3.2 Two Directions in the Architecture of Complexity

Many researchers have refuted the idea that the main evolutionary transitions occur in a linear sequence (McShea, 2001; McShea and Changizi, 2003; Michod, 1997; Michod and Herron, 2006; Turchin, 1977, and numerous references therein). Valentine (2000) reconsidered the simple and well-established fact that relatively few genes are needed to generate high levels of morphological complexity, and emphasized that such a complexity is generated in a combinatorial way, whereby each gene is used many times in many different combinations. Woese (1998, 2002) repeatedly turned to the idea of a collective mode of cellular evolution by considering that the diversity of cellular designs evolve simultaneously. Woese also emphasized that "only global invention arising in a *diverse* collection of primitive entities is capable of providing the requisite novelty" (Woese, 2002, p. 8746).

However, no biologist has attempted to apply these ideas to the entire diversity of the biological realm or to locate the layers of complexity on the basis of their structural, combinatorial, and emergent connections.

To make such a presentation in the context of referent conception, we first constructed a model with two (up-ward and orthogonal) directions in the architecture of complexity. This model includes five levels of complexity that correspond to sets of referents discussed earlier (Sect. 2.1). We concentrate on the three central levels because the lowest and highest levels can be partly associated with chemical and social types of evolution that demand special considerations. The model also shows two main trends (divergence and convergence) that are involved in the formation of biological complexity. The processes of divergence (multiplication, mixing, and differentiation) produce the diversity of referents of lower level of complexity (LLC), while those of convergence (recombination and integration) create referents of higher level of complexity (HLC). The model is required to satisfy the following conditions.

1. The common structural elements of referents are present at all levels of complexity and thus ensure succession and continuity.
2. The architecture of the model permits multiplication, mixing, and differentiation at every level of complexity to produce the material required for both diversity and integration.
3. The integration of referents of the LLC (rather than direct gradual transition) enables the transition between two adjacent levels of complexity to be accompanied via an increase in complexity.
4. Both forward and backward transitions between the levels have corresponding thresholds that control the probability of transitions.

Figure 2a represents the molecular, gene, cellular, organismal, and social levels of biological complexity as horizontal layers of referent transmutations. The two types of arrows in the figure indicate the relationships between divergence/convergence in producing a stepwise increase in complexity. This model is a greatly generalized fragment of the real world, and most of the referents that belong to a certain level of complexity are omitted from the fragment. This fragment is derived from the temporal and spatial distribution of referents; therefore, it does not illustrate the full diversity of variants of transitions in space and time.

A more detailed presentation of events that are characteristic of certain levels of complexity is shown in Fig. 2b. The divergent trends represent a combination of multiplication, mixing, and differentiation processes that occur at a given level of complexity. The diversity of referents can occur as a result of these processes. A part (possibly a very small part) of this diversity (of referents of the LLC) is involved in the converging trend of integration to produce the referents of the HLC. Different combinations of referents of the LLC ensure the initial diversity of referents of the HLC.

The architecture of the complexity presented herein in the framework of the referent concept (Fig. 2a and b) is rather heuristic, as are most widely accepted theories in biology, as every biological theory appears to be based upon the islands of experimental data and linking heuristic bridges. However, the

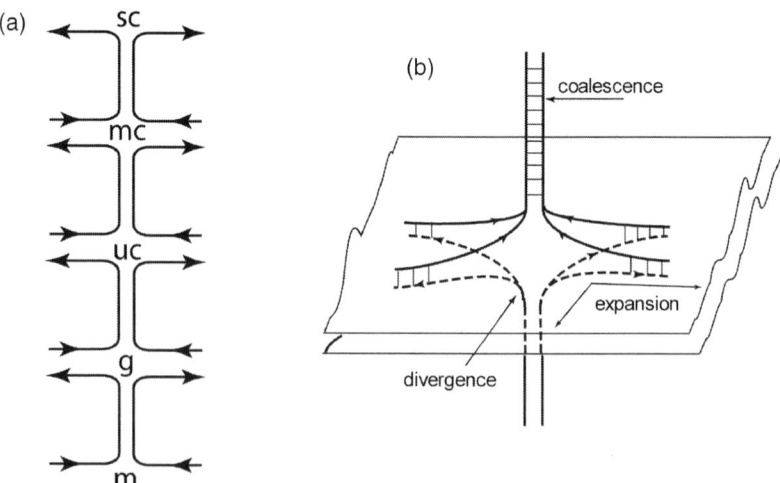

Fig. 2 a. Two-dimensional schema of referent-defined biological realm that depicts five levels of complexity and explains the relationships between divergent and convergent trends in complexity creation. m, molecular level; g, gene level; uc, level of unicellular organisms; mc, level of multicellular organisms; sc, social level; **b.** Detail representation of level-confined events. Diversity of referents appears as result of their multiplication and differentiation in the frame of the same level of complexity as it is depicted here by divergent arrows. Converging arrows symbolize the process of integration of elements to produce the referents of higher level

referent concept is broader than the STE paradigm because it includes a massive amount of data on microorganisms and non-organismal units that remain beyond the scope of the STE.

The core of our concept refers to those elements whose reality has been proven or can be proven. Important processes proposed for divergent/convergent trends, such as multiplication, mixing, differentiation, and integration, have been demonstrated in numerous experiments, and many can be observed with the naked eye. Of course, the integration is most intriguing because the novelty (as an acknowledged source of complexity) has its origin in the integration. Therefore, the referent concept states that the novelty emerges as a new entity due to the re-association of pre-existing entities. Some roots of this idea can be found in Alexander' publication (Alexander, 1920), while integration in dynamic systems was foretold by Turchin (1977), and many striking examples of integration in biology have been known for a long time: lichens of symbiotic origin (Schwendener, 1869), corals (see the references in Hatcher, 1997), eukaryotic organisms (Kozo-Poljansky, 1925; for a more comprehensive recent review, see Cavalier-Smith, 2004), allopolyploids and polyploid complexes in higher plants (see the review by Stebbins, 1950), and the multiple nature of the karyotypes (see the review by Ayala and Colluzi, 2005). Additional proof of integration can be found in the steadily increasing volume of information regarding the introgression of mtDNA in animals (Arnold, 1997; Harrison, 1990), as well as evidence of the reorganization and fusion of chromosomes in the *Arabidopsis* genome (Schoof et al., 2002) and other data on genome fusion and lateral gene transfer in the origin of eukaryotes that has been recently demonstrated and discussed (Omelchenko et al., 2003; Rivera and Lake, 2004; Simonson et al., 2005). Evidence for the chimerical origin of genetic code has been discussed elsewhere (Zhuravlev, 2002).

Taken together, these data supports the idea that the succession of level-specific events prescribed by the referent concept was important in the development of biological systems on Earth.

3.3 Transition Thresholds and the Irreversibility of Transition

In Fig. 2a, we used vast meta-sets to obtain the minimal number of layers of complexity that was desirable for the primary description. We also expected that in such generalized layers of complexity, the criterion for level segregation would be found. In fact, a clear structural difference between levels was found in this study; however, one further criterion related to the transition threshold must be considered here. It is of a probabilistic nature, and the value of the threshold is expected to increase during differentiation because of the intuitive expectation that the deepest differentiation produces the referents separated by the highest threshold. The integration of such divergent referents is related to the task of overcoming significant thresholds.

Let us make this evident via the example of transition from the gene level to the organismal level. Consider the transition that is represented by the integration (fusion) of five genes that each consist of 30 nucleotides that correspond to five very simple proteins of about 10 amino acid residues. The probability of the origin of unique sequences of nucleotides, p, (inverse value of the number of realizations) for every gene (in the context of genetic code) will be 4^{-30} for every separate gene (R_1–R_5); however, for a fusion of five genes, we have the product p_{fus}:

$$p_{\text{fus}} = \prod_{i=1}^{5} p = 4^{-5 \times 30}$$

The value 4^{-150} is vanishingly small. To overcome such a combinatorial threshold, a specific strategy of transition is required. We consider that this strategy actually exists because such combinatorially inconceivable transitions *have been realized* in terrestrial life.

It is necessary to emphasize that the probability of accomplishing reverse transitions out of the HLC references into the initial configuration of the LLC referents is also vanishingly small, as this transition requires the selection of a specific initial configuration from a combinatorially huge number of alternatives. Even if some examples of recapitulation are established, the reverse transformation is commonly simply impossible when a significant part of the LLC traits are lost after integration and differentiation at the HLC level. The evidence of the occurrence of such losses during referent transmutations can be obtained from the example of the very complicated fate of mitochondrial genes that originated within the α-proteobacteria. A number of ancestral bacterial genes have now been transferred from the mitochondrial to the nuclear genome at very specific conditions that appear to be irreproducible today (Andersson et al., 2003). The recent finding that sequence loss and cellular simplification are common modes of evolution (Kurland et al., 2006) can also be interpreted as evidence of such irreversibility.

The evidence of the irreversibility of transition discussed above is in close agreement with the thesis that "macroevolution is more than repeated rounds of microevolution" (for a history of this topic, see Erwin, 2000).

4 Formalization of the Referent Representation

4.1 Space of the Referent Signs

On the basis of the above-described ideas, we are going to construct such a description of referents in which sets of signs characterizing low-level referents are embedded into a set of signs characterizing high-level referents. A hierarchy of such embeddings will correspond to the hierarchy of referents.

Let us introduce N sets of symbols, S_i, $i = 1,...,N$, each of those includes p symbols $s_{i,k} \in S_i, k = 0, 1, ..., p - 1$ (p is a prime number, for further convenience). Let each symbol $s_{i,k}$ be interpreted as a sign of a referent of the level i. Let us introduce p^N of sequences of symbols $X = (s_0...s_{\gamma,k}...s_{N,k})^{(x)}$ being read left to right, where s_0 complements the symbol sets S_i, i.e., it is common for all symbol sequences X.

Let us introduce a hierarchy of partitions of the set $\{X\}$ to subsets $R_{\gamma,k}, 0 \leq \gamma \leq N$, so that each subset $R_{\gamma,k}$ includes all sequences coinciding (from the left) up to the symbol $s_{\gamma,k}$, inclusive, and only these ones. The $R_{\gamma,k}$ subsets make up a nesting hierarchy, $R_{\gamma+1,k} \subset R_{\gamma,k}, \cup R_{\gamma+1,k} = R_{\gamma,k}$. Let the hierarchy of the $R_{\gamma,k}$ subsets be matched to the hierarchy of referents according to the following rule. Let us match the referent of the "zero level" to the R_0-set consisting of all p^N sequences with the common (the first from the left) sign s_0 and matching, therefore, a space of the referent signs. The set R_0 is a conjugation of p non-intersecting subsets $R_{1,k}$, $k = 0,1,...,p-1$, each of p^{N-1} sequences with two coinciding signs: s_0 and s_1. Let us match to each of p subsets $R_{1,k}$ the referent of the first level. Then, let us match each of the $R_{1,k}$ subsets as a conjugation of p non-intersecting subsets $R_{2,k}$, each including p^{N-2} sequences, and let us match to each of p^2 subsets $R_{2,k}$ the referents of the second level. Further, let us continue this partitioning procedure until the level $\gamma = N$ on which each subset $R_{N,k}$ includes only one sequence. Thus, the sequences of the length N being the carriers of total sets of referent signs are associated with the most "complex" referents whose description demands for the largest set of individual signs. Referents of the lower level are identified by smaller sets of individual signs.

A class of objects (sequences) constructed in this way forms a space of referent signs (a RS-space) with a topology of embeddings. On the RS-space, a metrics could be introduced and defined as follows: a distance $d(X,Y)$ between sequences X and Y which coincide left up to the symbol s_{γ,k_γ} inclusive is equal to $p^{N-\gamma}$. The introduced distance satisfies to the strong triangular inequality, $d(X, Y) \leq max\{d(X,Z), d(X,Z)\}$, and consequently, the RS-space is an ultrametric space by construction. In the RS-space, there is an ultrametric sphere of the radius $p^{N-\gamma}$ corresponding to each referent of the γ-level.

For numerical description of the RS-space, let us use the field of p-adic numbers, Q_p, which is natural for ultrametric spaces. Each element of the quotient set Q_p/B_γ ($B_\gamma = (x : |x| \leq p^{N-\gamma})$, i.e., an ultrametric sphere of radius $p^{N-\gamma}$, is matched to the particular referent of the level γ (Fig. 3).

4.2 Evolutionary Equation

The evolutionary equation on the RS-space is constructed by the following way. Let us assume that the evolution of referents, i.e., variation of their distribution in the space of referent signs, is performed by means of "reproduction" of

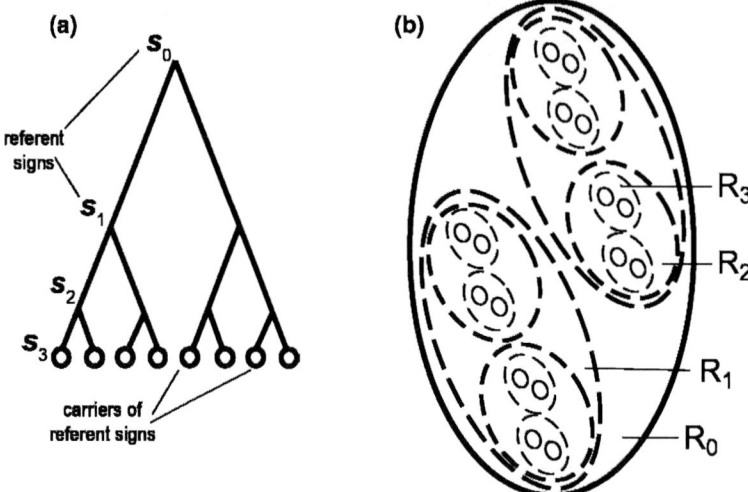

Fig. 3 a. Hierarchical tree-like ordering of the referent signs. For simplicity, a regularly branching tree of four levels is depicted. Each bottom node presents a carrier of the referent sings ($s_0 s_1 s_2 s_3 \ldots$). The complete set of nodes constitutes an ultrametric space of the referent signs. Vertices on the tree present both the ultrametric distances and the embedding role for different references. b. Illustrative presentation of a space of the referent signs with ultrametric topology. Each sphere is a particular set of carriers of referent signs: it presents a referent of particular level from the hierarchy of referent representation. Each smaller sphere is embedded into the larger one. Total hierarchy of embedded spheres (a whole tree) relates to zero "evolutionary temperature." At higher temperature, smallest spheres "dissolve" in larger one

carriers of the referent signs (sequences). Besides, let us introduce into the evolutions of referents a "randomness" assuming that random changes of signs (sequence symbols) are occurred via "reproduction" of carriers of the referent signs (sequences).

Let the sequence X be transformed into the sequence Y as the result of a random change of some sign in X. Let us assume that the probability of such change, $w(x|y) = w\left(|x-y|_p\right) = w(p^{N-\gamma})$, depends only on the ultrametric distance between the sequences, $p^{N-\gamma}$. As there is the hierarchy of distances between the sequences (and, accordingly, between the referents), there also exists the corresponding hierarchy of the probabilities of such random changes.

Using all this construction, it is then possible to write down a p-adic equation describing the biological evolution in the referent presentation. For the simplest approach, such equation has the following form (Avetisov, 2004; Avetisov and Zhuravlev, 2007):

$$\frac{\partial f(x,t)}{\partial t} = \int_{Q_p} w\left(|x-y|_p\right)[f(y) - f(x)]d\mu(y) + a\Omega\left(|x|_p\right)f(x) \qquad (1)$$

Here the real-valued function of the p-adic argument x and the time t, $f(x,t)$, is a distribution of carriers of the referent signs at instant t. An integral term in the right-side of Eq. 1 describes a Markovian random walk over the ultrametric RS-space (ultrametric diffusion). This random process models variation of the referent sign carriers by means of random changes of the signs themselves. The second term in the right-side of Eq. 1 describes "reproduction" of some set of carriers, $x \in Z_p$, namely those forming a sphere of unity radius with the centre $x = 0$ in the ultrametric RS-space.

All eigenvalues λ_j of the right-side operator in Eq. 1 excluding the maximal eigenvalues $\lambda_{max} = \lambda_\infty$ can be shown (Avetisov and Zhuravlev, 2007; Avetisov et al., 2002; Avetisov et al., 2003) to be negative. The maximal eigenvalue λ_{max} is positive or equal to zero, and, in the latter case, λ_{max} is the spectrum accumulation point. By $\lambda_{max} > 0$, the distribution $f(x,t)$ gets with time localized on the RS-space area, in which the reproducible carriers of signs are found. However, if $\lambda_{max} = 0$, the distribution $f(x,t)$ evolves to homogeneous distribution over the RS-space.

For our purposes, it is sufficient to estimate the conditions for the existence of the positive value of λ_{max}. Let us examine the case when the probability for random changes of the sign carriers decreases exponentially with the number of the hierarchy level on which these random changes take place: $w\left(|x-y|_p\right) = p^{-\alpha\gamma}, \alpha > 0$. Then, if $\alpha < 1$, the distribution of the sign carriers tends to be homogeneous. Since in this case, the distribution is not compact, no subsets of the referent signs and, consequently, the referents themselves may be distinguished in any bounded area of the RS-space. In other words, the referents have no identity, and referent representation lacks a hierarchical structure.

When $\alpha > 1$, the distribution of the sign carriers always tends to get localized over some bounded area of the RS-space. The larger is the value of the parameter a, the narrower is the distribution localization area. This area can be understood as a subset of the carrier signs and, consequently, of the referents corresponded to this subset, which has been evolutionary distinguished from others carriers of signs. In other words, the referents get identified on the RS-space.

4.3 Growing of the Referent Hierarchy

It is of interest, how the referent hierarchy grows in course of evolution. To describe this picture, it is convenient to introduce the notion of "evolutionary temperature" $T \sim T_c/\alpha$ (T_c is the "critical temperature," the meaning of which is clarified below). The value of the parameter α increases, the evolutionary temperature decreases.

Let the evolutionary temperature at the beginning be high, i.e., $T > T_c (0 < \alpha < 1)$. As it has been mentioned above, in this case the distribution of the sign carriers tends to be homogeneous on the RS-space, no subset of carriers is distinguished, all carriers could be referred to as referents of "zero" complexity level and, therefore, there is no referent hierarchy. If the evolutionary temperature T decreases, then, just below the value T_c, the distribution of the sign carriers is localized in some bounded area of the RS-space. Since the space is ultrametric, the area is an ultrametric sphere with a finite radius containing reproducible carriers of the "zero" level referents. Generally, just below the critical value T_c the RS-space "splits" into countable set of non-intersected areas (ultrametric spheres) which cover the whole space of the referent signs. The splitting of the RS-space into factorized areas (ultrametric spheres) presents all possible ways of evolutional realization of an event which, according to the referent representation, is associated with the appearance of the first level in a referent hierarchy. We emphasize that factorized areas present all potentialities for this "evolutionary jump." In course of the evolution itself, these potentialities could be implemented only partially.

Thus, just below the critical value T_c the RS-space changes stepwise its topology. Now it presents a set of factorized subspaces embedded into the RS-space. These subspaces correspond to the referents of the lowest hierarchical level, i.e., to the referents distinguished by most common characteristics in the referent representation.

Such a topology of the RS-space persists until the next threshold of the evolutionary temperature, $T_1 = T_c - \Delta_1$, will be reached. Just below this temperature threshold, each of the sign carrier subsets, already factorized at higher temperature T_c, "splits" again into a multitude of smaller subsets. In the ultrametric RS-space, this splitting relates to embedding of spheres of smaller radius into the spheres of larger one. Thus, below the "critical" temperature T_c (at which the first level arises in the referent hierarchy) there exists one more "critical" temperature $T_1 < T_c$ at which the topology of the RS-space changes stepwise again. Arisen at the first "evolutionary jump," the ultrametric spheres split into smaller ultrametric spheres embedded into the first ones, and the topology of the RS-space now reflects the two-level hierarchy of the referents.

During further decreasing of the evolutionary temperature, the third "critical temperature" is passed, and so on. Thereby, in each of the sign subsets factorized at the previous step, the subsets of the next level in the referent representation are factorized. Thus, while the evolutionary temperature is decreasing, a hierarchical tree of referents is growing by serial factorization of the referent signs (Fig. 3). This tree, as it has been mentioned above, reflects all potentialities for realization of sequential "evolutionary jumps" leading to emergence of the referent hierarchy. However, in the course of particular evolution only a part of these potentialities could be implemented, i.e., the referent presentation of the biological realm corresponds only to a subtree at the tree of all possible ways of "growing" of the referent hierarchy.

And one more interesting feature of the model (1) should be mentioned. Let us introduce "evolutionary barriers" $H(x,y)$ connecting them to the probability of random transitions $x \to y$ by the relation $w\left(|x-y|_p\right) \sim \exp\left[-H\left(|x-y|_p\right)/T\right]$. Then the above-described picture of "growing" of hierarchical referent tree can be shown (Avetisov and Zhuravlev, 2007; Avetisov et al., 2003) to arise when the barrier heights grows proportionally to the hierarchy level γ: $H\left(|x-y|_p\right) \sim \gamma$.

Such dependence means that the hierarchy of "evolutionary barriers" in the model (1) should possess a "scale-free" feature.

From this dependence it also follows that there takes place a simple relation between the ultrametric distance $d(x,y) = |x-y|_p = p^\gamma$ between the sign carriers in the RS-space and the height $H(x,y)$ of an "evolutionary barrier" dividing them: $H\left(|x-y|_p\right) \sim \ln(p^\gamma)$. Let us now mention that p^γ is the volume of ultrametric spheres of the radius p^γ. This volume is a measure, V_γ, of those subsets of signs which represent the referents of the level γ, and therefore $H_\gamma \sim \ln V_\gamma$. The last relation means that in the referent representation, the biological realm can be perceived as a hierarchically organized "scale-free" system of "small worlds."

5 Conclusions

In this study, the referent concept was proposed as a means to reconstruct the principal architecture of the biological realm and biological complexity. In pursuing this goal, new pictures of the distribution and connections of the "elementary units" of life, referents, were described. Every level of complexity includes the space where reproduction, mixing, and differentiation together lead to the origin of a rich diversity of referents (at the same level of complexity), while the recombination and integration of referents takes place to produce the referents of the HLC. The proposed hierarchical structure appears to be less contradictory than the linear or pyramidal structures presented previously, as no groups of biological referents are outside this hierarchy construction.

A hierarchical structure of referents can be described by a class of objects forming an ultrametric space. These objects can be associated with the sets of referent signs being ordered in such way that sets of signs of low-level referents are embedded into set of signs of high-level referents. The hierarchy of such embeddings would correspond to the referent hierarchy. For the space of referent signs constructed in such way, it is possible to introduce an evolutionary equation and to describe a "growing" of the hierarchical architecture of biological complexity.

Although the idea of "transitions" marking the integration of existing systems into a metasystem was suggested by Turchin as long ago as in 1977, it remains unpopular among biologists and is rarely referred to. Some scientists who are actively working on the problems of transition state and biological complexity are even convinced that "we should not expect any one model to cover everything" (Michot and Herron, 2006). In contrast, we believe that the referent approach is valid for the multi-purpose generalization of evolutionary events. We understand that such a wide-ranging concept is vulnerable to criticism, especially at its early stages of development. Nevertheless, the first evidence of the validity of the referent concept was demonstrated in this present chapter (the capacity of concept to comprise all the sets of the biological realm;

the demonstration of the specific and modular design of levels of complexity; a new view of the molecular and gene levels). We expect that the concept will be useful for solving the problem of "error catastrophe" (Avetisov and Zhuravlev, 2007), for the phylogenetic presentation of individual development (Zhuravlev et al., in preparation), and other evolutionary applications.

Acknowledgment The work was partly supported by the Program of Presidium of RAS "Origin and Evolution of Biosphere" Project No. 04-1-П25-035 FEB RAS and by grant NSH 6923-2006.4 "Leading Schools of Thought" of President of Russian Federation.

V. A. Avetisov is thankful to the COST-D27 program for the partial support of this work through the Russian-EU scientific collaboration in the field of the origin of life and evolution.

Yu. N. Zhuravlev is grateful to Prof. E. Ya. Frisman, and Dr. L. N. Vasiljeva for their helpful discussion of these issues, and to Dr. Elena Sundukova for her excellent assistance on the all stages of manuscript preparation.

References

Alexander, S. (1920) Space, Time, and Deity, vol. 2. Macmillan, London (cited by Korn 2005).
Andersson, S.G., Karlberg, O., Canback, B. and Kurland, C.G. (2003) On the origin of mitochondria: a genomics perspective. Phil. Trans. R. Soc. Lond. B 358, 165–177.
Arnold, M.L. (1997) Natural Hybridization and Evolution. Oxford University Press, New York.
Avetisov, V.A. (2004) Origin of biological homochirality: in search of evolutionary dynamics. In: G. Pályi, C. Zucchi and L. Calglioty (Eds), Progress in Biological Chirality. Elsevier, Amsterdam, pp. 3–12.
Avetisov, V.A., Bikulov, A.H., Kozyrev S.V. and Osipov, V.A. (2002) p-Adic models of ultrametric diffusion constrained by hierarchical energy landscapes. J. Phys. A: Math. Gen. 35, 177–189.
Avetisov, V.A., Bikulov, A.H. and Osipov, V.A. (2003) p-Adic description of characteristic relaxation in complex systems. J. Phys. A: Math. Gen. 36, 4239–4246.
Avetisov, V.A. and Zhuravlev Yu.N. (2007) An evolutionary interpretation of the p-adic ultrametric diffusion equation. Doklady Math. 75, 453–455.
Ayala, F.J. (2000) Debating Darwin. Biol. Phil. 15, 559–573.
Ayala, F.J. and Coluzzi, M. (2005) Chromosome speciation: humans, Drosophila, and mosquitoes. Proc. Natl Acad. Sci. 102 (Suppl. 1), 6535–6542.
Carroll, R.L. (2002) Evolution of the capacity to evolve. J. Evol. Biol. 15, 911–921.
Cavalier-Smith, T. (2004) Only six kingdoms of life. Proc. Roy. Soc. Lond. B 271, 1251–1262.
Changizi, M.A., McDannald, M.A. and Widders, D. (2002) Scaling of differentiation in networks: nervous systems, organisms, ant colonies, ecosystems, businesses, universities, cities, electronic circuits, and Legos. J. Theor. Biol. 218, 215–237.
Claverie, J.-M. (2001) What if there are only 30 000 human genes? Science 291, 1255–1257.
Drake, L.A., Choi1, K.-H., Ruiz, G.M. and Dobbs, F.C. (2001) Global redistribution of bacterioplankton and virioplankton communities. Biol. Invasions 3, 193–199.
Dupre, J. (2002) Hidden treasure in the Linnean hierarchy. Biology and Philosophy 17, 423–433.
Dobzhansky, T. (1937) Genetics and the Origin of Species. Columbia University Press, New York.
Eldredge, N. and Salthe, S.N. (1984) Hierarchy and evolution. Oxford Surv. Evol. Biol. 1, 184–208.

Ereshefsky, M. (2001) The poverty of the Linnaean hierarchy: a phylosophical study of biological taxonomy. Cambridge University Press, New York.

Erwin, D.H. (2000) Macroevolution is more than repeated rounds of microevolution. Evol. Dev. 2, 78–84.

Gould, S.J. (2002) The structure of evolutionary theory. Harvard University Press, Cambridge.

Haken, H. (1988) Information and Self-organization. A macroscopic Approach to Complex Systems. Second enlarged edition. URSS, Moskwa, Russian translation, 2000.

Harrison, R.G. (1990) Hybrid zones: windows on evolutionary process. Oxford Surv. Evol. Biol. 7, 69–128.

Hatcher, B.G. (1997) Coral reef ecosystems: how much greater is the whole than the sum of the parts? Coral Reefs 16(Suppl.), S77–S91.

Hey, J. (2001) The mind of the species problem. Trends Ecol. Evol. 16, 326–329.

Honma, N., Abe, K., Sato, M. and Takeda, H. (1998) Adaptive evolution of holon networks by an autonomous decentralized method. Appl. Math. Comp.. 91, 43–61.

Hutchinson, C.A., Peterson, S.N., Gill, S.R., Cline, R.T., White, O., Fraser, C.M., Smith, H.O. and Venter, J.C. (1999) Global transposon mutagenesis and a minimal mycoplasma genome. Science 286, 2165–2169.

Koestler, A. (1967) The Ghost in the Machine. Arkana. The Penguin Group, London.

Korn, R.W. (2002) Biological hierarchies, their birth, death and evolution by natural selection. Biol. Phil. 17, 199–221.

Korn, R.W. (2005) The Emergence Principle in Biological Hierarchies. Biol. Phil. 20, 137–151.

Kozo-Poljansky, B.M. (1925) New Principle in Biology. Studies on the Symbiogenetic Theory. Voronesh (in Russian).

Kurland, C.G., Collins, L.J. and Penny, D. (2006) Genomics and the irreducible nature of eukaryote cells. Science 312, 1011–1014.

Mayden, R.L. (1997) A hierarchy of species concepts: the denouement in the saga of the species problem. In: M.F. Claridge et al. (Eds), Species: the Units of Biodiversity. Chapman & Hall, London, pp. 381–424.

Mayr, E. (1982) The Growth of the Biological Thought. Harvard University Press, Belknap.

Mayr, E. and Provine, W.B. (1980) The Evolutionary Synthesis. Harvard University Press, Cambridge.

McShea, D.W. (2001) The minor transitions in hierarchical evolution and the question of a directional bias. J. Evol. Biol. 14, 502–518.

McShea, D.W. (2004) A Revised Darwinism. Biol. Phil. 19, 45–53.

McShea, D.W. and Changizi, M.A. (2003) Three puzzles in hierarchical evolution. Integr. Comp. Biol. 43, 74–81.

Michod, R.E. (1997) Cooperation and conflict in the evolution of individuality. 1. Multilevel selection of the organism. Am. Naturalist 149, 607–645.

Michod, R.E. and Herron, M.D. (2006) Cooperation and conflict during evolutionary transitions in individuality. J. Evol. Biol. 19, 1406–1409.

Noda, T., Sagara, H., Yen, A., Takada, A., Kida, H., Cheng, R.H. and Kawaoka, Y. (2006) Architecture of ribonucleoprotein complexes in influenza A virus particles. Nature 439, 490–492.

Oltvai, Z.N. and Barabási, A.-L. (2002) Life's complexity pyramid. Science 298, 763–764.

Omelchenko, M.V., Makarova, K.S., Wolf, Y.I., Rogozin, I.B. and Koonin, E.V. (2003) Evolution of mosaic operons by horizontal gene transfer and gene displacement in situ. Genome Biol. 4, R55.

Pattee, H.H. (1970) The problem of biological hierarchy. In: C.H. Waddington (Ed.), Towards a Theoretical Biology, Edinburgh University Press, Edinburgh, Vol. 3, pp. 117–136.

Pattee, H.H. (1995) Evolving self-reference: matter, symbols, and semantic closure. Comm. Cogn.-Artif. Intell. 12, 9–27.

Poli, R. (2001) The basic problem of the theory of levels of reality. Axiomathes 12, 261–283.

Rivera, M.C. and Lake, J.A. (2004) The ring of life provides evidence for a genome fusion origin of eukaryotes. Nature 431, 152–155.
Rocha, L.M. (2001) Evolution with material symbol systems. Biosystems 60, 95–121.
Salthe, S.N. (2004) The spontaneous origin of new levels in a scalar hierarchy. Entropy 6, 327–324.
Santelices, B. (2004) Mosaicism and chimerism as components of intraorganismal genetic heterogeneity. J. Evol. Biol. 17, 1187–1188.
Schoof, H., Zaccaria, P., Gundlach, H., Lemcke, K., Rudd, S., Kolesov, G., Arnold, R., Mewes, H.W. and Mayer, K.F. (2002) MIPS *Arabidopsis thaliana* Database (MAtDB): an integrated biological knowledge resource based on the first complete plant genome. Nucl. Acids Res. 30, 91–93.
Schwendener, S. (1869) Die Algentypen der Flechtengonidien. Basel.
Simon, H.A. (1962) The architecture of complexity: Hierarchic Systems. Proc. Am. Phil. Soc. 106, 467–482.
Simonson, A.B., Servin, J.A., Skophammer, R.G., Herbold, C.W., Rivera, M.C. and Lake, J.A. (2005) Decoding the genomic tree of life. Proc. Natl Acad. Sci. 102(Suppl. 1), 6608–6613.
Sites, J.W. and Marshall, J.C. (2003) Delimiting species: a Renaissance issue in systematic biology. Trends Ecol. Evol. 18, 462–470.
Smith, J.M. and Szathmáry, E. (1995) The Major Transitions in Evolution. Oxford University Press, Oxford.
Stebbins, G.L. (1950) Variation and Evolution in Plants. Columbia University Press, New York.
Suhre, K., Audic, S. and Claverie, J.M. (2005) Mimivirus gene promoters exhibit an unprecedented conservation among all eukaryotes. Proc. Natl Acad. Sci. 102, 14689–14693.
Turchin, V. (1977) The Phenomenon of Science. A cybernetic approach to human evolution. Columbia University Press, New York. Russian translation, 1993.
Valentine, J.W. (2000) Two genomic paths to the evolution of complexity in bodyplans. Paleobiology 26, 513–519.
Valentine, J.W. (2003) Architectures of biological complexity. Integr. Comp. Biol. 43, 99–103.
Valentine, J.W. and May, C.L. (1996) Hierarchies in biology and paleontology. Paleobiology 22, 23–33.
Vorontsov, N.N. (1999) The Development of Evolution Idea in Biology. Progress-Traditsia-Press, Moscow.
Woese, C. (1998) The universal ancestor. Proc. Natl Acad. Sci. 95, 6854–6859.
Woese, C.R. (2002) On the evolution of cells. Proc. Natl Acad. Sci. 99, 8742–8747.
Wommack, K.E. and Colwell, R.R. (2000) Virioplankton: viruses in aquatic ecosystems. Microbiol. Mol. Biol. Rev. 64, 69–114.
Wu, J. and David, J.L. (2002) A spatially explicit hierarchical approach to modeling complex ecological systems: theory and applications. Ecol. Modell. 153, 7–26.
Zimmer, C. (2006) Did DNA come from viruses? Science 312, 870.
Zhuravlev, Yu.N. (2002) Two rules of distribution of amino acids in the code table indicate chimeric nature of the genetic code. Dokl. Biochem. Biophys. 383, 85–87.

The Prebiotic Phase of the Origin of Life as Seen by a Physical Chemist

V. N. Parmon

Abstract It is shown that the main sufficient distinction of living matter from nonequilibrium abiogenic chemical systems seems to be the existence of a biological memory which allows the natural selection and thus an adaptive evolution of living systems. The known carriers of that biological memory in the currently existing terrestrial living systems are only the DNA molecules. However, the existence of a primitive "memory" without a DNA support is characteristic of autocatalytic systems having the properties of chemical mutation of autocatalysts. For this reason such autocatalytic systems can be considered as a real abiogenic predecessor of life on prebiotic Earth. The most interesting among the known autocatalytic systems with the "mutation" properties appears to be the Formosa reaction, i.e., autocatalytic aqueous synthesis of monosaccharides from formaldehyde in the presence of calcium hydroxide. One can expect that the Formosa reaction could initiate the above-mentioned prebiological evolution resulting in appearance of ribose which is known as a key element of both RNA and DNA molecules.

The first natural question that arises when physical chemists begin to discuss the origin of life is the possibility of a correct enough physicochemical definition of the phenomenon of life. Indeed, according to the opinion expressed since the second half of the nineteenth century by a great Marxist philosopher Friedrich Engels, the biological form of the motion of matter is qualitatively more complicated and is therefore not reducible to the simpler forms of the movement of matter, like chemical and physical ones (Engels, 1935). However, there is no doubt that any individually considered chemical processes in living organisms strictly obey all the laws of traditional physical chemistry.

What is the principal distinction between organic and inorganic matter? Unfortunately, we cannot offer a precise physicochemical definition of the

V. N. Parmon
Boreskov Institute of Catalysis, Novosibirsk, Russia
e-mail: Parmon@catalysis.ru

phenomenon of life, although there are dozens and even hundreds of various definitions of the phenomenon. Nevertheless, there are few extremely significant elements of such definitions which were given at different times by prominent specialists and are important in terms of physical chemistry.

According to Engels, who voiced the opinion of materialist scholars of the late nineteenth century, life is a mode of existence of proteins marked by *an exchange with the environment*. As time passed, this definition became considerably more specific. An important event was the establishment in the 1950s of the crucial role of DNA, which proved even more important than that of proteins. But as far as physical chemists were concerned, the key role belonged to the supplement provided by I. Prigogine and P. Glensdorf, who found that from the point of view of nonequilibrium thermodynamics, living organisms can be considered as self-organizing dissipative structures that exist only in conditions which are far from equilibrium due to the existence of reasonable flows of matter and energy through a system (Glensdorf and Prigozhin, 1973). In the 1960s, M. Eigen proved that the properties of living systems should be demonstrated by the so-called hypercycles on the basis of autocatalytic systems possessing a biological information storage and transmission mechanism (Eigen, 1971).

These points of view determine the foundation of the biochemical and biophysical approaches to the study of the phenomenon of life (Rubin, 1999). In the past decades, life was identified as a phenomenon coupled everywhere with the DNA and its possible evolutionary predecessor, RNA. Both classes of these complex biopolymers are capable of self-replication and, in consequence, the reproduction (or replication) of living organisms, and, because of errors of the replication, this leads to mutations that in the long run create the possibility of natural selection and the progressive or adaptive evolution of living species.

An analysis of these and many other definitions of the phenomenon of life prompts the conclusion that its principal physicochemical attributes include:

1. phase individuality of living systems;
2. the functioning of these systems via existence of an exchange of matter and energy with the environment;
3. the ability of the living systems for replication, that is, self-reproduction; and
4. the capacity of living systems for irreversible adaptive evolution (by means of the natural selection).

It is well known that the first attribute can be found in many equilibrium self-organizing systems in inorganic nature. A case in point is the spontaneous formation of emulsion micelles and supramolecular structures from molecules of large enough size.

The second attribute also does not belong exclusively to living systems. Indeed, any heterogeneous or microheterogeneous catalyst also exchanges matter and energy with its environment, because catalytic transformations take place only through the formation of catalytic intermediates, which emerge in the course of interaction of the active centers of the catalysts and the substance of initial chemical reagents (and products), provided there is a thermodynamic driving force for a resulting overall catalytic process. This force is constituted by the thermodynamic affinity of the catalyzed reaction, that is, its thermodynamic nonequilibricity.

The capacity for self-replication is also a well-known phenomenon in inorganic nature. It is demonstrated, for example, by crystalline solids, a negligibly small primer of which leads, in thermodynamically permitted conditions, to the growth of a phase identical to the primer phase. The capacity for self-replication is usually associated with the so-called *autocatalytic* properties. It is found in the above-mentioned micelles and microemulsions in conditions of their thermodynamic stability as the quantity of the high-molecular compound that forms them increases. Note that, naturally enough, neither micelles nor crystals nor autocatalytic systems can ever be placed in the category of living systems.

We see that the real qualitative difference between the phenomenon of life and inorganic nature is the capacity of living systems for adaptive irreversible evolution. Such evolution is a result of multistage natural selection and leads to ever more complex objects which retain all the attributes of the living objects listed above. On Earth, *Homo sapiens* is considered to be the crown of natural selection in the course of such evolution of living systems.

The capacity for irreversible adaptive evolution is organically determined by the "biological memory" of living systems, because any changes that appear as a result of mutations are built onto an already existing system and are fixed with the help of the genetic apparatus, which has a chemical nature. The possibility of reversion in the course of evolution is a rare exception which only confirms the rule, as is always the case. The unique role of DNA and RNA, with which the phenomenon of life (on Earth at least) is now closely associated, is determined by the fact that in the current conditions, these molecules can be considered as a material carrier of biological memory.

The uniqueness of DNA and RNA properties caused an extremely serious difficulty in all modern theories of the origin of life. Indeed, assuming that life is impossible without DNA and RNA, one has to recognize that the phenomenon of life on Earth appeared only when the first functioning molecule of this type appeared on our planet. Attempts to estimate the possibility of spontaneous self-assembly of such a complex biopolymer in a "broth" even saturated with all the necessary chemical components led to a vanishingly small value and, therefore, an unrealistically long time required for the formation of molecules of the most primitive biological information carriers, RNA, in the course of random chemical interactions. This is why even such dedicated materialists as biochemists are inclined to accept the theory of an extraterrestrial origin of life, for example, through the arrival of "seeds" of life, RNA or DNA molecules, with meteorites, in spite of an evident understanding that panspermia does not solve the problem and just moves it to another place.

The currently predominant theories of the natural origin of life on Earth (the theory of Oparin–Holdane–Bernal on the self-organization of high-molecular substance into coacervates; Miller's theory on the formation of the predecessors of biomolecules as a result of strong physical impacts such as UV irradiation or lightnings on the proatmosphere and the protoocean; Kaufmann's theory on the evolution-induced phase transition as molecules become more complex; etc.) do not even predict the natural emergence of RNA and DNA. Therefore, the main mystery of the origin of life boils down to the question of whence the first functioning RNA molecules, the most primitive known biological information storage and carriers, appeared.

In other words, in terms of chemistry or physical chemistry, the key issue when considering the origin of life is the possibility of finding an even simpler system than RNA or DNA that could retain their chemical prehistory and support an unidirectional, that is, progressive evolution of its properties through chemical changes in the carriers of the information about such a prehistory. Identification of such systems could solve the mystery of the origin of the first RNA and then DNA, since in the presence of progressively evolving chemical systems, RNA and DNA molecules could be a consequence of such evolution and not the origin of the phenomenon of life that is perceived as a much more complex event.

Curiously enough, in the last decades it was discovered that a capacity for monodirectional chemical evolution could be found among simple enough chemical systems that have no separate carriers of biological information.

Some studies (Ebeling et al., 1990; Parmon, 2001, 2002a,b) have shown that even in the simplest chemically reactive system, which contains mutually noninteracting autocatalysts, the phenomenon of natural selection leading to the adaptive irreversible evolution of autocatalysts inevitably emerges if the autocatalysts compete for the same type of "food" and the food niche becomes smaller.

It is very easy to explain why the "natural selection" appears in such systems. Assuming that the transformations of autocatalysts X_i proceed under the simplest scheme of their replication:

$$R + X_i \underset{k_{-i}}{\overset{k_i}{\rightleftarrows}} 2X_i,$$
$$X_i \overset{k_{ti}}{\rightarrow} P, \tag{1}$$

where R is an original substratum—a feedstock; P is a product of the death of the autocatalyst; and k_i, k_{-i} and k_{ti} are the rate constants of the relevant steps of the transformation.

When a preset concentration [R] of food is maintained in the system, it is easy to find the stationary concentrations $[X_i]$ of autocatalyst X_i by solving the stationary kinetic equation

$$\frac{d[X_i]}{dt} = k_i[R][X_i] - k_{-i}[X_i]^2 - k_{ti}[X_i] = 0.$$

The equation has two solutions:

$$\bar{X}_i^{(1)} = \frac{k_i}{k_{-i}}\left([R] - \frac{k_{ti}}{k_i}\right) \equiv \alpha_i([R] - R_{cri}),$$
$$\bar{X}_i^{(2)} = 0.|$$
(1)

The graphic dependence of these solutions upon [R] is presented in Fig.1.

Since only nonnegative solutions have a physical sense, point $[R] = k_{ti}/k_i \equiv R_{cri}$ is the "bifurcation point," in which the solution of the presented system of kinetic equations is branching. The specific characteristic of this point is that at $[R] < R_{cri}$) there is only one zero solution, $\bar{X}_i^{(2)} = 0$, while at $[R] > R_{cri}$, both solutions are possible: both the zero, $\bar{X}_i^{(2)} = 0$, and the positively defined one, $\bar{X}_i^{(1)} = \alpha_i([R] - R_{cri}$. It would be easy to show that at $[R] > R_{cri}$ only the positive solution $\bar{X}_i^{(1)}$ proves stable against possible random fluctuations of the X_i concentrations.

A specific property of the autocatalytic system is, however, that, having found itself on branch $\bar{X}_i^{(2)} = 0$ even in conditions $[R] > R_{cri}$, it is impossible to go over to the stable branch $\bar{X}_i^{(1)}$ unless there appears a primer of autocatalyst X_i, even in the form of just one molecule. In other words, when the food concentration falls below the critical level R_{cri}, the autocatalyst "dies out" completely. It cannot be restored even by increasing the food concentration in excess of value R_{cri}.

If the system has several, rather than one, types of autocatalysts X_i (i = 1,2,...) with differing of values $R_{cri} = k_{ti}/k_i$, the decrease in the food concentration below the critical values will cause a consecutive dying out of those autocatalysts for which R_{cri} proves higher than the value of the current food concentration (Fig. 2). If food concentration R falls, there occurs a

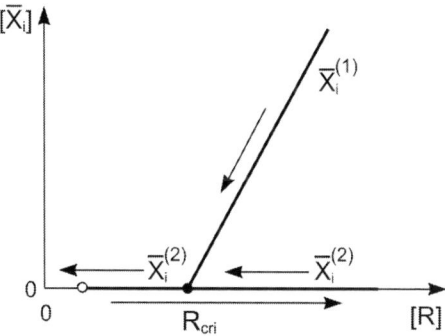

Fig. 1 Dependence of the stationary concentration of autocatalyst X_i on the concentration of food R. Point R_{cri} is the "bifurcation point" (bifurcation of the solutions of the equation). The *arrows* show the evolution of the stationary concentration of the autocatalyst under changes of food concentration in the vicinity of the bifurcation point. The *light-colored circle* on axis [R] designates the point after which the increase of food concentration is assumed

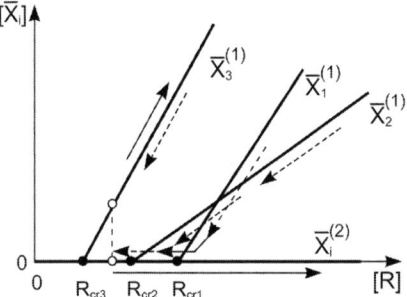

Fig. 2 Dependence of the stationary concentration of autocatalyst X_i on the concentration of food R in a system with several autocatalysts. After the food concentration falls to the value designated by the *light-colored circle*, the subsequent increase in the food concentration is able to restore only the "population" of autocatalyst X_3

consecutive and irreversible (in connection with the disappearance of "primers" even in the form of single molecules) dying out of all the autocatalysts characterized by the highest $R_{cri} = k_{ti}/k_i$ values. This means that in the system, there occurs a strictly unidirectional (towards the decrease in parameter R_{cri}) natural selection of the autocatalyst population, which corresponds to the presence of a biological memory prototype. It is easy to show (Parmon, 2001) that the physicochemical criterion of such selection proves to be quite natural for thermodynamics: in the course of the selection, there survive autocatalysts for which the maximum is their stationary chemical potential, not just their concentration or total mass. (Note that till now there is no definite opinion what the quantitative criteria of natural selection in biology are.)

A capacity for such natural selection also belongs to more complicated autocatalytic systems, which include additional reactions in the scheme of chemical transformations, e.g., nonautocatalytic metabolic and food assimilation reactions (see, e.g., Parmon, 2002a):

$$\begin{aligned} R &\rightleftarrows R_i \\ R_i + X_i &\rightleftarrows 2X_i \\ X_i &\rightleftarrows Y_i \\ Y_i &\rightarrow P \end{aligned} \quad (2)$$

where R_i are the intermediate products of the food assimilation and Y_i are the intermediates of the autocatalyst metabolism.

The same capacity for natural selection can also be observed among systems with slowed down autocatalysis, where autocatalytic replication takes place under the scheme

$$R + X_i \rightleftharpoons X_i + Z_i$$
$$Z_i \rightleftharpoons X_i \quad (3)$$
$$X_i \to P$$

through the formation of intermediate compounds Z_i, etc.

A capacity for natural selection with more complicated characteristics can be found among systems with nonideal replication under which autocatalyst X_i can generate not only an identical autocatalyst but also another autocatalyst X_j:

$$R + X_i \rightleftharpoons 2X_i$$
$$R + X_i \rightleftharpoons X_i + X_j$$
$$R + X_j \rightleftharpoons 2X_j \quad (4)$$
$$R + X_j \rightleftharpoons X_j + X_i$$

A significant outcome of this analysis is the conclusion that, under periodic decreases in the food concentration below the critical values, an adaptive (unidirectional) evolution of the surviving population through the accumulation of chemical "mutations" of autocatalysts X_i begins; that is, there are changes in the existing molecules of autocatalysts X_j which do not, however, lose their capacity for the autocatalytic replication. Thus, if the system has, apart from food R, other compounds capable of a reaction with the autocatalyst molecules, such mutation will be constituted by the formation of autocatalyst molecules modified by these compounds. In the case of the transformation scheme considered, such evolution can have only one direction: towards a decrease in factor R_{cri} in the newly emergent populations as compared to the populations that survived in the course of the earlier selection steps. In other words, this means that the evolution of mutating autocatalytic systems with natural selection is *irreversible*. There is no doubt that this assertion is equivalent to that made regarding the consolidation of acquired characters accumulated in the course of the previous natural selection.

The thermodynamically determined "target" and the end outcome of the development of the most primitive autocatalytic type (1) systems is the trend towards zero of factor

$$F_{ev} = R_{cri} = \frac{k_{ti}}{k_i},$$

that is, the trend towards zero of the ratio of the "death" and reproduction rate constants. A similar meaning can be found in the factor of evolution F_{ev} of more complex autocatalytic systems of type (2)–(4), which comprise additional steps. Competition for food results in the natural selection of the autocatalyst populations; only those populations that have the highest stationary chemical potential values under the preset flow of food are preserved and retain a capacity for a further progressive development.

It is significant that the specificity of the stability of stationary states of the autocatalytic systems under consideration prompts the conclusion that, for an autocatalyst with favorable values of factor F_{ev} to emerge and develop, only a single primer molecule of such an autocatalyst is sufficient. This indicates that, provided that the chemical evolution of autocatalysts has a determined direction, some of their properties may appear when the outcome of a random event becomes fixed. Such an event could be, e.g., the emergence and fixing of a certain optical activity in the autocatalyst populations (Gol'danskii and Kuz'min, 1989) from one optically active molecule of the autocatalyst that won the competition at an early enough stage of natural selection.

The conclusion that it is possible that the evolution-induced characteristics are fixed even in primitive autocatalytic systems is important for building possible physicochemical scenarios of the origin of life on Earth. Indeed, it becomes obvious that strong impacts upon the system (e.g., ionizing or UV radiation or electric discharges on the protoatmosphere model used in the Miller's classic experiments) could result in a significant, but short, deviation of the system from thermodynamic equilibrium, and therefore they were factors that initiated not the process of natural selection itself but mainly mutations (for example, by adding new substituents to the molecules of an existing autocatalyst) which create the possibility of the emergence of new objects from such a selection. The mutants with a lowered F_{ev} factor have better chances of survival. For this reason, it is improbable that the objects produced by two or more consecutive mutations, the first with an increase of factor F_{ev} and the second with a decrease, should survive too (that is, evolve to reach the stationary state of the system).

This assumption is in line with the theory that at the molecular level, a natural selection is possible only in the case of the "instantaneous" acquisition of additional favorable properties that are able to decrease the "evolution factor" F_{ev}. This means, specifically, that many of the enzymes adapted by evolving autocatalytic systems should have abiogenic but already functioning analogs of many prosthetic centers of currently known enzymes. This conclusion has recently been confirmed during attempts to artificially synthesize the functioning analogs of such complex redox enzymes as the oxygen-generating centers of photosystem II of natural photosynthesis and nonheme oxygenases (Elizarova et al., 2000, 2001).

Thus, the most interesting issue when considering possible scenarios for, and even modeling the origin of, life on Earth in abiogenic systems is which kind of abiogenic autocatalytic systems was able to function in a natural environment of the proto-Earth, thereby triggering the prebiotic natural selection.

In principle, such systems are known. Suitable ones for natural selection in the primordial broth include the reactions of synthesis of a broad range of monosaccharide structures in a water solution of formaldehyde, which is supposed to have been widely available on the proto-Earth. These reactions were discovered in 1861 by the Russian chemist A.M. Butlerov and subsequently became generally known under the name of the formosa reaction (formosa means a mixture of various monosaccharides; Khomenko et al., 1980):

$$n\text{CH}_2\text{O} \xrightarrow[\text{H}_2\text{O, Ca}^{2+},\text{OH}^-]{\text{room temperature}} (\text{CH}_2\text{O})_n.$$

In the presence of cations of calcium, magnesium and some other common inorganic bases, these reactions require the presence of a monosaccharide primer, have features of homogeneous autocatalytic synthesis (Khomenko et al., 1980) and take place in water solutions in conditions typical of the surface of the Earth. It is interesting that, according to Butlerov, autocatalytic sugar synthesis is not inhibited by chloride anions, which are most commonly found in natural water, although other anions do inhibit this process (Khomenko et al., 1980).

The possibility of artificial sugar synthesis under the formosa reaction was actively investigated in the 1970s in the Soviet Union and the United States with a view to creating reliable life support systems for very long space flights to Mars (see, e.g., Khomenko et al., 1980). However, this synthesis is still considered uncontrollable due to its autocatalytic character; the reaction produces sugars of different origin: trioses, tetroses, pentoses, hexoses, etc., with differing structures and optical isomerism (Fig. 3). Many C_5 and C_6 monosaccharides obtained under the formosa reaction are compounds that are typical of living organisms that have pronounced optical activity (see Fig. 3 and the structure of ribose in Fig. 4). It is especially important that one of such C_5-monosaccharides, D-ribose, is an irreplaceable RNA element.

It is assumed that the principal components of the Earth's protoatmosphere were methane, ammonia, carbon dioxide and water. It is also known that in a medium which contains, apart from formaldehyde, ammonia and other chemically active compounds, "mutations" of autocatalyst sugars occur easily. There is no doubt that the most probable routes of these mutations can be easily established with the help of experimental methods used in rapidly developing combinatorial chemistry.

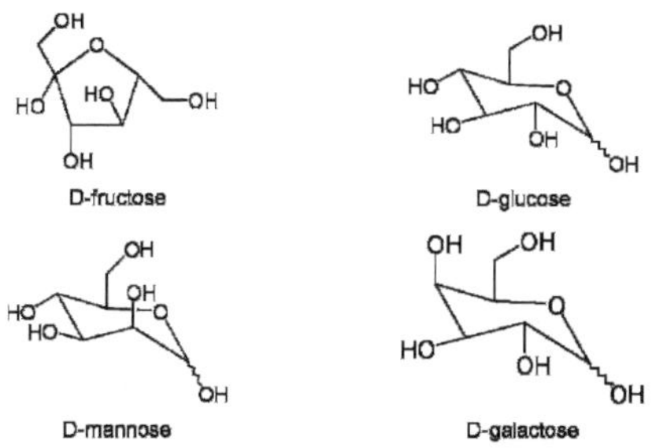

Fig. 3 Diversity of structures of some C_6 D-monosaccharides (hexoses) that are obtained from the formosa reaction and which can act as autocatalysts of this reaction (only "clockwise-rotating" isomers are shown)

D-fructose

D-glucose

D-mannose

D-galactose

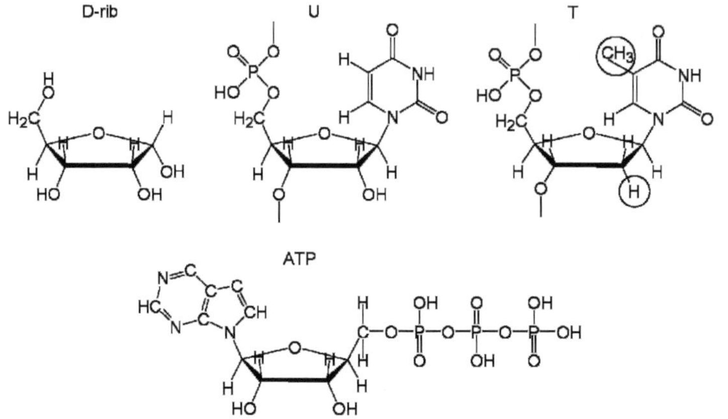

Fig. 4 Chemical formulae of the pentamerous monosaccharide D-ribose (D-rib), one of the product of autocatalytic synthesis in the formosa reaction, and, as an example, the residue of one of the ribonucleotides, uridylic acid (*U*) and its homologous DNA, thymine acid (*T*), as well as *ATP* which is the universal energy carrier in biological systems. The *circles* in the structure of *T* mark the evolutionary improvements of the biological memory elements under the evolution of RNA to DNA

Provided that the water solutions contain ions of phosphoric acid, phosphates of sugar could be formed; in the presence of nitrous bases, this could result in the appearance of universal energy carriers in biological systems, ATP molecules (Galimov, 2001) and RNA prototypes (Fig. 4) that retain the autocatalytic activity.

The possibility that the mechanism of modern biological memory originated in monosaccharides is also confirmed by the fact that, first, RNA is much less stable than DNA: RNA is easily decomposed by alkalis to mononucleotides (due to the presence of group 2'-OH), whereas the polynucleotide DNA strands improved in the course of evolution are stable in the same conditions; and, second, RNA usually comprises a single polynucleotide strand (Khimicheskaya entsiklopediya, 1992). Directly and selectively obtaining ribose diphosphates from glycoaldehyde phosphate implies that early life forms involved not only ribose-type pentamerous but also hexamerous sugars, which formed pyranosyl-RNA (Eschenmoser, 1999).

One can consider the formosa reaction as a possible candidate for the progressive natural selection of the living systems' predecessors in the natural water of ancient Earth, because the sugar synthesis through formaldehyde condensation (oligomerization) does not involve the formation of water molecules. This is why even the high-molecular products of such synthesis are not subject to hydrolysis, unlike polypeptides and proteins. This is extremely essential for synthesizing oligomers in lean solutions: due to thermodynamic reasons, the hydrolysis of proteins during the latter's synthesis from amino acids does not promote the synthesis of such biopolymers in diluted water solutions.

Small amounts of formaldehyde required for the autocatalytic sugar synthesis with subsequent natural selection could form not only under a strong "physical impact" upon the Earth's protoatmosphere (electric discharges or UV radiation), but also during thermocatalytic gas transformations in the reducing protoatmosphere, which had a high methane content when the gases came into contact with a lava substance with a temperature of 700°C or more. This substance, which has a high nickel and iron content in the reducing environment, has pronounced catalytic properties for thermocatalytic steam methane conversion to synthesis-gas (a mixture of hydrogen and carbon monoxide) and/or formaldehyde:

$$CH_4 + H_2O \xrightarrow[Ni,Fe,lava]{>700\,°C} CH_2O + 2H_2.$$

The products of similar thermocatalytic transformations could include other highly reactive "building blocks" for gradually evolving and increasingly complicated autocatalytic systems, e.g., cyanic acid obtained in the course of the thermocatalytic methane–ammonia reaction. The most debatable issue here is the reason why nucleic acids, which are the necessary components and carriers of the biological information of modern organic objects, were involved in the processes of natural selection. It could be a natural consequence of the evolutionary addition of nitrogen- and phosphate-containing fragments to the most primitive autocatalytic oligosaccharide systems already functioning in the "protobroth," which had passed the early stages of their evolution under temperatures that were too high for proteins and nucleic acids.

Below, there is a list of several general questions that can be answered with the help of physicochemical analysis of the origin of life phenomenon on the basis of the above kinetic and thermodynamic analysis of the "natural" selection of autocatalytic systems.

Prevalent opinion	The findings of analysis
Progressive evolution through natural selection mutations can take place in the presence of DNA and/or RNA only	Prebiological progressive evolution through natural selection can take place without DNA and RNA
The "prebiotic" broth was "fatty" (that is, had a high content of the original compounds necessary for the synthesis of biomolecules)	To initiate prebiotic natural selection, the broth must be "lean"
A living organism is inevitably a "dissipative structure" (chemical transformations of the metabolic cycle should have some nonlinear stages which must be kinetically irreversible; Rubin, 1999)	At the early stages of evolution, the predecessors of living organisms need not have been "dissipative structures" (only the stage of "death" must be kinetically irreversible; Parmon, 2002a)

(continued)

Prevalent opinion	The findings of analysis
Even primitive prebiotic structures in nature must have been formed from complex organic molecules	The presence of the simplest compound, formaldehyde, was enough to initiate natural selection
Even the first prebiotic processes are a polycondensation of those organic molecules whose condensation in water solution is thermodynamically obstructed	The formation of sugar from formaldehyde in water encounters no thermodynamic obstruction

The physicochemical position under discussion makes it possible to simplify many points of the origin of life that are otherwise difficult to explain. Finally, this analysis allows the formulation of the principal elements of the physicochemical definition of the phenomenon of life without mentioning the uniqueness of DNA or RNA: *life is a phase-isolated form of existence of functioning autocatalysts capable of chemical mutations, which underwent prolonged evolution owing to natural selection.*

One can hope that these ideas may prove useful in the discussion of the most obscure part of prebiotic evolution at the time when life on Earth originated.

Based on the meager data on the kinetics of the formosa reaction, one can assume that in the absence of live "eaters" of mono- and oligosaccharide molecules produced by autocatalytic synthesis, the prebiotic phase of natural selection before the appearance of the first RNA was not excessively long in terms of geologic time, occupying only tens of millions rather than billions of years after the cooling of the Earth's surface and the appearance of natural liquid water. Obviously, the chemical evolution of autocatalytic systems in the conditions on Earth was inevitable. This is why the prototypes of living organisms could appear and disappear repeatedly, and the trend of their natural selection was determined by the chemical composition of the Earth's surface and was bound to produce more or less similar life forms based on RNA and then DNA. In this way the main features of life on our planet are predetermined.

Indeed, one cannot exclude that the appearance of monosaccharide ribose needed for the formation of the first RNA was not the consequence of the above spoken prototype of the "natural selection" in a "community" of autocatalysts. This can be the case in the situation when there are some still unknown ways of selective enough synthesis of ribose from formaldehyde and/or its simplest derivatives like C_2–C_3 monosaccharides.

Recently, we have carried out a wide research which was aimed to find results of the "natural selection" in the formosa system in a stationary flow-through reactor with nearly ideal mode of operation (Simonov, A.N., Pestunova, O.P., Parmon, V.N., accepted for publication). Unfortunately, we failed to reach such formaldehyde concentrations at which the selection starts. It has been found that for a change in the selectivity of the formosa reaction to be achieved, one has to either cool the system (and thus to diminish the rate of the formosa reaction to consider the reaction as nearly dead) or diminish the starting concentration of formaldehyde below the level of sensitivity of liquid chromatography.

However, simultaneously we have found experimentally that there is the ability of a quite selective synthesis of ribose from formaldehyde and C_2, C_3-monosaccharides in the presence of heterogeneous catalysts like, e.g., natural mineral apatite, synthetic hydroxoapatite, mineral vivianit or homogeneous phosphates (see Pestunova et al.'s "Prebiotic Carbohydrates and Their Derivates," this book). Note that the presence of phosphates in these catalytic systems could be able to facilitate a phosphorylation of the sugars and, as a result, the formation of the first proto-RNA. Anyhow, the consequence of such synthesis would be the start of the same progressive evolution of RNA on the molecular level in the "RNA world."

References

Ebeling, W., Engel, A. and Feistel, R. (1990) Physik der Evolutionarozesse. Akademie-Verlag, Berlin.
Eigen, M. (1971) "Molekulare Selbstorganisation und Evolution" (Self organization of matter and the evolution of biological macro molecules). Naturwissenschaften 58(10), 465–523.
Elizarova, G.L., Zhidomirov, G.M. and Parmon, V.N. (2000) Hydroxides of transition metals as artificial catalysts for oxidation of water to dioxygen. Catal. Today 58, 71–88.
Elizarova, G.L., Matvienko, L.G., Kuzmin, A.O., Savinova, E.R. and Parmon, V.N. (2001) Copper and iron hydroxides as new catalysts for redox reactions in aqueous solutions. Mendeleev Commun. 11(1), 15–17.
Engels, F. (1935) Dialektik der Natur. Chemical etiology of nucleic acid structure. Marx-Engels Inst. Publ., Moscow.
Eschenmoser, F. (1999) Science 284, 2118–2124.
Galimov, E.M. (2001) Fenomen Zhizni (Phenomenon of Life). URSS Publisher, Moscow (in Russ.).
Glensdorf, P. and Prigozhin, I. (1973) Thermodynamic Theory of Structure, Stability, and Fluctuations. Mir, Moscow (in Russ.).
Gol'danskii, V.I. and Kuz'min, V.V. (1989) Spontaneous disturbances in mirror symmetry in nature and origin of life. Usp. Fiz. Nauk 157(1), 3–50.
Khimicheskaya entsiklopediya (Chemical Encyclopedia) (1992) Bol'shaya Rossiiskaya Entsiklopedia, Moscow, vol. 3, p. 297 (in Russ.).
Khomenko, T.I., Sakharov, M.M. and Golovina, O.A. (1980) Hydrocarbon synthesis from formaldehyde. Usp. Khim. 49(6), 1079–1105 (in Russ.).
Parmon, V.N. (2001) Natural selection in a homogeneous system with noninteracting autocatalyst "populations." Doklady Phys. Chem. 377(4), 91–95.
Parmon, V.N. (2002a) Physicochemical driving forces and the pattern of selection and evolution of prebiotic autocatalytic systems. Russ. J. Phys. Chem. 76(1), 126–133.
Parmon, V.N. (2002b) The origin of life: The prebiotic phase. Herald Russ. Acad. Sci. 72(6), 592–598.
Rubin, A.B. (1999) Biofizika (Biophysics). Knizhnyi Dom "Universitet", Moscow, vols. 1 and 2 (in Russ.).

Prebiotic Carbohydrates and Their Derivates

O. P. Pestunova, A. N. Simonov, V. N. Snytnikov, and V. N. Parmon

Abstract The most significant experimental results on the putative synthesis of various carbohydrates and their derivates from simple substrates in plausible prebiotic conditions are summarized and discussed. The synthesis of monosaccharides from formaldehyde and lower carbohydrates (glycolaldehyde, glyceraldehyde, dihydroxyacetone) can be catalyzed by different compounds such as lead, phosphate and borate ions and several natural minerals. Lower carbohydrates can be directly formed in aqueous formaldehyde solutions under the action of UV-irradiation. The possible role of carbohydrates and their derivates in the chemical evolution and development of presumable abiogenic metabolism is illustrated as well.

1 Introduction

An abiogenic evolution of chemical compounds on the early prebiotic Earth was, undoubtedly, the first step of the origin of life. Most likely such evolution started from the simplest substances like H_2, CO, H_2O, CH_2O and HCN which are widely distributed in the universe (Hudson and Moore, 2004; Litvak, 1972). In the course of evolution, these substances were converted into more complex organic compounds such as amino acids, carbohydrates, carboxylic acids and some others which constituted the basis of life.

The presence of aqueous medium could be the most favorable factor for the prebiotic chemical evolution to start. The first reactions which yielded complex organic compounds from the simple ones might be catalyzed by inorganic substances like metal oxides and salts. There are a number of alternative theories, with an origin of life based, for example, on proteins (de Duve, 1991), lipids (Segre and Lancet, 1997), inorganic crystals (Cairns-

---

O. P. Pestunova
Department of Nontraditional Catalytic Processes, Boreskov Institute
of Catalysis, Novosibirsk, Russia
e-mail: oxanap@catalysis.ru

Smith, 1982; Hartman, 1998), FeS–H_2S autotrophy (Wächtershäuser, 1992), prebiotic organic synthesis under hydrothermal conditions (Simoneit, 2004), methane-hydrate deposits (Ostrovskii and Kadyshevich, 2007) or extraterrestrial sources of life (postponing its origin) (Crick, 1981). However, some experts suppose carbohydrates to be the major source of the energy-driving prebiotic metabolism and chemical evolution (Tolstoguzov, 2004; Weber, 1997, 2001). Actually, carbohydrates play an inestimable role in the organic life. Monosaccharides and their derivates constitute the building blocks of various biomolecules like DNA and RNA, ATF, cellulose, chitin and starch which are indispensable for the living organisms. From the point of abiogenic chemical evolution, carbohydrates are supposed to be very important as well. The simplest monosaccharides—glycolaldehyde, glyceraldehyde and dihydroxyacetone—can serve as substrates for further synthesis of amino acids, polyols, aldehydes and carboxylic acids. Higher monosaccharides—hexoses and pentoses—can also be converted into various organic substances including heterocycles.

Among all prebiotic carbohydrates the main emphasis is placed on ribose. Indeed, the RNA-world (Gesteland and Atkins, 1993) is the most reasoned hypothesis on the prebiotic chemical evolution and the origin of life. Recently it has been experimentally proved that some ribonucleic acids possess an enzymatic-like activity (Altman, 1989; Cech et al., 1981). According to the model, the first molecules of RNA must be present for the chemical evolution to start (Joyce, 1991).

Thus, the main problem of prebiotic chemistry is the feasible synthesis of the RNA components: D-ribose, pyrimidine or purine bases and phosphoric acid and their selective assembly into ribonucleotides and then into a sequence of RNA.

The major obstacle in such RNA synthesis is the availability of D-ribose. Besides, ribose is rather unstable (Larralde et al., 1995). For these reasons some experts even suggest that the backbone of the first analogs of modern RNAs did not contain this carbohydrate.

In this paper we focus on possible prebiotic routes to synthesis of carbohydrates, especially ribose, and their derivates.

To create a model of the prebiotic synthesis of any biosubstance one should determine the starting point—the first substrates and catalysts. According to the common knowledge (Hudson and Moore, 2004; Litvak, 1972), formaldehyde (CH_2O), water, ammonia (NH_3) and cyanhydric acid (HCN) are the most possible starting prebiotic substrates. The inorganic catalysts like magnesium, calcium, iron and lead oxides, phosphates, borates can be supposed to be available.

2 The Formose Reaction

In 1861, an eminent Russian chemist A.M. Butlerov discovered that boiling of the aqueous formaldehyde solution in the presence of calcium and barium oxides results in the formation of a complex mixture of various carbohydrates

(Khomenko et al., 1980). Later the catalytic activity of numerous inorganic and organic bases for that reaction was demonstrated. The blend of the carbohydrates derived from formaldehyde was called "formose" and the process for the moment is known as the Butlerov or formose reaction. The general equation of the formose reaction is

$$n\text{CH}_2\text{O} \xrightarrow{\text{catalyst}} (\text{CH}_2\text{O})_n \qquad (1)$$

The conditions, substrates and catalysts required for the process to proceed are very simple. Water and formaldehyde constitute a sufficient part of prebiotic substances, while magnesium and calcium oxide are among the most abundant compounds.

Actually, the incubation of an aqueous formaldehyde solution in the presence of alkaline catalysts at the elevated temperature gives start for several parallel processes based on the alkaline catalysis in addition to the carbohydrate synthesis itself. These are the Cannizzaro reaction, the Lobry de Bruyn Alberda van Ekenstein reaction, retroaldol cleavage, saccharine rearrangement and oxidation of carbohydrates by oxygen.

Thus, the "formose" constitutes a very complex mixture of different carbohydrates (including branched), polyols and organic acids. This gave rise to doubts in the very role of the formose reaction in the prebiotic chemical evolution.

The mechanism of the formose reaction was intensively studied during the twentieth century. However, the general intimate process pathway was not proposed till recently.

In spite of the evident autocatalytic character of the formose reaction, till 1980s all investigators were sure that the process was not initiated by any carbohydrate. The autocatalytic stage of the process was assumed to start from the condensation of formaldehyde into glycolaldehyde—the simplest carbohydrate. Several schemes of this pretended reaction were proposed (Harsch et al., 1984; Mizuno and Weiss, 1974); however, the impossibility of this process was clearly demonstrated earlier (Socha et al., 1981). This fact laid down an additional condition for the formose reaction to pass in prebiotic conditions, i.e., the availability of any simplest carbohydrate needed to initiate the process.

Apart from the stage of the initiation of the formose reaction, the similar further routes of the monosaccharides formation were proposed by different investigators. The common scheme of the monosaccharides formation proposed by Mizuno and Weiss (1974) is presented in Fig. 1. The growth of the carbohydrates chain is governed by numerous aldol-condensation reactions of lower carbohydrates with formaldehyde and one another.

Breslow (1959) and then de Bruijn et al. (1986) were the first to assume the retroaldol cleavage of higher monosaccharides to be the reason of autocatalysis of the formose reaction.

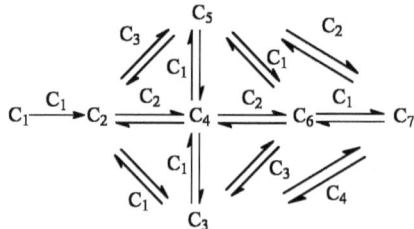

Fig. 1. Scheme of the formose reaction (Mizuno and Weiss, 1974).
C_1– formaldehyde,
C_2– glycolaldehyde,
C_n– carbohydrate with n carbon atoms

In the last few years, thorough investigation of the mechanism of the formose reaction was performed by Simonov et al. (2007a). The extensive evidence on the kinetics of the formose reaction has shown that the initiation of the formose reaction by higher monosaccharides (C_5–C_6) is due to their cleavage with the formation of two lower carbohydrates (Eq. 2). The key role in this stage is played by endiol complexes of higher monosaccharides with calcium ions. The carbohydrate chain growth is governed by aldol condensation of lower carbohydrates (Eqs. 3 and 4). The autocatalytic nature is caused by the retroaldol cleavage (2) of higher monosaccharides (C_4–C_5) formed at stages (3) and (4). The scheme of the reaction can be described by analogy with the chain mechanism with the confluent branching (2)–(4).

$$C_iH_{2i}O_i \underset{-2H_2O}{\overset{+Ca(OH)_2}{\rightleftarrows}} [C_iH_{2i-2}O_i^*Ca] \underset{-Ca(OH)_2}{\overset{+2H_2O}{\longrightarrow}} C_{i-j}H_{2(i-j)}O_{i-j} + C_jH_{2j}O_j, \qquad (2)$$

$$C_jH_{2j}O_j + CH_2O \xrightarrow{Ca(OH)_2} C_{j+1}H_{2(j+1)}O_{j+1}, \qquad (3)$$

$$C_nH_{2n}O_n + C_mH_{2m}O_m \xrightarrow{Ca(OH)_2} C_{n+m}H_{2(n+m)}O_{n+m} \qquad (4)$$

where $i = 4$–7, $j = 2$–4; $n, m = 2$–5; $n + m < 8$.

Thus, the formation of monosaccharides from formaldehyde can only start in the presence of a carbohydrate initiator. Glycolaldehyde is the simplest and most active one. However, the possibility of its formation and presence on prebiotic Earth is not evident.

3 Prebiotic Synthesis of Lower (C_2–C_3) Carbohydrates

Recently, astronomers have discovered a frigid reservoir of glycolaldehyde molecules in a cloud of gas and dust near the center of our galaxy (Saxton, 2004). Such clouds are considered to be the raw material for the formation of new stars and planets. The detection of cold glycolaldehyde indicates that a considerable quantity of this simple carbohydrate exists at extremely low temperatures in the universe. Thus, the beginning of the chemical evolution on a new planet might get started in the dust of interstellar clouds.

Some experimental results were obtained for the low-temperature reaction of water with atomic carbon (Ahmed et al., 1983; Flanagan et al., 1992), which was identified in the interstellar space. The interaction of C with H_2O at 77 K yields hydroxymethylene which rearranges to formaldehyde. A subsequent nucleophilic addition of hydroxymethylene to formaldehyde with hydrogen transfer generates glycolaldehyde. Actually, the interaction of atomic carbon with water generates not only glycolaldehyde but a mixture of straight-chain aldoses with up to five carbon atoms. The main products are glycolaldehyde, glyceraldehyde, erythrose and threose. Aldopentoses are formed as a minor fraction. Thus, the condensation of extraterrestrial carbon atoms on an ice surface may generate not only glycolaldehyde, but other carbohydrates including ribose as well.

Ultraviolet light is recognized as the most abundant energy source for organic synthesis on the prebiotic Earth. Many attempts of simulated prebiotic synthesis of carbohydrates were made using UV-irradiation.

Low-temperature results are known for photolysis of formaldehyde in argon and carbon monoxide matrices (Molina et al., 1978; Sodeau and Lee, 1978). Glycolaldehyde is formed when formaldehyde is subjected to the UV-light action ($220 < \lambda < 410$ nm) at 12 K.

The condensation of formaldehyde was observed in aqueous formaldehyde solutions subjected to UV-irradiation. The first results were obtained by Baly (1924), more recent report was made by Shigemasa et al. (1977). The studies were carried out using a highly concentrated formaldehyde solution (up to 8 M) in the presence of highly concentrated inorganics like NaOH or $NaHCO_3$ (0.62–1.63 M). The UV-irradiation of such solutions leads to the formation of polyols—pentaerythritol and 2-hydroxymethyl glycerol. However, the first stage of the reaction is the photoinduced formation of glycolaldehyde from formaldehyde. The results of the irradiation of more dilute formaldehyde solutions (0.1 M) were almost similar (Schwartz and de Graaf, 1993). Irie (1989) reported highly selective synthesis of ethylene glycol and pentaerythritol via γ-irradiation of aqueous formaldehyde in the presence of various inorganic bases.

The reasons for the formation of polyols rather than carbohydrates were most likely the presence of concentrated inorganic salts and the strong alkaline medium. The maximum of the single adsorption band of methyleneglycol (the primary form of formaldehyde in aqueous solutions) locates at approximately 200 nm, while all alkaline solutions absorb UV-radiation at $\lambda < 240$ nm. Hence, in the experiments in the presence of inorganics, the mercury lamp radiation seems to affect mostly the base catalyst, i.e., the OH^- anions, and lead mostly to the formation of hydroxyl radical. This process may be followed by the acyloin condensation, aldol condensation and crossed-Cannizzaro reaction (Irie, 1989) yielding polyols.

The UV-irradiation of a pure neutral aqueous formaldehyde solution has been studied recently (Pestunova et al., 2005). It appears to result predominantly in the decomposition of formaldehyde into gaseous carbon oxides, hydrogen and light hydrocarbons. At the same time, glycolaldehyde and glyceraldehyde are formed in the liquid phase with the registered maximum yields of 4.2 and 0.18%, respectively. Thus, these results indicate the possibility of photoinitiated formation of low carbohydrates at standard conditions in pure aqueous solutions. Note, that neither extremely low temperatures, nor extraterrestrial substrates, nor energy sources are needed for the formation of glycolaldehyde from formaldehyde.

4 Selective Prebiotic Synthesis of Carbohydrates

As it was already mentioned, ribose can be generated from formaldehyde under the catalytic action of $Ca(OH)_2$. Unfortunately, the whole array of straight-chain and branched oligomers of $(CH_2O)_n$ is obtained where $n = 2$, 3, 4, 5, 6... and the maximal content of ribose is 1%. Thus, the classical autocatalytic Butlerov reaction seems to be judged as a very unlikely prebiotic source of ribose.

Zubay (1998) has reported that the formose reaction in the presence of magnesium hydroxide suspension at pH 9.4, Pb(II) salts and catalytic amounts of any intermediate of the prebiotic pentose pathway (for example, dihydroxyacetone) yields aldopentoses with up to 30% of the final product. Unfortunately, the analysis in this work was performed by thin-layer chromatography which is not very reliable for the identification of components of complex mixtures. It was suggested that Pb^{2+} ions catalyze the synthesis of glycolaldehyde from formaldehyde in the presence of dihydroxyacetone. Zubay has accepted the mechanism of the catalytic cycle as it was proposed earlier by Langenbeck (1954). However, we suppose that the catalytic effects of Pb^{2+} and Ca^{2+} cations are analogous and at the reaction conditions used by Zubay, the classical autocatalytic formose reaction takes place. In this case, the selective formation of aldopentoses is impossible.

The most efficient way to increase the selectivity of the formose reaction is deceleration of the disproportionation and retroaldol cleavage reactions. It can be easily achieved by decreasing the pH. The rate of retroaldol cleavage reactions can be decreased by the replacement of calcium by other cations as well. Since the retroaldol cleavage, which is the source of lower carbohydrates in the formose system, is excluded, the synthesis of monosaccharides from formaldehyde must be performed in the presence of considerable amounts of lower C_2–C_3 carbohydrates.

4.1 Carbohydrates Synthesis Catalyzed by Natural Minerals

The data on the activity of natural minerals in the carbohydrates formation from formaldehyde are important for understanding possible ways of the putative synthesis of monosaccharides in prebiotic conditions.

The earliest (unfortunately, doubtful) results were obtained by Mayer and Jaschke (1960), who reported that calcium carbonate catalyzes formaldehyde oligomerization with selective yield of ribulose. Its identification was made by paper chromatography, which is doubtful. However, the catalytic activity of $CaCO_3$ in the formose reaction was confirmed by Reid and Orgel (1967). Gabel and Ponnamperruma (1967) reported that minerals kaolinite and illite, as well as γ-alumina, are capable of catalyzing the reaction at neutral pH. Analysis by paper chromatography suggested the production of hexoses, pentoses, tetroses and trioses at the formaldehyde concentrations as low as 0.01 M. The catalytic activity of a large number of natural minerals has been tested by Cairns-Smith et al. (1972) on refluxing 0.13 M formaldehyde. Quartz, calcite and galena were found to catalyze the reaction. Schwartz and de Graaf (1993) have also performed a screening of the catalytic activity of a considerable number of natural minerals. Dilute solutions of formaldehyde undergo the autocatalytic formose reaction when heated in the presence of montmorillonite (pH 6.5–7.1), alumina (pH 6.3), calcite (pH 7.0–7.2) and synthetic hydroxylapatite (pH 6.0). Schwartz and de Graaf (1993) claim that the reaction neither requires the presence of dissolved calcium ions or a strong base, nor depends on the presence of carbohydrates. No particular selectivity in the reaction has been observed and all tested minerals produced the same complex pattern of the products.

4.2 Selective Synthesis of Carbohydrates Phosphates

One of the most original examples of selective synthesis of straight-chain carbohydrate derivatives from formaldehyde and glycolaldehyde derivate—glycolaldehyde phosphate—was reported by Müller et al. (1990).

A possible way for the glycolaldehyde phosphate synthesis was reported by Krishnamurty et al. (1999). Amidotriphosphate in the presence of magnesium cations easily phosphorylates glycolaldehyde in the aqueous solution at near neutral pH. Amidotriphosphate can be formed by the reaction of cyclotriphosphate and ammonia. The problem of the formation of cyclotriphosphate was amply discussed in literature (see, e.g., Arrhenius et al., 1997).

The reaction of glycolaldehyde phosphate in an alkaline medium leads to the formation of racemic mixture of hexose-2,4,6-triphosphates in ca. 80% overall yield with allose-2,4,6-triphosphate as the dominant product. In the presence of formaldehyde, glycolaldehyde phosphate aldolizes to a mixture of glyceraldehyde-2-phosphate, aldotetrose- and aldopentose-2,4-diphophates and aldohexose-2,4,6-triphosphates in which ribose-2,4-diphosphate is the main component.

Such synthesis excludes the formation of branched carbohydrates and ketoses which dominate among autocatalytic formose reaction products.

To avoid the use of highly alkaline medium it was suggested to use common expanding sheet structure metal hydroxide minerals that pose a high affinity for inorganic phosphates (Pitsch et al., 1995). Glycolaldehyde phosphate, sorbed from the solution into the interlayer of metal hydroxide minerals, condenses into racemic aldotetrose-2,4-diphosphates and aldohexose-2,4,6-triphosphates. In the presence of glycerlaldehyde-2,4-diphosphate under otherwise similar conditions, aldopentose-2,4-diphosphates also form, but only as a small fraction of products.

4.3 Co-condensation of Lower Carbohydrates and Formladehyde

The extensive study of the condensation of formaldehyde with lower carbohydrates and co-condensation of lower carbohydrates catalyzed by heterogeneous natural minerals and inorganic salts as well as by homogeneous phosphates and borates was performed recently by Simonov et al. (2007b). Natural minerals apatite $Ca_5(OH, F, Cl)(PO_4)_3$ and vivianite $Fe_3(PO_4)_2$, synthetic hydroxylapatite $Ca_5(OH)(PO_4)_3$, calcium carbonate $CaCO_3$ and calcium phosphate $Ca_3(PO_4)_2$ were used as heterogeneous catalysts for the condensation of formaldehyde and lower carbohydrates at neutral pH. For all catalysts, the pathway of formation of carbohydrates appeared to be the same: formaldehyde and lower carbohydrates condense with one another resulting in a chain growth and formation of few carbohydrates with 3-pentulose and erythrulose being the main products. No autocatalytic behavior of the system was observed. The homogeneous condensation of lower carbohydrates with formaldehyde in the presence of phosphates (Na_2HPO_4 + KH_2PO_4), borate ($Na_2B_4O_7$) and pyrophosphate ($Na_4P_2O_7$) passes even more effectively than in the presence of heterogeneous catalysts. 3-Pentulose can be transformed into aldopentoses in the course of some rearrangement reactions which proceed easily in a slightly alkaline medium.

One of the latest striking results on the synthesis of ribose catalyzed by borates were reported by Ricardo et al. (2004) and Benner (2004). Borate is known to form complexes with organic molecules which have 1,2-dihydroxyl groups. Thus, Ricardo et al. (2004) and Benner (2004) demonstrated that borates form diglyceraldehyde–borate complexes and highly non-reactive complexes with two molecules of aldopentoses in their cyclic form stabilizing the carbohydrates.

In the presence of $Ca(OH)_2$, solution of glycolaldehyde and glyceraldehyde turns rapidly to brown. But when the same incubation is done in the presence of the natural borate minerals colemanite, ulexite and kernite (Kawakami, 2001; Moody, 1976), the solution does not turn brown even after months, and ribose accumulates in the reaction mixture.

The results by Simonov et al. (2007b) seem to approve the unique stabilizing ability of borates. So, the condensation of 5 mM glycolaldehyde and 5 mM glyceraldehyde in the presence of Na_2HPO_4 and KH_2PO_4 leads to the formation of erythrose (0.14 mM), threose (0.16 mM), ribose (0.03 mM), arabinose (0.02 mM) and fructose (0.03 mM). On the other hand the same condensation catalyzed by $Na_2B_4O_7$ yields much higher amounts of carbohydrates: erythrose (0.31 mM), threose (0.63 mM), ribose (0.1 mM), arabinose (0.1 mM) and fructose (0.11 mM).

Magnesium oxide MgO, which seems to be one of the most abundant prebiotic oxides, appears to be the most efficient catalyst for the condensation of glycolaldehyde and glyceraldehyde. The interaction of lower carbohydrates leads to the formation of fructose (18%), sorbose (23%), ribose (10%) and arabinose (8%) within few hours (our unpublished results).

Alternatively, the direct aqueous aldol reactions of glycolaldehyde and glyceraldehyde are known to be catalyzed by a zinc–amino acid complex under conditions that are compatible with the expected prebiotic environment (Kofoed et al., 2005). The reaction of glycoaldehyde and glyceraldehyde in the presence of the Zn-proline complex gives rise to pentoses (62%) with ribose accounting for about 30% of the pentoses. Ribose and other pentoses are formed by aldol reactions and are stable in aqueous media at room temperature. Kofoed et al. (2005) suggests the catalytically active Zn–amino acid complex available in the prebiotic environment.

The mentioned experimental data allow the assumption that ribose might be present in reasonable amounts in an early stage of the evolution under expected prebiotic conditions. Putative routes to further stabilization procedures of the carbohydrate were also described.

4.4 Putative Prebiotic Synthesis of Carbohydrates from Formaldehyde

All described possible prebiotic routes to carbohydrate synthesis start from the composition of either formaldehyde with lower carbohydrate or just lower carbohydrates assuming that the presence of the latter is obvious. As it was considered earlier, glycolaldehyde might be formed in terrestrial prebiotic conditions under the action of UV-irradiation on aqueous formaldehyde solution. The further condensation of formaldehyde and glycolaldehyde in the presence of various catalysts leads to the monosaccharide formation. An attempt to combine these two processes was reported by Simonov et al. (2007b).

The first stage was the irradiation of formaldehyde solution in the presence of homogeneous phosphates, and the second stage was a reasonable heating of the irradiated mixture for the acceleration of the catalytic reaction. After the irradiation, glycolaldehyde and glyceraldehyde were formed in the solution as expected. Then the catalytic reaction started. 3-Pentulose and erythrulose were formed from one simplest substrate, formaldehyde, in the course of the described combination of photochemical and catalytic processes in neutral aqueous solution in the presence of dissolved phosphates. Unfortunately, the conditions with the minimal formaldehyde concentration, in comparison with glycolaldehyde and glyceraldehyde concentrations, to synthesize ribose were still not achieved by Simonov et al. (2007b).

5 Important Prebiotic Organics from Carbohydrates

The prebiotic role of monosaccharides seems not to be limited only to the synthesis of the RNA precursors and other monosaccharide-based biological molecules. The presence of alcoholic and carboxylic groups in the structure of monosaccharides makes them convenient for disproportionation reactions which can provide free energy for the synthesis of other prebiotic compounds. In addition, carbohydrates can be substrates for synthesis of amino acids, polyols, carboxylic acids and heterocycles.

5.1 Synthesis of Heterocycles from Carbohydrates and Ammonia

The possibility of the synthesis of different imidazoles, pyrazines and pyridines from carbohydrates and ammonia is well known (Grimmet, 1965; Kort, 1970). In general, the interaction of carbohydrates with ammonia results in the formation of a complex mixture of imidazoles, while metal salts are known to catalyze the reaction. The initial step of that process involves dehydrogenation of an aldose. The dehydrogenation product reacts with ammonia and formaldehyde to yield imidazole. A number of pyrazine and pyridine compounds are also formed. The formation of these heterocycles is connected with the transformation of, e.g., fructose first to the fructosylamine and then to glucoseamine. The latter condenses to form 2,5-substituted pyrazine.

These processes require rather high temperatures (up to 200°C) and go slowly under milder conditions. However, the model discussed below takes into account the possibility of the nitrogen heterocycles formation from carbohydrates and ammonia.

5.2 A "Sugar Model" by A.L. Weber

The data from a number of investigations by A.L. Weber (Weber, 1998 and ref. therein) resulted in the creation of the so-called "sugar model" (Weber, 2001). The following deduction by Weber (2000) was declared basing on the comparison of redox and kinetic properties of carbon groups in order to identify the optimal biosubstrate: "Sugars are the optimal biosynthetic carbon substrate of aqueous life. Carbohydrates contain the maximum number of biosynthetically useful high energy electrons/carbon atom while still containing a single carbonyl group needed to kinetically facilitate their conversion to useful biosynthetic intermediates."

In addition to common substrates like formaldehyde, glycolaldehyde and ammonia, the "sugar model" also requires thiols which act as catalysts for the synthesis of amino acids (Weber, 1998).

Fig. 2 shows the reactions of the sugar model. The pathway of the model starts by aldol condensation of glycolaldehyde with formaldehyde to give trioses and tetroses (Harsch et al., 1984). Next, β-dehydration of the trioses and tetroses generates the α-ketoaldehydes, pyruvaldehyde and 3-C-hydroxymethyl-pyruvaldehyde (Feather and Harris, 1973). In the presence of ammonia and a thiol, the redox rearrangement of α-ketoaldehydes yields alanine and homoserine thioesters that either hydrolyze to amino acids or oligomerize to peptides (Weber, 1985, 1998). The energy values in Fig. 2 show that the condensation of formaldehyde and glycolaldehyde is energetically favorable but not irreversible. Both sugar dehydration to α-ketoaldehydes and ketoaldehyde conversion to amino acids are irreversible. The aldol condensation of glycolaldehyde with other aldehyde and ketone products of the "sugar model" pathway could lead to the synthesis of other types of amino acid thioesters.

Figure 2 also shows the side reactions. They result in the synthesis of (a) larger pentose or hexose sugars by aldol condensations (Harsch et al., 1984), (b) amines by Amadori rearrangements (Kort, 1970), (c) imidazoles, pyrazines and pyridines (Grimmet, 1965; Kort, 1970), (d) acetaldehyde (Grimmett, 1965), (e) pyruvate (Weber, 2001) and (f) lactate thioester (Weber, 1984a,b).

The next step in the development of the "sugar model" was the creation of possible autocatalytic system in the context of the prebiotic model under discussion.

First, imidazole and pyridine products are possible autocatalysts, since they were shown to catalyze the formose reaction (Gutsche et al., 1967). Second, ammonia and amines (including amino acids) were shown to catalyze the formose reaction too and, in addition, a subsequent conversion of sugars to

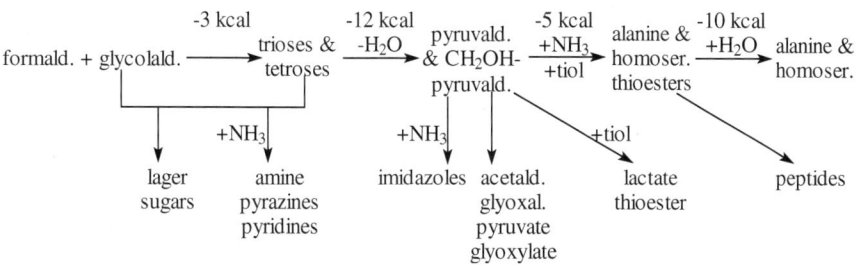

Fig. 2 The "sugar model" by A.L. Weber

carbonyl-containing products. The amine-catalyzed reactions yield glyceraldehyde, dihydroxyacetone, erythrose, threose, erythrulose, pyruvaldehyde, acetaldehyde, glyoxal, pyruvate, glyoxylate and several unidentified carbonyl products. Since alanine is synthesized from glycolaldehyde and formaldehyde via pyruvaldehyde (Weber, 1998), the ability of alanine to catalyze the conversion of glycolaldehyde and formaldehyde to pyruvaldehyde indicates that the "sugar model" is capable of autocatalysis.

Thus, the "sugar model" by Weber (2001) exhibits several essential characteristics of a plausible "origin of life" process. First, the model uses simple carbon substrates—formaldehyde and glycolaldehyde which are undoubtedly prebiotic compounds. Second, the "sugar model" can be regarded as autocatalytic, while the "one-pot" nature of the model facilitates the autocatalysis by having the reaction intermediates and catalytic products made in close proximity to one another. Besides, the model generates amino acid thioesters for the peptide synthesis and consequently it can develop a more sophisticated type of sequence-dependent catalysis based on polypeptide replication (Lee et al., 1997; Yao et al., 1998). Finally, since the model converts sugar substrates to "biochemicals" by carbon redox disproportionation (Weber, 1997), chemical transformations of the "sugar model" could develop into modern metabolism (Weber, 2001).

5.3 Synthesis of Cytidine Ribonucleotides on Sugar Phosphate

Although various paths of the syntheses of the nucleic acid bases exist and the possible ribose prebiotic synthesis has been just discussed, no prebiotically plausible methods for attaching pyrimidine bases to ribose to give nucleosides have been described.

No prebiotic condensation of uracil or cytosine with ribose to give nucleosides is known. The equilibrium constant for the reaction of ribose with uracil to form uridine has been calculated to be in the region of 10^{-3} M^{-1} (Kolb et al., 1994). An alternative route to the pyrimidine nucleosides involving a stepwise assembly of the nitrogenous base on the sugar has been demonstrated by Sanchez and Orgel (1970) but a stereochemical impasse was encountered.

Unique results concerning this problem were reported by Ingar et al. (2003). A process in which the base is assembled in stages on a sugar phosphate can produce cytidine nucleotides. The sequential action of cyanamide and cyanoacetylene on arabinose-3-phosphate produces cytidine-2′,3′-cyclophosphate and arabinocytidine-3′-phosphate (Fig. 3). Both cyanamide and cyanoacetylene are presumed to be prebiotic compounds and can be produced in reasonable yields (Orgel, 2002; Schimpl et al., 1965). The fact that this conversion proceeds readily in an aqueous solution at near-neutral pH values suggests that this chemistry is prebiotically plausible.

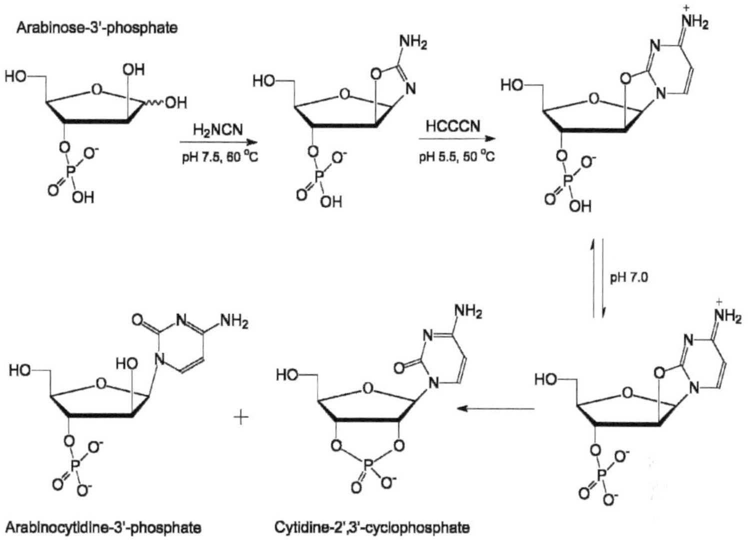

Fig. 3 Synthesis of cytidine-2′,3′-cyclophosphate and arabinocytidine-3′-phosphate

6 Conclusions

This overview summarizes few most significant results on the putative synthesis of various carbohydrates, in particular ribose and their derivates, from simple substrates in plausible prebiotic conditions. The described routes of the synthesis of monosaccharides can be formally divided into two groups. The first group unites synthesis of carbohydrates in the interstellar space initiated by UV-irradiation or cosmic radiation. The known substrates for these processes are atomic carbon, water and formaldehyde. The second group of possible scenarios of prebiotic carbohydrates synthesis embodies the catalytic processes in the aqueous formaldehyde solutions. In general these processes are the formose reaction and its various modifications based on the mechanism of the aldol-condensation reactions. The catalytic synthesis of monosaccharides seems to be more plausible. One cannot expect reasonable yields of carbohydrates for the processes of the first group because of the presence of free radicals.

The synthesis of biologically important monosaccharides from formaldehyde, first of all ribose, with increased yields can be catalyzed in plausible prebiotic conditions by different compounds: lead, phosphate and borate ions, several natural minerals ($CaCO_3$, apatites, MgO and others). Higher yields can be achieved when formaldehyde interacts with lower carbohydrates (glycolaldehyde, glyceraldehyde, dihydroxyacetone), which can be directly formed in formaldehyde aqueous solutions under the action of UV-irradiation.

The possible role of carbohydrates and their derivates in the chemical evolution and development of presumable abiogenic metabolism not only as substrates for other "biochemical" substances but also as a source of energy is illustrated as well.

Acknowledgment The financial support of RFBR (Grant No. 05-03-32862), program of Presidium of RAS "Origin and evolution of biosphere" (Grant RNP.2.1.1.1969) and "Scientific Schools of Russia" (Grant No. N.Sh. 6526.2006.3) are gratefully acknowledged.

References

Ahmed, S.N., McKee, M.L. and Shevlin, P.B. (1983) J. Am. Chem. Soc. 105, 3942–3947.
Altman, S. (1989) Adv. Enzymol. Relat. Areas Mol. Biol. 62, 1.
Arrhenius, G., Sales, B., Mojzsis, S. and Lee, T. (1997) J. Theor. Biol. 187, 503.
Baly, E.C.C. (1924) Ind. Eng. Chem. 16, 1016–1018
Benner, S.A. (2004) Acc. Chem. Res. 37, 784–797.
Breslow, R. (1959) Tetrahedron Lett. 1, C22–C26.
de Bruijn, J.M., Kieboom, A.P.G. and van Bekkum, H. (1986) J. Carbohydr. Chem. 5, 561–569.
Cairns-Smith, A.G. (1982) Genetic Takeover and the Mineral Origins of Life. Cambridge University Press, Cambridge.
Cairns-Smith, A.G., Ingram, P. and Walker, G.L. (1972) J. Theor. Biol. 35, 601–604.
Cech, R.R., Zaug, A.J. and Grabowski, P.J. (1981) Cell. 27, 487.
Crick, F. (1981) Life Itself. It's Origin and Nature. Simon and Schuster, New York.
de Duve, C. (1991). Blueprint for a Cell: The Nature and Origin of Life. Patterson, New York.
Feather, M.S. and Harris, J.F. (1973) Dehydration reactions of carbohydrates. In: R.S. Tipson and D. Horton (Eds), Advances in Carbohydrate Chemistry and Biochemistry. Academic Press, New York, 28, pp. 161–224.
Flanagan, G., Ahmed, S.N. and Shevlin, P.B. (1992) J. Am. Chem. Soc. 114, 3892–3896.
Gabel, N.W. and Ponnamperuma, C. (1967) Nature 216, 453–455.
Gesteland R.F. and Atkins J.F. (Eds) (1993) The RNA World. Cold Spring Harbor Laboratory, Cold Spring Harbor, NY.
Grimmet, M.R. (1965) Rev. Pure Appl. Chem. 15, 101–108.
Gutsche, C.D., Redmore, D., Buriks, R.S., Nowotny, K., Grassner, H. and Armbruster, C.W. (1967) J. Am. Chem. Soc. 89, 1235–1245.
Harsch, G., Bauer, H. and Voelter, W. (1984) Kinetik, Liebigs Ann. Chem. 4, 623–635.
Hartman, H. (1998). Orig. Life Evol. Biosph. 28, 515–521.
Hudson, R.L. and Moore, M.H. (2004) The formation, destruction, and spectra of extraterrestrial biological and prebiological molecules. Abstract 35th COSPAR Scientific Assembly, Paris, France (CD-ROM, COSPAR04-A-03541).
Ingar, A.A., Luke, R.W.A., Hayter, B.R. and Sutherland, J.D. (2003) ChemBioChem 4, 504–507.
Irie, A. (1989) Carbohydr. Res. 190, 23–28.
Joyce, G.F. (1991) New Biol. 3, 399.
Kawakami, T. (2001) J. Metamorph. Geol. 19, 61–75.
Khomenko, T.I., Sakharov, M.M. and Golovina, O.A. (1980) Russ. Chem. Rev. 49, 570–584.
Kofoed, J., Reymond, J. and Darbre, T. (2005) Org. Biomol. Chem. 3, 1850–1855.
Kolb, V.M., Dworkin, J.P. and Miller, S.L. (1994) J. Mol. Evol. 38, 549–557.
Kort, M.J. (1970) Reactions of free sugars with aqueous ammonia. In: R.S. Tipson and D. Horton (Eds) Advances in Carbohydrate Chemistry and Biochemistry. Academic Press, New York, 25, pp. 311–349.
Krishnamurty, R., Arrhenius, G. and Eschenmoser, A. (1999) Orig. Life Evol. Biosph. 29, 333–354.

Langenbeck, W. (1954) Angew. Chem. 66, 151.
Larralde, R., Robertson, M.P. and Miller, S.L. (1995) Proc. Natl Acad. Sci. USA. 92, 8158–8160.
Lee, D.H., Severin, K., Yokobayashi, Y. and Ghadiri, M.R. (1997) Nature 390, 591–594.
Litvak, M.M. (1972) Non-equilibrium processes in interstellar molecules. In: T.R. Carson, and M.J. Roberts (Eds), Atoms and Molecules in Astrophysics. Academic Press, London and New York, p. 201.
Mayer, R. and Jaschke, L. (1960) Lieb. Ann. Chem. 635, 145–153.
Mizuno, T. and Weiss, A. (1974) Adv. Carbohydr. Chem. Biochem. 29, 173.
Molina, L.T., Tang, K.Y., Sodeau, J.R. and Lee, E.K.C. (1978) J. Phys. Chem. 82, 2575–2578.
Moody, J.B. (1976) Lithos 9, 125–138.
Müller, D., Pitsch, S., Kittaka, A., Wagner, E., Wintner, C.E. and Eschenmoser, A. (1990) Helv. Chim. Acta 73, 1410.
Orgel, L.E. (2002) Orig. Life Evol. Biosph. 32, 279–281.
Ostrovskii, V.E. and Kadyshevich E.A. (2007) Phisics-Uspekhi 50, 175—196.
Pestunova, O., Simonov, A., Snytnikov, V., Stoyanovsky, V. and Parmon, V. (2005) Adv. Space Res. 36(2), 214–219.
Pitsch, S., Eschenmoser, A., Gedulin, B., Hui, S. and Arrhenius, G. (1995) Orig. Life Evol. Biosph. 25, 297–334.
Reid, C. and Orgel, L.E. (1967) Nature 216, 455.
Ricardo, A., Carrigan, M.A., Olcot, A.N. and Benner, S.A. (2004) Science 303, 196.
Sanchez, R.A. and Orgel, L.E. (1970) J. Mol. Biol. 47, 531–543.
Saxton, B. (2004) Cold Sugar in Space Provides Clue to the Molecular Origin of Life. http://www.nrao.edu/pr/2004/coldsugar/.
Schimpl, A., Lemmon, R.M. and Calvin, M. (1965) Science 147, 149–150.
Schwartz, A.W. and de Graaf, R.M. (1993) Mol. Evol. 36, 101–106.
Segre, D. and Lancet, D. (1997). Mutually catalytic amphiphiles: simulated chemical evolution and implications to exobiology. In: J. Chela-Flores and F. Raulin (Eds) Exobiology: Matter, Energy, and Information in the Origin and Evolution of Life in the Universe, Proc. the 5th Trieste Conf. on Chemical Evolution. Kluwer Academic Publishers, Trieste, pp. 123–131.
Shigemasa, Y., Matsuda, Y., Sakazawa, C. et al. (1977) Bull. Chem. Soc. Jpn 50, 222–226.
Simoneit, B.R.T. (2004) Adv. Space Res. 33, 88–94.
Simonov, A.N., Pestunova, O.P., Matvienko, L.G. and Parmon, V.N. (2007a) Kinet. Catal. 48(2), 245–254.
Simonov, A.N., Pestunova, O.P., Matvienko, L.G., Snytnikov, V.N., Snytnikova, O.A., Tsentalovich, Yu.P. and Parmon, V.N. (2007b) Possible prebiotic synthesis of monosaccharides from formaldehyde in presence of phosphates. Adv. Space Res. doi:10.1016/j.asr.2007.08.002
Socha, R.F., Weiss, A. and Sakharov, M.M. (1981) J. Catal. 67, 207–217.
Sodeau, J.R. and Lee, E.K.C. (1978) Chem. Phys. Lett. 57(1), 71–74.
Tolstoguzov, V. (2004) Orig. Life Evol. Biosph. 34, 571–597.
Wächtershäuser, G. (1992). Prog. Biophys. Mol. Biol. 58, 85–201.
Weber, A.L. (1984a) J. Mol. Evol. 20, 157–166.
Weber, A.L. (1984b) Orig. Life Evol. Biosph. 15, 17–27.
Weber, A.L. (1985) J. Mol. Evol. 21, 351–355.
Weber, A.L. (1997) J. Mol. Evol. 44, 354–360.
Weber, A.L. (1998) Orig. Life Evol. Biosph. 28, 259–270.
Weber, A.L. (2000) Orig. Life Evol. Biosph. 30, 33–43.
Weber, A.L. (2001) Orig. Life Evol. Biosph. 31, 71–86.
Yao, S., Ghosh, I., Zutshi, R. and Chmielewski, J. (1998) Nature 396, 447–450.
Zubay, G. (1998) Orig. Life Evol. Biosph. 28, 13–26.

Theoretical and Computer Modeling of Evolution of Autocatalytic Systems in a Flow Reactor

S. I. Bartsev and V. V. Mezhevikin

Abstract A chemoautotrophic concept of the initial stages of chemical prebiotic evolution, which eliminates key difficulties in the problem of life origin and permits experimental tests, is proposed. The concept leads to an important statement—organisms emerging (out of the Earth and/or inside an experimental reactor) have to be based on biochemical bases, different from those occurring on our planet. According to the concept the predecessor of living beings has to be sufficiently simple to provide non-zero probability of self-assembly during a short (in geological or cosmic scale) time. In addition the predecessor has to be capable of autocatalysis, and further complexification (i.e., evolution). A theoretical model of a multivariate oligomeric autocatalyst coupled with a phase-separated particle is presented. This model, possessing non-genomic inheritance, describes a version of the "metabolism first" approach to life origin. Conducted computer simulation shows the origin of an autocatalytic oligomeric phase-separated system to be possible at reasonable values of the kinetic parameters of involved chemical reactions in a small-scale flow reactor.

1 Introduction

1.1 The "Only" Statements

A concept of life as a phenomenon ought to be based on well-grounded assumptions about the *sufficient* conditions of life origin. But *only* experiments pertaining to life origin can meet these demands. Confirmation, or at least support for, the initial conditions which are *sufficient* for life to originate can be obtained *only* after successful completion of the experiment.

S. I. Bartsev
Institute of Biophysics SB RAS, Theoretical Biophysics Department,
e-mail: bartsev@yandex.ru

However organization of the experiment has to provide at least *necessary* conditions for starting pre-biotic chemical evolution and the origin of life. Since we have no information on extraterrestrial life forms the *necessary* conditions can be taken *only* from general ideas about life in general and possible ways of its arising. So let us consider the main properties of living beings.

1.2 On the Main Properties of Living Beings

At the beginning a seemingly tautological but key statement is suggested: the main property of living *beings* is their ability *to be*. Logical consequences from this statement are many; however the consideration of life as an established phenomenon distinct from life in the process of formation seems to have different emphasis (Bartsev, 2004; Bartsev et al., 2001). In this chapter the arising consequences relevant to the stage of life are considered.

So ability "to be" means

1. To be separate from an environment (e.g., cell, organism). The simplest example of this property realization are phase-separated systems—coacervate droplets, micelles, inorganic catalytic compartments etc.
2. To extend its own existence as long as possible in spite of external destructive influences. The way this property is realized is by a combination of self-support and self-reproduction. The simplest example of self-reproduction in non-living systems is autocatalysis. Self-support and self-reproduction are not possible without consuming free energy from outside.

So phase-separability and an autocatalytic nature are two properties that are inherent to all living and some non-living systems and which have to underlie the emergence of life.

This knowledge alone is insufficient to mount experiments on the life origin. It makes sense only if the known problems of the life origin are eliminated, at least hypothetically.

1.3 Key Problem Is Due to Attempts to Explain the Exact Origin of Earth's Life Form

A key problem arises if one wishes to build a cause–effect relation (or exactly reproduce in experiment) the origin of the Earth's life. On one hand it is easy to evaluate that the mean time of spontaneous self-assembling of the simplest living cell of the type presently existing on Earth from organic compounds under mostly favorable conditions would be greater by many orders of magnitude than the age of the Universe. On the other hand it is known that life on the Earth arose almost immediately (on the geological scale) after the ocean condensed (Martin and Russell, 2002).

This contradiction has to be solved (at least hypothetically) before beginning the experiment, and before searching for extraterrestrial life. It can be seen that the later problems of chirality (Gleiser and Thorarinson, 2006; Goldberg, 2007), the genetic code (Altstein, 1987; Feigin, 1987; Nelson et al., 2000; Rasmussen et al., 2002), and the specific chemical composition of cells are similar to the key problem. Two mutually amplifying presuppositions capable of eliminating the contradiction can be suggested.

1.4 Presuppositions Capable of Eliminating the Problem

Presuppositions which are presented in this section are based on well-known facts and ideas. However, their simultaneous usage allows us to come to constructive moves to the experiment.

Presupposition 1 (continuity principle). Modern cells are the result of gradual evolution from simple pre-biotic non-living system (Carny and Gazit, 2005; Martin and Russell, 2002; Pross, 2004; Rasmussen et al., 2002).

Presupposition 2 (many-to-one type of structure-function mapping). The same function (including catalysis) can be performed by many different structures (Fontana and Shuster, 1998; Fox and Dose, 1977; Schuster, 2000).

These presuppositions have three implications:

1. Life can exist on different chemical bases, and an earthly life form is only one of a large number of possible realizations. Therefore it is not necessary to reproduce exactly specific properties of Earth's living organisms.
2. The origin of a living cell must have arisen from some predecessor of such a simple structure that the possibility of its self-assembly was not negligible (Martin and Russell, 2002).
3. A "fully-fledged" living cell is the result of evolution from a simple predecessor driven by "natural selection" that acted even at the stage of pre-biotic chemical evolution. So, in principle, an initial chemical base of the predecessor can essentially differ in its essentials from one of the "mature" life forms (Cody, 2004; Cody et al., 2000).

Summarizing namely multiplicity of variants of chemical evolution starting, and simplicity of the living cell predecessor give a hope to successful experimental modeling of initial stages of pre-biotic chemical evolution leading to life origin.

1.5 On the Nature of the Predecessor

It seems logical to suppose that the existence and evolution of the predecessor of living systems was based on molecules of a simple chemical nature and non-genomic inheritance (Martin and Russell, 2002; Oba et al., 2005; Pohorille and Deamer, 2001; Pross, 2004). This allows us to set aside the implementation of the very complex function of matrix duplication in the initial stage of chemical evolution.

In accordance with main properties of living beings a predecessor has to be a phase-separated autocatalytic system. In addition phase separation is necessary for effective selection of more suited autocatalytic systems (Feistel et al., 1980). Many types of phase-separated systems (PSS) self-formatting in the course of different physical processes are known (Fox and Dose, 1972; Rasi et al., 2004; Segre et al., 2001).

Phase-separated systems of micellar type can be used for the modeling stage of the origin of metabolism. In the course of their growth such systems can reach a critical size and then, due to mechanical instability, divide, i.e., undergo primitive self-reproduction. It was shown that the particles themselves can operate as autocatalytic systems (Segre and Lancet, 2000; Segre et al., 2001).

An ability to reproduce (autocatalyze) is a necessary, but not sufficient, property for enabling possible chemical evolution from the predecessor. Structures of simple chemical auto-catalysts do not possess sufficient degrees of freedom to produce different variants of auto-catalysts. Therefore simple chemical auto-catalysts are not capable of further complication or evolution.

A possible variant of an autocatalytic system capable of evolution is a statistical chemical autocatalytic system (Segre et al., 1998) in which an ensemble of different catalysts cross-catalyze the syntheses of the other catalysts. Thereby the total effect is the reproduction of all components of the system providing self-replication in mutually catalyzing sets. However, it is not clear how coordination in this population of heterogeneous catalytic reactions can be provided. Besides, such a system can appear only if the origin of the complete population of the catalysts happens simultaneously to provide a closed chain of reproduction—an eventuality that has to be considered highly improbable.

The hypothesis that the primary autocatalyst has to be a linear oligomeric molecule capable of catalyzing polymerization of monomers from which it is assembled (Bartsev and Mezhevikin, 2005, 2006) seems to be well grounded. In this case only one event of self-assembly is required for initiating autocatalytic processes. This event is provided by only one type of chemical reaction—polymerization.

Generally properties of a linear oligomer depend upon the length of chain and its monomer sequence. While simple oligomers are not capable of matrix duplication, under appropriate conditions, linear oligomerase can randomly synthesize large numbers of different linear oligomers. These oligomers can perform different catalytic activity and some fraction of them can be auto-oligomerases supporting oligomerization process.

In parallel, at the initial stage of chemical evolution, membranes can be assembled from amphiphiles presented in the environment of PSS. Amphiphiles can be produced by different physical–chemical processes. Micelles can be formed at low concentration ($\sim 10^{-8}$ M) of amphiphiles possessing high hydrophobicity of the tail (Ruckenstein and Nagarajan, 1976). The presence of hydrophobic nuclei can start formation of micelles at negligible concentrations of amphiphiles. Due to phase effects such PSS can accumulate different substances. At the accumulation inside them of appropriate monomers, different linear molecules possessing oligomerase activity may be assembled. If oligomers capable of synthesizing appropriate amphiphiles are present in this pool of oligomers, then a phase-separated system (PSS) possessing primitive metabolism can emerge. A conventional scheme of primitive metabolism arising is shown in Color Plate 1 (see p. 393). However there is a question what component of pre-biotic metabolism—autocatalytic oligomerase or PSS has to arise first.

As the kinetic characteristics of oligomers of the given type are constant, it follows that a multivariate oligomeric autocatalytic system cannot evolve. However the background of various catalytic activities can promote the emergence of more complex monomers for more effective auto-oligomerases of a new type. These new auto-oligomerases can be the kernel of a new multivariate oligomeric autocatalytic system forcing out the predecessor. It means that any multivariate oligomeric autocatalytic system permanently stands on the verge of the risk of generating their own "grave digger".

This permanent forcing out can be completed after specific matrix copying occurs. Only then can biological evolution gradually replace chemical evolution. So, figuratively speaking, chemical evolution represents by itself a selection of "potential" biochemistries. It is important to note that this scheme does not assume an obligatory emergence of protein–nuclein variant of life at all. Life in principle can be based on a different biochemistry, repeating nevertheless its key properties (Bartsev, 2004).

A general suggesting concept can be applied to description of the first stages of pre-biotic evolution in the surroundings of underwater hydrothermal vents. They can provide relatively stable flow of simple high-energy molecules and different organic compounds for initial abiogenous synthesis (Braun and Libchaber, 2004; Koonin and Martin, 2005; Martin and Russell, 2002). Considered pre-biotic system is of chemoautotrophic type since relatively complex organic compounds are necessary as building material at the stage of self-assembling of the predecessor. Very important property of hydrothermal vents is the flow of a medium which can serve as an analog of flow reactor where "free living" phase-separated autocatalytic systems are subjected to almost Darwin's selection.

For conducting experiments on pre-biotic stages of life origin the choice of chemical compounds is essential. Since exact local conditions and the time of life origin are not known the obligatory request to reproduce early Earth's conditions seems to not be absolutely correct. There are some "degrees of freedom" giving possibility to vary initial conditions in the scope of relatively reliable data. It is well known that substances of various natures can demonstrate various catalytic activities (Carny and Gazit, 2005; Ferris et al., 1996; Gilbert, 1986; Klotz et al., 1971; Laszlo, 1999; Segre et al., 2001). There are no grounds to deny possible catalytic activity of short linear hetero-oligomers, which can consist of very simple monomers producing by hydrothermal vents (Braun and Libchaber, 2004). Taking this into account allows us to introduce into computer experiments a flow of substances of different chemical nature.

2 Description of Computer Model and Conditions of Real Experiment

A computer model was built to estimate the possibility and conditions of the origin of PSSs in which processes providing their dynamic stability are coupled with auto-oligomerase reaction. This model simulates, at the first approximation, the selection of phase-separated oligomer autocatalytic systems and allows evaluation of conditions under which the process of the selection of the most adaptable PSS takes place. Considered in the model is a flow reactor with changeable rates of dilution which allows the provision of the optimum conditions for selection and identification of oligomers possessing autopolymerase activity (Bartsev et al., 2001).

Complete schemes of reactions inside and outside the phase-separated systems are presented in Figs. 1 and 2. An extended explanation of these schemes and assumptions are considered below.

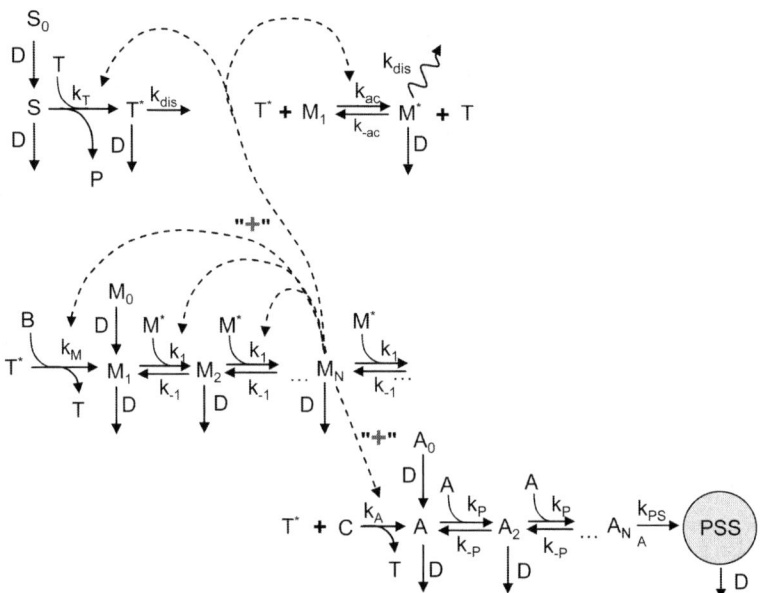

Fig. 1 Reactions inside the zone of origin, but outside PSS. Notation: D, flow rate; S_0, initial concentration of high-energy substrate; S, concentration of the substrate inside the reactor; T, concentration of carrier; T^*, concentration of activated carrier; M_0, initial concentration of monomer; M_1, concentration of monomer inside reactor; M^*, concentration of activated monomer; M_i, concentration of oligomer of i-th length; A_0, initial concentration of amphiphiles; A, concentration of amphiphiles inside reactor; A_i, concentration of aggregate, consisting of i amphiphiles; B, C, concentrations of some non-organic and simple organic compounds which can be used for synthesis of monomers and amphiphiles; k_{dis}, constant of dissipation. *Dash curves* present activation. Other notations are clear from the figure

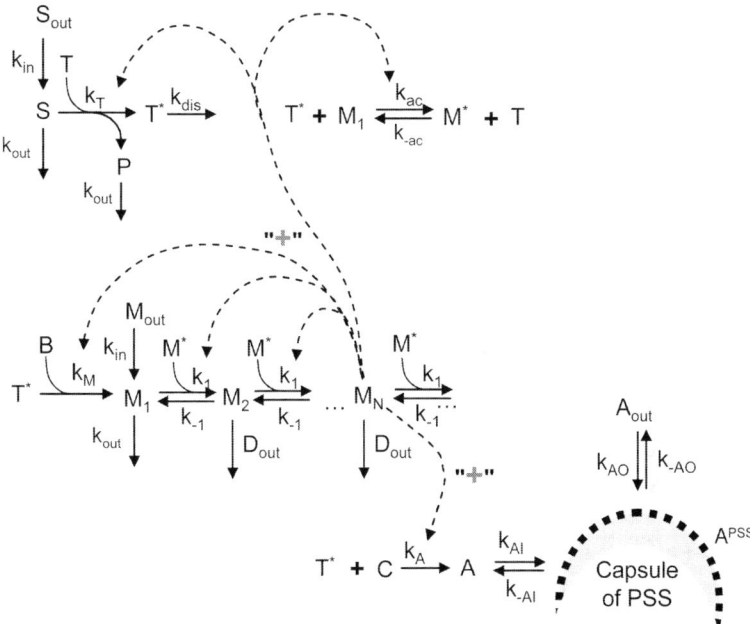

Fig. 2 Reactions inside PSS. Notation: S_{out}, concentration of high-energy substrate outside the PSS; S, concentration of the substrate inside the PSS; T, concentration of carrier; T^*, concentration of activated carrier; M_{out}, concentration of monomer outside the PSS; M_1, concentration of monomer inside reactor; M^*, concentration of activated monomer; M_i, concentration of oligomer of i-th length; A_{out}, concentration of amphiphiles outside the PSS; A, concentration of amphiphiles inside PSS; A^{PSS}, amount of amphiphiles in PSS capsule; B, C, concentrations of some non-organic and simple organic compounds which can be used for synthesis of monomers and amphiphiles; k_{dis}, constant of dissipation; k_{in}, constant rate of the substrate incoming from outside; k_{out}, constant rate of the substrate leakage from PSS; D_{out}, constant rate of the oligomers leakage due to temporary disintegration of PSSs capsule. Other notations are clear from the figure

Details of computer model.

1. There is "a zone of origin" where high-energy substances (substances including hydrogen, phosphate or another high energetic groups), various inorganic substances (e.g., CO_2, NH_3, etc.), amphiphiles and simple organic monomers are supplied. This zone is organized as a flow reactor of ideal mixing.
2. A high-energy substrate transfers energy to a carrier, which provides chemical activation of monomers.
3. Phase separated systems (PSS) are formed when the amount of bounded amphiphiles exceeds some critical value.
4. Activated monomers are capable of spontaneous and reversible binding to form linear oligomers. Some fraction of these oligomers is capable

of catalyzing synthesis of amphiphiles. Another fraction is capable of monomer synthesis. And some fraction of oligomers is auto-oligomerases.
5. Internal concentration of monomers inside PSS is higher than outside due to inter-phases redistribution, and the polymerization reaction is shifted to oligomer synthesis.
6. The increase of PSS in size occurs due to embedding of amphiphiles, synthesized both inside PSS, and coming from the outside.
7. Since oligomers for performing catalytic functions have to possess a tertiary three-dimensional structure then an assumption on terminating oligomer chain elongation after its folding is accepted.

3 Results of the Computer Simulation and Discussion

In principle two variants of the initial stages of pre-biological chemical evolution are possible in the reactor. According to the first variant, the auto-oligomerase reaction appears first in the flow reactor. This reaction forms the background of varied catalytic activities in particular promoting the syntheses of amphiphiles. When concentration of amphiphiles reaches a certain threshold, then phase-separated systems also containing auto-oligomerases within appear. As water is often a product of polymerization then syntheses of oligomers in aqueous solution is difficult. And since the water activity is reduced inside the phase-separated particles, it is possible to expect an increase in the rate of oligomer synthesis. In addition the presence of phase boundaries can lead to increase in concentrations of reagents inside PSS, and by this to increase in auto-oligomerase reaction rates, and to the growing of PSS itself. Further chemical evolution and its substitution by biological evolution will occur inside PSS selected under conditions in the flow reactor.

According to the second variant if the amount of amphiphiles in the flow reactor is sufficient for PSS production, then they can self-multiply by micellar autocatalysis. As in the first variant, inside PSS conditions provide acceleration of chemical reactions and shift polymerization reactions toward the generation of the longest average oligomer chain. Then auto-oligomeraze reaction onsets inside PSS. Further possible stages comply with the first variant.

Computer simulation shows depending on kinetic constants both variants of the events are possible. Fig. 3 demonstrates the "hump" of typical oligomer length distribution inside PSS. It means that autocatalytic process takes place inside PSS. This simulation was conducted at reasonable parameters of reaction and flow rates in the reactor.

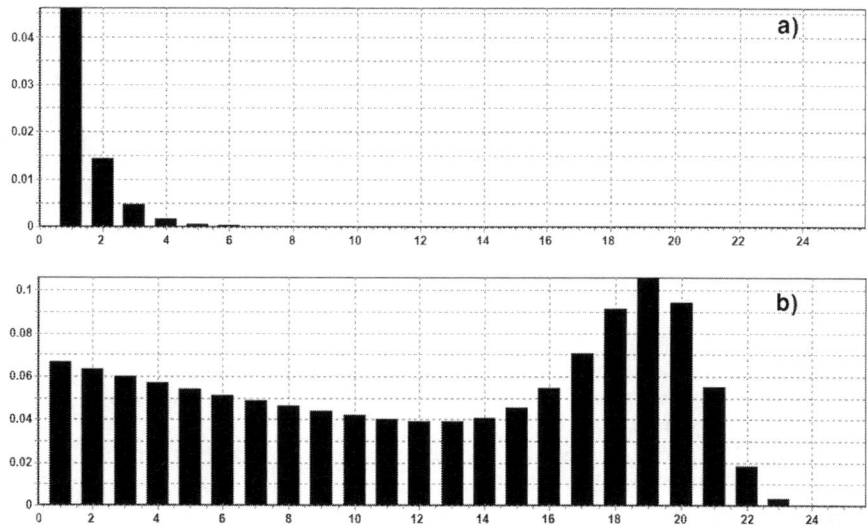

Fig. 3 Computer simulation of steady state oligomer distribution with respect to their length: (**a**), inside the flow reactor; (**b**), inside a phase-separated system compartment

4 Conclusions

1. Addressing the oligomers of the simplest chemical nature, and the renunciation of considerations of matrix duplication at initial stages of chemical evolution, allows a simpler and more probable scenario of pro-biont metabolism origin. It gives possibility to put the objectives for experimental modeling of initial stages of evolution.
2. The model considered here allows a preliminary estimation of necessary parameters of autocatalytic synthesis of oligomers, and the coupling to syntheses of inphase-separated systems. Preliminary estimations show that the origin of autocatalytic oligomeric PSS is possible at reasonable values of kinetics parameters of the chemical reactions involved in flow reactor processes. The signature of auto-oligomerase reaction is the presence of "a hump" on the distribution of oligomers with respect to the length of their chains.
3. It is supposed that at the initial stages of chemical evolution the variants of the most successful compositions of amphiphiles and oligomers of different chemical nature, i.e., variants of possible future biochemistries, are selected to create conditions that encourage matrix duplication. Following this chemical evolution gives gradually the place to biological evolution.

Acknowledgment This work was conducted with the support of RAS Program No. 18.2 "Biosphere Origin and Evolution" and RFBR Grant No. 06-04-49016.

References

Altstein, A.D. (1987) The origin of the genetic system: the progene hypothesis. Mol. Biol. 21(2), 309–322 (in Russian).
Bartsev, S.I. (2004) Essence of life and multiformity of its realization: expected signatures of life. Adv. Space Res. 33(8), 1313–1317.
Bartsev, S.I. and Mezhevikin, V.V. (2005) Pre-biotic stage of life origin under non-photosynthetic conditions. Adv. Space Res. 35(9), 1643–1647.
Bartsev, S.I. and Mezhevikin, V.V. (2006) On the theoretical and experimental modeling of initial states of metabolism formation in prebiotic systems. Paleontol. J. 40(4), S536–S542.
Bartsev, S.I., Mezhevikin, V.V. and Okhonin, V.A. (2001) Life as a set of matter transformation cycles: ecological attributes of life. Adv. Space Res. 28(4), 607–612.
Braun, D. and A. Libchaber (2004) Thermal force approach to molecular evolution: Physical Biology 1, 1–8.
Carny, O. and Gazit, E. (2005) A model for the role of short self-assembled peptides in the very early stages of the origin of life. FASEB J. 19, 1051–1055.
Cody, G.D. (2004) Transition metal sulfides and the origin of metabolism. Ann. Rev. Earth Planet. Sci. 32, 569–599.
Cody, G.D., Bactor, N.Z., Filley, T.R., Hazen, R.M., Scott, J.H., Sharma, A., Yoder, H.S., Jr. (2000) Primordial carbonylated iron-sulfur components and the synthesis of pyruvate. Science 289(5483), 1337–1340.
Feigin, A.M. (1987) On a possibility of origin of nucleic acids on the basis of polymers with a simpler structure. J. Evol. Biochem. Physiol. 23(4), 417–422.
Feistel, R., Romanovsky, Yu.M. and Vasil'ev, V.A. (1980) Evolution of Eigen's hyper-cycle taking place in coacervate. Biophysics 25(5), 882–887 (in Russian).
Ferris, J.P., Hill, A.R., Jr., Liu, R. and Orgel, L.E. (1996) Synthesis of long prebiotic oligomers on mineral surfaces. Nature 381, 59–61.
Fontana, W. and Shuster, P. (1998) Shaping space: the possible and the attainable in RNA genotype–phenotype mapping. J. Theor. Biol. 194, 491–515.
Fox, S.W. and Dose, K. (1977) Molecular Evolution and the Origin of Life. Marcel Dekker, New York., p. 370.
Gilbert, W. (1986) The RNA world. Nature 319, 618.
Gleiser, M. and Thorarinson, J. (2006) Prebiotic homochirality as a critical phenomenon. Orig. Life Evol. Biosph. 36, 501–505.
Goldberg, S.I. (2007) Enantiomeric enrichment on the prebiotic earth. Orig. Life Evol. Biosph. 37, 55–60.
Klotz, I.M., Royer, G.P. and Scarpa, I.S. (1971) Synthetic derivatives of polyethyleneimine with enzyme-like catalytic activity (synzymes). PNAS 68(2), 263–264.
Koonin, E.V. and W. Martin, (2005) On the origin of genomes and cells within inorganic compartments. Trends Genet. 21, 647–654.
Laszlo, P. (1999) Catalysis of organic reactions by inorganic solids. Pure Appl. Chem. 62(10), 2027–2030.
Martin, W and Russell, M.J. (2002) On origins of cells: a hypothesis for the evolutionary transitions from abiotic geochemistry to chemoautotrophic prokaryotes, and from prokaryotes to nucleated cells. Phil. Trans. R. Soc. London, B. 358, 27–85.
Nelson, K.E., Levy, M. and Miller, S.L. (2000) Peptide nucleic acids rather than RNA may have been the first genetic molecule. PNAS 97, 3868–3871.
Oba, T., Fukushima, J, Maruyama, M., Iwamoto, R. and Ikehara, K. (2005) Catalytic activities of [GADV]-peptides. Orig. Life Evol. Biosph. 34, 447–460.
Pohorille, A. and Deamer, D. (2001) Artificial cells: prospects for biotechnology. Trends Biotechnol. 20, 123–128.
Pross, A. (2004) Causation and the origin of Life. Metabolism or replication first? Orig. Life Evol. Biosph. 34, 307–321.

Rasi, S., Mavelli, F. and Luisi, P.L. (2004) Matrix effect in oleate micelles–vesicles transformation. Orig. Life Evol. Biosph. 34, 215–224.

Rasmussen, S., Chen, L., Stadler, B.M.R. and Stadler, P.F. (2002) Proto-organism kinetics: evolutionary dynamics of lipid aggregates with genes and metabolism. Santa Fe Institute, Working Paper, No. 02-10-054.

Ruckenstein, E. and Nagarajan, R. (1976) On critical concentrations in micellar solutions. J. Colloid Interface Sci. 57(2), 388–390.

Schuster, P. (2000) Molecular insight into the evolution of phenotypes. Santa Fe Institute, Working Paper No. 00-02-013.

Segre, D. and Lancet, D. (2000) Composing life. EMBO Reports 1(3), 217–222.

Segre, D., Lancet, D., Kedem, O. and Pilpel, Y. (1998) Graded autocatalysis replication domain (GARD): kinetic analysis of self-replication in mutually catalytic sets. Orig. Life Evol. Biosph. 28, 501–514.

Segre, D., Ben-Eli, D., Deamer, D.W. and Lancet, D. (2001) The lipid world. Orig. Life Evol. Biosph. 31, 119–145.

RNA World: First Steps Towards Functional Molecules

A.V. Lutay, M.A. Zenkova, and V.V. Vlassov

Abstract This review is focused on the evolution of prebiotic RNA molecules—from the first nucleotide's origin towards the supramolecular complexes of RNA molecules, possessing the catalytic activities. On the assumption of the numerous experimental data known, the principal stages of the RNA World's evolution are described in terms of the molecular complexity. Special attention is given to our study of the metal-catalyzed nonenzymatic cleavage/ligation reaction, which provides a mechanism of how the RNA molecules could elongate and diversify.

Studies of the last decades provided evidences that "there was once an RNA world" (Orgel, 2004). The natural selection of such complicated molecule as RNA is a low-probability outcome for the primitive system, so there are several hypotheses that some polymers of another type could forerun RNA (Joyce and Orgel, 1999; Orgel, 2003). Since the primordial catalytic RNAs from that era have been replaced by protein enzymes, the exact reconstruction of a hypothesized "RNA world" is difficult, though one can verify experimentally some crucial features of RNA world (for reviews see Dworkin et al., 2003; Orgel, 2003).

This chapter focuses on the nonenzymatic reactions that could provide the possibility of emergence of large functionally active RNA molecules required for the development of RNA world. The following three sections describe the main stages of the RNA world establishment according to their titles.

1 First RNA Monomers: Puzzle of Isomers

The chemical steps required for the emergence of RNA oligomers include successive reactions of condensation increasing the molecular complexity (Fig. 1).

A.V. Lutay
Institute of Chemical Biology and Fundamental Medicine SB RAS, Novosibirsk, Russia
e-mail: lutay_av@niboch.nsc.ru

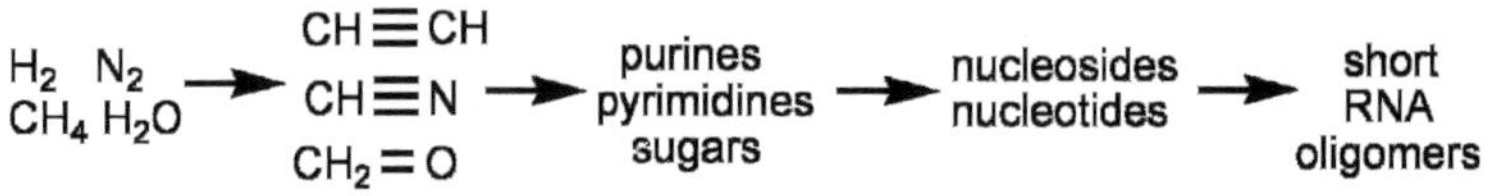

$$\begin{matrix}H_2 & N_2 \\ CH_4 & H_2O\end{matrix} \longrightarrow \begin{matrix}CH\equiv CH \\ CH\equiv N \\ CH_2=O\end{matrix} \longrightarrow \begin{matrix}\text{purines} \\ \text{pyrimidines} \\ \text{sugars}\end{matrix} \longrightarrow \begin{matrix}\text{nucleosides} \\ \text{nucleotides}\end{matrix} \longrightarrow \begin{matrix}\text{short} \\ \text{RNA} \\ \text{oligomers}\end{matrix}$$

Fig. 1 The prebiotic pathway leading to the emergence of the first RNA molecules

The first stage was the generation of organic molecules from simple precursors such as primitive gases that might have constituted the primal atmosphere. A variety of energy sources like heat from sparks, chemical exudates from ocean vents and solar radiation were all capable of generating the key prebiotic molecules. The high energy favours the formation of the multiple bonded forms $-C\equiv C-$, $-C\equiv N$, $>C=O$ that are capable of further polymerization, producing a wide variety of organic molecules.

Polymerization of formaldehyde in the presence of simple mineral catalysts—the formose reaction—is a unique, cyclic autocatalytic process that takes place in aqueous solution and converts a very simple substrate, formaldehyde, to a mixture of complex molecules, many of which are important biochemicals. The first product of polymerization is glycoaldehyde, which is later converted to glyceraldehyde and a variety of tetrose, pentose and hexose sugars. However, it had not been possible to channel this reaction to the synthesis of any particular sugar, and ribose usually has been a notoriously minor product. Eschenmoser and his co-workers showed that the pattern of products could be greatly simplified if glycoaldehyde and glyceraldehyde were replaced by their monophosphates. The condensation of glycoaldehyde phosphate with glyceraldehyde phosphate on the mineral surfaces results in more effective formation of ribose-2,4-diphosphate, but it has another disposition of the phosphate groups than the natural ribose–phosphate backbone in RNA molecule (Mueller et al., 1990; Pitsch et al., 1995).

The prebiotic purine synthesis is concerned with the polymerization process of HCN, which can proceed in a wide range of conditions, yielding substantial amounts of purines. The reaction of cyanoacetylene or cyanoacetaldehyde with urea and the reaction of cyanoacetylene with cyanate in eutectic solution are the most plausible prebiotic routes to cytosine. Uracil is formed from cytosine by hydrolysis, and this has been proposed as a prebiotic synthesis. In contrast to purines that react with D-ribose with the formation of nucleosides, no direct synthesis of pyrimidine nucleosides from ribose and uracil or cytosine has been reported, although some variants of the indirect synthesis have been proposed (see Orgel, 2004 for the references). The mechanism of selection of nucleotides with "correct" β-glycosidic linkage remains unknown.

The last structural unit to be mentioned is the phosphodiester bond by which all nucleosides are linked together in the RNA structure. Like today, phosphorus on the primitive Earth was likely present mostly as various phosphates. Nucleosides could be converted to a complex mixture of products containing one or more phosphate groups by heating at moderate temperatures with phosphate salts and urea. Some progress has been achieved in attempts to direct this solid-state reaction to the synthesis of 5′-phosphorylated nucleosides, since the latter are considered to be the most probable nucleotide precursors. While most efforts have been devoted to the phosphorylation of preformed nucleosides, some plausible routes with phosphorylated sugars have also been demonstrated. Whatever phosphate was incorporated at the stage of nucleoside or ribose syntheses or it appeared later via some phosphorylation reactions, it is clear that the primitive condensation reactions between phosphorylated nucleosides could produce a mixture of isomeric phosphodiester bonds.

Thus many RNA-like molecules (including various isomers) could have been synthesized in the reactions of prebiotic synthesis and there is no clear conception how the RNA molecules, containing 3′,5′-phosphodiester bonds, D-ribose and β-glycosidic bonds, might have been selected. Detailed consideration of the mentioned problems can be found in the reviews (Dworkin et al., 2003; Orgel, 2004).

2 Prebiotic Synthesis of RNA Oligomers

Polymerization of nucleotides in aqueous solution must necessarily make use of external activating agents. Since many highly reactive compounds (e.g. hydrocyanic acid, cyanoacetylene, cyanoamide, formamide, different aldehydes) were likely involved in the initial syntheses, it is natural to suppose that many nucleophiles were available and the phosphate group could be activated under these conditions. Activated forms of phosphate group that were likely involved in prebiotic chemistry of RNA include phosphoramidates (e.g. phosphorimidazolides; Lohrmann and Orgel, 1977), polyphosphates (Lohrmann, 1975) and 2′,3′-cyclic phosphates (Lutay et al., 2005; Renz et al., 1971; Verlander et al., 1973) (Fig. 2a–c).

A number of investigations were focused on two possible types of prebiotic polymerization reactions: template-free polymerization and template-directed synthesis (Fig. 2d and e). In the first case, the meeting of RNA fragments and the total efficiency of the reaction is determined by diffusion. Since various molecules can react with activated phosphates, it is unlikely that template-independent polymerization in solution had been able to yield significant contribution to the synthesis of RNA although it was shown that a number of activated nucleotides are able to polymerize in the absence of template upon drying (Sleeper et al., 1978; Verlander and Orgel, 1974), freezing (Kanavarioti et al., 2001; Monnard et al., 2003; Renz et al., 1971; Trinks et al., 2005) or adsorption on the surface of minerals (Acevedo and Orgel, 1986).

Formation of double-stranded complexes by means of Watson–Crick base pairing could be another mechanism for the RNA contiguity. The template brings complementary activated monomers or short oligomers together, providing proximity of specific reactive groups, and facilitates their ligation.

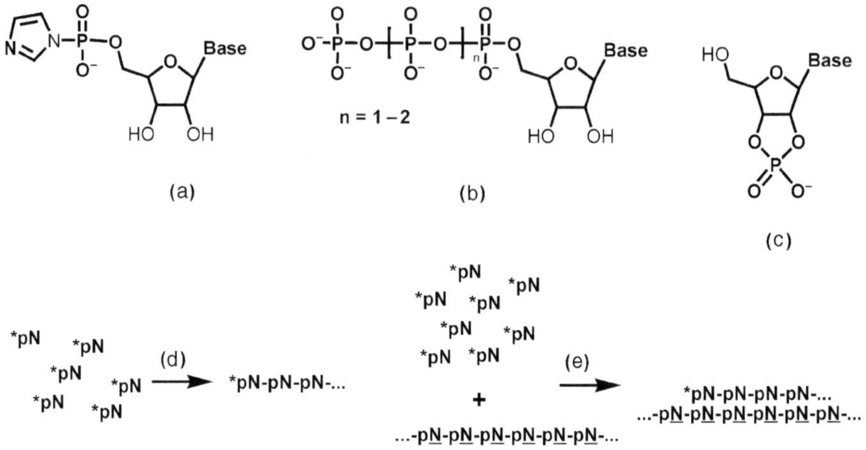

Fig. 2 The most probable prebiotic activated RNA monomers: phosphorimidazolides (**a**), tri- or tetraphosphates (**b**), 2′,3′-cyclic phosphates (**c**). Polymerization in the absence of templates (**d**). Template-directed synthesis (**e**)

Metal ions are capable of coordinating with the oxygens of phosphate groups, which could accelerate the reaction and affect the ratio of the resulting isomers. The ratio of 3′,5′-/2′,5′-phosphodiester linkages can be expected to depend on the nature of the nucleoside base and the activating group. In some cases this ratio can be as high as four (Prabahar and Ferris, 1997), in other cases the regiospecificity favours the formation of 2′,5′-phosphodiester bonds (Ertem and Ferris, 1997). Selective production of the 3′,5′-phosphodiester linkages has never been observed. Oligonucleotides, containing both 2′,5′- and 3′,5′-phosphodiester linkages, could serve as a template or a primer in the template-directed synthesis that results in the double-stranded RNA complex formation (Sawai et al., 1998, 2006).

An imperfection of the template-directed synthesis model lies in the distinguished efficiencies of nucleotide incorporation depending on the opposite nucleotide. For example, a pair of adjacent A residue in the template is an almost complete barrier to further synthesis, and wobbling pairing of G opposite U leads to extensive misincorporation of G. The results of sequence dependence investigations have shown that a wide variety of different RNA sequences can be copied, but not all (Joyce, 1987).

The L-enantiomers of activated nucleotides are efficient inhibitors of template-directed synthesis using the naturally occurring D-enantiomers. Since any prebiotic synthesis produces racemates, one more challenge in any scheme for polynucleotide replication from plausibly prebiotic monomeric substrates is an enantiomeric cross-inhibition (Joyce, 1984).

The most efficient synthesis of RNA oligonucleotides was achieved by polymerization of 5′-phosphorimidazolides of the montmorillonite surface.

Repeated addition of new portions of phosphorimidazolides led to synthesis of RNA oligomers as long as 35 to 40-mers (Ferris, 2002).

It is not known if phosphorimidazolides could have been presented in large amounts on the primitive Earth, and no similar reaction with phosphorimidazolides (or any other phosphoramidates) involvement can be found in any living systems today. Thus, there can be no doubt that described mineral-catalyzed reactions are relevant to chemical evolution. Emerged oligonucleotides could form double-stranded regions by means of Watson–Crick base pairing. Nonenzymatic ligation reactions proceeding in those complexes seem to be the most probable mechanism of RNA elongation, as template-assisted polymerization of activated monomers is less effective due to some sequence restrictions.

3 Emergence of Catalytic RNAs

After the synthesis of RNA oligomers the next key event in the RNA world evolution is the emergence of RNA molecules long enough to possess the catalytic activity. At this stage the only function of the first ribozymes was catalysis of RNA replication, the building blocks of which could be either monomers or short RNA oligomers. The development of "in vitro selection" and "in vitro evolution" strategies allowed to obtain ribozymes with predetermined activity; thus significant progress was achieved in understanding ribozyme's potential in phosphodiester bond formation reactions.

In most studies the derived ligase ribozymes could catalyze phosphodiester bond formation between substrate molecule and ribozyme itself (i.e. in *cis*), making the catalyst discharged for the subsequent reactions, what does not simulate prebiotic evolution. Thus, the ribozymes with *trans* activity are of special interest.

The first ligase ribozyme (class I ligase ribozyme) was selected from a large pool of 220-nt RNA molecules and catalyzed the ligation of 3'-hydroxyl of substrate molecule onto 5'-triphosphate group of enzyme with formation of 3',5'-phosphodiester bond (Bartel and Szostak, 1993). Its modified form was 93-nt long and could append several NTPs in template-dependent manner. In vitro evolution of that ribozyme with random RNA domain appended at its 3'-end allowed to evolve an enzyme, catalyzing polymerization of NTPs on external template. Novel ribozyme containing approximately 200 nt was insensitive to nucleotide sequences of template and primer and could append up to 14 nucleotides (Johnston et al., 2001).

In independent investigations several other ribozymes with 3',5'-ligase activity have been evolved. These include catalytic domains smaller than that of class I ribozyme. It is interesting that the L1 ligase can be activated by an oligonucleotide, protein effector or small molecule to act as an allosteric ribozyme, suggesting a possible mechanism for primitive phenotypic regulation (Ellington and Szostak, 1990; Robertson and Ellington, 2001). The R3C ribozyme has been engineered to function in an intermolecular reaction format with one substrate containing the attacking 3′-hydroxyl and the second substrate containing the 5′-triphosphate (Rogers and Joyce, 2001). Later R3C ribozyme was redesigned for the development of self-replicating system based on the ribozyme that catalyzes the assembly of additional copies of itself through RNA-catalyzed RNA ligation reaction (Paul and Joyce, 2002).

Variants of the hc ligase ribozyme, which catalyzes ligation of the 3′-end of an RNA substrate to the 5′-end of the ribozyme, were utilized to evolve a ribozyme that catalyzes ligation reactions of an external RNA template. The evolved ribozyme has a length of about 300 nucleotides and catalyzes joining of an oligonucleotide 3′-OH group to the 5′-triphosphate of an RNA hairpin molecule. The enzyme also demonstrates predominately sequence-independent mechanism for substrate recognition and can catalyze the addition of NTP onto the 3′-end of an oligonucleotide primer in a template-dependent manner (McGinness and Joyce, 2002).

The highest length obtained in the polymerization of activated nucleotides is about 40 nt, which is 2–3 times smaller than the minimal length of the evolved ligase ribozymes. But it seems that the prebiotic ribozyme might have been a little bit smaller (60–100 nt). One great opportunity is hidden in supramolecular complexes of small RNA molecules that could possess the catalytic activity. It was shown that truncated and fragmented derivatives of the hairpin ribozyme can catalyze ligation of a wide variety of RNA molecules to a given sequence in frozen solution (Vlassov et al., 2004). Another example is the binary form of hammerhead ribozyme, consisting of two parts and possessing catalytic activity similar to that of the native ribozyme (Kuznetsova et al., 2004; Vorobjeva et al., 2006).

The progress of in vitro selection and in vitro evolution methods facilitates the evaluation of the most important aspects of prebiotic ribozymes, namely, catalytic efficiency and probability of emergence. It is evident that, despite the low efficiency of obtained ribozymes, more active RNA polymerases acting in *trans* might be developed. Thus, progress in the minimal ribozyme design should be expected.

4 RNA Recombination: Elongation and Diversity

The recombination reaction can provide formation of novel RNA molecules. Chetverin observed emergence of novel sequences in the colony of RNA molecules, exponentially amplified by Qb-replicase, and suggested that similar recombination reactions might have proceeded in RNA world, providing its sequence diversity (Chetverin, 1999).

RNA molecules contain double-stranded regions of various lengths, formed by partially complementary sequences. Since the single-stranded regions are much more susceptible to transesterification reaction, cleavage can occur in such a way that in the nick 2′,3′-cyclic phosphates and 5′-hydroxyl groups are located. The intramolecular transesterification reaction is accelerated in the presence of many compounds (metal ions, organic bases, etc.). The transformation pathway of the new-born 2′,3′-cyclic phosphate, as well as any activated phosphate, depends on the pH and availability of nucleophilic molecules present in the system. At pH close to neutral the time of 2′,3′-cyclic phosphate half-life is long enough to permit interaction with other available RNA molecules. Thus RNA fragments with 5′-hydroxyl group and 2′,3′-cyclic phosphate, obtained as a result of RNA cleavage, could cross-interact with each other forming new phosphodiester linkages and yielding new RNA molecules. Therefore 2′,3′-cyclic phosphate could be viewed as an activated phosphate emerging in a customary process without any other activating groups.

It was shown that binding of oligoadenylates with 2′,3′-cyclic phosphate terminus to polyuridylic acid template in the presence of ethylenediamine leads to the formation of dimers (24%) and trimers (5%). The formed phosphodiester bonds were predominately 2′,5′-isomers (Usher and McHale, 1976).

Recently, we have investigated the nonenzymatic template-directed ligation of oligonucleotides containing 2′,3′-cyclic phosphate (Lutay et al., 2006). Ligation of the oligonucleotides occurred even in the absence of metal ions and its efficiency demonstrated an escalating pH–yield profile within a pH interval of 6.0–8.8, which suggests base catalysis of the reaction proceeding via increasing of nucleophilic properties of the attacking 5′;-hydroxyl group, as was shown earlier (Renz et al., 1971). Divalent cations (Mg^{2+}, Mn^{2+}, Zn^{2+}, Pb^{2+}) accelerated the reaction in pH-dependent manner, which is explained by the following events: (1) activation of hydroxyl groups via the coordination of the attacking oxygen and increase of its nucleophilicity, (2) stabilization of the developing negative charge on the leaving oxygen group, (3) activation of phosphate centre atom for nucleophilic attack through coordination to the nonbridging oxygen (Dahm et al., 1993; Hampel and Cowan, 1997; Pyle, 2002; Steitz and Steitz, 1993). At pH 7.2 the efficiency of the metal ions as catalysts inversely correlated with the pK_a values of the metal-bound water molecules. However, the catalytic potential of the metal ion is determined by solubility of its metal ion hydroxide; maximal yields were obtained at the highest possible concentrations of the metal ions in solution. At pH 8.8 insoluble metal hydroxides are formed resulting in the decrease of the ligation product yield, except for Mg^{2+} ions, which showed the best catalytic properties, yielding up to 15% of the ligation product (Fig. 3).

At least 95% of the isolated products of nonenzymatic ligation were resistant to ribonuclease T1 under the conditions where 3′,5′-bonds are completely cleaved by the enzyme. The result of comparative T1-probing can be attributed to the 2′,5′-bond formation in the nonenzymatic ligation products (Lutay et al., 2006).

Similar ligation reaction was shown to proceed in vivo within peach latent mosaic viroid (PLMVd). Self-ligation occurs when the 5′-hydroxyl and the 2′,3′-cyclic phosphate termini produced by the hammerhead ribozyme self-cleavage

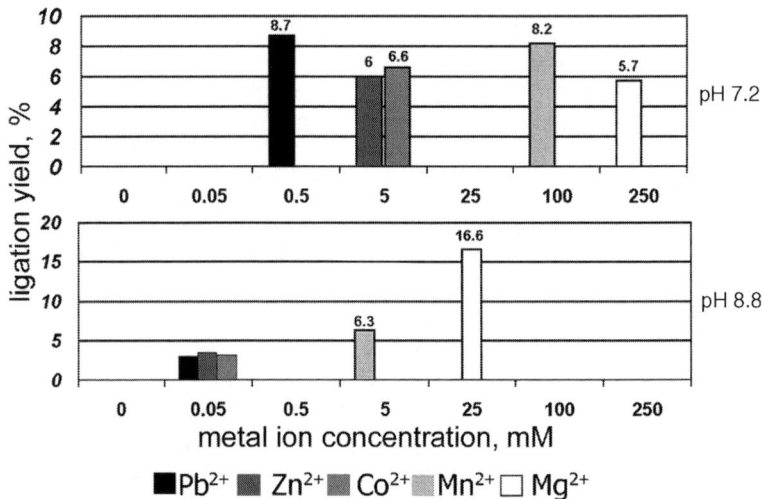

Fig. 3 The highest ligation yields obtained in the presence of metal ions at pH 7.2 (*above*) and 8.8 (*below*). For each metal ion the maximal ligation yield is indicated *above* the bar at the concentration used. Oligonucleotide complex was incubated in the presence of metal ions (50 μM–250 mM) at pH 7.2 and 8.8 for 3 days at 37 °C

of the viroid RNA are juxtaposed by the viroid rod-like structure. The formed linkages were almost solely 2′,5′ (>96%) and were shown to stabilize the replicational circular templates. Since the reverse transcriptase is able to read through a 2′,5′-linkage with the efficacy more than 50%, their presence in the viroid structure does not constitute an important obstacle to polymerase progression and results in a significant advantage in terms of viroid viability, protecting the viroid integrity (Cote et al., 2001).

The explanation of the observed stereospecificity concerns the geometry of A-type RNA helix conformation. The rate of the transesterification reaction depends on the probability of the in-line conformation in the transition state, when forming O–P and cleaving P–O bonds are placed on the opposite sides of the phosphorus atom. In the presence of complementary strand, 2′,5′-linkages are 10^3 times less stable than the 3′,5′-linkages. Similarly, the rate of the ligation reaction depends on the probability of the in-line attack by the 5′-oxygen atom and the type of formed linkage is determined by the direction of attack—whether 2′- or 3′-oxygen atom is on the opposite side. Since the transesterification reaction of 2′,5′-phosphodiester bonds, resulting in the formation of 2′,3′-cyclic phosphate, is in equilibrium with the ligation reaction, one can conclude that the terminal 2′,3′-cyclic phosphate participates in reaction in single spatial orientation, when 5′-attacking oxygen is in in-line conformation with 3′-cleaving O–P bond of cyclic phosphate. It means that the equilibrium (phosphodiester bond vs. 2′,3′-cyclic phosphate and 5′-OH) involves the unstable 2′,5′-phosphodiester bond.

Due to this conclusion the scientific interest to the possible role of 2′,3′-cyclic phosphates in nonenzymatic template-directed synthesis of RNA decreased, but ligation reaction could also proceed in the complex with catalytic nucleic acid, and the orientation of reacting groups depends on the structure of the complex.

In vitro selection procedure was used to obtain deoxyribozymes that catalyze ligation of two RNA oligonucleotides, bearing 2′,3′-cyclic phosphate and 5′-hydroxyl group. At least 96% of the formed linkages were 2′,5′ (Flynn-Charlebois et al., 2003). On the contrary, two RNA oligonucleotides, one of which contained 2′,3′-cyclic phosphate, were ligated by fragmented hairpin ribozyme derivatives with formation of 3′,5′-phosphodiester bonds (Vlassov et al., 2004).

Presented data suggest that RNA cleavage yielding oligonucleotides with 2′,3′-cyclic phosphates, followed by reaction of these oligonucleotides with each other, could provide a source of new RNA molecules under prebiotic conditions. We investigated two-step cleavage/ligation reaction of RNA oligonucleotides catalyzed by metal ions. The oligonucleotides were designed so as to form concatemeric pseudo-duplex complex with dangling tetraadenylate tails, not involved in Watson–Crick base pairing (Fig. 4).

Magnesium ions were shown to catalyze cleavage of phosphodiester bonds within single-stranded RNA sequences (namely, A_4), followed by the ligation reaction. The ligation products contain not only dimers, but also detectable amounts of trimers.

An increase in pH from 7.0 to 9.5 weakens RNA duplex stability and thus enhances the degradation of phosphodiester bonds. This process can result in a decrease in the complexes in which the template-directed ligation can occur. Nevertheless, the yield of ligation products demonstrates an escalating profile within chosen pH range. The almost linear accumulation of the ligation products is observed at pH 9.5 during 7 days incubation at 25°C up to maximal total yield of 5%. At 37°C the yield of the ligation products increases during 24 h up to 3% and then diminishes almost completely by 72 h. According to the profile of the degradation, the disappearance of the ligation product is caused mainly, if not entirely, by cleavage of the phosphodiester bond formed upon

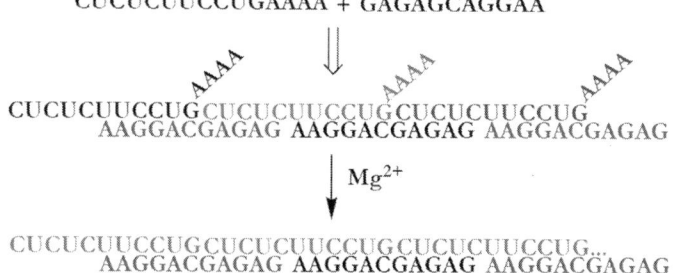

Fig. 4 Mg^{2+}-catalyzed combined cleavage/ligation reaction proceeding in the complex of two oligoribonucleotides and leading to the emergence of new RNA sequences. Different molecules are marked with *different colour intensities*

ligation, suggesting its 2',5'-isomeric form. Though the formed phosphodiester bond is liable to metal ion-catalyzed degradation, the resistance of the other linkages has been increased due to stabilization of the double-stranded complex caused by growth of oligonucleotide length. Formation of the ligation product is also observed at lower temperatures (15 and 20°C), where its life-time seems to be longer than a week.

It is natural to consider that the Watson–Crick base pairing existed under prebiotic conditions as primary essential mechanism of RNA–RNA interactions. Since the RNA molecules are able to form complexes with tertiary structure, including bulges and loops, it seems that many phosphodiester bonds should be accessible for the cleavage reaction. Thus, described cleavage/ligation reactions could also proceed, catalyzed with ubiquitous metal ions. Co-existence of numerous RNAs might provide catalysis via formation of some sophisticated complexes (as in the case of binary ribozymes), suggesting the hypothetic network of the supramolecular interactions, in which one RNA molecule might have been involved in many catalytic reactions.

5 Concluding Remarks

The RNA world model of the origin of life begins with the production of small molecules, which then combine to form ribonucleic acids. Small molecules are unfeasible to promote specific synthesis of RNA structural constituents, thus we should consider the existence of many RNA-like molecules representing all possible isomers of RNA. No mechanism of selection has been proved yet.

RNA monomers were likely to possess an activated phosphate group and could couple in condensation reactions, of which the mineral-catalyzed condensation is the most efficient. Under that conditions ribonucleic acids up to 10 monomers in length could be formed. These new molecules might be a source of RNA catalysts directly or via formation of supramolecular complexes.

The combined cleavage/ligation reactions might contribute to production of a variety of new RNAs in a recombination-like process and yield large RNAs needed for the development of the RNA world.

Acknowledgment This work was supported by grants from FCSTP (RI012/001/254), RAS programs "Origin of life and evolution of the biosphere" and "Molecular and cellular biology."

References

Acevedo, O.L. and Orgel, L.E. (1986) Template-directed oligonucleotide ligation on hydroxylapatite. Nature 321, 790–792.
Bartel, D.P. and Szostak, J.W. (1993) Isolation of new ribozymes from a large pool of random sequences. Science 261, 1411–1418.
Chetverin, A.B. (1999) The puzzle of RNA recombination. FEBS Lett. 460, 1–5.

Cote, F., Levesque, D. and Perreault, J.P. (2001) Natural 2′,5′-phosphodiester bonds found at the ligation sites of peach latent mosaic viroid. J. Virol. 75, 19–25.

Dahm, S.C., Derrick, W.B. and Uhlenbeck, O.C. (1993) Evidence for the role of solvated metal hydroxide in the hammerhead cleavage mechanism. Biochemistry 32, 13040–13045.

Dworkin, J.P., Lazcano, A. and Miller, S.L. (2003) The roads to and from the RNA world. J. Theor. Biol. 222, 127–134.

Ellington, A.D. and Szostak, J.W. (1990) In vitro selection of RNA molecules that bind specific ligands. Nature 346, 818–822.

Ertem, G. and Ferris, J.P. (1997) Template-directed synthesis using the heterogeneous templates produced by montmorillonite catalysis. A possible bridge between the prebiotic and RNA worlds. J. Am. Chem. Soc. 119, 7197–7201.

Ferris, J.P. (2002) Montmorillonite catalysis of 30–50 mer oligonucleotides: laboratory demonstration of potential steps in the origin of the RNA world. Orig. Life Evol. Biosph. 32, 311–332.

Flynn-Charlebois, A., Wang, Y., Prior, T.K., Rashid, I., Hoadley, K.A., Coppins, R.L., Wolf, A.C. and Silverman, S.K. (2003) Deoxyribozymes with 2′-5′ RNA ligase activity. J. Am. Chem. Soc. 125, 2444–2454.

Hampel, A. and Cowan, J.A. (1997) A unique mechanism for RNA catalysis: the role of metal cofactors in hairpin ribozyme cleavage. Chem. Biol. 4, 513–517.

Johnston, W.K., Unrau, P.J., Lawrence, M.S., Glasner, M.E. and Bartel, D.P. (2001) RNA-catalyzed RNA polymerization: accurate and general RNA-templated primer extension. Science 292, 1319–1325.

Joyce, G.F. (1984) Non-enzymatic template-directed synthesis of RNA copolymers. Orig. Life 14, 613–620.

Joyce, G.F. (1987) Nonenzymatic template-directed synthesis of informational macromolecules. Cold Spring Harb. Symp. Quant. Biol. 52, 41–51.

Joyce, G.F. and Orgel, L.E. (1999) Prospects for understanding the origin of the RNA world. In: R.F. Gesteland, T.R. Cech and J.F. Atkins (Eds), The RNA World. Cold Spring Harbor Press, Cold Spring Harbor, pp. 49–78.

Kanavarioti, A., Monnard, P.A. and Deamer, D.W. (2001) Eutectic phases in ice facilitate nonenzymatic nucleic acid synthesis. Astrobiology 1, 271–281.

Kuznetsova, M., Novopashina, D., Repkova, M., Venyaminova, A. and Vlassov, V. (2004) Binary hammerhead ribozymes with high cleavage activity. Nucleosides Nucleotides Nucleic Acids 23, 1037–1042.

Lohrmann, R. (1975) Formation of nucleoside 5′-polyphosphates from nucleotides and trimetaphosphate. J. Mol. Evol. 6, 237–252.

Lohrmann, R. and Orgel, L.E. (1977) Reactions of adenosine 5′phosphorimidazolide with adenosine analogs on a polyuridylic acid template. J. Mol. Biol. 113, 193–198.

Lutay, A.V., Chernolovskaya, E.L., Zenkova, M.A. and Vlassov, V.V. (2005) Nonenzymatic template-dependent ligation of 2′,3′-cyclic phosphate-containing oligonucleotides catalyzed by metal ions. Doklady Biochem. Biophys. 401, 163–166.

Lutay, A.V., Chernolovskaya, E.L., Zenkova, M.A. and Vlassov, V.V. (2006) The nonenzymatic template-directed ligation of oligonucleotides. Biogeosciences 3, 243–249.

McGinness, K.E. and Joyce, G.F. (2002) RNA-catalyzed RNA ligation on an external RNA template. Chem. Biol. 9, 297–307.

Monnard, P.A., Kanavarioti, A. and Deamer, D.W. (2003) Eutectic phase polymerization of activated ribonucleotide mixtures yields quasi-equimolar incorporation of purine and pyrimidine nucleobases. J. Am. Chem. Soc. 125, 13734–13740.

Mueller, D., Pitsch, S., Kittaka, A., Wagner, E., Wintner, C.E. and Eschenmoser, A. (1990) Chemistry of alpha aminonitriles. Aldomerization of glycoaldehyde phosphate to racemic hexose 2,4,6-triphosphates and (in presence of formaldehyde) racemic pentose 2,4-diphosphates: rac-allose 2,4,6-triphosphate and racemic ribose 2,4-diphosphate are the main reaction products. Helv. Chim. Acta 73, 1410–1468.

Orgel, L.E. (2003) Some consequences of the RNA world hypothesis. Orig. Life Evol. Biosph. 33, 211–218.

Orgel, L.E. (2004) Prebiotic chemistry and the origin of the RNA world. Crit. Rev. Biochem. Mol. Biol. 39, 99–123.

Paul, N. and Joyce, G.F. (2002) Inaugural article: a self-replicating ligase ribozyme. Proc. Natl Acad. Sci. USA 99, 12733–12740.

Pitsch, S., Eschenmoser, A., Gedulin, B., Hui, S. and Arrhenius, G. (1995) Mineral induced formation of sugar phosphates. Orig. Life Evol. Biosph. 25, 297–334.

Prabahar, K.J. and Ferris, J.P. (1997) Adenine derivatives as phosphate-activating groups for the regioselective formation of 3′,5′-linked oligoadenylates on montmorillonite: possible phosphate-activating groups for the prebiotic synthesis of RNA. J. Am. Chem. Soc. 119, 4330–4337.

Pyle, A.M. (2002) Metal ions in the structure and function of RNA. J. Biol. Inorg. Chem. 7, 679-690.

Renz, M., Lohrmann, R. and Orgel, L.E. (1971) Catalysts for the polymerization of adenosine cyclic 2′,3′-phosphate on a poly (U) template. Biochim. Biophys. Acta 240, 463–471.

Robertson, M.P. and Ellington, A.D. (2001) In vitro selection of nucleoprotein enzymes. Nat. Biotechnol. 19, 650–655.

Rogers, J. and Joyce, G.F. (2001) The effect of cytidine on the structure and function of an RNA ligase ribozyme. RNA 7, 395–404.

Sawai, H., Totsuka, S., Yamamoto, K. and Ozaki, H. (1998) Non-enzymatic, template-directed ligation of 2′-5′ oligoribonucleotides. Joining of a template and a ligator strand. Nucleic Acids Res. 26, 2995–3000.

Sawai, H., Wada, M., Kouda, T. and Nakamura Ozaki, A. (2006) Nonenzymatic ligation of short-chained 2′-5′- or 3′-5′-linked oligoribonucleotides on 2′-5′- or 3′-5′-linked complementary templates. Chembiochem 7, 605–611.

Sleeper, H.L., Lohrmann, R. and Orgel, L.E. (1978) Formation of the imidazolides of dinucleotides under potentially prebiotic conditions. J. Mol. Evol. 11, 87–93.

Steitz, T.A. and Steitz, J.A. (1993) A general two-metal-ion mechanism for catalytic RNA. Proc. Natl Acad. Sci. USA 90, 6498–6502.

Trinks, H., Schroder, W. and Biebricher, C.K. (2005) Ice and the origin of life. Orig. Life Evol. Biosph. 35, 429–445.

Usher, D.A. and McHale, A.H. (1976) Nonenzymic joining of oligoadenylates on a polyuridylic acid template. Science 192, 53–54.

Verlander, M.S. and Orgel, L.E. (1974) Analysis of high molecular weight material from the polymerization of adenosine cyclic 2′, 3′-phosphate. J. Mol. Evol. 3, 115–120.

Verlander, M.S., Lohrmann, R. and Orgel, L.E. (1973) Catalysts for the self-polymerization of adenosine cyclic 2′, 3′-phosphate. J. Mol. Evol. 2, 303–316.

Vlassov, A.V., Johnston, B.H., Landweber, L.F. and Kazakov, S.A. (2004) Ligation activity of fragmented ribozymes in frozen solution: implications for the RNA world. Nucleic Acids Res. 32, 2966–2974.

Vorobjeva, M.A., Zenkova, M.A., Venyaminova, A.G. and Vlassov, V.V. (2006) Binary hammerhead ribozymes with improved catalytic activity. Oligonucleotides 16, 241–254.

Trans Hammerhead Ribozyme: Ligation vs. Cleavage

M. A. Vorobjeva, A. S. Privalova, A. G. Venyaminova, and V. V. Vlassov

Abstract *Trans* hammerhead ribozymes are considered from the point of view of the RNA world theory. An attempt was made to reconstruct the "ancestors" of contemporary hammerhead ribozymes with improved ligation activity by altering the sequence of *trans* hammerhead motifs. RNA ligation activity of *trans* hammerheads was shown to be affected significantly by minor changes in non-conservative regions. In particular, introduction of heptanucleotide bulge into stem III led to the 10-fold increase of ligation rate constant. At that, RNA cleavage was predominant activity in all cases. It was shown that *trans* hammerhead ribozyme can assemble from two separate short oligoribonucleotides upon binding to RNA substrate. This binary hammerhead ribozyme possesses a higher RNA-cleaving activity than its full-length analog. It can be assumed that such self-assembling multi-subunit catalytic RNAs could exist at early stages of the prebiotic evolution.

1 Introduction

The discovery of catalytic RNAs (ribozymes) capable of performing the site-specific phosphodiester bond cleavage in the absence of proteins (Cech et al., 1981; Guerrier-Takada et al., 1983) provided a paradigm shift in molecular biology and led to the hypothesis of RNA world which is generally accepted nowadays as origin-of-life scenario (see reviews: Joyce, 2002; Spirin, 2002; Muller, 2006). According to this theory, first RNA molecules were formed from abiogenous ribonucleotides or their activated derivatives. As it was discovered by Chetverin et al. (Chetverina et al., 1999), spontaneous rearrangements of RNA sequences can occur in the absence of any catalysts or cofactors. In the early RNA world, this reaction could be a driving force for the elongation

M. A. Vorobjeva
Institute of Chemical Biology and Fundamental Medicine, Novosibirsk, Russia
e-mail : kuzn@niboch.nsc.ru

of oligoribonucleotides, formation of polyribonucleotides and complication of their structure. As a result of such random recombinations, different catalytic RNA molecules could arise. The emergence of complementary RNA replication was a breakthrough that provided an amplification of RNA sequences and gave rise to the evolution of the RNA world. It is likely that in the early RNA world the complementary replication was carried out by ribozyme ligases using short oligonucleotides as substrates (Chen et al., 2007) and ribozyme polymerases appeared at the next stage of evolution. Nowadays, it has been clearly shown that *in vitro* selected RNA molecules are capable of catalyzing a wide variety of reactions, including the RNA cleavage, RNA ligation, RNA phosphorylation, and RNA polymerization (Chen et al., 2007; Joyce, 2002). This evidences the possibility of the existence of self-sustaining RNA-based prebiotic system where RNA molecules were both the genetic material and catalysts which performed all reactions necessary for autonomous functioning and evolution of the RNA world. It can be assumed that later ribozymes of the RNA world could adopt amino acids which are readily created by prebiotic chemistries, as cofactors to assist catalysis (Szatmary, 1999). Due to a greater catalytic versatility of mino acids, such "ribonucleopeptides" gained an advantage over all-RNA ribozymes, and their evolution led to the appearance of protein enzymes. At the late stages of this RNA–protein world, DNA could appear as more stable and safe genetic material which provided much larger genomes. As a result, DNA substituted for RNA as coding molecule, and proteins displaced ribozymes as better catalysts.

All above considerations lead to the conclusion that the RNA ligation and cleavage were of great importance in the early RNA world, since they provided an extension of RNA molecules and complication of their structure *via* RNA recombination. In the ligation reaction, catalytic RNA plays two roles: a template responsible for specific recognition and positioning of RNA substrates, and a catalyst necessary for the formation of phosphodiester bonds. Small natural ribozymes, such as hairpin, hammerhead and VS ribozymes can catalyze both RNA cleavage and RNA ligation reactions (Tanner, 1999) and can be considered as "descendants" of ancient RNA catalysts because of their relatively small size and high catalytic activity. The simplest among them, hammerhead catalytic motif was discovered in highly divergent organisms and obtained independently by in vitro selection (Conaty et al., 1999; Salehi-Ashtiani and Szostak, 2001) which is evident for the early evolution of hammerhead ribozymes. Natural *cis* hammerhead ribozymes can be divided into a non-conservative "substrate" part and a "ribozyme" part containing most of the conservative ribonucleotides. The resulting *trans* hammerhead ribozyme is the most simple nature-derived catalytic RNA, and we assume that hammerhead-like *trans* ribozymes could probably exist in the early RNA world. The aim of our work was to investigate the regularities of RNA ligation and cleavage by *trans* hammerhead ribozyme as an example of processes that took place during the RNA world evolution.

2 RNA Ligation by *Trans* Hammerhead Ribozymes

A lot of artificial ribozymes capable of site-specific RNA cleavage were designed on the basis of hammerhead motif (see reviews: Amarzguioui and Prydz, 1998; Citti and Rainaldi, 2005; Grassi et al., 2004; Peracchi, 2004; Schubert and Kurreck, 2004), but the ligation activity of these ribozymes is studied rather poor. It was shown that the rate constant for the reaction of RNA ligation (which has been measured for a few hammerhead sequences) is, in average, 0.01 min^{-1}, while the rate constant for the cleavage is about 1 min^{-1} (Hertel and Uhlenbeck, 1995; Hertel et al., 1994; Stage-Zimmermann and Uhlenbeck, 1998). It was proposed that the strong preference for the cleavage reaction is caused by the flexibility of complex of hammerhead ribozyme with two ligation substrates providing an entropic advantage for cleavage (Hertel et al., 1995). This suggestion was further proved by design of "conformationally restricted" *cis* hammerhead ribozyme, where ribozyme and ligation substrate were jointed covalently by a disulfide crosslink (Blount and Uhlenbeck, 2002; Stage-Zimmermann and Uhlenbeck, 2001). The modification made the complex more rigid, giving 50-fold increase in the ligation efficiency. Another possibility to improve the ligation activity by reducing the flexibility of ribozyme–substrates complex could be the introduction of auxiliary elements imitating loops or bulges in stem I of natural hammerhead ribozymes. It is known that tertiary interactions between these elements and loop II stabilize the active conformation of the complex formed by ribozyme and cleavage substrate, thus enhancing the cleavage activity (Burke and Greathouse, 2005; Penedo et al., 2004; Przybilski and Hammann, 2006; Weinberg and Rossi, 2005; Westholf, 2007). It can be assumed that such loop–loop interactions could as well stabilize the complex of ribozyme with ligation substrates and improve the ligation activity. Indeed, it has been shown recently that the presence of loop I in *cis* hammerhead ribozyme from satellite tobacco ringspot virus (sTRSV) increases the efficiency of RNA ligation (Nelson et al., 2005).

We supposed that during the evolution of hammerhead ribozyme, its structure changed so that the balance between cleavage and ligation shifted towards cleavage. In the present study, we examined the possibility to reconstruct the "ancestors" of contemporary hammerhead ribozymes with improved ligation activity by altering the sequence of hammerhead motif. Two variants of *trans* hammerhead ribozymes were designed which could potentially possess an improved ligation activity. In the first variant, a semi-conservative U$_7$ in the catalytic core was replaced by cytidine, adenosine or guanosine (Fig. 1a). According to (Simorre et al., 1997) these mutations would disrupt U$_4$–U$_7$ base pair in a complex of ribozyme with substrates of ligation and change its structure; the authors supposed that this conformational change can potentially be favorable for the RNA ligation. The second variant represents hammerhead ribozyme containing 7-nt bulge in stem I (**Rz-bulge-I**, Fig. 1b). Since all ribozymes designed in this work are based on sTRSV hammerhead motif, the

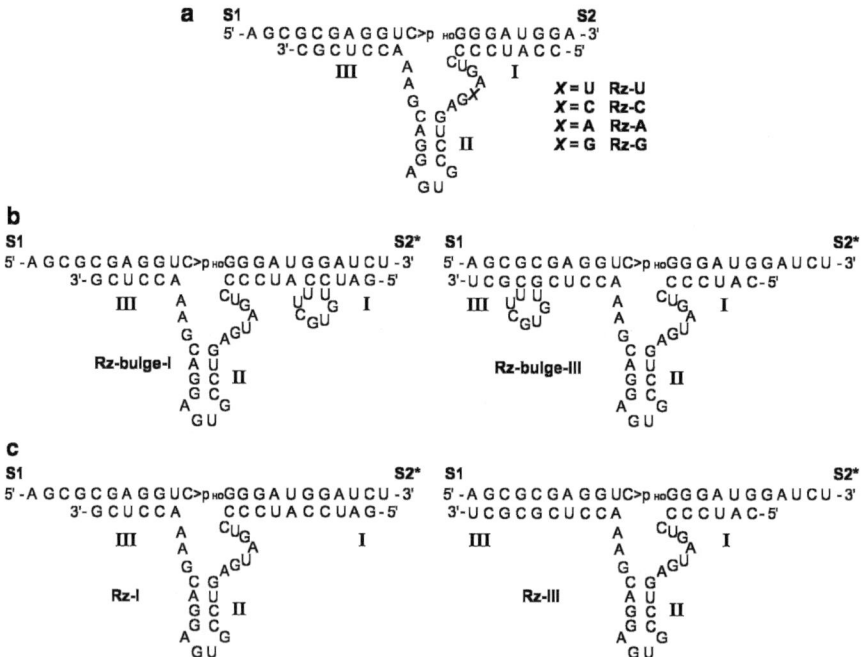

Fig. 1 The variants of *trans* hammerhead ribozymes in complexes with substrates of ligation. **a.** Hammerhead ribozymes containing single nucleotide mutation in position 7. **b.** Hammerhead ribozymes containing 7-nucleotide bulge in stem I (**Rz-bulge-I**) or in stem III (**Rz-bulge-III**). **c.** Asymmetric hammerhead ribozymes with lengthened stem I (**Rz-I**) or stem III (**Rz-III**): S1, S2, S2*—substrates of ligation

sequence of bulge corresponds to loop I in natural sTRSV ribozyme. We assumed that the bulge would increase the ligation efficiency, in analogy with *cis* ribozymes (Nelson et al., 2005). We also wonder if the same bulge introduced in stem III could affect the ligation activity of ribozyme (**Rz-bulge-III**, Fig. 1b).

All ribozymes were tested on their ability to ligate two oligoribonucleotides: 5′-substrate, bearing 2′,3′-cyclophosphate (**S1**) and 3′-substrate with 3′-OH group (**S2** or **S2***). Among ribozymes containing point mutation in position 7, only ribozyme **Rz-C** ligated RNA more efficiently than the control non-modified ribozyme **Rz-U**; **Rz-A** demonstrated the same efficiency as the control ribozyme, and **Rz-G** was less efficient (Fig. 2). One can suggest that the $U_7 \rightarrow G_7$ mutation disrupts strongly the structure of the catalytic core. Ribozyme **Rz-bulge-III** demonstrated the highest ligation efficiency: the fraction of product at equilibrium was 4 vs. 0.8 % for **Rz-U**. Cleavage constants were also determined for the designed ribozymes (Table 1). It was shown that introduction of purine residues in position 7 (**Rz-A**, **Rz-G**) decreases significantly the cleavage constant, while replacement of U_7 by cytidine results in the same cleavage rate. Interestingly, these data are not consistent with the results of mutagenesis in

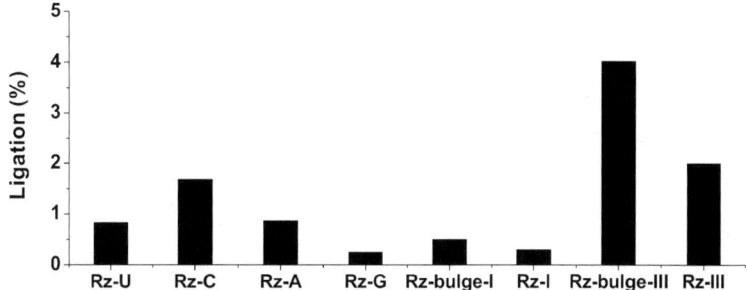

Fig. 2 The equilibrium extent of RNA ligation by different variants of hammerhead ribozymes. Reaction conditions: 15°C, 50 mM Tris-HCl, pH 7.5, 10 mM MgCl$_2$, 5 µM ribozyme, 50 nM [^{32}P]-labeled **S1**, 5 µM **S2** or **S2***

position 7 obtained by Ruffner et al. (1990): ribozymes containing purines in position 7 were more active than their C$_7$-containing analog. The rate constants for ligation were calculated from cleavage constants and equilibrium amount of the ligated product (S_{eq}), using Eq. 1 (Hertel et al., 1994):

$$K_{eq} = k_{cleav}/k_{lig} = 1/S_{eq} \tag{1}$$

Among ribozymes with mutations in position 7, A$_7$ and G$_7$-containing ribozymes demonstrated a lower ligation rate than control **Rz-U**. Since their cleavage activity is also lower, one can conclude that purine residues in position 7 cause a significant distortion of the ribozyme structure. Ribozyme **Rz-bulge-I**

Table 1 RNA cleavage and ligation parameters for hammerhead ribozymes designed in this work

Ribozyme	k_{cleav}, (min^{-1})	S_{eq}	K_{eq}	k_{lig}, (min^{-1})
RzU	5.0	0.0083	120.5	0.042
RzC	5.0	0.0168	59.5	0.084
RzA	1.8	0.0087	114.9	0.016
RzG	1	0.0025	400	0.0025
Rz-bulge-I	0.3	0.005	200	0.0015
Rz-I	0.15	0.0030	333.3	0.0005
Rz-bulge-III	11.1	0.0402	24.9	0.45
Rz-III	6.1	0.0200	50	0.12

Cleavage conditions: 15°C, 50 mM Tris-HCl, pH 7.5, 10 mM MgCl$_2$, 10 nM [^{32}P]-labeled cleavage substrates, 10 µM ribozyme. Ligation conditions: 15°C, 50 mM Tris-HCl, pH 7.5, 10 mM MgCl$_2$, 5 °M ribozyme, 50 nM [^{32}P]-labeled **S1**, 5 µM **S2**. S_{eq} – fraction of ligation product at equilibrium, K_{eq} – equilibrium constant for the cleavage-ligation reaction. The relative experimental error was ≤20%.
Cleavage substrates: 5'-AGCGCGAGGUCGGGAUGGA (for **Rz-U, Rz-C, Rz-A,** and **Rz-G**) or 5'-AGCGCGAGGUCGGGAUGGAUCU (for **Rz-bulge-I, Rz-I, Rz-bulge-III, and Rz-III**). Ligation substrates are presented at Fig. 1.

with loop I-imitating bulge was found to be less efficient than control ribozyme **Rz-U** both in ligation and in cleavage reactions. This unexpected loss of activity can be attributed to the asymmetry of its substrate-binding domains. The bulge has to be placed six nucleotides apart form catalytic core to resemble the wild-type structure, therefore the overall length of stem I in **Rz-bulge-I** is 10 base pairs. According to (Hendry and McCall, 1996), lengthening of stem I can slow the exchange between the active and inactive conformations of the ribozyme–substrate complex and inhibit the cleavage. We assume that 10 bp stem I of **Rz-bulge-I** turned out to be long enough to produce the same effect. It is noteworthy that **Rz-bulge-III**, where the same bulge is placed in stem III, and stem I contains only 6 bp, displayed the maximal cleavage and ligation efficiency. To estimate directly the influence of the flanking arms asymmetry on catalytic properties, two control asymmetric ribozymes were designed, **Rz-I** and **Rz-III** (Fig. 2c) with the same asymmetry of substrate-binding domains as in **Rz-bulge-I** and **Rz-bulge-III**, respectively. Ribozyme **Rz-I** with lengthened stem I displayed lower catalytic activity both in ligation and in cleavage reactions than its bulge-containing analog **Rz-bulge-I**. These results clearly evidence that the loss of activity is caused solely by the lengthening of stem I, and the introduction of 7-nt bulge into stem I, though positively influence catalytic activity, cannot entirely compensate this effect. On the other hand, the catalytic activity of ribozyme **Rz-III** with lengthened stem III was closer to that of control **Rz-U**, and significantly lower as compared to **Rz-bulge-III**. Thus, 7-nt bulge in stem III is responsible for the increase of cleavage activity.

It was also of interest to determine the nature of phosphodiester bond formed upon ligation. The ligation product was gel-purified and hydrolyzed by nuclease P1 which is known to cleave only $3'-5'$ bonds. If linkage between G and C is $2'-5'$, no cleavage would proceed at this site. It was shown that all linkages in oligoribonucleotide obtained by ligation were hydrolyzed by P1 nuclease, thus we concluded that natural $3'-5'$ bond is a product of ligation.

Thus, ligation activity of *trans* hammerhead ribozymes can be improved by changing the nucleotide sequence in non-conservative positions. However, equilibrium for the hammerhead reaction is still shifted towards cleavage.

3 Binary Hammerhead Ribozymes

It can be assumed that in the early RNA world, catalytic RNAs were represented by multi-subunit complexes made of several short oligonucleotides. We have shown that hammerhead ribozyme can be divided into two separate partially complementary oligonucleotides capable to form catalytically active structure without tetraloop II after binding to RNA substrate (Fig. 3) (Kuznetsova et al., 2003, 2004; Vorobjeva et al., 2005a,b; Vorobjeva et al., 2006).

Under single turnover conditions this binary ribozyme cleaves RNA with the same efficiency as full-length analog. It is notable that under multiple turnover

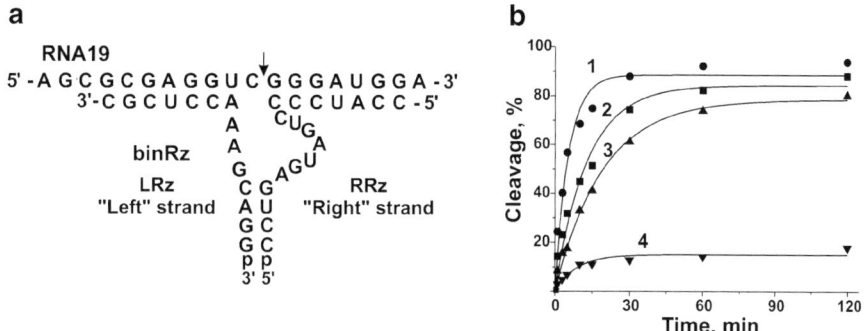

Fig. 3 RNA cleavage by binary hammerhead ribozyme **binRz** and its full-length analog Rz-U under multiple turnover conditions. Kinetic curves are numbered as follows: 1: [**RRz**] = 0.1 μM, [**LRz**] = 0.01 μM (circles); 2: [**RRz**] = 0.01 μM, [**LRz**] = 0.1 μM (squares); 3: [**Rz-U**] = 0.01 μM (triangles); 4: [**RRz**] = [**LRz**] = 0.01 μM (reverse triangles). Reaction conditions: 37°C; 50 mM Tris-HCl, pH 7.5, 10 mM MgCl$_2$, 0.1 μM [^{32}P]-labeled **RNA19**

conditions catalytic activity of binary ribozyme can be altered by varying the ratio of ribozyme chains, and an increase in the concentration of any strand resulted in a significant increase in the cleavage efficiency, at that binary hammerhead ribozyme becomes more effective than full-length one (Fig. 3b). These properties of binary ribozyme could be attributed to the mechanism of its interaction with RNA substrate. Apparently, formation of the binary ribozyme-substrate complex includes two steps: (1) binding of one strand to RNA, and (2) association of the second strand with pre-formed complex. This suggestion was proved by a gel-shift assay of ribozyme-substrate complex formation. To monitor directly the complex association, non-cleavable analog of RNA substrate was used, bearing 2'-*O*-methyl cytidine in the cleavage site (**RNA19^m**). It was shown that two-component complex of RNA and one of the strands forms first, and then the second strand joins this complex (Fig. 4).

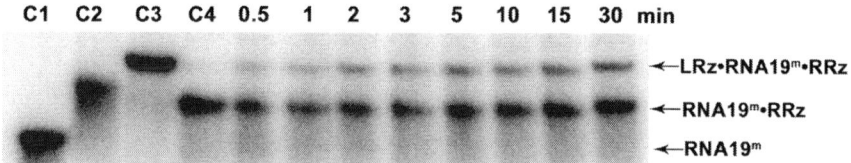

Fig. 4 Assembling of three-component complex of binary ribozyme and non-cleavable substrate **RNA19^m** (autoradiograph of non-denaturing 15% polyacrylamide gel). Complex association was monitored at 25°C under the same buffer conditions as those for ribozyme cleavage (50 mM Tris-HCl, pH 7.5, 10 mM MgCl$_2$), at 10 nM **LRz**, 50 nM **RRz**, and 50 nM [^{32}P]-**RNA19^m**. Controls: C1—**RNA19^m**, C2—**LRz·RNA19^m** complex, C3—**LRz·RNA19^m·RRz** complex, C4—**RRz·RNA19^m** complex. Lanes 0.5,..., 15—non-cleavable RNA substrate incubated with **LRz** and **RRz** for the indicated time

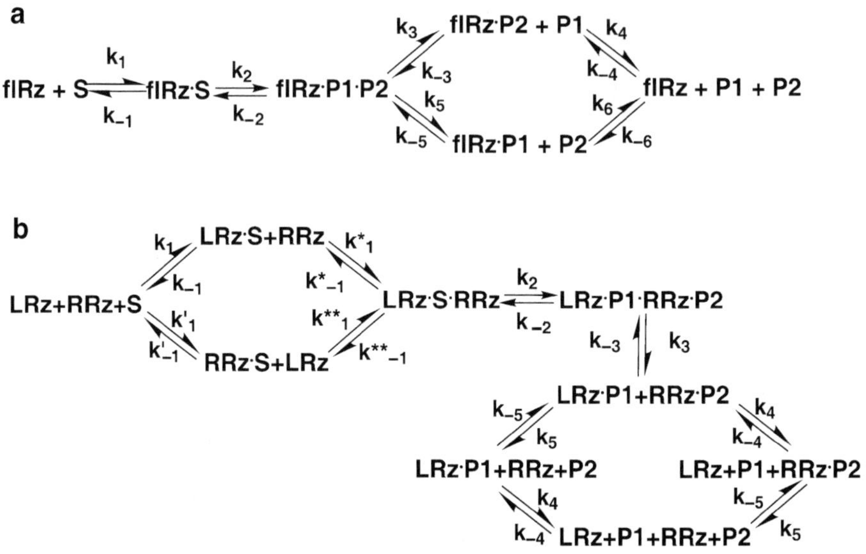

a

$$flRz + S \underset{k_{-1}}{\overset{k_1}{\rightleftarrows}} flRz \cdot S \underset{k_{-2}}{\overset{k_2}{\rightleftarrows}} flRz \cdot P1 \cdot P2$$

with branches:
- $\overset{k_3}{\underset{k_{-3}}{}}$ flRz·P2 + P1 $\overset{k_4}{\underset{k_{-4}}{}}$ flRz + P1 + P2
- $\overset{k_5}{\underset{k_{-5}}{}}$ flRz·P1 + P2 $\overset{k_6}{\underset{k_{-6}}{}}$ flRz + P1 + P2

b

$$LRz + RRz + S \rightleftarrows \begin{array}{c} LRz \cdot S + RRz \\ RRz \cdot S + LRz \end{array} \rightleftarrows LRz \cdot S \cdot RRz \underset{k_{-2}}{\overset{k_2}{\rightleftarrows}} LRz \cdot P1 \cdot RRz \cdot P2$$

with rate constants $k_1, k_{-1}, k^*_1, k^*_{-1}, k'_1, k'_{-1}, k^{**}_1, k^{**}_{-1}$

$\underset{k_{-3}}{\overset{k_3}{\updownarrow}}$

branches to:
- $\overset{k_{-5}}{\underset{k_5}{}}$ LRz·P1 + RRz·P2 $\overset{k_4}{\underset{k_{-4}}{}}$ LRz + P1 + RRz·P2
- LRz·P1 + RRz + P2 $\overset{k_5}{\underset{k_{-5}}{}}$
- $\overset{k_4}{\underset{k_{-4}}{}}$ LRz + P1 + RRz + P2 $\overset{k_5}{}$

Fig. 5 Kinetic schemes of RNA cleavage by full-length (**a**) and binary (**b**) hammerhead ribozymes. Abbreviations: **flRz**, full-length ribozyme; **LRz**, "left" strand of binary ribozyme; **RRz**, "right" strand of binary ribozyme; **S**, RNA substrate; **P1**, 5′-cleavage product; **P2**, 3′-cleavage product

Thus, different catalytic properties of binary and full-length ribozymes are caused by distinctive features of their interaction with RNA substrate and cleavage products. To reveal these differences and to investigate in detail the properties of binary catalytic RNA, we compared catalytic cycles of binary and full-length ribozymes.

To propose a catalytic cycle of RNA cleavage by binary ribozyme, we based on a kinetic scheme for full-length hammerhead ribozyme (Fedor and Uhlenbeck, 1992; Hertel and Uhlenbeck, 1995, Hertel et al., 1994), including substrate binding, cleavage and release of both products by different ways as well as reverse reactions (Fig. 5a). Binary ribozyme is built of two separate oligonucleotides, and ribozyme-substrate association, as we discussed above, is a two-stage process. Products dissociation step also seems to be different in catalytic cycles of full-length and binary ribozymes. After the phosphodiester bond cleavage, some of non-canonical base pairs between nucleotides of catalytic core that stabilize stem II in ribozyme–substrate complex are loss (Simorre et al., 1997), and duplex II, which is less stable than 7-bp duplexes I and III, dissociates in the first instance. After that two complexes of ribozyme strands with correspondent cleavage products would dissociate independently. Based on these considerations, we proposed a scheme for the catalytic cycle of binary ribozyme (Fig. 5b). We have shown previously that the rate constants for phosphodiester bond cleavage (k_2) are similar for binary and full-length

Table 2 Rate constants for the key stages of catalytic cycles of binary and full-length hammerhead ribozymes

Ribozyme–substrate association	Association rate constant ($\mu M^{-1} \cdot min^{-1}$)
LRz + RNA19^m	28 ± 3
RRz + RNA19^m	17 ± 1
LRz·RNA19^m + RRz	34 ± 7
RRz·RNA19^m + LRz	44 ± 7
Rz–U + RNA19^m	36 ± 8
Dissociation of ribozyme–substrate complex	Dissociation rate constant (min^{-1})
Rz–U·RNA19^m → Rz–U + RNA19^m	0.012 ± 0.002
LRz·RNA19^m·RRz → LRz + RNA19^m + RRz	0.020 ± 0.003
Products dissociation	Dissociation rate constant (min^{-1})
Rz-U·P2·P1 → Rz-U·P2 + P1	0.20 ± 0.04
Rz-U·P1 → Rz-U + P1	0.18 ± 0.03
LRz·P1 → LRz + P1	–
Rz-U·P2·P1 → Rz-U·P1 + P2	0.016 ± 0.005
Rz-U·P2 → Rz-U + P2	0.020 ± 0.001
RRz·P2 → RRz + P2	0.05 ± 0.005

^{a} Dissociation is so fast that LRz·P1 complex is undetectable even after 10 s of incubation. All reactions were performed at 25°C, in a buffer containing 50 mM Tris-HCl, pH 7.5, and 10 mM MgCl$_2$.

ribozymes (21.5 and 15.9 min^{-1} at 25°C, respectively) (Kuznetsova et al., 2004; Vorobjeva et al., 2006). Thus, to distinguish the differences between catalytic cycles of full-length and binary ribozymes, association and dissociation stages were studied (Table 2).

Association rate constants for assembling of ribozyme–substrate complexes were found to be of the same order of magnitude for full-length and binary ribozymes, and similar to constants measured in (Hertel et al., 1994) for full-length hammerhead ribozymes. It is worth noting that the binding of one ribozyme strand to an RNA substrate accelerates the binding of another one: twofold increase in the association rate constant is observed for both **RRz** and **LRz**. At that the "left" strand of binary ribozyme binds the substrate and the pre-formed complex faster than the "right" strand.

To estimate the rate of dissociation of ribozyme–RNA complexes, a large excess of unlabeled substrate was added to a pre-formed complex of labeled substrate and ribozyme, and the rates of displacement by non-labeled RNA were monitored.

It was shown that the complex of full-length ribozyme **Rz-U** with RNA is two times as stable as the complex of binary ribozyme (Table 2). However, after the formation of ribozyme–substrate complex two reactions are possible: the reverse dissociation reaction or the cleavage of phosphodiester bond. As it was mentioned above, the rate of cleavage for binary ribozyme is 21.5 min^{-1}, i.e., three orders of magnitude higher than the rate of dissociation. So, a decrease in the complex stability in the case of binary ribozyme would not influence significantly the efficiency of RNA cleavage.

Another characteristic distinction in catalytic cycles of binary and full-length ribozymes is the step of product release. Under multiple turnover conditions, this stage is often rate-limiting and determines the number of ribozyme turnovers and the efficiency of RNA cleavage. In the case of binary ribozyme, complex with two cleavage products was not observed even at large excess of ribozyme components, so we were able to measure only the rates of dissociation of each product from the corresponding strand of binary ribozyme. Both cleavage products dissociated from the components of binary ribozyme markedly faster than from full-length analog: the rate of dissociation of 3′-product **P2** from the "right" strand is twofold higher than from complexes with full-length ribozyme **Rz-U·P1·P2** and **Rz-U·P2** (Table 2). Moreover, 5′-product **P1** released from the "left" strand of binary ribozyme so quickly that we could not measure the dissociation rate: **LRz·P1** complex was not observed after 10 s of incubation. Under the same conditions **Rz-U·P1** complex was detected even after 5 min of incubation. Taken together, these results clearly indicate that the rate of release of the 5′-product is much higher in the case of binary ribozyme.

As it can be concluded from the kinetic description of catalytic cycles, the assembling of binary ribozyme from two oligomers on RNA substrate proceeds as fast as ribozyme-substrate complex formation in the case of its full-length analog. At the same time, the binary ribozyme is characterized by a faster release of cleavage products.

Thus, binary hammerhead ribozyme possesses an improved ratio of association/dissociation rates in the catalytic cycle, as compared to its full-length analog. Obtained results are in a good agreement with RNA cleavage experiments under multiple turnover conditions (Fig. 3b). The use of the "left" strand as deficient component at equal concentrations of "right" strand and substrate results in the maximal cleavage efficiency, which is significantly higher than efficiency of the full-length ribozyme. As it can be concluded from our kinetic studies, this effect is provided by several factors: (1) fast rate of association of **LRz** with pre-formed complex **RRz·S**; (2) independent dissociation of cleavage products from the components of binary ribozyme; (3) fast release of the "left" ribozyme strand from the complex with cleavage product.

Binary hammerhead ribozyme can be considered as the simplest RNA endonuclease with high efficiency and specificity. We hypothesize that at the early stages of RNA world catalytic RNAs could emerge not only from spontaneous RNA recombinations, but also from occasional assembling of short oligoribonucleotides into catalytically active complexes.

4 Conclusion

To conclude, hammerhead ribozymes can be considered as "molecular fossils" of RNA world because of their small size, structural simplicity and high catalytic activity, as well as the fact that hammerhead motifs were discovered in highly divergent organisms. The data described above shows that ligation activity of *trans* hammerhead ribozymes can be significantly enhanced by altering the nucleotide sequence of non-conservative regions, at that RNA cleavage is the main catalytic activity of these ribozymes. It is likely that both activities of "ancient" hammerhead ribozymes were employed in early RNA world, providing the extension of RNA molecules and recombination of their fragments. It was demonstrated that highly active hammerhead-like structure can assemble from separate catalytically inactive oligonucleotides. Detailed kinetic comparison of binary and full-length ribozymes revealed that binary ribozymes possess better ratio of association/dissociation rates in catalytic cycle. It can be assumed that such self-assembling multi-subunit catalytic RNAs could exist at early stages of pre-biological evolution, acting as specific and efficient RNA endonucleases. Since both ribozyme components were in close proximity, bound to the same RNA template, they might be ligated by some other ribozymes, with the formation of joint full-length hammerhead ribozyme. It is worth noting that binary hammerhead ribozyme can catalyze the RNA cleavage, but not RNA ligation, because the formation of four-component complex between two ribozyme components and two ligation substrates is almost impossible. In contrast, full-length hammerhead ribozymes are able to catalyze both cleavage and ligation reactions, and this extended functionality could be an advantage over the binary analogs, that provided the transition from binary to full-length ribozymes during the natural selection.

Acknowledgment This work was supported by RFBR (project No. 05-04-48341), RAS Program "Origin of Life and Evolution of the Biosphere", and Lavrentiev Competition of Young Scientists' projects of SB RAS.

References

Amarzguioui, M. and Prydz, H. (1998) Hammerhead ribozyme design and application. Cell. Mol. Life Sci. 54, 1175–1202.
Blount, K.F. and Uhlenbeck, O.C. (2002) Internal equilibrium of the hammerhead ribozyme is altered by the length of certain covalent cross-links. Biochemistry 41, 6834–6841.
Burke, D.H. and Greathouse, S.T. (2005). Low-magnesium, *trans*-cleavage activity by type III, tertiary stabilized hammerhead ribozymes with stem I discontinuities. BMC Biochemistry 6, 14.
Cech, T.R., Zaug, A.J. and Grabowski, P.J. (1981). *In vitro* splicing of the ribosomal RNA precursor of *Tetrahymena*: involvement of a guanosine nucleotide in the excision of the intervening sequence. Cell 27, 487–496.
Chen, X., Li, N. and Ellington, A.D. (2007) Ribozyme catalysis of metabolism in the RNA world. Chem. Biodivers. 4, 633–655.
Chetverina, H.V., Demidenko, A.A., Ugarov, V.I. and Chetverin, A.B. (1999) Spontaneous rearrangements in RNA sequences. FEBS Lett. 450, 89–94.
Citti, L. and Rainaldi, G. (2005) Synthetic hammerhead ribozymes as therapeutic tools to control disease genes. Curr. Gene Ther. 5, 11–24.

Conaty, J., Hendry, P. and Lockett, T.J. (1999) Selected classes of minimised hammerhead ribozymes have very high cleavage rates at low Mg^{2+} concentration. Nucleic Acids Res. 27, 2400–2407.
Fedor, M.J. and Uhlenbeck, O.C. (1992) Kinetics of intermolecular cleavage by hammerhead ribozymes. Biochemistry 31, 12042–12054.
Grassi, G., Dawson, P., Guarneri, G., Kandolf, R. and Grassi, M. (2004) Therapeutic potential of hammerhead ribozymes in the treatment of hyper-proliferative diseases. Curr. Pharm. Biotechnol. 5, 369–386.
Guerrier-Takada, C., Gardiner, K., Marsh, T., Pace, N. and Altman, S. (1983) The RNA moiety of ribonuclease P is the catalytic subunit of the enzyme. Cell 35, 849–857.
Hendry, P. and McCall, M. (1996) Unexpected anisotropy in substrate cleavage rates by asymmetric hammerhead ribozymes. Nucleic Acids Res. 24, 2679–2684.
Hertel, K.J. and Uhlenbeck, O.C. (1995) The internal equilibrium of the hammerhead ribozyme reaction. Biochemistry 34, 1744–1749.
Hertel, K.J., Herschlag, D. and Uhlenbeck, O.C. (1994) A kinetic and thermodynamic framework for the hammerhead ribozyme reaction. Biochemistry 33, 3374–3385.
Joyce, G.F. (2002) The antiquity of RNA-based evolution. Nature 418, 214–221.
Kuznetsova, M., Fokina, A., Lukin, M., Repkova, M., Venyaminova, A. and Vlassov, V. (2003) Catalytic DNA and RNA for targeting *MDR1* mRNA. Nucleosides Nucleotides Nucleic Acids 23, 1521–1523.
Kuznetsova, M., Novopashina, D., Repkova, M., Venyaminova, A. and Vlassov, V. (2004) Binary hammerhead ribozymes with high cleavage activity. Nucleosides Nucleotides Nucleic Acids 24, 1037–1043.
Muller, U.F. (2006) Re-creating an RNA world. Cell. Mol. Life Sci. 63, 1278–1293.
Nelson, J.A., Shepotinovskaya, I. and Uhlenbeck, O.C. (2005) Hammerheads derived from sTRSV show enhanced cleavage and ligation rate constants. Biochemistry 44, 14577-14585.
Penedo, J.C., Wilson, T.J., Jayasena, S.D., Khvorova, A. and Lilley, D.M. (2004) Folding of the natural hammerhead ribozyme is enhanced by interaction of auxiliary elements. RNA. 10, 880–888.
Peracchi, A. (2004) Prospects for antiviral ribozymes and deoxyribozymes. Rev. Med. Virol. 14, 47–64.
Przybilski, R. and Hammann, C. (2006) The hammerhead ribozyme structure brought in line. ChemBioChem 7, 1641–1644.
Ruffner, D.E., Stormo, G.D. and Uhlenbeck, O.C. (1990) Sequence requirements of the hammerhead RNA self-cleavage reaction. Biochemistry 29, 10695–10702.
Salehi-Ashtiani, K. and Szostak, J.W. (2001) *In vitro* evolution suggests multiple origins for the hammerhead ribozyme. Nature 414, 82–84.
Schubert, S. and Kurreck, J. (2004) Ribozyme- and deoxyribozyme-strategies for medical applications. Curr. Drug Targets 5, 667–681.
Simorre, J.P., Legault, P., Hangar, A.B., Michiels, P. and Pardi, A. (1997) A conformational change in the catalytic core of the hammerhead ribozyme upon cleavage of an RNA substrate. Biochemistry 36, 518–525.
Spirin, A.S. (2002) Omnipotent RNA. FEBS Lett. 530, 4–8.
Stage-Zimmermann, T.K. and Uhlenbeck, O.C. (1998) Hammerhead ribozymes kinetics. RNA 4, 875–889.
Stage-Zimmermann, T.K. and Uhlenbeck, O.C. (2001) A covalent crosslink converts the hammerhead ribozyme from a ribonuclease to an RNA ligase. Nat. Struct. Biol. 8, 863–867.
Szatmary, E. (1999) The origin of the genetic code. Amino acids as cofactors in an RNA world. Trends Genet. 15, 223–229.
Tanner, N.K. (1999) Ribozymes: the characteristics and properties of catalytic RNAs. FEMS Microbiol Rev. 23, 257–275.

Vorobjeva, M., Gusseva, E., Repkova, M., Kovalev, N., Zenkova, M., Venyaminova, A. and Vlassov, V. (2005a) Modified binary hammerhead ribozymes with high catalytic activity. Nucleosides Nucleotides Nucleic Acids 25, 1105–1109.

Vorobjeva, M.A., Gusseva, E.V., Venyaminova, A.G. and Vlassov, V.V. (2005b) Binary catalytic RNA. International Workshop "Biosphere Origin and Evolution". 74–75. Abstracts.

Vorobjeva, M., Zenkova, M., Venyaminova, A. and Vlassov, V. (2006) Binary hammerhead ribozymes with improved catalytic activity. Oligonucleotides 16, 239–252.

Weinberg, M.S. and Rossi, J.J. (2005) Comparative single-turnover kinetic analyses of trans-cleaving hammerhead ribozymes with naturally derived non-conserved sequence motifs. FEBS Lett. 579, 1619–1624.

Westholf, E. (2007) A tale in molecular recognition: the hammerhead ribozyme. J. Mol. Recogn. 20, 1–3.

Paradoxical Bistate Status of a Prebiotic Microsystem: Universal Predecessor of Life

V. N. Kompanichenko

Abstract A prebiotic organic microsystem has a chance to be transformed into a simplest living unit only through the intermediate stage—acquisition of the specific *bistate* status. The status is a result of uncompleted bifurcate transition of a chemical system from the initial state into advanced state under oscillating nonequilibrium conditions. In the case of the balance between the tendencies to the forward and reverse transitions, the system oscillates around the highest point of bifurcation displacing to the initial and advanced states by turns. It acquires the paradoxical way of organization—"stabilized instability." The remarkable characteristic of such organization is that the principally unstable point of bifurcation is "incorporated" between two opposite but equal forces. This way of organization allows the prebiotic microsystem to maintain the following properties, which are at the foundation of life: incessant inner fluctuations and re-arrangement of molecules; integrity through cooperative events; exchange by matter and energy with the environment; latent biforked structure consisting of two interrelated co-structures; natural dichotomy at the end of cycle of the existence, etc. The aim of the suggesting experimental research is to obtain bistate prebiotic microsystems, which are able to evolve to life. During the experiments various prebiotic models should be explored at the state of bifurcate transition and under oscillating conditions in experimental chamber.

1 Introduction

Since the classical works by A. Oparin, various organic microsystems are considered as possible prebiotic models: coacervates (A. Oparin), RNA-World macromolecules (D. Gilbert, G. Joice), liposomes (D. Deamer, P. Luisi), proteinoide microspheres (S. Fox, K. Dose), marigranules (M. Ventilla,

---

V. N. Kompanichenko
Institute for Complex Analysis at Birobidzhan (Russia)
e-mail: kompanv@yandex.ru
University of California at Santa Cruz, Department of Chemistry and Biochemistry (USA)
e-mail: vladk@soe.ucsc.edu

F. Egami), aromatic World molecules (P. Ehrenfreund), etc. These microsystems demonstrate some signs of the internal and external activity that are relevant for living processes. For instance, RNA-World molecules are able to self-replicate, proteinoides possess catalytic activity, coacervates may grow and selectively assimilate substance. However, these initial signs of activity in the microsystems are not converted into the corresponding self-maintaining processes (metabolism, natural self-replication), which are characteristic of life. We have to conclude that there exists is a principal gap separating the prebiotic models and living cells.

The gap between non-living prebiotic microsystems and living microorganisms shows us that they are fundamentally different types of natural systems. If we comprehend the systemic difference between prebiotic microsystems and cells, we will get a good chance to understand a nature of the gap and outline the way to life. This thesis was a starting-point for the elaborated initial version of the systemic approach to the origin of life (Kompanichenko, 2004). It is aimed to comparison of biological and non-biological types of natural systems basing on their fundamental properties, i.e. in fact comparison of the biological and non-biological properties. The major result of the initial version is that any kind of prebiotic microsystems has a chance to be transformed into a simplest living unit only through acquiring the *bistate status* that is an intermediate stage between non-life and life. So, according to the author's conception the general way to life lies through initial transformation of a prebiotic organic microsystem (its kind should be discussed) into bistate prebiotic microsystem, and then the last one into simplest living unit (probiont). The article is devoted to advanced consideration of the intermediate stage of the origin of life process—a prebiotic microsystem having bistate status.

2 Fundamental Properties of Biological Systems

About 230 properties (features) of biological systems were distinguished by several tens competent scientists around the whole world in the book "Fundamentals of Life" (2002). The book is the latest summarizing in this field of fundamental biology. It contains 78 short definitions of life and 25 selected fundamental papers. More or less definite properties of life are substantiated in 64 short definitions and papers. The properties suggested by the different authors are often identical or similar to each other. According to the carried out generalization, 31 fundamental properties of biological systems can be compiled on the basis of this data (Kompanichenko, 2003, 2004). Only 19 of them can be considered as the unique fundamental properties of biosystems, which are not peculiar to any other natural system. The remaining 12 are attributed to the non-unique fundamental properties. These or similar features sometimes display themselves in few non-biological systems, although they are devoid of any biological specificity.

The unique fundamental properties can be united into four key, or integrated properties, which are expressed an essence of the biological organization. They are the following.

1. The ability of a living organism to concentrate free energy and information by means of extraction from the environment.
2. The ability for the intensified counteraction to external influences.
3. The ability for expedient behavior, or the expedient character of interaction with the environment.
4. Regular self-renovation at different levels, including renovation of cellular structures, self-reproduction of organisms, and evolution of species.

In fact, the unique fundamental properties emphasize the strict barrier separating animate and inanimate parts of nature, while the non-unique ones can be considered as the connecting thread between them. During the origin of life process on the Earth the unique properties for the first time appeared, unlike the non-unique ones that transited from the maternal geological medium and acquired the biological specificity. Therefore, the non-unique properties may serve as a "load-star" to characterize conditions in the geological Cradle of Life, especially in immediate temporal proximity to the emergence of the simplest living units. From this point of view, the following three non-unique fundamental properties of biological systems are the most peculiar: (1) thermodynamic and chemical nonequilibrium; (2) integrity through cooperative events (auto-organization of molecules, emergent properties); (3) capability for self-organization (self-maintenance). These properties were distinguished as the most fundamental in biological systems by 17 scientists. They indicate specific nonequilibrium processes that were characteristic for the geological cradle of life on the Earth. The necessity of nonequilibrium conditions is remarkable. This requirement relates the origin of life process with the wide range of events investigating within the framework of irreversible thermodynamics, including first of all the theory of dissipative structures (Nicolis and Prigogine 1977; Prigogine, 2003; Prigogine and Stengers 1984) and synergetics (Haken, 1978, 2003).

3 Stabilized Bifurcation as a Starting Point of Life

There exist two different mechanisms of evolution of natural systems. The first one consists in gradual accumulation of small changes in a system. Thus, importance of this mechanism for biological evolution was emphasized by C. Darwin. The second revolutionary mechanism leads to radical transformation in a system's structure. The mechanism acts through the bifurcate transition of a system. A bifurcation occurs when a system cannot develop further in the current conditions. The general scheme of a bifurcate transition is the following (Fig. 1): initial period of stable existence of a system—pre-bifurcate period (its structure become unstable due to rising fluctuations)—critical, or bifurcation point (highest transformation of a system, reorganization of the

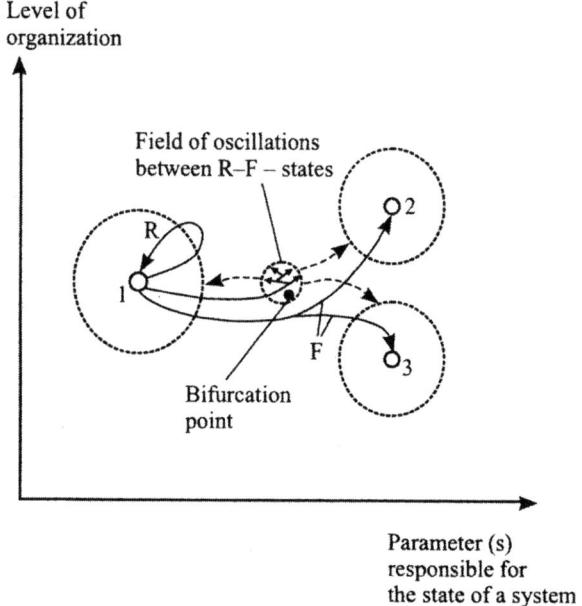

Fig. 1 Principal scheme of the appearance of a bistate system simultaneously existing in the two opposite states. 3—field of the stable prebifarcate (recurrent) state of a system; 2,3—fields of the stable postbifitrcate (forward) state of a system. The *arrows* with a *continuous line* indicate: R—a schematic way of the recurrent transition of a system into the prebifurcate state 1; F—schematic ways of the forward transition of a system into the postbifitrcate states 2–3. The arrows with the *dotted line* indicate a potential (force) striving of a system to move into one of these states. The *circle* around the bifurcation point represent the field where the tendencies for the transition to recurrent and forward transitions are in balance

structure and choose new way of development)—the post-bifurcate period (fading away of fluctuations)—the nest period of stable development. After the bifurcation a system may follow plenty of permissible ways of development, but they can be united into three principal trends: full destruction; simplification (degradation); complication through self-organization.

Around the bifurcation point a chemical system possesses some specific properties, which are not peculiar to it during a period of stable existence. Let us pay attention to striking analogies between these bifurcate or critical properties and the regularities that are at the foundation of biological organization.

3.1 Main Properties of a Chemical System Close to the Bifurcation Point

1. *Sharp heterogeneity with the intensive counter processes along and against the gradients.* Multitude high gradients (concentration, pressure, temperature,

electric field, and other kinds) appear throughout in a chemical system that approaches to the critical point of transition.
2. *Continuous waves and re-arrangement of molecules.* The particles are organized by fluctuations through cooperative events and synergism. A competition between the fluctuations continuously changes their radiuses of correlation.
3. *Integrity through cooperative processes.* At the critical point the dominant fluctuation integrates all particles because its radius of correlation expands to the entire system. The particles are perceptive each other on macroscopic distances due to long-range links and cooperative effect.
4. *Open system: uninterrupted exchange by matter and energy with the surroundings.* Powerful fluctuations arising around the critical point open the system and maintain the exchange by matter and energy, as well as by simple structural information, with the surroundings.

3.2 The Corresponding Characteristics of a Living Cell

1. *Heterogeneous structure and the counter processes of synthesis and destruction.* Distinct heterogeneity of a cell is displayed as in ionic-molecular and as in structural levels. It is maintained by extraordinary complex circulation of the spontaneous processes (decreasing the gradients) and the non-spontaneous ones (increasing them) providing simultaneously synthesis and destruction (assimilation and dissimilation).
2. *Continuous dynamic processes and re-arrangement of molecules/atoms.* A cell is an assembly-line system in which manifold dynamic processes proceed uninterruptedly. Permanent chemical nonequilibrium between a cell and the outside world supports continuity of the processes.
3. *Integrity based on cooperative interaction.* Integrity of a cell is maintained by the organization and regulation of cellular functions on different hierarchical levels. The organization of entire cell appears in its organic process of division, coordinate movement, etc.
4. *Open system: continuous exchange of a cell by matter, energy, and information with the environment.* Existence of any organism is maintained through the various exchange processes with the outside world. An organism is highly sensitive to external changes and responds upon them on different hierarchical levels.

The analogies between these two sets of the properties are obvious. Each of four critical properties is distinctly reflected in the vital processes. Although four listed characteristics of living cells do not express the unique essence of life, they are in its background. Absence of even one of these characteristics would make life impossible. Meanwhile, the critical properties are temporal characteristic of a chemical system. They appear since the beginning of the bifurcate transformation and disappear with the completed transition to the advanced stable state. The following thesis is formulated on the basis of these facts: *Life arose of the bifurcate state* that inhered to still indeterminate kind of organic microsystems.

This thesis has led to the paradoxical situation: the bifurcate state should be considered as a starting-point of living systems, but the bifurcation point is unstable in principle. A system cannot be at the bifurcation point for a long time. But nature on the Earth found the way to prolong the bifurcate state of the initial prebiotic microsystems about four billion years ago. The permissible opportunity for this way is investigated in the next chapter.

4 Bistate System as Prototype of a Living Organism: Theoretical Substantiation

The formulated thesis brings us to the consideration of the rare boundary events arising between the irreversible and reversible bifurcate transitions. This is the case, when conditions in the outside world oscillate and constrain a chemical system to balance as well. These events have not been investigated from the origin of life point of view. The situation with the starting point of life is contradictory: close to the bifurcation point a chemical system is unstable, but maintains the critical properties, while far from the point it is stable, but loses the properties. It seems this contradiction can be resolved in only if a chemical system oscillates around the bifurcation point, developing to the advanced state and to the initial one. The balanced oscillations provide relative stability of the system, while the "caught"-inside unstable bifurcation point serves as a permanent source of dynamic transformations. There appears the paradoxical state "stabilized instability" that is a background of living systems, in accordance with the author's conception. On this basis the following thesis can be formulated as a hypothesis: the critical properties in a prebiotic microsystem can be kept for a long time *through regular oscillations around the bifurcation point*. This principal opportunity to maintain the critical properties allowed the original prebiotic microsystems to evolve to life on the early Earth. The thesis will be theoretically corroborated below. The way to its future experimental proof will be outlined as well.

The considered case is shown on Fig. 1. Let us assume that a chemical system begins to transit from the initial state 1 into the advanced states 2 or 3 through the bifurcation point. The transition is initiated by significant changes in the external parameters. There exist three possible scenarios of behavior for a system that has been obtained the impulse for transition. The first one implies the impulse is insufficient for the system to reach the bifurcation point. In this case, the system returns to the initial field 1, according to the schematic trend R. The second scenario: the impulse is superfluous (more than sufficient) to overcome the bifurcation point. In the long run, the system selects a position in the

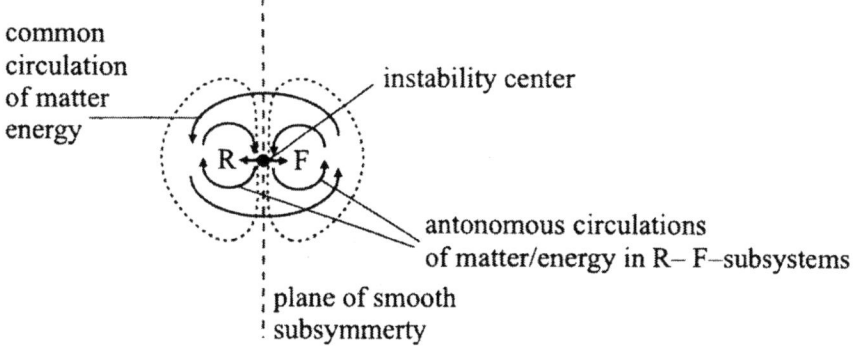

Fig. 2 Principal scheme of a bistate system

fields 2 or 3, in accordance with the schematic trends F (or goes to ruin if the change is too strong). In case the intermediate value of the impulse the system stays swinging between the opposite and equal forces for a while. This third scenario implies possible acquisition of temporal stability of the system. Displacement of the system into forward state can be compensated by the adequate reverse change of the external conditions, and visa versa. This theoretically substantiated type of natural systems was called *bistate system*, or shortly *bisystem* (Kompanichenko, 2004). Its principal scheme is shown on the Fig. 2. Both opposite initial and advanced states are nonequilibrium: they are attracted to the bifurcation point and balance each other. The bistate system begins to develop now to the initial (recurrent) now advanced (forward) states, correspondingly these nonequilibrium states become dominant by turns. The oscillating balance between the R-state (reverse force) and F-state (forward force) is an indispensable condition for stability of a bistate system, otherwise the system would leave the bifurcate area and loose its bistate status. Real stability of a bisystem supports through the cooperative events and self-organization that exist near by the bifurcation point. The bifurcation point itself is a principal point of instability of a bistate system. The instability maintains due to high fluctuations that also peculiar to bifurcate state of a system. A continuous "opposition-interaction" between the instability and the equilibrium leads to a continuous rearrangement of molecules, structures and circulative physical–chemical processes within bisystems.

Although existence of the bistate type of natural systems is based theoretically, this notion allows us to get satisfactory principal responses to the following questions concerning the biological way of organization.

1. Why a cell normally fissions at the end of its own existence? This property follows the paradoxical way of organization of a bisystem, i.e. simultaneous repulsion the nonequilibrium co-structures from the central instability point and integration through maintenance of relative equilibrium between them.

Maintenance of the relative equilibrium on the background of repulsive forces should inevitably lead to the final division into two sub-identical components. Figuratively speaking, normal cycle of existence of a bisystem represents its gradual "sawing" combined with the tendency to formation of the sub-identical dichotomous units.

2. Why advanced forms of life possess obvious plane of smooth subsymmetry dividing pair limbs, eyes, ears, cerebral hemispheres, etc? According to the elaborated conception, these characteristics are the developed consequence of the *latent biforked structure* of a bisystem. It consists of two autonomous co-structures (Fig. 2). The development of a bistate system from the central point of instability simultaneously into the opposite directions should result in its specific biforked macrostructure divided by plane of smooth subsymmetry The plane reflects the virtual zone of the relative equilibrium between the forces striving to outstretch the bisystem simultaneously "ahead and back."

3. Why various inner processes in a living organism have oscillating character, for instance, oscillations of flagellums in independent eucariotic cells or hierarchical biophysical rhythms in advanced forms of life? *Oscillating character of existence* is a principal property of a bistate system related with periodic change of the dominant states. Oscillations of all inner processes support stability of a bisystem. Ceasing of the oscillations constrains a bisystem to leave the bifurcate area with the following irreversible transition into one of the stable equilibrium states. Oscillations between the polar forces sustain the internal tension that is a source of incessant dynamic processes and intensive rearrangement of molecules in a bistate system.

4. What is a cause of mutability and heredity of life forms? According to the theory of dissipative structures and synergetics, each forward transition over the bifurcation point (from the initial into advanced state) brings some accidental changes in a chemical system. During the back transition the system may return only into the altered initial state due to the happened changes. So, the reverse transition can be considered as a way directed to conservation of the previous state of a chemical system. In this context, regular oscillations around the bifurcation point occurring in a bistate chemical system step by step work towards accumulation of the accidental changes that provide its ability to evolve. The forward transitions initiate new transformations in a bisystem (mutability), while the reverse ones retain its state (heredity).

As it was based, a bistate system may acquire relative stability in the oscillating thermodynamic and physical–chemical external conditions. Rewording, it must undergo the regular influences from the outside world. Unlike an ocean, significant thermodynamic and physical–chemical fluctuations are regularly generated in hydrothermal systems where they are generated by the contradictory interaction between rising hydrodynamic and descending lithostatic pressures. It follows from the conception that hydrothermal systems with the adjoining areas in ocean and terrestrial groundwater systems were the most suitable geological cradle of life on the Earth.

5 Conclusion

1. The results of the executed experimental efforts to transform various kinds of prebiotic microsystems into really living units were negative because of the missing intermediate step—acquisition of the bistate status. The following permissible succession of transformation of the kind is suggested:

   Prebiotic microsystem ("monostate") → Bistate microsystem
   → Living probiont

   Unlike initial prebiotic model, a living probiont possesses dynamic self-sustaining heterogeneity and responds to instability by continuous internal reorganization.
2. Experimental getting of prebiotic bistate microsystems represents an important aim of science. In order to produce a useful model of the bistate system, synthetic prebiotic organic microsystems should be explored; first, close to the bifurcation point of transition, and second, under oscillating conditions in the experimental chamber. To select the most suitable kind of prebiotic models, all of them should be explored under these conditions. In fact, this way consists in investigation of organic microsystems "on the verge of survival," with the attempt to prolong their bifurcate state by the adequate changes of conditions in the experimental system. Increase of instability in the bistate system due to spontaneous displacement into the stable forward state might become balanced with an adequate change of the external conditions in the reverse direction, and visa versa.
3. Bistate systems possess ability for evolution and complication by their own nature. This means they are able to generate motive power under certain conditions and naturally turn the biosynthesis process. So, initial probionts could be composed of very simple components, with the following complicating synthesis of the both classes of biopolymers inside the evolving probiont.
4. Traces of volcanic and associated hydrothermal events on other planets and satellites are favorable signs to suppose that life might take place there. In this context, even space bodies like Moon and Io should not be excluded out of the program of life search.

References

Fundamentals of Life (2002) G. Palyi, C. Zucci and L. Caglioti (Eds), Elsevier SAS, Paris.
Haken, H. (1978) Synergetics. Springer-Verlag, Berlin, New York.
Haken Hermann (2003) Special issue. Nonlinear Phenomena in Complex Systems. Vol. 5, N4.

Kompanichenko, V.N. (2003) Distinctive properties of biological systems: the all-round comparison with other natural systems. Front. Perspect. 12(1), 23–35.

Kompanichenko, V.N. (2004) Systemic approach to the origin of life. Front. Perspect. 13(1), 22–40.

Nicolis, G. and Prigogine, I. (1977) Self-organization in nonequilibrium systems. Wiley, New York.

Prigogine, I. and Stengers, I. (1984). Order out of chaos. Bantam, New York.

Prigogine, I. (2003) Advances in chemical physics. Wiley, New York, Vol. 127.

Part III
Archaen–Proterozoic Ecosystems: Their Interaction and Contemporary Analogues

The Ancient Anoxic Biosphere Was Not As We Know It

A. E. Fallick, V. A. Melezhik, and B. M. Simonson

Abstract A previously unreported observation is that whereas c. post-2000 Ma sedimentary rocks commonly contain microbially mediated, ^{13}C-depleted carbonate concretions, such features are largely absent in older rocks. Excluding banded iron formations, δ^{13}C of sedimentary and diagenetic carbonates of the older rocks are ∼ 0 ± 3‰ VPDB and are not characteristic of microbially recycled organic matter. We propose that c. pre-2000 Ma the biosphere operated in a different mode compared to the modern. Organic matter remineralisation was predominantly in the anoxic water column and sediment–water interface and rarely within sediments. Isotopically distinctive CO_2 and CH_4 readily escaped to atmosphere leaving no vestige as diagenetic carbonates. Around c. 2000 Ma, in response to the now oxic near-surface conditions, the anaerobic recycling microorganisms retreated deep into sediments to escape poisonous dioxygen. Redox gradients developed in the sedimentary column and abundant ^{12}C-rich carbonate concretions then formed using recycled organic carbon, as observed in the Phanerozoic world.

1 Introduction

If one were to seek unequivocal evidence for biological activity in the Earth's deep subsurface (i.e. significantly below the sediment–water interface) what, in the absence of obviously active or once-active organisms, would be demanded? Ideally, we propose, one would look for integrated petrographic/fabric, mineralogical, chemical and isotopic features which in aggregate compel us to deduce that biological agents have been active in modifying the sediment relatively soon after burial. Such features exist at the present day: they are called *concretions*, defined as a mass or nodule of mineral matter usually differing in

A.E. Fallick
Scottish Universities Environmental Research Centre, Glasgow, G75 0QF, Scotland
e-mail: T.Fallick@suerc.gla.ac.uk

chemical composition from the host rocks. Concretions of calcite, dolomite, silica, Fe- and Mn-oxide, Fe-sulphide, phosphate and gypsum are known; they may vary in size from a fraction of a centimetre to a few metres and in form from regular spheroidal and oblate nodules, through burrow-fills, to laterally extensive beds (Sellés-Martínez, 1996).

Concretions are abundant in post-Precambrian rocks and attract interest because they stand out from their host rocks. Their chemical and isotopic composition can be used for tracking the burial (diagenetic) history of sediments and subsurface microbial life. In most cases, concretions formed by diffusion from a small volume of concentrated microbial activity in sediments after they were deposited (Raiswell et al., 1993), e.g. during diagenesis (Coleman, 1993). Constituents for their growth (HCO_3^-, HS^-) are supplied via processes of microbial decomposition of organic matter (OM) deposited in sediments. At the present day, the chemical and isotopic composition of *carbonate concretions* is the result of degradation of OM proceeding via a succession of processes during sediment burial, each mediated by a specific microbial group (e.g. Irwin et al., 1977) including

$$CH_2O + O_2 \rightarrow \mathbf{CO_2} + H_2O, \tag{1}$$

$$2CH_2O + SO_4^{2-} \rightarrow \mathbf{2CO_2} + S^{2-} + 2H_2O, \tag{2}$$

$$2CH_2O \rightarrow CH_4 + \mathbf{CO_2}, \tag{3}$$

$$R.COOH \rightarrow RH + \mathbf{CO_2}, \tag{4}$$

where CH_2O is used generically for organic matter and R is an aliphatic or aromatic hydrocarbon group.

These and other reactions characterise at least four depth-related zones in the sequence of sediment burial, starting with (Eq. 1) microbial oxidation, followed by (Eq. 2) bacterial sulphate reduction, (Eq. 3) microbial fermentation and (Eq. 4) abiotic decarboxylation. Carbon dioxide ($\mathbf{CO_2}$, Eqs. 1–4) is the common product for all these processes. The CO_2 produced in the burial sequence dissolves readily in pore water, thus increasing the concentration of bicarbonate whose carbon isotopic composition is very different from that of marine reservoir bicarbonate ($0 \pm 3‰$). During burial, sediments pass successively through different zones within which OM is being altered and CO_2 produced.

Bicarbonate activities sufficient to cause carbonate super-saturation can be reached, thus causing precipitation of diagenetic carbonates in the form of cement and concretions. Specific microbial groups decomposing OM can be constrained via isotopic composition of postdepositional carbonate cement and concretions. Carbonates precipitated from zone 1 (Eq. 1) are commonly isotopically light (down to −27‰). Carbonates of similar isotopic composition form in zone Eq. 2 mediated by sulphate-reducing bacteria (SRB). In contrast, zone Eq. 3 carbonates, mediated by *fermenters*, commonly show significant enrichment in ^{13}C (up to +15‰). Moreover, methane (CH$_4$) produced in zone Eq. 3 can be oxidised to CO$_2$ with δ^{13}C down to –70‰. Therefore, in general, carbonate minerals formed in zone Eq. 3 may show a wide range in δ^{13}C (–50 to 15‰) (e.g. Watson et al., 1995). Zone Eq. 4 carbonates are marked by return to more narrowly constrained δ^{13}C (e.g. –15‰; Irwin et al., 1977). Unstable primary carbonate minerals (δ^{13}C = –3 to +3‰) can be dissolved in all the above-mentioned zones, thus adding bicarbonate with different isotopic compositions and potentially masking the above-described isotopic trends.

In the modern system, the recycling of OM produced by primary productivity is remarkably efficient. The ratio of CaCO$_3$ to organic carbon exported from the top 100 m of the open-ocean water column is calculated as 0.06 ± 0.03 on a molar basis (Sarmiento et al., 2002). This is reconciled with the ratio of around 4 for carbon sequestered long term in sediments (e.g. Schidlowski, 1987) by the estimation that only of the order of 0.1% of annual organic matter production ends up buried in surface sediment (e.g. Tissot and Welte, 1984). The proportion of buried OM which is recycled in sediments after burial has been estimated (Neruchev, 1982; Romankevich, 1977): in shelf regions the amount of OM is reduced by a factor of 2 and in the deep ocean by a factor of 4. This provides the feedstock for the abundant concretions observed in post-Precambrian rocks using organically derived, ^{12}C- and ^{13}C-rich bicarbonate (Coleman and Raiswell, 1981, 1993; Curtis et al., 1986; De Craen et al., 1999; Hennessy and Knauth, 1985; Irwin et al., 1977; Mortimer and Coleman, 1997; Mozley and Burns, 1993; Pye et al., 1990; Raiswell, 1987).

In this article, we explore trends in the temporal distribution and isotopic composition of postsedimentary carbonates and concretions in Archaean and Palaeoproterozoic rocks for tracking ancient microbial life involved in remineralisation of OM specified by reactions like Eqs. 1–3. This approach has benefited from an earlier discussion with John Hayes who independently and simultaneously explored the idea of the migration of the anaerobes deeper in sedimentary column when the terrestrial hydrosphere had passed from anoxic to oxic conditions. Although the approaches of the two groups differ in details and in the proposed cause-and-effect relationships, we all agree upon a radical change of remineralisation of OM during this transition, whose precise onset is yet to be established.

2 δ^{13}C$_{carb}$ of Early Precambrian Carbonates

Whilst the mean isotopic composition of the deep Earth (including the oceanic and continental mantle) carbon is in the range –5 to –7‰ (Deines, 1992; Kyser, 1986), the sedimentary marine carbonate record for much of the Earth's history suggests that primary $\delta^{13}C$ was within a few per mil of 0‰ (see review by Ripperdan, 2001); this is usually attributed to a relatively constant burial ratio of carbonate carbon to reduced carbon (e.g. Schidlowski, 1987). Excluding samples from banded iron formations (BIF), $\delta^{13}C$ values for sedimentary and diagenetic carbonates in the time interval 3700 to 2060 Ma with very few exceptions are near 0 ± 3‰ (Fig. 1), corroborating that diagenetic shifts of carbon isotope ratios are systematically very low for Archaean carbonates (Abell et al., 1985; Schidlowski et al., 1979; Veizer et al., 1989, 1990). Primary micrite and late sparite show similar isotopic signatures and even calcite disseminated in "black schists" (C_{org} = 4%) is depleted in ^{13}C by only 2‰ compared to carbonate minerals precipitated from coeval seawater (Beukes et al., 1990). Numerous whole-rock isotopic analyses obtained from 2700 Ma stromatolitic carbonates in the Belingwe Greenstone Belt (Abell et al., 1985) show a surprisingly tight cluster around 0‰.

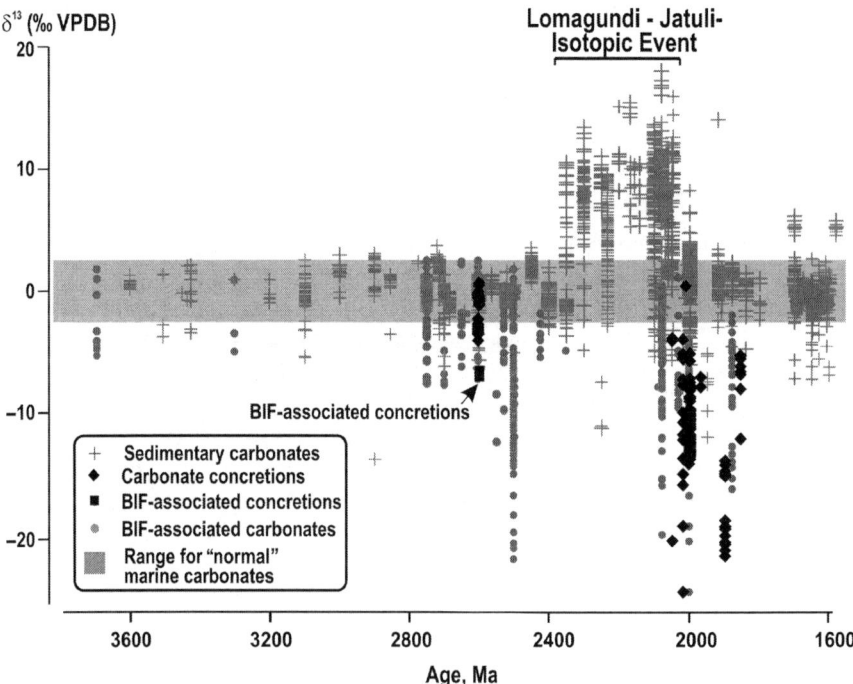

Fig. 1 Variation of carbon isotopes in sedimentary and diagenetic carbonates through the early Precambrian. Sources for isotopic data are from Melezhik and Fallick (1996), Melezhik et al. (1999) and Shields and Veizer (2002) and references therein; additional data are from Perry et al. (1973), Thode and Goodwin (1983), Abell et al. (1985), Baur et al. (1985) and Beukes et al. (1990). All databases have been screened against metamorphic overprints, non-marine and hydrothermal carbonates

3 Diagenetic Concretions in the Early Precambrian

The scarcity of diagenetic carbonate concretions in the Archaean has been acknowledged long ago (e.g. Pettijohn, 1940). A targeted search for carbonate concretions over several decades (Melezhik, 1992) and extensive carbonate concretion bibliography, listing 700 articles, (Dietrich, 1997) suggest that there is also a general lack of carbonate concretions in Early Palaeoproterozoic sedimentary formations older than 2060 Ma.

Similarly, sulphide concretions have rarely been reported from the Archaean (Krapež et al., 2003; Smyk and Watkinson, 1990). The Hamersley basin represents an exception. The concretions are widespread in the Mt. McRae and the shalier parts of the Wittenoom formations. They are best developed in the Roy Hill Shale of the Jeerinah Formation, which is older than 2630 ± 5 Ma (a SHRIMP date from a tuff right near its top; Hassler, 1993). However, even those Archaean sulphide occurrences, which have well preserved primary sedimentological features and biogenicity (Donnelly and Crick, 1992; Grassineau et al., 2002; Shen et al., 2001), very often do not show any concretional development of sulphides.

If BIF-associated concretions are excluded, there are only very few places from which carbonate concretions are known or have been reported. This includes the Boomplaas Formation constrained to c. 2642 to 2588 Ma (Altermann and Nelson, 1998; Nelson et al., 1999), in the Kaapvaal Craton, in South Africa, and the Wittenoom and Mt. McRae formations in the Pilbara Craton of Australia (Krapež et al., 2003; Simonson et al., 1993a,b) deposited roughly from 2600 to 2500 Ma (Trendall et al., 2004).

The first abundant development of diagenetic ^{13}C-depleted carbonates occur at c. 2000 Ma in the aftermath of the Jatulian-Lomagundi positive isotopic excursion of ^{13}C/^{12}C in sedimentary carbonates (reviewed in Melezhik et al., 1999) They coincide in time with the first widespread development of carbonate concretions (Fig. 1).

3.1 Isotopic Composition of Archaean Concretions

Our study involved concretions from both currently known areas. These are the Boomplaas Formation in the Kaapvaal Craton and the Wittenoom Formation in Pilbara Craton. The Boomplaas concretions are calcitic in composition and occur as flattened lenses (Color Plate 2a, see p. 394) in 1–2 m thick thinly laminated black shales. The host shales show a moderate differential compaction with respect to the concretions. The concretions exhibit a vague zoning enhanced by recent weathering. Marginal parts contain pyrite (Color Plate 2b, see p. 394). δ^{13}C ranges between −1.2 and +0.7‰. This approximates to δ^{13}C of coeval marine carbonates (Sumner, 2000). Microcored isotopic transects suggest no or very limited growth-associated δ^{13}C variations as concretionary cores are isotopically heavier by at most ca.

1‰ (Table 1, Color Plate 2b, see p. 394). Hence there is no indication that concretions incorporated a sizable amount of CO_2 derived through microbial remineralisation of OM as specified by reactions 1–3. $\delta^{34}S$ of concretion-hosted pyrite tightly clusters around zero averaging $+0.7 \pm 0.7$ (1σ) (Table 2), thus there is no clear evidence suggesting that microbial sulphate reduction was involved in the formation of these sulphides.

Samples analysed from the Wittenoom Formation include dolomite and calcite concretions and several layers of sedimentary carbonates (Table 1 and 3). Carbonate concretions are widespread in the Paraburdoo and especially Bee Gorge members of the formation (Simonson et al., 1993a). They range up to 75 cm long by 15 cm thick and appear as massive or concretionary layered but mostly laminated. The vast majority of the concretions have remained in situ since their formation (Color Plate 2c and d, see p. 394) whereas a minority were pene-contemporaneously reworked by high-energy sediment gravity flows, most notably a major eruptive event that formed the Main Tuff Interval (Hassler, 1993). This indicates that concretions originally formed very close to the water–sediment interface. Associated carbonate beds display lumpy lamination, roll-up and other structures related to

Table 1 Isotopic profiles through selected concretions based on microcoring

Sample #	Profile	Distance (cm)	δ^{13} C (‰) VPDB	δ^{18} O (‰) VSMOW
Palaeoproterozoic calcite concentrations and host rock				
Onega Basin, Kondopoga Fm., c. 1900 Ma				
Greywacke-hosted, lensoidal, zoned concretion (#K1–05)				
a	Core	5.5	−19.3	14.6
b		4.5	−18.8	14.5
c		4	−19.3	14.5
d		3.1	−19.5	14.5
e		2.3	−20.6	14.6
f		1.5	−21.3	15.3
g		0.8	−21.8	15.6
h	Rim	0	−15.2	16.2
Pechenga Greenstone Belt, Pilgujärvi Sedimentary Fm., 2004 ± 9 Ma (Re-Os)				
Greywacke-hosted, lensoidal, zoned concretion (#2900/537)				
n	Core	7.0	−5.2	21.3
m		6.5	−10.8	15.0
l		6.0	−13.6	12.1
k		5.5	−8.9	17.0
j		5.0	−9.8	15.3
g		3.5	−5.8	20.8
f		3.0	−10.5	14.4
e		2.5	−12.4	13.4
d		2.0	−12.8	12.1
c		1.5	−13.1	12.2
b		1.0	−13.2	12.0
a	Rim	0.5	−13.0	11.8

Table 1 (continued)

Sample #	Profile	Distance (cm)	δ^{13} C (‰) VPDB	δ^{18} O (‰) VSMOW
Late Archaecan carbonate concretions				
Kaapvaal Craton, Schmidsdriff Subroup, Boomplaas Formation, ca. 2600 Ma				
Shale-hosted, lensoidal, calcite concretion (#SA 11)				
d	Core	2.5	−1.1	19.0
c		2.0	−1.1	19.4
b		1.5	−1.1	19.3
a	Rim	1.0	−1.0	19.7
Shale-hosted, lensoidal, calcite concretion (#SA 12)				
i	Core	2.3	−0.8	19.1
h		2.0	−0.9	19.1
g		1.5	−0.9	19.2
f		1.0	−0.9	19.1
c	Rim	0.5	−1.0	19.1
Shale-hosted, lensoidal, calcite concretion (#SA 13)				
e	Core	2.3	+0.4	20.2
d		2.0	+0.7	20.2
c		1.5	−0.4	17.7
b		1.0	−0.7	17.6
a	Rim	0.5	−0.6	18.4
Shale-hosted, lensoidal, calcite concretion (#SA 17)				
p	Core	3.2	−1.2	18.4
o		2.8	−0.7	19.5
m+n		2.3	−1.0	18.6
l		1.5	−1.1	18.7
k	Rim	1.0	−1.2	18.1
Pilbara Craton, Hamersley Group, Wittenoom Fm., c. 2603 ± 7 Ma				
Shale-hosted, reworked, calcite concretion (#96255)				
a	Top	4.0	−2.7	20.9
b		3.0	−3.2	20.4
c		2.0	−4.1	18.5
d		1.6	−3.6	19.9
f	Core	1.0	−3.3	19.5
g		0.8	−3.2	19.6
h		0.5	−3.2	19.5
i	Base	0.0	−3.4	18.8
Pilbara Craton, Hamersley Group, Wittenoom Fm., c. 2603 ± 7 Ma				
Marl-hosted dolomite concretionary bed preserving roll-up structure (#92011)				
f	Top	6.0	−0.2	17.0
e	Core	3.5	−0.3	17.4
d	Base	0.0	−0.1	16.5
BIF-associated, shale-hosted, dolomite concretion (#88273)				
a	Top	4.5	−6.5	23.7
b		4.0	−6.5	24.1
c		3.5	−6.5	24.2
d		3.0	−6.9	24.3
e	Core	2.5	−6.7	24.3
f		2.0	−6.8	24.2

Table 1 (continued)

Sample #	Profile	Distance (cm)	δ^{13} C (‰) VPDB	δ^{18} O (‰) VSMOW
g		1.5	−6.7	24.3
h		0.8	−6.8	24.3
i	Base	0.0	−6.7	24.4

Concretion specimens were slabbed, polished, scanned and examined macroscopically. Polished 150 μm thick sections were made to provide a core-to-rim coverage. They were scanned and examined in transmitted and reflected light. The same sections were examined with backscatter-mode scanning electron microscopy in order to study authigenic carbonates. Ca, Mg, Fe and Mn core-to-rim trends were semi-quantitatively recorded in untreated samples at low vacuum by electron-microprobe analysis of thin sections. Back-scattered electron microscopy and microprobe analyses were performed using the scanning electron microscopy facility at the Geological Survey of Norway, which consists of a 1450VP electron microscope from LEO Electron Microscopy Ltd equipped with X-ray detection systems Energy 400 and Wave 500 from Oxford Instruments. The microscope is capable of running prepared or conducting samples under high-vacuum conditions as well as untreated samples in lower vacuum. The electron detection systems include a secondary electron detector for topographic contrast and two back-scattered electron detectors, primarily for atomic number contrast. The X-ray detection system includes an energy-dispersive spectrometer and a single wavelength dispersive spectrometer for trace element analysis.

Submilligram microsamples were acquired for oxygen and carbon isotope analyses using an Ulrike Medenbach microcorer, which enables core diameters from <50 μm to several millimeters to be obtained with a maximum drilling depth of 1 mm. The microsamples were cored from 150 μm thick polished sections which have been examined by the back-scattered electron microscopy and microprobe analyses. Oxygen and carbon isotope analyses of whole-rock and microcore samples were performed at the Scottish Universities Environmental Research Centre, Glasgow, using the phosphoric acid method of McCrea (1950) as modified by Rosenbaum and Sheppard (1986) for operation at 70°C. Oxygen isotope data for dolomites were corrected using the fractionation factor 1.0099 recommended by Rosenbaum and Sheppard (1986). Carbon and oxygen isotope ratios in carbonate constituents of the whole-rock and microcore samples were measured on an AP 2003 mass spectrometer. Analyses were calibrated against NBS 19, and precision (1σ) for both isotope ratios is better than ± 0.2‰. The carbon isotope data are reported relative to VPDB whereas the oxygen isotope data are relative to VSMOW.

soft-sediment deformation (Color Plate 2e, see p. 394). Some concretions trapped roll-up structures (Color Plate 2f, see p. 394). It has been proposed that the roll-up structures are a result of soft-sediment deformation of laminated mud which was originally stabilised by cohesive microbial mats built by non-photosynthetic, heterotrophic or chemoautotrophic microbes below wave base (Simonson and Carney, 1999).

Bulk analyses of both in situ and reworked calcite concretions show similar δ^{13}C ranging between −3.5 and −2.3‰. Hosting marls have indistinguishable ratios fluctuating between −3.5 and −3.0‰ though they are slightly enriched in ^{18}O (Tables 1 and 3). A bottom–top microcored transect through the redeposited concretion reveals a fluctuation between −4.1 and −2.7‰. The central region is relatively depleted in ^{13}C (Color Plate 2g, see p. 394). Dolomitic

Table 2 Isotopic composition of microcored sulphides from Boomplaas carbonate concretions

Sample #	$\delta^{34}S$ (‰ CDT)
Concretion # SA 11	
1	+0.4
3	+0.1
4	−0.4
5	0.0
6	+0.6
7	+1.8
8	+0.6
Concretion #SA 17	
9	+1.4
10	+1.6
Concretion #SA 17A	
11	+1.4
12	+0.3
Mean	+0.7 ± 0.7 (1σ)

Individual cores were laser-combusted to SO_2 (Fallick et al., 1992; Kelley and Fallick, 1990) which was cryogenically purified and isotopically assayed on an in-line dual inlet gas source mass spectrometer (SIRA II). A pyrite laser correction of −1‰ was applied (see Kellley and Fallick, 1990). Combusting cores is more difficult to control than for polished blocks and there is some loss in precision which we estimate as 0.7‰ (1σ) from analyses of standards. Therefore, there is little evidence of geological scatter in the analyses of the cores.

concretions and concretionary beds with roll-up structure and their host marls exhibit $\delta^{13}C$ close to zero (Color Plate 2h, see p. 394). Similar to the calcite concretions, the host marl is slightly enriched in ^{18}O (Table 3). The overall data suggest that a very limited ^{13}C-depleted, organically derived source was involved in concretionary calcite. No ^{13}C-depleted source was involved in concretionary dolomite whose $\delta^{13}C$ is close to zero. In contrast, a dolomite concretion collected in close proximity to the Jimblebar iron ore deposit (Blockley, 1990) shows significant ^{13}C depletion with $\delta^{13}C$ averaging around −6.7‰ though without any significant fluctuation along a bottom-to-top profile (Color Plate 2i, see p. 394). Such values are a typical feature of BIF-associated carbonates throughout the Archaean (Fig. 1)

3.2 Isotopic composition of Palaeoproterozoic concretions

The first abundant appearance of carbonate concretions is dated to c. 2050 Ma (Color Plate 4a, see p. 396) (Melezhik, 1992). From c. 2000 Ma they became widespread and formed in diverse depositional settings extending from marine (Color Plate 4a–e, see p. 396) to subaerial (Color Plate 4f, see p. 396) and

Table 3 Isotopic composition of Precambran concretions based on whole-rock analyses

Sample #	$\delta^{13}C$(‰) VPDB	$\delta^{18}O$ (‰) VSMOW
Palaeoproterozoic calcite concretions		
Onega Basin, Kondopoga Fm., ca 1900 Ma		
Greywacke-hosted, lensoidal, zoned concretion		
K1–05	−20.1	14.9
K2–05	−20.8	14.8
Greywacke-hosted, calcite concretionary layers		
Kn1–02	−13.9	13.8
Kn2–02	−14.1	13.2
Kn3–02	−14.3	13.9
Kn4–02	−15.0	13.6
Kn5–02	−15.0	13.6
Pechenga Greenstone Belt, Pilgujärvi		
Volcanic Fm., 1970 ± 5 Ma (U–Pb zircon)		
Felsic tuff-hosted, spherical, zoned concretions		
1	−7.1	11.8
2	−7.9	9.6
Pechenga Greenstone Belt, Pilgujärvi		
Sedimentary Fm., 2004 ± 9 Ma (Re–Os)		
Calcite cement in greywackes		
2900/502	−10.7	13.7
2900/512	−11.2	13.4
2900/564	−11.2	12.7
R2900/565.7	−10.7	14.3
R2900/566	−11.1	12.7
R2900/567	−10.3	12.4
R2900/568	−7.2	25.8
2900/587	−13.8	13.0
2900/743	−10.9	13.1
2900/758	−10.6	13.2
2900/770	−9.1	13.1
2900/787	−11.9	13.2
2900/798	−8.8	13.4
2900/811	−10.9	13.5
2900/822	−10.2	13.5
2900/860	−9.9	13.5
2900/869	−11.3	13.4
Greywacke-hosted, spherical, zoned concretions		
2900/448	−11.8	13.7
2900/453	−12.3	13.2
2900/472	−10.1	13.2
2900/473	−10.2	12.8
2900/508.5	−12.6	13.1
2900/517	−13.4	13.0
2900/549	−12.7	13.6
299/561.8	−7.5	11.6
2900/563.9	−8.2	13.0

Table 3 (continued)

Sample #	$\delta^{13}C(‰)$ VPDB	$\delta^{18}O$ (‰) VSMOW
K2900/562	−7.8	13.5
K2900/562.3	−7.8	13.4
K2900/562.9	−7.9	12.8
Greywacke-hosted, spherical, zoned concretions		
K2900/565.7	−10.6	13.8
K2900/567	−11.1	12.9
2900/568	−11.2	13.9
K2900/568.4	−10.1	13.1
K2900/578.2	−9.9	13.8
K2900/580.3	−9.8	13.2
K2900/580.4	−10.3	12.1
2900/686	−13.5	13.1
2900/699.7	−10.6	14.0
K2900/704	−12.1	12.7
Greywacke-hosted, lensoidal, zoned conretions		
2900/535a	−13.1	12.3
2900/535b	−14.2	10.8
2900/535c	−13.2	12.3
2900/524a	−13.3	13.4
2900/590	−10.9	12.9
2900/592	−12.5	13.1
2900/659	−11.4	12.7
2900/665	−12.5	13.1
2900/668	−12.2	13.1
2900/682	−12.5	12.9
2900/683	−13.1	13.0
2900/721	−13.3	12.3
2400/248	−11.3	11.2
2900/600	−9.7	12.6
2400/248	−11.3	11.2
Greywacke-hosted concretionary beds		
2900/802	−10.0	12.3
2900/816	−10.5	11.1
2900/831	−11.3	13.4
2900/844	−9.8	13.4
2900/844a	−9.7	12.1
2900/878	−12.2	13.3
2900/926	−12.0	13.7
2900/926a	−12.0	13.6
2900/995	−11.3	13.4
2900/1009	−8.7	12.7
2900/1011	−8.7	12.4
2900/1030	−8.1	13.1
2900/1044	−8.8	13.1
2900/1052	−10.4	13.2
2900/1075	−12.5	12.7
2798/600	−9.7	11.8

Table 3 (continued)

Sample #	δ^{13}C(‰) VPDB	δ^{18}O (‰) VSMOW
Imandra/Varzuga Greenstone Belt, Il'mozero Sedimentary Fm., c. 2050		
Greywacke-hosted, lensoidal, calcite concretions		
496	−3.9	13.4
877	−4.1	12.8
3027	−20.5	13.3
Late Archaean carbonate concretions		
Pilbara Craton, Hamersley Group, Wittenoom Fm., ca., 2603 ± 7 Ma		
Shale-hosted, in situ, lensoidal, calcite concretion		
92026	−2.3	22.6
Shale-hosted, in situ, lensoidal, calcite concretion		
92489-A	−2.8	21.3
Shale-hosted, reworked, calcite concretion		
96255	−3.5	20.3
Marl-hosted, reworked, calcite concretions Host marl		
92848b	−3.5	21.9
92848d	−3.1	21.0
Marl-hosted, reworked, calcite concretion Host marl		
92848h	−3.0	20.4
Concretion		
92848c	−2.3	19.8
92848e	−2.3	19.3
92848f	−2.9	19.6
92848g	−3.3	20.1
Marl-hosted dolomite concretion preserving roll-up structure		
Host marl		
92011a	−0.4	17.9
92011c	−0.5	18.2
92011g	−0.6	18.3
92011h	−0.5	18.2
92011i	−0.5	18.3
Concretion preserving roll-up structure		
92011b	−0.3	17.5
BIF-associated, shale-hosted, dolomite concretion affected by recent weathering		
88273	−6.9	31.6

lacustrine (Color Plate 6a–g, see p. 398) environments (Melezhik et al., 1998; Winter and Knauth, 1992). Importantly, this period was marked by formation of large arrays of other diagenetic products including Fe- and Mn-oxides, Mn- and Fe carbonates (siderite and rhodochrosite), phosphate and sulphide concretions (Melezhik, 1992).

Palaeoproterozoic concretions isotopically analysed in our study are from the Fennoscandian Shield. Concretions that occur immediately in the aftermath of the Lomagundi-Jatuli event have been reported from the Il'mozero Sedimentary Formation in the Imandra/Varzuga Greenstone Belt (Melezhik and Fallick, 1996). They formed in volcaniclastic greywacke deposited in a setting transitional from fluvial and deltaic to marine (Melezhik, 1992). Concretions from marine setting have the lowest $\delta^{13}C$ (–20.5‰) and are associated with phosphate nodules.

Concretions formed in the 2004 Ma, c. 1000 m thick Pilgujärvi Sedimentary Formation of the Pechenga Greenstone Belt (e.g. Melezhik et al., 1998) are voluminous and occur as zoned lenses (Color Plate 2a and b, see p. 394), massive beds and lenses (Color Plate 4d, see p. 396) and zoned irregular spheres (Color Plate 4e, see p. 396). Host rocks are turbiditic, volcaniclastic greywackes deposited in a marine environment. Bulk analysis of 51 concretions shows $\delta^{13}C$ ranging from –14.2 to –7.5‰. $\delta^{13}C$ correlates neither with the concretion morphology nor with the stratigraphy. Diagenetic calcite cement in the hosted marine greywackes shows similar $\delta^{13}C$ varying between –13.8 and –7.2‰ (Table 3). Calcite concretions in 1970 Ma, subaerially deposited, felsic tuffs of the overlying volcanic formation have similar $\delta^{13}C$ at around –7‰ (Table 3). An isotopic transect through zoned, lensoidal concretion exhibits $\delta^{13}C$ irregularly fluctuating between –13.6 and –5.2‰ (Color Plate 4b, see p. 396) though showing a strong positive correlation between $\delta^{13}C$ and $\delta^{18}O$ ($r = +0.99$), apparently suggesting a physical mixture of two end members.

The c. 1900 Ma Kondopoga Formation of the Onega Basin contains abundant calcite and siderite concretions formed in a lacustrine environment. The concretions are found in c. 100 m thick, volcaniclastic, greywacke unit deposited from turbidity currents. Concretions occur in various forms. Laterally continuous beds, 5–7 cm thick, dominate (Color Plate 6a, see p. 398). Many such beds show sedimentary boudinage (Color Plate 6b, see p. 398). Lensoidal concretions commonly occur as a series of laterally linked bodies (Color Plate 6c, see p. 398). Many exhibit a considerable differential compaction (Color Plate 6d, see p. 398) suggesting formation in very early diagenesis. In all cases concretions formed within coarse sediments comprising lower part of Bouma cycles. Some sandy ripples and channels underwent early cementation by carbonates, causing the concretion's shape to mimic bedding surfaces (Color Plate 6e, see p. 398). In bedding parallel sections, such concretions appear as either roughly aligned (Color Plate 6f, see p. 398) or sub-perpendicular (Color Plate 6g, see p. 398) sausage-shaped bodies. In places, concretions show concentric zoning. Numerous, large-scale, high-energy, erosional channels are filled with debris including redeposited carbonate concretions (Color Plate 6g, see p. 398). All features suggest that carbonate concretions formed extensively and that cementation occurred during early stages of diagenesis. Both bulk and microcore-based analyses show low $\delta^{13}C$ values fluctuating between –21.8 and –13.9‰ (Tables 1 and 3).

All Palaeoproterozoic concretions involved in this study as well as previously published (e.g. Winter and Knauth, 1992) are significantly depleted in ^{13}C, which clearly indicates the involvement of CO_2 derived through remineralisation of OM via either bacterial oxidation or bacterial sulphate reduction.

4 C_{org} Recycling

4.1 Prior to c. 2000 Ma

The scarcity of carbonate concretions in the pre-2000 Ma geological record can be explained neither by biased sampling nor by preservation potential. The latter does not seem to be a problem, because carbonate concretions are known even from rocks subjected to high-temperature amphibolite facies metamorphism and partial melting (Color Plate 6h, see p. 398). Because concretions form laterally extensive or spheroidal bodies, they tend to catch the eye in the field (Color Plate 6i, see p. 398): therefore it is highly unlikely that their scarcity in the pre-2000 Ma geological record is an artefact of biased sampling.

The Late Archaean concretions from known occurrences have $\delta^{13}C$ either fluctuating close to zero and thus indistinguishable from coeval seawater bicarbonate (Boomplaas data) or only slightly depleted in ^{13}C (Wittenoom data). The scarcity of concretions and the $\delta^{13}C$ values of early Precambrian carbonates do not support a sizable contribution of carbon related to postdepositional bacterial degradation of organic matter, an observation that represents a fundamental difference with the post-2000 Ma world.

This enigma can be explained if prior to c. 2000 Ma, because of an essentially anoxic atmosphere/hydrosphere, the water column and sediment water interface were colonised by carbon recycling organisms. The organic carbon in this world predominantly originated from protozoans—for the most part small, buoyant, high surface-area-to-volume entities not rich in strongly refractory, unreactive carbon biopolymers. OM recycling did not take place to any significant extent in the deeper sediments, and the remineralisation byproducts of isotopically distinctive CO_2 and CH_4 readily escaped to the water column and atmosphere (Pavlov et al., 2003) where rapid mixing diluted away their isotopic fingerprints. In the sediment column, there were no redox fronts, no redox gradient zones, no major OM recycling, no concretionary growth using recycled carbon. The rock-forming processes can be characterised by one-stage sedimentogenesis (cf. Strakhov, 1962) with negligible formation of diagenetic minerals caused by remineralisation of OM. Exceptionally rare late Archaean carbonate concretions have $\delta^{13}C$ close to coeval marine bicarbonate (Fig. 1). This suggests that they have also formed either in close proximity to sediment–water interface (slightly depleted in ^{13}C Wittenoom concretions) or on the seafloor (isotopically indistinguishable from coeval seawater Boomplass concretions) where their isotopic composition was greatly influenced by exchange with the larger marine bicarbonate reservoir.

The preferential metabolism in the water column also explains the rarity of Early Precambrian sulphide concretions. If sulphate and SRBs were available, OM was oxidised to produce sulphide (HS^-) in the water column. This HS^- may form Fe-sulphide (e.g. pyrite) by reacting with dissolved Fe^{2+}. Such pyrite particles gravitationally settled to the seafloor, occurring in the geological record either as disseminated or layered form, but not as concretions. The latter require formation in the sedimentary column. Moreover, the pyrite produced in such open system should have a limited fractionation of the sulphur isotopes which is, in fact, a typical feature of the Archaean (Strauss, 2003). Although there are exceptions (e.g. Bekker et al., 2004; Shen et al., 2001) they are very few and Archaean concretions show fractionation of the sulphur isotopes only at a fine scale.

4.2 After c. 2000 Ma

Abundant diagenetic ^{13}C-depleted concretions (Color Plates 4 and 6, see p. 396 and 398, respectively) and diagenetic carbonates occurring at c. 2000 Ma (Fig. 1) are evidence of the establishment of biologically induced diffusion gradients in the water and sediment columns, following significantly increased atmospheric oxygen levels (Baker and Fallick, 1989; Bekker et al., 2004; Karhu and Holland, 1996). Escaping the poison dioxygen, the obligate anaerobic recycling organisms retreated deep into sediments where significant recycling of OM took place from then on. This led to the establishment of two-stage sedimentogenesis in which burial diagenesis involved the reaction of various oxidising agents with OM and the formation of a variety of diagenetic concretions (Color Plates 4 and 6, see p. 396 and 398, respectively) with incorporation of organically derived, ^{13}C-depleted CO_2 into diagenetically formed carbonates, similar to the Phanerozoic world.

5 BIF-Associated Carbonates

Within the context of the ideas presented above, BIF carbonates and sulphides are clearly anomalous, suggesting unusual circumstances of formation. Although sedimentological evidence relevant to primary depositional environment of BIFs is well-preserved in the Hamersley basin and a number of other iron formations (Clout and Simonson, 2005; Simonson, 2003), the chemistry of BIFs is highly contentious, and mineralogy is mostly diagenetic or metamorphic in origin (Krapež et al., 2003). Secondly, there is no consensus over the ^{13}C-depleted source for BIF carbonates: Beukes et al. (1990) postulated a hydrothermal input and stratified water column, with thermal decarboxylation occurring later during burial; Kaufman et al. (1990) suggested low-temperature biological processes. Fermentative metabolism or anaerobic respiration occurring near the sediment–water interface was invoked by Baur et al. (1985) whereas Walker (1984) supported "suboxic diagenesis," albeit high original concentrations of OM must be postulated. We note that carbon isotopic and sedimentological patterns (e.g. sulphide concretions) of BIFs are suggestive of OM recycling similar to that developed at around 2000 Ma, when the free dioxygen concentration in seawater had become substantial. Perhaps this accords with the idea of localised "oases" such that short-term oxygenated basins might have punctuated the prevailing oxygen-free world described above (Baur et al., 1985; Huston and Logan, 2004; Thode and Goodwin, 1983), a world that was to change irrevocably at around 2060 Ma.

6 OM Recycling and the Lomagundi-Jatuli Isotopic Event

An intriguing yet unexplained feature of the 2300(?) to 2060 Ma Lomagundi-Jatuli carbon isotope positive excursion(s) is that there is no evidence for a subsequent compensating low $\delta^{13}C$ excursion (see, e.g. the representation in Karhu and Holland, 1996). Transient low $\delta^{13}C$ excursions have been identified as possible characteristic features of extinction events (Magaritz, 1989) and of ventilation or turnover of deep anoxic basins (e.g. Aharon and Liew, 1992). Such an excursion might be anticipated to follow an episode of enhanced organic carbon burial as recycling and remineralisation proceed. However, the lack of a compensating negative excursion emerges as a predictive consequence of the hypothesis, developed above, of the creation of a new locus for OM recycling. Since the active biomass reservoir within the sediments and crust endures, there is no net return of oxidised OM of low $\delta^{13}C$ to the ocean–atmosphere system to become recorded by carbonate $\delta^{13}C$. Proponents of a "deep biosphere" have a new insight into possible early stages and causative mechanisms of their favoured concept.

Hayes and Waldbauer (2006) have interpreted the Lomagundi-Jatuli excursion differently, namely as an indicator of shifting methanogenesis from water column and the water–sediment interface deeper into the sedimentary column at around 2400 Ma as a response to increasing SO_4^{2-} and O_2 availability in the ocean. The resulting effect of this step function in microbial ecosystem dynamics was formation of ^{13}C-rich diagenetic carbonates with the absence of congruent variations in the OM isotopic record and without requiring enhanced burial of OM.

This model somewhat echoes our earlier hypothesis, suggesting that the cause for the Lomagundi-Jatuli ^{13}C-rich carbonates was penecontemporaneous recycling of OM in cyanobacterial mats with the production and consequent loss of CO_2 and CH_4 in usual shallow-water depositional settings (Melezhik et al., 1999). However, our subsequent search for the record of ^{13}C-depleted end-member of methanogenetic processes in Jatuli carbonates was unsuccessful (Melezhik and Fallick, 2003), thus leaving the methanogenetic model unproved. However, interestingly, both the Hayes and Waldbauer's approach and the "concretionary" approach explored in this contribution agree that there was an important biological change as a response to the development of the O_2-rich hydrosphere. One remaining issue to be resolved is the precise timing of onset of the event.

Acknowledgement The idea presented above has emerged in 2004 (Melezhik and Fallick, 2004) and it was first discussed publicly during the 2004 Goldschmidt Conference with M.L. Coleman, S.D. Golding and H.D. Holland. Since then, we have elaborated our model from stimulating debates with J.M. Hayes, A. Lepland, L.R. Kump and H. Strauss. Field work by BMS was supported by grants from the National Geographic Society. Field trip to South Africa by VAM was supported by the Geological Survey of Norway. The isotope analyses were performed at the Scottish Universities Environmental Research Centre supported by the Consortium of Scottish Universities and the Natural Environment Research Council. The article represents contribution to the International Programme "Biosphere Origin and Evolution" and to the Fennoscandian Arctic Russia—Drilling Early Earth Project within the framework of the International Continental Scientific Drilling Program.

References

Abell, P.I., McClory, J., Martin, A. and Nisbet, E.G. (1985) Archaean stromatolites from the Ngesi Group, Belingwe Greenstone Belt, Zimbabwe; preservations and stable isotopes—preliminary results. Precambrian Res. 27, 357–383.

Aharon, P. and Liew, T.C. (1992) An assessment of the Precambrian/Cambrian transition events on the basis of carbon isotope records. In: M. Schidlowski, S. Golubic, M.M. Kimberley, D.M. McKirdy and P.A. Trudinger (Eds), Early Organic Evolution: Implications for Mineral and Energy Resources. Springer, Berlin, pp. 212–223.

Altermann, W. and Nelson, D.R. (1998) Sedimentation rates, basin analysis and regional correlations of three Neoarchaean and Palaeoproterozoic sub-basins of the Kaapvaal craton as inferred from precise U–Pb zircon ages from volcaniclastic sediments. Sediment. Geol. 120, 225–256.

Baker, A.J. and Fallick, A.E. (1989) Evidence from Lewisian limestones for isotopically heavy carbon in two-thousand-million-year-old sea water. Nature 337, 352–354.

Baur, M.E., Hayes, J.M., Studley, S.A. and Walter, M.R. (1985) Millimeter-scale variations of stable isotope abundances in carbonates from banded iron-formations in the Hamersley Group of Western Australia. Econ. Geol. 80, 270–282.

Bekker, A., Holland, H.D., Wang, P.L., Rumble III, D., Stein, H.J., Hannah, J.L., Coetzee, L.L. and Beukes, N.J. (2004) Dating the rise of atmospheric oxygen. Nature 427, 117–120.

Beukes, N.J., Klein, C., Kaufman, A. and Hayes, J.M. (1990) Carbonate petrography, kerogen distribution, and carbon and oxygen isotope variations in an Early Proterozoic transition from limestone to iron-formation deposition, Transvaal Supergroup, South Africa. Econ. Geol. 85, 663–690.

Blockley (1990) Iron ore. Geology and Mineral Resources of Western Australia. Geol. Surv. West. Austral. Memoir 3, pp. 679–692.

Clout, J.M.F. and Simonson, B.M. (2005) Precambrian iron formations and iron formation-hosted iron ore deposits. In: J.W. Hedenquist, J.F.H. Thompson, R.J. Goldfarb and J.P. Richards (Eds), Economic Geology One Hundredth Anniversary Volume 1905–2005. Soc. Econom. Geol., Littleton, CO, pp. 643–679.

Coleman, M.L. (1993) Microbial processes: controls on the shape and composition of carbonate concretions. Mar. Geol. 113, 127–140.

Coleman, M.L. and Raiswell, R. (1981) Carbon, oxygen and sulphur isotope variations in concretions from the upper Lias of N.E. England. Geochim. Cosmochim. Acta 45, 329-340.

Coleman, M.L. and Raiswell, R. (1993) Microbial mineralization of organic matter: mechanism of self-organization and inferred rates of precipitation of diagenetic minerals. Philos. Trans. R. Soc. Lond. A 344, 69–87.

Curtis, C.D., Coleman, M.L. and Love, L.G. (1986) Pore water evolution during sediment burial from isotopic and mineral chemistry of calcite, dolomite and siderite concretions. Geochim. Cosmochim. Acta 50, 2321–2334.

De Craen, M., Swennen, R., Keppens, E.M., Macaulay, C.I. and Kiriakoulakis, K. (1999) Bacterially meditated formation of carbonate concretions in the Oligocene Boom Clay of Northern Belgium. J. Sediment. Res. 69, 1098–1106.

Deines, P. (1992) Mantle carbon: concentration, mode of occurrence, and isotopic composition. In: M. Schidlowski, S. Golubic, M.M. Kimberley, D.M. McKirdy and P.A. Trudinger (Eds), Early Organic Evolution: Implications for Mineral and Energy Resource. Springer, Berlin, pp. 133–146.

Dietrich, R.V. (1997) Carbonate Concretions: A Bibliography. http://www.cst.cmich.edu/users/dieter1rv/concretions/, 64 pp.

Donnelly, T.H. and Crick, I.H. (1992) Biological and abiological sulfate reduction in two Northern Australian Proterozoic basins. In: M. Schidlowski, S. Golubic, M.M. Kimberley, D.M. McKirdy and P.A. Trudinger (Eds), Early Organic Evolution: Implications for Mineral and Energy Resource. Springer, Berlin, pp. 398–407.

Fallick, A.E., McConville, P., Boyce, A.J., Burgess, R. and Kelley, S.P. (1992) Laser microprobe stable isotope measurements on geological materials: some experimental considerations (with special reference to $\delta^{34}S$ in sulphides). Chem. Geol. 101, 53–61.

Grassineau, N.V., Nisbet, E.G., Fowler, C.M.R., Bickle, M.J., Lowry, D., Chapman, H.J., Mattey, D.P., Abell, P., Yong, J. and Martin, A. (2002) Stable isotopes in the Archaean Belingwe Belt, Zimbabwe; evidence for a diverse microbial mat ecology. In: C.M. Fowler, C.J. Ebinger and C.J. Hawkesworth (Eds), The Early Earth: Physical, Chemical and Biological Development. Geol. Soc. Spec. Publ. 199, pp. 309–328.

Hassler, S.W. (1993) Depositional history of the Main Tuff Interval of the Wittenoom Formation, late Archean–Early Proterozoic Hamersley Group, Western Australia. Precambrian Res. 60, 337–359.

Hayes, J.M. and Waldbauer, J.R. (2006) The carbon cycle and associated redox processes through time. Philos. Trans. R. Soc. Lond. B 361, 931–950.

Hennessy, J. and Knauth, L.P. (1985) Isotopic variations in dolomite concretions from the Monterey Formation, California. J. Sediment. Petrol. 55, 120–130.

Huston, D.L. and Logan, G.A. (2004) Barite, BIFs and bugs: evidence for the evolution of the Earth's early hydrosphere. Earth Planet. Sci. Lett. 220, 41–45.

Irwin, H., Curtis, C. and Coleman, M. (1977) Isotopic evidence for source of diagenetic carbonates formed during burial of organic rich sediments. Nature 269, 209–213.

Karhu, J.A. and Holland, H.D. (1996) Carbon isotopes and the rise of atmospheric oxygen. Geology 24, 867–879.

Kaufman, A.J., Hayes, J.M. and Klein, C. (1990) Primary and diagenetic controls of isotopic compositions of iron-formation carbonates. Geochim. Cosmochim. Acta 54, 3461–3473.

Kelley, S.P. and Fallick, A.E. (1990) High precision spatially resolved analysis of $\delta^{34}S$ in sulphides using a laser extraction technique. Geochim. Cosmochim. Acta 54, 883–888.

Krapež, B., Barley, M.E. and Pickard, A.L. (2003) Hydrothermal and resedimented origins of the precursor sediments to banded iron formation: sedimentological evidence from the Early Palaeoproterozoic Brockman Supersequence of Western Australia. Sedimentology 50, 979–1011.

Kyser, T.K. (1986) Stable isotope variations in the mantle. Rev. Mineral. 16, 141–164.

Magaritz, M. (1989) ^{13}C minima follow extinction events: a clue to faunal radiation. Geology 17, 337–340.

McCrea, J.M. (1950) On the isotopic geochemistry of carbonates and a paleotemperature scale. J. Chem. Phys. 18, 849–857.

Melezhik, V.A. (1992) Palaeoproterozoic Sedimentary and Rock-Forming Basins of the Fennoscandian Shield. Nauka (Science), Leningrad (in Russ.).

Melezhik, V.A. and Fallick, A.E. (1996) A widespread positive $\delta^{13}C_{carb}$ anomaly at around 2.33–2.06 ga on the Fennoscandian Shield: a paradox. Terra Nova 8, 141–157.

Melezhik, V.A. and Fallick, A.E. (2003) $\delta^{13}C$ and $\delta^{18}O$ variations in primary and secondary carbonate phases: several contrasting examples from Palaeoproterozoic ^{13}C-rich dolostones. Chem. Geol. 201, 213–228.

Melezhik, V.A. and Fallick, A.E. (2004) Concretions tell that ancient bugs lived differently. Goldschmidt Conference, Copenhagen, Denmark, June 5–11, 2004. Geochim. Cosmochim Acta, Abstract Volume, A796.

Melezhik, V.A., Grinenko, L.N. and Fallick, A.E. (1998) 2000 Ma sulphide concretions from the 'Productive' Formation of the Pechenga Greenstone Belt, NW Russia: genetic history based on morphological and isotopic evidence. Chem. Geol. 148, 61–94.

Melezhik, V.A., Fallick, A.E., Medvedev, P.V. and Makarikhin, V.V. (1999) Extreme $^{13}C_{carb}$ enrichment in ca. 2.0 Ga magnesite-stromatolite-dolomite-'red beds' association in a global context: a case for the world-wide signal enhanced by a local environment. Earth Sci. Rev. 48, 71–120.

Mortimer, R.J.G. and Coleman, M. (1997) Microbial influence on the oxygen isotopic composition of diagenetic siderite. Geochim. Cosmochim. Acta 61, 1705–1711.

Mozley, P.S. and Burns, S.J. (1993) Oxygen and carbon isotopic composition of marine carbonate concretions: an overview. J. Sediment. Petrol. 63, 73–83.

Nelson, D.R., Trendall, A.F. and Altermann, W. (1999) Chronological correlations between the Pilbara and Kaapvaal cratons. Precambrian Res. 97, 165–189.

Neruchev, S.G. (1982) Uranium and Life in the Earth History. Nedra, Moscow, 208 pp (in Russ.).

Pavlov, A.A., Hurtgen, M.T., Kasting, J.F. and Arthur, M.A. (2003) Methane-rich Proterozoic atmosphere? Geology 31, 87–90.

Perry, E.C., Jr, Tan, F.C. and Morey, G.B. (1973) Geology and stable isotope geochemistry of the Biwabik Iron Formation, Northern Minnesota In: Precambrian iron-formations of the world. Soc. Econ. Geol. Bull. 68, 1110–1125.

Pettijohn, F.J. (1940) Archean metaconcretions of Thunder Lake, Ontario. Geol. Soc. Am. Bull. 51, 1841–1850.

Pye, K., Dickson, J.A.D., Schiavon, N., Coleman, M.L. and Cox, M. (1990) Formation of siderite–Mg-calcite–iron sulphide concretions in intertidal marsh and sandflat sediments, north Norfolk, England. Sedimentology 37, 325–343.

Raiswell, R. (1987) Non-steady state microbiological diagenesis and the origin of concretions and nodular limestones. In: J.D. Marshall (Ed.), Diagenesis of Sedimentary Sequences. Geol. Soc. Spec. Publ. 36, pp. 41–54.

Raiswell, R., Whaler, K., Dean, S., Coleman, M.L. and Briggs, D.E.G. (1993) A simple three-dimensional model of diffusion-with-precipitation applied to localised pyrite formation in framboids, fossils and detrital iron minerals. In: R.J. Parkes, P. Westbroek and J.W. Leeuw (Eds), Marine Sediments, Burial, Pore Water Chemistry, Microbiology and Diagenesis. Marine Geol. 113, pp. 89–100.

Ripperdan, R.L. (2001) Stratigraphic variation in marine carbonate carbon isotope ratios. Rev. Mineral. 43, 637–662.

Romankevich, E.A. (1977) Geochemistry of Organic Matter in the Ocean. Nauka (Science), Moscow, 265 pp (in Russ.).

Rosenbaum, J.M. and Sheppard, S.M.F. (1986) An isotopic study of siderites, dolomites and ankerites at high temperatures. Geochim. Cosmochim. Acta 50, 1147–1159.

Sarmiento, J.L., Dunne, J., Gnadadesikan, A., Key, R.M., Matsumoto, K. and Slater, R. (2002) A new estimate of the $CaCO_3$ to organic carbon export ratio. Global Biogeochem. Cycles 16, 1107.

Schidlowski, M. (1987) Application of stable carbon isotopes to early biogeochemical evolution on Earth. Earth Planet. Sci. Annu. Rev. 15, 47–72.

Schidlowski, M., Appel, P.W.U., Eichmann, R. and Junge, C.E. (1979) Carbon isotope geochemistry of the 3.7×10^9 yr old Isua sediments, West Greenland; implications for the Archaean carbon and oxygen cycles. Geochim. Cosmochim. Acta 43, 189–200.

Sellés-Martínez, J. (1996) Concretion morphology, classification and genesis. Earth Sci. Rev. 41, 177–210.

Shen, Y., Buick, R. and Canfield, D.E. (2001) Isotopic evidence for microbial sulphate reduction in the early Archaean era. Nature 410, 77–81.

Shields, G. and Veizer, J. (2002) Precambrian marine carbonate isotope database: version 1.1. Geochem. Geophys. Geosyst. G3(3), 1–12.

Simonson, B.M. (2003) Origin and evolution of large Precambrian iron formations. In: M. Chan and A. Archer (Eds), Extreme Depositional Environments: Mega End Members in Geologic Time. Geol. Soc. Am. Spec. Pap. 370, pp. 231–244.

Simonson, B.M. and Carney, K.E. (1999) Roll-up structures: evidence of in situ microbial mats in late Archean deep shelf environments. Palaois 14, 13–14.

Simonson, B.M., Hassler, S.W. and Schubel, K.A. (1993a) Lithology and proposed revisions in stratigraphic nomenclature of the Wittenoom Formation (Dolomite) and overlying formations, Hamersley Group, Western Australia. West. Austral. Geol. Surv. Rep. 34, Prof. Pap. 65–79.

Simonson, B.M., Schubel, K.A. and Hassler, S.W. (1993b) Carbonate sedimentology of the early Precambrian Hamersley Group of Western Australia. Precambrian Res. 60, 287–335.

Smyk, M.C. and Watkinson, D.H. (1990) Sulphide remobilization in Archean volcano-sedimentary rocks and its significance in Proterozoic silver vein genesis, Cobalt, Ontario. Can. J. Earth. Sci. 27, 1170–1181.

Strakhov, N.M. (1962) Stages of development of exospheres and sedimentary rock-forming processes in the Earth history. Proc. Acad. Sci. USSR, Geol. Series 12, 3–22 (in Russ.).

Strauss, H. (2003) Sulphur isotopes and the early Archaean sulphur cycle. Precambrian Res. 126, 349–361.

Sumner, D.Y. (2000) Autotrophic fixation of HCO_3^- caused calcite precipitation in 2.5 Ga. Geological Society of America, 2000 Annual Meeting. Geol. Soc. Am. Abstr. Programs 32, 257.

Thode, H.G. and Goodwin, A.M. (1983) Further sulphur and carbon isotope studies of late Archaean iron-formations of the Canadian Shield and the rise of sulfate reducing bacteria. Precambrian Res. 20, 337–356.

Tissot, B.P. and Welte, D.H. (1984) Petroleum Formation and Occurrence, 2nd ed. Springer, Berlin.

Trendall, A.F., Compston, W., Nelson, D.R., De Laeter, J.R. and Bennett, V.C. (2004) SHRIMP zircon ages constraining the depositional chronology of the Hamersley Group, Western Australia. Aust. J. Earth Sci. 51, 621–644.

Veizer, J., Hoefs, J., Lowe, D.R. and Thurston, P.C. (1989) Geochemistry of Precambrian carbonates: II. Archean greenstone belts and Archean seawater. Geochim. Cosmochim. Acta 53, 859–871.

Veizer, J., Clayton, R.N., Hinton, R.W., Von Brunn, V., Mason, T.R., Buck, S.G. and Hoefs, J. (1990) Geochemistry of Precambrian carbonates: III. Shelf seas and non-marine environments of the Archean. Geochim. Cosmochim. Acta 54, 2717–2729.

Walker, J.C.G. (1984) Suboxic diagenesis in banded iron formation. Nature 309, 340–342.

Watson, R.S., Trewin, N.H. and Fallick, A.E. (1995) The formation of carbonate cement in the Forth and Balmoral Fields, northern North Sea: a case for biodegradation, carbonate cementation and oil leakage during early burial In: A.J. Hartley and D.J. Prosser (Eds), Characterization of Deep Marine Clastic Systems. Geol. Soc. Spec. Publ. 94, pp. 177–200.

Winter, B.L and Knauth, L.P. (1992) Stable isotope geochemistry of early Proterozoic carbonate concretions in the Animikie Group of the Lake Superior region; evidence for anaerobic bacterial processes. Precambrian Res. 54, 131–151.

Evolutionary Aspects of Geochemical Activity of Microbial Mats in Lakes and Hydrotherms of Baikal Rift Zone

B. B. Namsaraev, V. M. Gorlenko, Z. B. Namsaraev, D. D. Barkhutova, L. P. Kozyreva, O. P. Dagurova, and A. V. Tatarinov

Abstract The long-term investigations of structure and biogeochemical activity of phototrophic microbial mats and non-phototrophic biofilms of hydrotherms, freshwater lake, soda and saline lakes located in Baikal rift zone were carried out. Microbial mats, especially phototrophic, are high productive systems with prevalence of production above destruction. Participation of the microbial community in travertine deposition as a model of the Precambrian era carbonaceous stromatolite formation (travertine) was shown on the Garga spring. Cyanobacterial mats of soda lakes and hydrotherms of Zabaikalie and Mongolia (the Baikal rift zone) can serve as model systems, which imitate conditions of existence of biological communities in the Precambrian era in the region of volcanic activity and the ancient soda ocean.

1 Introduction

Microbial mats are concerned to be the most ancient biological systems (Zavarzin, 1984). Actualistic paleontology data show that microbial mats played an important role in biogeochemical cycles during the Precambrian era, formation of stromatolites and formation of the modern atmosphere of the Earth (Rozanov, 2002). The distribution of modern microbial mats is restricted to ecosystems where the elevated temperature or high salinity eliminates almost all eukaryotic organisms. These are hydrotherms, hypersaline lagoons and soda lakes.

The Baikal rift zone and Transbaikalian zone of mesozoic tectonic activity are characterized by a variety of extreme environments (Fig. 1).

Microbial mats of these environments exhibit a great range of types of mat-building microbial communities: oxygenic mats built by cyanobacteria

---

B.B. Namsaraev
Institute of General and Experimental Biology, Siberian Branch of Russian Academy of Sciences, Buryat State University Ulan-Ude, Russia
e-mail: bair_n@mail.ru

Fig. 1 Location scheme of studied hydrotherms and soda and saline lakes (Zabaikalie and Mongolia regions). Hydrotherms: *1* Alla, *2* Kuchiger, *3* Bolshaya Rechka, *4* Garga, *5* Seiya, *6* Uro, *7* Umkhei, *8* Zmeinaya, *9* Sukhaya Zagza, *10* Khoito Gol, *11* Shumak, *12* Zhoigon, *13* Khuzhirta. Soda and saline lakes: *14* Khilganta, *15* Gorbunka, *16* Solenoe, *17* Barun Torey, *18* Dabas Nuur, *19* Verkhnee Beloe, *20* Niznee Beloe, *21* Khotontyn Nuur, *22* Shara Burdyin Nuur

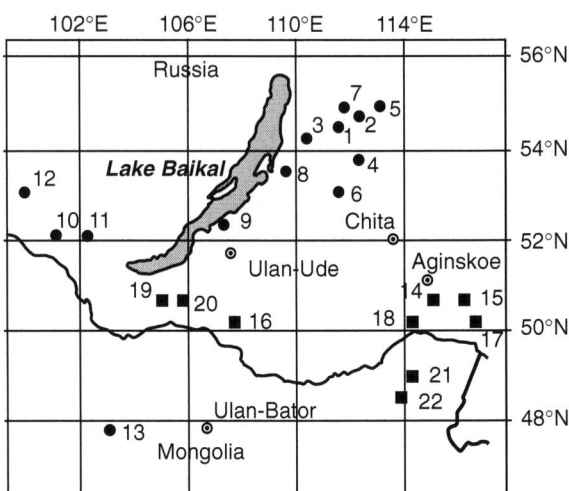

("cyanobacterial mats"), mats built by purple and green anoxygenic phototrophic bacteria ("purple and green mats") and mats built by colorless sulfur bacteria ("sulfuric mats") (Gorlenko et al., 1999) (Table 1).

Table 1 Types of mats and characteristic of development places

Type	Dominants	Development places	T°C	Salinity g l^{-1}	pH
Cyanobacterial	*Phormidium, Oscillatoria, Synechococcus, Mastigocladus*	Hydrotherms	25–68	0.2–1.2	7.8–10.4
Cyanobacterial	*Microcoleus, Phormidium, Oscillatoria, Arthrospira*	Alkaline lakes	−4–43	3–390	9.0–10.6
Purple	*Ectothiorhodospira, Allochromatium, Thiocapsa, Rhodovulum*	Hydrotherms	25–66	0.2–1.2	7.8–10.4
Purple	*Ectothiorhodospira, Marichromatium, Allochromatium, Thiocapsa, Rhodovulum*	Alkaline lakes	−4–43	3–390	9.0–10.6
Green	*Chloroflexus*	Hydrotherms	25–70	0.2–1.2	7.8–10.4
Sulfuric	*Thiothrix, Beggiatoa, Thiophisa, Thioploca*	Hydrotherms	25–56	0.2–1.2	7.8–10.4
		Lake Baikal	3.4–6	0.1	7.5–8.6

Knowledge of these mat types provides evidence of the possible types of mats which may have been important at different times.

The purpose of this article is to present findings on geochemical activity of different types of microbial mats from our long-term (1994–2005) research.

Methods of hydrochemical, microbiological and biogeochemical researches are published in separate papers (Gorlenko et al., 1999).

2 Hot Spring Phototrophic Microbial Mats

Hot springs of the Baikal rift zone belong to nitric, carbonic and methane types. The domination of gases in gas composition of springs was as follows: nitrogen dominates in hot springs of Barguzin valley and Mongolia, carbon dioxide in hot springs of East Sayan Mountains, methane in hot springs of the Tunka valley.

A combination of temperature, light and sulfide presence favors the formation of microbial mats in these springs (Brock, 1967; Gorlenko and Bonch-Osmolovskaya, 1989; Gorlenko et al., 1985). The temperature of water at the outflow of the investigated springs is from 25 to 79 °C; the highest temperature was on griffon of the Alla spring. The salinity of the investigated springs is low (0.2–1.2 g l^{-1}), pH 7.8–10.4. Sulfide concentrations are 0–31 mg l^{-1}. Thus, the microbial communities of alkaline hydrotherms of the Baikal Rift Zone are exposed to the combined influence of extreme factors: high temperature, high pH and, in some cases, high sulfide concentration.

Anoxygenic mats built by purple bacteria are observed in springs Kucheger, Umkhey and Uro. These mats are 0.2–1.5 mm thick. Cyanobacteria, aerobic and anaerobic destructors are also present in mats. Green mat dominated by *Chloroflexus aurantiacus* develops in the Alla spring in the zone of mixing of sulfide-containing alkaline thermal waters with pH 9.0–9.9 and river waters with pH 8.3. The thickness of this mat is 1.0–1.5 mm. These mats have high evolutionary significance showing that mats may be built solely by anoxygenic phototrophic bacteria. Mats of this type may have preceded the development of cyanobacterial mats in the anoxic early Precambrian Earth (Ward et al., 1989).

The most microbial mats of the Baikal rift zone hot springs are those produced by cyanobacteria. Most mats are 0.5–6 cm thick. In the Seiya spring the cyanobacterial mats accumulate up to 13 cm in thickness. The principal microbial components are cyanobacteria of genera *Phormidium, Oscillatoria* and *Synechococcus*. In some hot springs cyanobacteria of genus *Mastigocladus* are found.

In alkaline hydrotherms phototrophic communities appear at lower temperature than in slightly alkaline and neutral hydrotherms. For comparison, in the hot spring Uro with pH 8.8 cyanobacterial mats grow at temperatures lower than 64 °C, whereas in the slightly alkaline Octopus spring with pH 8 cyanobacterial mat appears at temperature as high as 73 °C (Brock, 1967). The

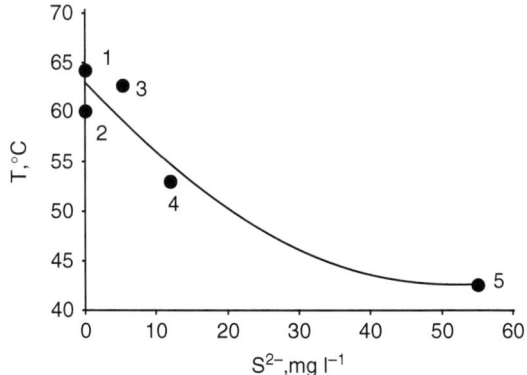

Fig. 2 Restriction of the top limit of microbial mats development in alkaline hydrotherms by the combined influence of temperature and sulfide concentration. *1* Uro spring, *2* Garga spring, *3,4* Bolsherechenskii spring, *5* mineralizing spring Paoha (USA)

inverse relationship between the highest temperature limit of microbial mats development and the concentration of sulfide in spring water is observed (Fig. 2).

Another important feature of alkaline hydrotherms is the domination of cyanobacteria in microbial mat development in hot springs with sulfide concentration more than 1 mg l^{-1}. It was considered earlier that in springs with such concentration of sulfide the mats are formed exclusively by anoxygenic phototrophic bacteria (Castenholz, 1976, 1977; Ward et al., 1989). Our research has shown that this rule cannot be applied to alkaline hydrotherms. In the pH–T diagram based on literature and our data, areas of distribution of various types of microbial communities are indicated at the concentration of sulfide higher than 1 mg l^{-1} (Fig. 3). Anoxygenic phototrophic bacteria dominate over slightly acidic and neutral conditions, but in alkaline conditions, as it has been shown by us, cyanobacteria dominate.

This phenomenon can be explained by a decrease in hydrogen sulfide toxicity for an increase in pH (Howsley and Pearson, 1979). In the diagram pH–S^{2-} it is shown that there is a consecutive change in types of communities during increase of pH in hydrogen sulfide-containing hydrotherms at temperature about 60 °C (Fig. 4).

At pH lower than 7 the majority of hydrogen sulfide molecules are not dissociated and are capable of penetrating easily through a cellular wall. The community dominated by *Chloroflexus aurantiacus* develops. At pH higher than 7 less toxic hydrosulfide ion starts to dominate and cyanobacteria appear in the structure of microbial mat. They form a layer under top layer of *Chloroflexus aurantiacus* which protects cyanobacteria from the influence of high concentration of hydrogen sulfide dissolved in water above the mat. At pH higher than 8.5 all hydrogen sulfide molecules are transformed to a hydrosulfide ion and, as it has been shown by us, cyanobacteria dominate in the microbial community. *Chloroflexus aurantiacus* reaches significant number in microbial mats at pH higher than 8.5. The investigated cultures are capable of photoautotrophic growth, using sulfide as an electron donor, but prefer to grow photoheterotrophically and probably function as primary destructors in the community.

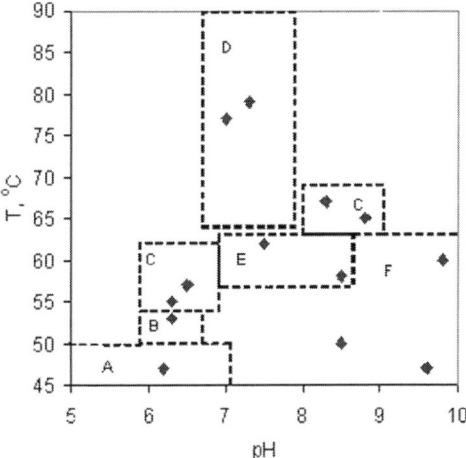

Fig. 3 Distribution of various types of microbial communities on diagram pH–T (in the diagram the data about communities developing under sulfide concentration more than 1 mg l^{-1} in springs are used). Microbial mats with domination of anoxygenic phototrophic bacteria are: A *Chlorobium tepidum* (Castenholz, 1977; Ward et al., 1989); B *Chromatium tepidum* (Madigan et al., 1989; Ward et al., 1989); C *Chloroflexus aurantiacus* (Giovannoni et al., 1987; Madigan et al., 1989; Skirnisdottir et al., 2000; Ward et al., 1989; our data); D cyanobacterial mats with domination of *Chloroflexus aurantiacus* in the upper layer and domination of cyanobacteria in the lower layer (Gorlenko et al., 1999; Jorgensen and Nelson, 1988; Ward et al., 1989); E biofilms of Thermotrix and Aquifex (Caldwell et al., 1976; Gorlenko and Bonch-Osmolovskaya, 1989; Skirnisdottir et al., 2000); F cyanobacterial mats with domination of cyanobacteria (Bauld and Brock, 1973; Ward et al., 1989; our data)

Fig. 4 Distribution of various types of microbial communities on the diagram pH–S^{2-} (communities developing at temperature of about 60 °C are taken into account). Designations of communities are specified in Fig. 3. Ratio between hydrogen sulfide and hydrogen sulfide ion at various pH: *I* H$_2$S > HS$^-$; *II* H$_2$S < HS$^-$; *III* HS$^-$ only. The *bold line* specifies border distributions of microbial mats (our data)

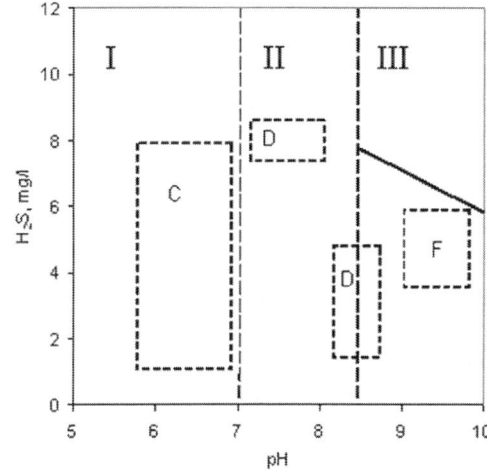

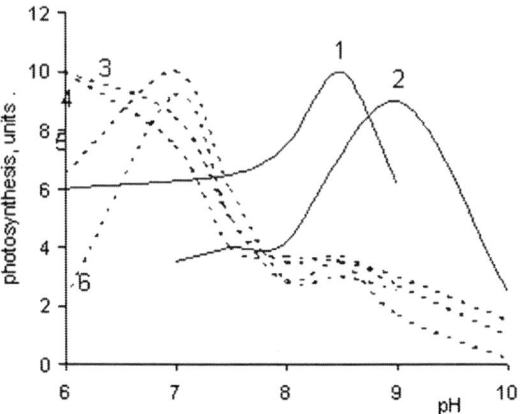

Fig. 5 Dependence on photosynthesis rates of phototrophic bacteria, isolated from the Bolsherechenskii spring (pH 9.8). Cyanobacteria: 1 *Synechococcus elongatus*, 2 *Synechococcus minuscula*. Anoxygenic photosynthetic bacteria: 3 *Blastohloris viridis*, 4 *Rubrivivax gelatinosus*, 5 *Rhodomicrobium vannielii*, 6 *Rhodopseudomonas palustris*

Comparison of pH optimum for cyanobacteria, anoxygenic phototrophic bacteria and chemotrophic bacteria isolated from the Bolsherechenskii spring indicates that cyanobacteria dominating in microbial mat are more adapted to high pH, than anoxygenic phototrophic bacteria. Cyanobacteria, which raise pH during oxygenic photosynthesis, are alkaliphiles, whereas anoxygenic phototrophic and chemotrophic bacteria are neutrophilic and alkalitolerant (Fig. 5).

An increase in stability of a hydrosulfide ion under alkaline conditions also can cause the absence of mass development of thermophilic sulfur bacteria at temperature about 70 °C and pH higher than 8.5.

Microbial communities of alkaline hydrotherms have high productivity comparable with other highly productive ecosystems where eukaryotic organisms dominate. The maximal content of chlorophyll *a* is 892 mg m^{-2}. The rate of oxygenic photosynthesis reaches 3.5 g C m^{-2}d^{-1}, that is comparable to rates of oxygenic photosynthesis in the Termophilnii spring, Russia, Kamchatka (2.3 g C m^{-2}d^{-1}) and Octopus spring, USA, Yellowstone (4 g C m^{-2}d^{-1}) and also in the mat of the Spencer Gulf, South Australia (6.13 g C m^{-2}d^{-1}) (Bauld, 1984; Gorlenko et al., 1999; Revsbech and Ward, 1984). The rate of anoxygenic photosynthesis (5.5 g C m^{-2}d^{-1}) reaches high values in microbial mats even at the dominance of cyanobacteria in microbial mat. It can be explained by switching cyanobacteria to anoxygenic photosynthesis. The maximal rate of dark fixation of CO_2 is 5.5 g C m^{-2}d^{-1}. Maximal total production reaches 21 g C m^{-2}d^{-1}. The isotope analysis data show that cyanobacteria and anoxygenic phototrophic bacteria of hydrotherms use volcanic and atmospheric CO_2 ($\delta^{13}C$ are from –5.9 to – ‰).

The combination of high temperature, high pH and high sulfide influences the intensity of biogeochemical processes in microbial mats. The productivity of microbial mats in springs Garga, Seiya and Uro with low sulfide concentrations is very high; and the maximum of primary production is at temperature 45–50 °C. In the Bolsherechenskii spring with high sulfide concentration and highest pH the maximal productivity is observed at temperatures 33–39 °C and values of productivity are lower than in non-sulfide springs.

Participation of the microbial community in travertine deposition as a model of the Precambrian era carbonaceous stromatolite formation was studied on the Garga spring. The maximum thickness of the travertine is 2.5 m. Travertine is carbonate-calcium in composition and have high content of SiO_2 (3.61%) and MnO (1.27%). The travertine age is 19,245–25,725 (Middle–Late Pleistocene) (Plyusnin et al., 2000). Hydrothermal water belongs to sulfate-sodium type and contains relatively low concentration of calcium and carbonate, therefore the fact of formation from such waters of calcium-carbonate travertine represents significant interest (Borisenko et al., 1976). In opinion of Plyusnin and coauthors (2000), the formation of travertine could not occur during a process of decompression of carbon dioxide at an outflow because the concentrations of carbon dioxide carbonate, bicarbonate and calcium ions are too low. According to our observations, the travertine deposition is closely associated with the activity of microbial mat. The highest activity of this process is measured in zones of microbial mat with excess of primary production above destruction and constant access of hot spring water. During decrease of the hot spring water flow the destruction process (0.89 g C $m^{-2}d^{-1}$) exceeds considerably the primary production (0.24 g C $m^{-2}d^{-1}$), the microbial mat dehydrates and forms a thin gradually collapsing crust; lithification of microbial mat does not occur.

Samples of dehydrated microbial mat, surface and deeper layers of travertine are investigated using scanning electronic microscopy. Microfossils (filaments and spheres) are found in travertine samples and microbial mat (Fig. 6).

Among filaments the samples with diameter from 2.5 up to 4.5 μm (30 units, 60%) dominate. Filaments with thickness of up to 2.5 μm are found in lower numbers (17 units, 34%). Filaments with thickness of up to 10 μm are found only in three cases (6%). Diameter of the spheres is about 10 μm. The highest number of microfossils was revealed in the dried mat (28 filaments and 67 spheres). In lower levels of travertine the number of microfossils decreases probably due to destruction during diagenesis (Fig. 7).

Travertine deposit of the Garga spring at the initial stage of formation is represented by gel organic argillo carbonated substrate which then is transformed into calcites 1, 2 and 3. Formation of calcite grains is located around

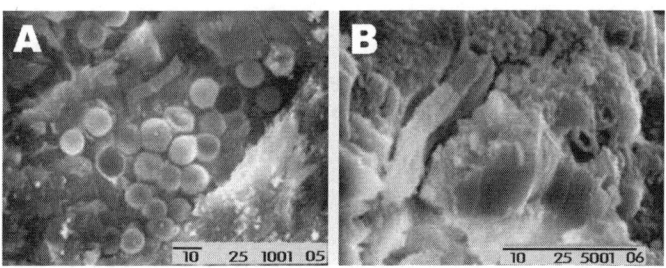

Fig. 6 Microfossils of Garga spring: (**A**) dehydrated mat; (**B**) surface layer of travertine

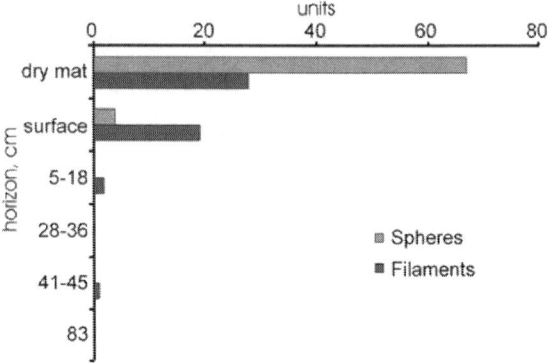

Fig. 7 Changes of compound and number of microfossils in cyanobacterial mat and travertine of Garga spring

pelitomorphic organic argillo carbonated centers with bacterial basis. Fragments of cyanobacteria are basically found in connection with relict particles of gel organic argillo carbonated substrate in calcite grains 1 and partly 2.

Comparison of neutral carbonic springs with high content of calcium and carbonates located in the Baikal rift zone and the sulfate-sodium Garga spring with low concentrations of calcium and carbonates shows that the microbial community plays an important role in the travertine deposition in the Garga spring. In the process of oxygenic photosynthesis an alkaline geochemical barrier is formed and precipitation of calcium carbonate occurs on it. Colloidal systems with high content of bioorganic substance and bicarbonate true solutions are formed in the processes of bacterial production and destruction of organic matter. At later stages bicarbonate solution becomes the centers of crystallization of calcite.

At pH > 8.4 the travertine deposition does not occur due to the absence of calcium. Under these conditions siliceous deposit (geyserite) is formed around Bolsherechenskii, Alla and Uro hot springs. These systems can be regarded as modern analogs of the Precambrian fossil-bearing siliceous stromatolites.

3 Non-phototrophic Biofilms

In the absence of light or at temperatures excluding the growth of phototrophic microorganisms (>73 °C) microbial communities are based on the activity of chemolithoautotrophic aerobic and anaerobic bacteria.

In the Frolikha Bay of the Lake Baikal, underwater vents were studied (Namsaraev et al., 1994). These waters are characterized by increased concentrations of K, Na, Ca ions, lowered concentrations of Fe, Mn, Ni, Zn ions and an enormously high concentration of Cl ion. A sulfur biofilm dominated by *Thioploca* develops in this area (Namsaraev and Zemskaya, 2000). Gentle biofilms are located mosaically and have thickness of 0.1-0.3 mm. The rate of dark CO_2 fixation in underwater mats reaches 24.8 mg C $m^{-2}d^{-1}$, chemosynthesis reaches 13.7 mg C m^{-2}d^{-1} and methane oxidation reaches 0.23 ml CH$_4$ dm^{-3}d^{-1}. High activity of methanogenic and methane-oxidizing bacteria in the microbial community of the Frolikha Bay influences the biogeochemical cycles in this area. The values of δ^{13}C of the benthic animals of this area are from –59.5 to 72.3‰. These data indicate that they use carbon of biogenic methane.

Sulfur streamers are found in high-sulfide slightly alkaline warm springs Alla, Kucheger, Khoito-Gol and Zmeinaya. Streamers are dominated by colorless sulfur bacteria of genera *Thiotrix, Beggiatoa, Thiophisa* and *Thiobacillus*.

4 Microbial Mats of Saline and Soda Lakes

Other systems where microbial mats develop are saline and soda lakes of the Transbaikalian zone of mesozoic tectonic activity. Salinity of investigated lakes varies from 3 to 390 g l^{-1}. The lakes are alkaline with pH from 9.0 to 10.6. Waters are sulfate-chloride, sulfate and carbonate types. The temperature of water in the lakes in the summer warms up to 28–30 °C, in shallow lagoons of lakes up to an ambient temperature (sometimes above 40 °C).

Anoxygenic phototrophic bacteria of genera *Ectothiorhodospira, Allochromatium, Thiocapsa, Rhodovulum* dominate in purple mats of coastal lagoons of soda lakes Khilganta, Gorbunka, Solenoe (Kyran), Barun Torey, Dabas Nuur, Verkhnee and Nizhnee Beloe, Khotontyn Nuur and Shara Burdyin Nuur.

Thin (2–3 mm) cyanobacterial microbial mats are observed in most of the soda lakes. Thick (1.5–2.0 cm) layered cyanobacterial mat is formed only in the shallow soda Lake Khilganta (Gerasimenko et al., 2003). Six vertical layers are revealed (Table 2).

Table 2 Vertical structure of the Khilganta cyanobacterial mat

Zone	Layer (mm)	Color	Components
1	0–2	Dark green	Cyanobacteria (*Microcoleus chthonoplastes, Aphanothece salina, Oscillatoria* sp., *Phormidium molle*), green algae *Chlorella minutissima* , diatoms, purple bacteria, minerals
2	2–4.5	Purple and brown	Purple bacteria (*Ectothiorhodospira, Allochromatium, Thiocapsa, Rhodovulum*), cyanobacteria, green filamentous bacteria *Chloroflexus* sp., minerals
3	4.5–7.5	Brown-green	Cyanobacteria, purple bacteria, minerals
4	7.5–10.5	Grey	Minerals, fragments of cyanobacteria
5	10.5–12.5	Light brown	Fragments of cyanobacteria
6	12.5–18.5	Dark grey	Bacteria-destructors, lysed cells of cyanobacteria

The first has dark green color by filamentous cyanobacteria domination. The next is the layer of development of the purple bacteria. In the third layer cyanobacteria and anoxygenic phototrophic bacteria are observed in equal numbers. In the fourth layer minerals and remnants of filamentous bacteria are revealed. The fifth layer is formed by lysed filaments of cyanobacterium *Microcoleus*. The last layer is a zone of sulfate reduction with dark grey color.

Isotopic data show that primary producers in microbial mats of soda and saline lakes use carbon of microbial origin ($\delta^{13}C$ from -15.6 to $-24.2‰$). Microbial mats of shallow soda lakes have productivity of about $0.1-3.9$ g C m^{-2}d^{-1}. Dark CO_2 fixation reaches $0.02-0.26$ g C m^{-2}d^{-1}. The highest total photosynthesis $1.4-3.9$ g C m^{-2}d^{-1} is measured in the Lake Khilganta cyanobacterial mat. Oxygenic photosynthesis is found in the upper zone of the mat where its part of the total photosynthetic production reaches 25%. In lower layers the photosynthetic production is carried out entirely by anoxygenic photosynthesis. That the rate of anoxygenic photosynthesis is high in all layers indicates participation of cyanobacteria in the process of anoxygenic photosynthesis by using photosystem I. The microbial community of the Lake Khilganta may be concerned as a modern analog of ancient sulfide environments where the evolution of cyanobacteria occurred.

Deposition of minerals in the mat of the Lake Khilganta is found. Aragonite, calcite and calcium phosphate precipitated in zones of development of oxygenic cyanobacteria and anoxygenic photobacteria. Deposition of sulfide basically occurred in the sulfate reduction zone.

5 Terminal Destruction Processes in Microbial Mats

High rates of terminal destruction processes are revealed in microbial mats of hot springs. The maximal value of sulfate reduction reaches 5.5 g S m^{-2}d^{-1}. The rate of methane formation is low, no more than 1.5 mg C m^{-2}d^{-1}. The rate of organic matter mineralization in the course of sulfate reduction has the value of two to three orders of magnitude higher than that characterizing the methane formation. The high rate of sulfate reduction (up to 5.5g S m^{-2}d^{-1}) results in a high activity of sulfur cycle's microorganisms (anoxygenic phototrophic bacteria, sulfur-oxidizing bacteria) even in the absence of dissolved sulfide. In cyanobacterial mats from Spencer Gulf (South Australia) the maximal sulfate reduction rate is 3.5 g S m^{-2}d^{-1}, in Termophilnii spring it is 1.4 g S m^{-2}d^{-1} (Gorlenko et al., 1989; Skyring et al., 1983).

In microbial mats of soda lakes sulfate reduction is also higher than methane formation. Therefore in these mats most part of the organic matter is mineralized during the sulfate reduction.

In the Lake Baikal mat most part of the organic matter (10.3 mg C m^{-2}d^{-1}) is used for formation of methane, whereas reduction of sulfates takes only 0.15 mg C m^{-2}d^{-1}.

6 Conclusion

It is shown that microbial mats of hot springs, soda lakes and freshwater of the Lake Baikal of the Baikal rift zone represent different types of mats which may have been important at different times of the Precambrian era. Modern cyanobacterial mats, purple and green mats and sulfur biofilms are independent cooperative communities consisting of different functional groups of producents and destructors. In cyanobacterial mats the basic producers are filamentous and coccoid cyanobacteria capable of oxy- and anoxygenic photosynthesis and of fixing molecular nitrogen. Purple and green mats are formed by anoxygenic phototrophic bacteria in hydrotherms with a high sulfide content and in soda lakes. In underwater hydrotherms in the absence of light and at high sulfide concentration the streamers of colorless sulfur bacteria are found. These bacteria are capable of using biogenic and volcanic hydrogen sulfide.

Earlier the different microbial mats have been described in hydrotherms, shallow soda and salt lakes and marine lagoons (Zavarzin, 1984). Comparative analysis of literature and our data shows that the microbial communities in microbial mats of hydrotherms and shallow soda and salt lakes are similar in structural and functional characteristics (Bauld, 1984; Gorlenko et al., 1989; Ward et al., 1989).

In these systems the high activity of microorganisms participating in the organic matter production and destruction is measured. Cyanobacterial and purple microbial mats of investigated environments of the Baikal rift zone are high productive systems with dominance of production above destruction (Table 3).

Sulfate-reducing bacteria play a key role in the terminal destruction processes at alkaline pH in a wide range of salinity and temperature in hydrotherms and soda lakes. In microbial mats of freshwater of the Lake Baikal the major terminal destruction process is methane formation. Methane-oxidizing and sulfur-oxidizing bacteria form effective filter on the way of volcanic and biogenic gases.

Phototrophic, lithotrophic and heterotrophic bacteria of microbial mats participate in the formation of biogenic minerals, first of all carbonates, phosphates and sulfides. Precipitation of minerals during the formation of geochemical

Table 3 Intensity of microbial processes in microbial mats of water ecosystems of the Baikal rift zone

Water ecosystems	Photosynthesis ($g\,C\,m^{-2}d^{-1}$)		Chemo-synthesis ($g\,C\,m^{-2}d^{-1}$)	Sulfate reduction ($g\,S\,m^{-2}d^{-1}$)	Methane formation ($\mu l\,CH_4\,m^{-2}d^{-1}$)
	Oxygenic	Anoxygenic			
Hydrotherms	0.002–3.65	0.004–5.48	–	0.006–5.5	0.28–1560
Soda lakes	0.04–3.86	0.003–3.28	–	0.01–0.12	3.5–80.7
Lake Baikal	–	–	1.4	0.014	12000

barriers by microbial communities of hot springs and soda lakes allows considering these communities as modern analogs of ancient stromatolites.

Acknowledgment This work was supported by Programs of the Presidium of the Russian Academy of Sciences "Origin of life and evolution of biosphere" and "Molecular and cellular biology," grants of the Siberian Branch of the Russian Academy of Sciences 24, grant of the Russian Foundation for Basic Research 05-04-97215, grants of "Russian Universities" 07.01.474 and Program of Russian Federation Department of Education and Science "Development of higher school educational potential (2006–2008)."

References

Bauld, J. (1984) Microbial mats in Shark Bay and Spencer Gulf. In: Y.Cohen, R.W. Castenholz and H.O. Halvorson (Eds), Microbial Mats: Stromatolites. Alan R. Liss, New York, pp. 39–58.
Bauld, J. and Brock, T.D. (1973) Ecological studies of *Chloroflexus*, a gliding photosynthetic bacterium. Arch. Microbiol. 92, 267–284.
Borisenko, I.M., Ochirov, Yu.Ch. and Suslenkova, R.M. (1976) The travertine compound from sediments of the some mineral springs of Transbaikalia. Proc. Buryatia Geol. Inst. Ulan-Ude 7(15), 36–52.
Brock, T.D. (1967) Microorganisms adapted to high temperatures. Nature 214, 882–885.
Caldwell, D.E., Caldwell, S.J. and Laycock, J.P. (1976) *Thermotrix thioparus* gen. nov. sp. nov. a facultatively anaerobic facultative chemolithotroph living at neutral pH and high temperature. Can. J. Microbiol. 22, 1509–1517.
Castenholz, R.W. (1976) The effect of sulfide on the blue-green algae of hot springs. I. New Zealand and Iceland. J. Phycol. 12, 54–68.
Castenholz, R.W. (1977) The effect of sulfide on the blue-green algae of hot springs. II. Yellowstone National Park. Microb. Ecol. 3, 79–105.
Gerasimenko, L.M., Mityushina, L.L. and Namsaraev, B.B. (2003) The mats *Microcoleus* from alkaliphilic and halophilic communities. Microbiology 72, 84–93 (in Russian).
Giovannoni, S.J., Revsbech, N.P., Ward, D.M. and Castenholz R.W. (1987) Obligately phototrophic *Chloroflexus*: primary production in anaerobic hot spring microbial mats. Arch. Microbiol. 147, 80–87.
Gorlenko, V.M. and Bonch-Osmolovskaya, E.A. (1989) The microbial mats formation and activity of production and destruction processes. In: Calder Microorganisms. Nauka, Moscow, pp. 53-64.
Gorlenko, V.M., Kompantseva, E.I. and Puchkova, N.N. (1985) The influence of temperature on the distribution of phototrophic bacteria in the hot springs. Microbiology 54, 848–853 (in Russian).
Gorlenko, V.M., Namsaraev, B.B., Kulyrova, A.V., Zavarzina, D.G. and Zhilina, T.N. (1999) The activity of sulfate-reducing bacteria in the sediments of the soda lakes in Southeastern Transbaikal Region. Microbiology 68, 580–585.
Howsley, R. and Pearson, H.W. (1979) pH dependent sulfide toxicity to oxygenic photosynthesis in cyanobacteria. FEMS Microb. Lett. 6, 287–292.
Jorgensen, B.B. and Nelson, D.C. (1988) Bacterial zonation, photosynthesis and spectral light distribution in hot spring microbial mats of Iceland. Microb. Ecol. 16, 133–148.
Madigan, M.T., Takigiku, R., Lee, R.G., Gest, R. and Hayes, J.M. (1989) Carbon isotope fractionation by thermophilic phototrophic sulfur bacteria: evidence for autotrophic growth in natural populations. Appl. Environ. Microbiol. 55, 639–644.

Namsaraev, B.B. and Zemskaya, T.I. (2000) Microbial processes of carbon circulation in bottom sediments of Lake Baikal. Publishing House of Siberian Branch of RAS, Novosibirsk.

Namsaraev, B.B., Dulov, L.E., Dubinina, G.A., Zemskaya, T.I., Granina, L.Z. and Karabanov, E.V. (1994) The participation of bacteria in processes of synthesis and destruction of the organic matter in microbic mats the lake Baikal. Microbiology 63, 345–352.

Plyusnin, A.M., Suzdalnitskii, A.P., Adushinov, A.A. and Mironov A.G. (2000) Formation of travertine from carbonated and nitric hydrotherms in the Baikal rift zone. Geol. Geophys. 41, 546–552 (in Russian).

Revsbech, N.P. and Ward, D.M. (1984) Microelectrode studies of interstitial water chemistry and photosynthetic activity in a hot spring microbial mat. Appl. Environ. Microbiol. 48, 270–275.

Rozanov, A.Yu. (Ed.) (2002) Bacteriological Paleontology. PIN RAS, Moscow.

Skirnisdottir, S., Hreggvidsson, G.O., Hjorleifsdottir, S., Marteinsson, V.T., Petursdottir, S.K., Holst, O. and Kristjansson J.K. (2000) Influence of sulfide and temperature on species composition and community structure of hot spring microbial mats. Appl. Environ. Microbiol. 66, 2835–2841.

Skyring, G.M., Chambers, L.A., Bauld, J. (1983) Sulfate reduction in sediments colonized by cyanobacteria, Spencer Gulf, South Australia. Aust. J. Mar. Freshw. Res., 34, 359–374.

Ward, D.M., Weller, R., Shiea, J., Castenholz, R.W. and Cohen, Y (1989) Hot spring microbial mats: anoxygenic and oxygenic mats of possible evolutionary significance. In: Y. Cohen and E. Rosenberg (Eds), Microbial Mats: Physiological Ecology of Benthic Microbial Communities. ASM, Washington, pp. 3–15.

Zavarzin, G.A. (1984) Bacteria and composition of atmosphere. Nauka, Moscow.

On the Concept for the Organization of the Modern Biosphere in the Terrestrial Subsurface

A. A. Oborin, L. M. Rubinstein, and V. T. Khmurchik

Abstract There are strong indications that microbial life is widespread at depth in the crust of the Earth. The proposed concept suggests a subdivision of the biosphere in the terrestrial subsurface into three zones. The zone of the hydrocarbon-oxidizing bacterial filter is situated at depths of up to 1000 m. The biota of this zone utilizes methane and other hydrocarbons. The zone of the carbon dioxide–hydrogen bacterial filter is situated at even greater depths (3500–7000 m). The biota of this zone utilizes H_2 and CO_2 of mantle fluids. The zone of naphthidiobiosis could appear between these two filters, and the activity of its biota could result in the formation of primary hydrocarbons. These zones are closely interrelated in their functions and represent an important part of the biosphere. The most recent evaluations of the subsurface biota are presented.

1 Introduction

The subsurface microbiology was distinguished as an independent scientific direction in the environmental microbiology over 15 years ago, and microbiological investigations of the lithosphere showed that rather diverse and active bacterial life occurred at depths being available for the exploration (Fliermans and Balkwill, 1989; Ivanov, 1991). The influence of different geological factors on the subsurface microbiota was examined, and it was determined that microbiological properties correlated to geological, hydrological, and geochemical properties of the lithosphere (Amy and Haldeman, 1996; Chapelle, 1993; Ehrlich, 1996). At the same time a great number of fundamental research issues remain undisclosed including the structure and pecularities of the subsurface microbial communities and the evolution of the subsurface microbial world.

V. T. Khmurchik
Institute of Ecology and Genetics of Microorganisms of Russian Academy of Sciences, Perm, Russia
e-mail: khmurchik@iegm.ru

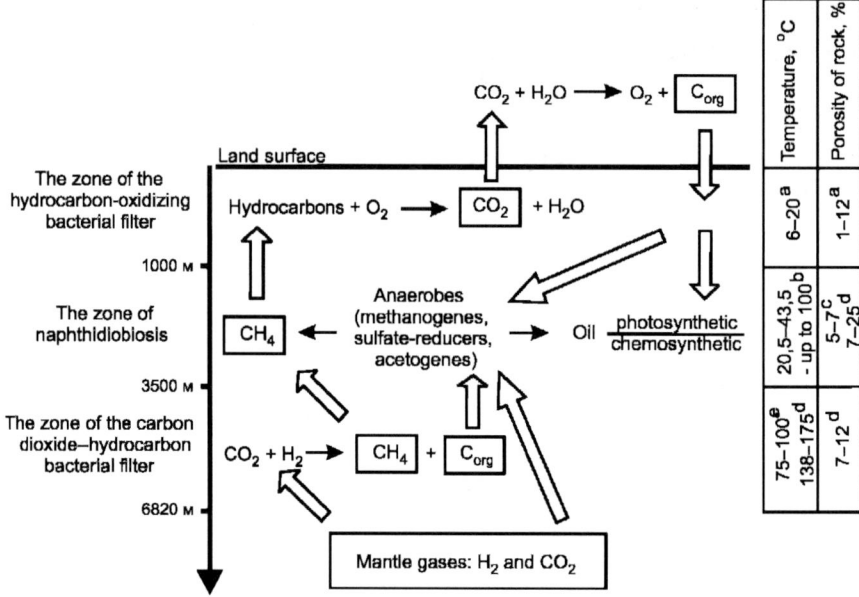

Fig. 1 The structural organization of subsurface biota and fluxes of carbon substances in terrestrial subsurface ([a]Popov, 1985; [b]Bars and Zaydelson, 1973; [c]Sharonov, 1971; [d]Esipko et al., 2001; [e]Bashkova, 2001)

We suggest the concept for the structural organization of the modern biosphere in the terrestrial subsurface. It is known that the lithosphere consists of two regions: oceanic, which comprises about ¾$_{parts}$ of the lithosphere's area, and terrestrial one. The terrestrial region in turn could be subdivided into two types: geosyncline type and platform type. It is known that the biosphere could penetrate into all these types and regions of the lithosphere. The proposed concept tends to describe the structural organization of subsurface biota within rocks of the terrestrial lithosphere of the platform type only and supposes that the biota could be subdivided into three zones that are closely interrelated in their functions: the zone of the hydrocarbon-oxidizing bacterial filter at the upper part of the lithosphere, the zone of the carbon dioxide–hydrogen bacterial filter at the super deep part, and the zone of naphthidiobiosis that appears to be formed between them (Fig. 1).

2 The Zone of the Hydrocarbon-Oxidizing Bacterial Filter

The upper part of the subsurface biota occupies the depth down to 1000 m. It was termed "the hydrocarbon-oxidizing bacterial filter" (Mogilevsky et al., 1979). This filter is represented by the unique trophic natural system of microorganisms that includes biotopes of synthrophic microbiological communities of aerobic, chemolithotrophic, and heterotrophic microorganisms, and is based on the activity of hydrocarbon-oxidizing bacteria. Hydrocarbon-oxidizing bacteria assimilate methane and its homologues and finally oxidize them to carbon dioxide and water. They provide most aminoacids as exometabolites, and—under cellular lysis—all essential components (proteins, lipids, carbohydrates) for heterotrophically grown bacteria (Oborin and Stadnik, 1996; Zavarzin, 1972). Carbon dioxide is in turn a growth substrate for all autotrophic microorganisms. While utilizing hydrocarbons that are "exhausted" from the earth's entrails the microorganisms of the filter sustain the hydrocarbon-free conditions of the earth's atmosphere. The activity of the hydrocarbon-oxidizing bacterial filter renders significant effect on the environments: it causes the sharp changes in the migration capability of most elements of chalkophile group and elements of variable valency—Fe, Mn, Cu, Ti, Pb, Zn, S, the essential elements of carbonate rocks—Ca, Mg, Ba, Sr, the elements of clay minerals—Si, Al, and the elements in complex organo-mineral compounds—V, Ni, Co, etc. Most alterations occur in carbonate rocks: recrystallization and dolomitization of limestone, pyritization, silicification, reduction of ferric iron to ferrous iron, deconsolidation of rocks, and the formation of secondary porosity therein. Chloritization, serpentinization, kaolinization, montmorillonization, and also the localization de novo of sulfide minerals (pyrite, chalkopyrite, sphalerite, galenite, and cinnabar) are observed in the terrigenous rocks. The formation of new additional sulfide and magnetic minerals provides the alterations in magnetic field and electric conductivity of rocks, and the creation of geophysical abnormalities above the hydrocarbon deposits (Oborin et al., 1996). The integral geochemical activity of the Earth's bacterial filter biocenosis is manifested in the alteration in the parameters of oil-gas-prospecting probation, in the formation of litho-, phyto-, and aerocosmic abnormalities over the oil-gas deposits (Galkin et al., 2006).

The activity of the hydrocarbon-oxidizing bacterial filter depends on the surface and the sun via photosynthesis: almost all oxygen on the Earth is produced via photosynthesis; heterotrophic bacteria could utilize the input of organic matter from the surface. As the hydrocarbon-oxidizing bacterial filter and the surface biota are interrelated in their substrates and products one could suggest the coevolution of them. The issue of the evaluation of this filter role in the involvement of "juvenile" carbon of deep fluids into a biogenous cycle and in the enrichment of the rocks with organic substances seems to be topical as well.

3 The Zone of the CO_2–H_2 Bacterial Filter

The lower part of the subsurface biota occupies the depth from 3500 m to more than 7000 m below surface, but the proper depth of penetration of the biosphere into the lithosphere (so-called "the lower boundary of the life") remains unknown. As it is largely comprised of bacteria possessing chemolithotrophic type of metabolism that actively uptake hydrogen and carbon dioxide of mantle fluids (Gold, 1993), we term it "the carbon dioxide-hydrogen bacterial filter". The detection of the lower boundary of the biosphere in the terrestrial subsurface is the first and most important scientific problem (Ivanov, 1991). It was supposed that the upper temperature limit of bacterial life may well be within the region of 110–150 °C (Stetter et al., 1990). Then, the lower boundary of the biosphere could be located in depth between 5 and 10 km in most areas of the crust (Gold, 1992). Our investigations of core samples taken from the Ural and the Tyumen superdeep wells revealed viable hydrogen-, carbon dioxide-, and carbon monoxide utilizing bacteria in them (Ilarionov et al., 1996; Oborin et al., 1999). So, at present the reliable determined depth of propagation of the biosphere could comprise 6820 m (Ilarionov et al., 1996).

The dispersed organic matter in this zone could be formed at the sacrifice of chemosynthetic activity of the biota (Pedersen, 1997; Stevens, 1997). As the biota does not depend on input of photosynthetic organic matter from the surface it could be a model for how ecosystems functioned before the evolution of photosynthesis (Stevens, 1997). Moreover, it was supposed that the place for the origin of life could be a deep subterranean igneous rock environment where the early life could survive cosmological events, which sterilized the surface of the planet (Pedersen, 1997; Stevens, 1997). The biota of the carbon dioxide–hydrogen bacterial filter uses as its energy source various chemical imbalances that the outgassing process creates as gases and liquids stream up through the rocks, so one could suggest the dependence of the evolution of the subsurface biota on the evolution of the lithosphere and the presence of adaptations of the biota to high temperature similar to that observed in microflora of hydrothermal abyssal vents in the oceans. However, further microbiological investigations are required.

4 The Zone of Naphthidiobiosis

Our concept supposes that at depth from 1000 to 3500 m (up to 5000 m in several cases) below surface the zone of naphthidiobiosis is situated. Naphthidiobiosis is a process of microbiological synthesis of gaseous and liquid hydrocarbons. Major microorganisms that are involved into the biosynthesis of hydrocarbons appear to be methanogenic, sulfate-reducing, and probably acetogenic ones. Vernadsky attracted the attention to the possibility of direct microbiological synthesis of oil hydrocarbons (Vernadsky, 1954), and as long ago as 1930s he stated, "One of the regular scientific issues appears to be the elucidation of whether those organisms that gave the origin to... oil and bitumen... are preserved; as we know, they [oil and bitumen-*auth*.] are not created merely by physical-chemical processes" (Vernadsky, 1987, p. 57). Subsequently this suggestion was supported in field and laboratory experiments (Bagaeva, 1997; Bagaeva and Zolotukhina, 1994; Simoneit et al., 2004; Slobodkin and Bonch-Osmolovskaya, 1994). Carbon and hydrogen sources for naphthidio-generating biota could be organic substances of photosynthetic origin, migrated organic matter of the area of the carbon dioxide–hydrogen bacterial filter, and inorganic gases (CO_2, H_2) of a migration stream from the Earth's entrails. It is the authors' opinion that this area does not spread everywhere in the lithosphere, but its distribution is associated with deep-seated joints and fractures and so has a local character.

Vernadsky supposed that the geological process of formation of oil fields is proceeding at present, "But, indeed, simultaneously the generation of oils, coals, bitumens etc. takes place in newest rocks, in the subsurface biosphere at present" (Vernadsky, 1987, p. 249). The origin of oil and gas fields remains highly topical. According to the concept, atoms of carbon in hydrocarbon molecules could originate from both sources—photosynthetic organic matter and mantle fluids, and the biota of the zone of naphthidiobiosis is involved into hydrocarbon generation that could explain the low-temperature character of the process and the relationship of oil fields with the fracture disturbances of the earth crust (Oborin et al., 2005). It is the authors' opinion that the formation of oil fields is a very slow process, which is limited by many insufficiently studied factors of the environment.

5 Evaluation of the Subsurface Biota

All living organisms represent a large geological force (Vernadsky, 1987), and so the evaluation of the subsurface biota in terms of mass and volumes is one of the main scientific issues. It was supposed that total mass of the subsurface biota is comparable to all the living mass at the surface (Gold, 1992). As a part of the International Ocean Drilling Program, investigations of subseafloor sediments taken from the depth up to 900 m below sea floor level revealed that their biota built up 10% in the mass of the surface biota (Parkes et al., 2000). We should point out that this appraisal is related to the subsurface biota into the oceanic region only, not the lithosphere as a whole. The opinion of some investigators is that microbial life at subsoil levels is much higher in mass and volume than that in soils (Vorobyova et al., 1997). We studied the microbiota in subsurface waters of the terrestrial lithosphere of the platform type only. According to our calculations, the total mass of free-living bacteria in aquifers at the depth up to 2.2 km could comprise about 2% of mass of the surface biota (Oborin et al., 2004). However, further investigations are required.

6 Conclusion

Numerous investigations of the lithosphere confirm the existence of diverse microbial world at depths being available for the exploration, and the results of the activity of the subsurface biota possess a global character. It was supposed that total mass of the subsurface biota is comparable to all the living mass at the surface (Gold, 1992). To authors' calculations on the total mass of free-living bacteria in aquifers at the depth up to 2.2 km could comprise about 2% of the mass of the surface biota (Oborin et al., 2004). According to the concept the biosphere in the terrestrial subsurface could be subdivided into the three zones. The zone of the hydrocarbon-oxidizing bacterial filter is located at the depth up to 1000 m. The biota of this zone utilizes hydrocarbons that are exhausted from the deep levels of the lithosphere providing the hydrocarbon-free conditions for the earth's atmosphere. The zone of the carbon dioxide–hydrogen bacterial filter is situated at the super deep depth (3500–7000 m). The biota of this zone uptakes H_2 and CO_2 of mantle fluids and synthesizes methane and organic matter. The zone of naphthidiobiosis could appear between this two filters, but its distribution is associated with deep-seated joints and fractures, and so the zone does not spread everywhere in the lithosphere. The biota of the zone could use the organic matter of photosynthetic origin, the migrated organic matter of the zone of the carbon dioxide–hydrogen bacterial filter, and gases of a migration stream from the Earth's entrails; its activity could result in the formation of the primary oil hydrocarbons. These three zones are closely interrelated in their functions, represent the important part of the biosphere, and could coevolve with the lithosphere and the surface biota.

Acknowledgment This work is supported by the Program of the Presidium of RAS "Scientific fundamentals of biodiversity preservation in Russia" and RFBR 07 – 05 – 00541a grant .

References

Amy, P.S. and Haldeman, D.L. (1996) The Microbiology of the Terrestrial Deep Subsurface. CRC Lewis Publications, Boca Raton.
Bagaeva, T.V. (1997) The ability of sulfate-reducing bacteria of various taxonomic groups to synthesize extracellular hydrocarbons. Mikrobiologiya 66, 666–668.
Bagaeva, T.V. and Zolotukhina, L.M. (1994) The formation of hydrocarbons by sulfate reducing bacteria during the growth in chemolithoheterotrophic conditions. Mikrobiologiya 63, 993–995.
Bars, E.A. and Zaydelson, M.I. (1973) Hydrogeological Conditions for the Forming of Oil and Gas Field at Volga-Ural Basin. Nedra, Moscow.
Bashkova, S.E. (2001) The temperature regimen in the main oil and gas fields of Russia at the depth of 5 km. In: M.B. Keller and A.V. Lipilin (Eds), Assessment Criteria for Oil-and-Gas Content Below Commercially Developed Horizons and Prioritization of Geological-Prospecting Work. KamNIIKIGS, Perm, T. 1, pp. 30–40.
Chapelle, F.H. (1993) Ground-Water Microbiology and Geochemistry. Wiley, New York.
Ehrlich, H.L. (1996) Geomicrobiology. Marcel Dekker, New York.
Esipko, O.A., Gorbachev, V.I. and Sokolova, T.N. (2001) The results of geophysical studies at the Tyumen Superdeep Well. In: M.B. Keller and A.V. Lipilin (Eds), Assessment Criteria for Oil-and-Gas Content Below Commercially Developed Horizons and Prioritization of Geological-Prospecting Work. KamNIIKIGS, Perm, T. 2, pp. 27–52.
Fliermans, C.B. and Balkwill, D.L. (1989) Life in the terrestrial deep subsurface. BioScience 39, 370–377.

Galkin, V.I., Oborin, A.A. and Khmurchik, V.T. (2006) The fundamentals of theory, methodology, and geological-economical efficiency of biogeochemical methods of oil and gas deposits searching. Sci. Ind. 1, 28–29.

Gold, T. (1992) The deep, hot biosphere. Proc. Natl Acad. Sci. USA 89, 6045–6049.

Gold, T. (1993) The origin of methane in the crust of the Earth. In: D.G. Howell (Ed.), The Future of Energy Gases. USGS Professional Paper 1570. US Government Printing Office, Washington.

Ilarionov, S.A., Oborin, A.A., Seleznyov, I.A., Khmurchik, V.T., Rubinstein, L.M., Bakhareva, E.Z. and Titova, A.V. (1996) Microbiological study of core samples from the Tyumen Superdeep Well. In: V.B. Mazur (Ed.) The Tyumen Superdeep Well (0-7502 m). Results of Boring and Research. Nedra, Perm, pp. 294–296.

Ivanov, M.V. (1991) Problems of subsurface microbiology in connection with drilling of superdeep holes. Sovetskaya Geol. 8, 10–13.

Mogilevsky, G.A., Bogdanova, V.M. and Kichatova, S. (1979) Bacterial filter in the zone of oil and gas deposits, its peculiarities and methods of investigation. In: Geochemical Method of Oil and Gas Search and the Problems of Nuclear Geology. Nedra, Moscow, pp. 210–246.

Oborin, A.A. and Stadnik, E.V. (1996) Oil-and-Gas Prospecting Geomicrobiology. UB RAS, Ekaterinburg.

Oborin, A.A., Ilarionov, S.A., Seleznyov, I.A. and Khmurchik, V.T. (1999) Microbiological investigations at the Ural Superdeep Well. In: B.N. Khakhaev and A.F. Morozov (Eds), Results of Boring and Research at the Ural Superdeep Well (SD-4). Nedra, Yaroslavl, pp. 354–360.

Oborin, A.A., Rubinstein, L.M., Khmurchik, V.T. and Churilova, N.S. (2004) The Concept for the Organization of the Subsurface Biosphere. UB RAS, Ekaterinburg.

Oborin, A.A., Rubinstein, L.M. and Khmurchik, V.T. (2005) The role of subsurface microbiota in carbon flows of the upper part of lithosphere. Geol. Geophys. Develop. Oil Gas Deposits 9–10, 34–36.

Parkes, J.R., Cragg, B.A. and Wellsbury, P. (2000) Recent studies on bacterial populations and processes in subseafloor sediments: a review. Hydrogeol. J. 8, 11–28.

Pedersen, K. (1997) Microbial life in deep granitic rock. FEMS Microbiol. Rev. 20, 399–414.

Popov, V.G. (1985) Hydrogeochemistry and Hydrodynamics of Pre-Ural. Nauka, Moscow.

Sharonov, L.V. (1971) The Formation of Oil and Gas Fields at the North of Volga-Ural Basin. PKI, Perm.

Simoneit, B.R.T., Lein, A.Yu., Peresypkin, V.I. and Osipov, G.A. (2004) Composition and origin of hydrothermal petroleum and associated lipids in the sulfide deposits of the Rainbow Field (Mid-Atlantic Ridge at 36N). Geochim. Cosmochim. Acta 68, 2275–2294.

Slobodkin, A.I. and Bonch-Osmolovskaya, E.A. (1994) The growth and producing of methabolites by extreme-thermophilic archaea *Desulfurococcus*. Mikrobiologiya 63, 981–986.

Stetter, K.O., Fiala, G., Huber, G., Huber, R. and Segerer A. (1990) Hyperthermophilic microorganisms. FEMS Microbiol. Rev. 75, 117–124.

Stevens, T. (1997) Lithoautotrophy in the subsurface. FEMS Microbiol. Rev. 20, 327–337.

Vernadsky, V.I. (1954) Essays on geochemistry. In: Selected Works. USSR Academy of Sciences, Moscow.

Vernadsky, V.I. (1987) The Chemical Structure of the Earth's Biosphere and Its Environment. Nauka, Moscow.

Vorobyova, E., Soina, V., Gorlenko, M., Minkovskaya, N., Zalinova, N., Mamukelashvili, A., Gilichinsky, D., Rivkina, E. and Vishnivetskaya, T. (1997) The deep cold biosphere: facts and hypothesis. FEMS Microbiol. Rev. 20, 277–290.

Zavarzin, G.A. (1972) Lithotrophic Microorganisms. Nauka, Moscow.

Biomineralization and Evolution. Coevolution of the Mineral and Biological Worlds

I. S. Barskov

Abstract The concept of coevolution as a process of irreversible changes in the composition, structure, and function of two or more co-existing systems of different origin resulting from the exchange of the matter, energy, and information is applicable to a broad range of processes in nature. This chapter considers biomineralization, one of the fundamental biospheric phenomena, as an example of coevolution of the biological and mineralogical worlds.

1 Introduction

The evolution of the biosphere is a process of dynamic interactions of at least three of its components (three natural systems): living matter, non-living matter, and the space in which they exist. Irreversible changes in these systems, including their content (taxonomic content of the biota, composition of the atmosphere and geosphere, positions of continents and oceans), constitute their inherent evolution. Biological evolution is thought to be stochastic, and its results are unpredictable. Evolution of abiotic systems is determined by their previous states. Changes in each system are not autonomous, but connected with the evolution of other systems influencing their future, i.e. contributing to coevolution.

2 Definition of Coevolution

The first definition of coevolution is connected to the phenomenon of reciprocal evolutionary changes in two interacting species or groups of species, when change in the genetic composition of one species (or group) leads to a genetic

---

I. S. Barskov
Paleontological Institute of the Russian Academy of Science, Moscow, Russia
e-mail: barskov@paleo.ru

change in another (Ehrlich and Raven, 1964). Presently, the concept of coevolution extends far beyond the study of two or several species and biological systems in general and is used in a wider sense, although remaining within ecology and biosphere evolution studies. For instance, while discussing today's ecological crisis Danilov-Danilian et al. (2005, p. 140) wrote: "...this term (meaning coevolution – I.B.) successfully describes a wide spectrum of phenomena related to the evolution of any interacting systems or elements of the same system." Moissejev (1997), discussing present day interactions between man and biosphere, uses the term "coevolution" to describe these interactions and equates it to the term "stable development." According to this viewpoint, coevolution is favourable to both interacting systems and allows their trouble-free and progressive development. The academic program "Origin and Evolution of the Biosphere," by containing a section on "Coevolution of Abiotic and Biotic Events," expands this term to the past states of the biosphere and geospheres of the Earth. Thus, the term "coevolution" is understood as a collection of general rules of the evolution of nature and requires a definition describing general rules. In the philosophical sense, the essence and purpose of the existence of everything that exists is the interaction of co-existing entities, i.e. coevolution. Nothing can exist without such an interaction. Interaction in a general sense is an exchange of matter, energy, and information between systems. Given that, coevolution in general may be defined as a process of irreversible change in the composition, structure, and function of co-existing natural systems and their parts due to the enhancement (acceleration, modification) of the exchange of matter, energy, and information between systems. The hierarchy of biological and non-biological systems comprising the biosphere allows their evolution to be studied at various hierarchical levels. The study of these processes constitutes the basis of various fields of modern science.

The interaction between the biological molecules which are products of metabolism (either normal or pathological) of organisms and ions of the environment that results in the formation of the "secondary" mineral world is studied by the science of biological mineralization (biomineralization). Classical ecology (currently often referred to as bioecology) studies interactions on another hierarchical level, i.e. interactions of organisms and the environment. Complex relationships between communities of organisms and their environment (biotope), including their mutual changes constituting microevolution, are studied by biogeocoenology.

The investigations of global tendencies in the mutual effects of the Earth's biota at different stages of its development and of the Earth's geospheres, which form a global ecotop of the biota, are studied by global ecology or biospherology. At present, when the influence of human activity on all natural systems: biota, lithosphere, hydrosphere, atmosphere, and on man himself as a special systemic entity (social, religious, cultural, and informational) approaches a permissible limit, new fields of coevolutionary studies: geoecology and ecological geology, problems of stable development in the relationship between man and the biosphere, genetic ecology, cultural ecology, etc. have arisen. Each of these fields, including the first definition, has its own understanding of coevolution and studies particular patterns of the coevolutionary process, but each of them in the end can be reduced to the above definition of coevolution. Processes of the origin of life, which are described by the hypothesis of astrocataclysm (Snytnikov, "Astrocatalysis Hypothesis for Origin of Life Problem," this book), are interactions between the simultaneously born systems of the ancestral mineral and biological worlds and are also processes of coevolution.

3 Biomineralization Is Coevolution of the Mineral and Biological Worlds

This paper discusses biological mineralization, among the most important aspects of coevolution for understanding the history of the biosphere. Biological mineralization is one of the fundamental processes in nature occurring at the molecular level and resulting in the formation of the biogenic mineral world, i.e. of the secondary mineral nature. The appearance and evolution of the process of biomineralization that can be traced in the geological (and possibly in the pre-geological) past profoundly affected the evolution of the biological world and development and change of all of the inorganic (abiological) envelopes (geospheres) of our planet.

Coevolution of the biological and mineral worlds is the most striking example of the environment-creating function of living matter.

Presently, biomineralization studies extended far beyond the framework of classical biology and natural history into theory (molecular biology, genetic, and biophysics) and practical use (orthopaedics, dentistry, technology, etc.). Major issues of biomineralization and their implications for the Earth's history are discussed below.

3.1 Modern Understanding and Types of Biomineralization

Rigorous studies of biomineralization in recent decades significantly extended the original, apparent understanding of biomineralization as a process of secretion of minerals by living organisms. In the modern understanding, biomineralization is a process when biological molecules and supramolecular cellular and extracellular structures forming during metabolism act as mediators, templates, and matrices for mineral compounds. The processes of biomineralization occur in conditions, which are very distinct from the conditions of inorganic mineralization, i.e. in "normal" temperature ranges, in a water medium, with the participation of ions of water-soluble compounds.

Some 80 minerals are currently known to be formed through the metabolism of organisms. Theoretically any mineral can be created by biomolecules because the number of possible spatial conformations of biological molecules (mediators of mineralization) greatly exceeds 230 Fedorov's groups of crystalline symmetry that describe a possible arrangement of particles inside crystals of any mineral.

In most reviews of biomineralization (Dove et al., 2004; Lowenstam and Weiner, 1989; Mann, 2001) two major types of biomineralization are recognized. These are the types first recognized by Lowenstam (1981): (1) induced and (2) matrix mediated. Recent studies have significantly contributed to the understanding of the types of biomineralization. It has been shown that natural and synthetic biopolymers, their supramolecular structures, biofilms, and also simple organic compounds can work as mediators and templates of mineral formation from non-saturated water solution, which is not observed in abiogenic mineralization. This process has been referred to as biomimetics (Sarikaya, 1994). Biomimetic technology is thought to have immense potential for producing minerals with predetermined qualities, and generally belongs to the field of coevolution of the noosphere and mineral world and leads to the creation of a third mineral kingdom, which is controlled by man. It seems logical to assume that this process was performed by nature much earlier than it was discovered by man and was a significant factor in the biosphere.

Recently, with the discovery and study of nanobacteria, an additional mineralization phenomenon has become known. Nanobacteria are microorganisms with a diameter less than 0.2 μm. Formerly, small mineralized objects referred to as nanobacteria had been discovered in various rocks and on the surface of some minerals (Folk, 1992, 1993) and were interpreted as fossilized remains of extremely small organisms, one tenth of the size and one thousandth of the volume, of any known bacteria. In 1989 the existence of living nanobacteria, isolated from kidney stones, human blood, cow blood, and commercial blood serum preparations, was confirmed (Kajander and Ciftçioglu, 1998). They form the tiny mineral structures of which apatite is composed. Many critics have argued that they are smaller than the minimum possible size for a living cell, and some have argued that the tiny structures in question were formed by abiological processes (Cisar et al., 2000; Kirkland et al., 1999), and that there is no evidence for the existence of any life forms which are only 10% of the size of the smallest microbes known on Earth.

Although it is not yet universally accepted that nanobacteria are living organisms, numerous workers have referred to numerous examples of nanobacteria on Earth. Minuscule 25–200 nm calcite and aragonite particles, of spheroidal or oval shape, have been found in hot springs carbonates, hardgrounds and other carbonate sediments, recent and Precambrian stromatolite and ovoid limestone (Folk, 1992, 1993) and have been interpreted as nanobacteria. Structures resembling nanobacteria are found in Martian meteorites (Benzerara et al., 2003; Folk and Lynch, 1998; Hoover et al., 2004; McKay et al., 1996) and Precambrian phosphorites (Zhegallo et al., 2000). There is a hypothesis that manganese and iron ores have also been formed by nanobacteria (Chafetz et al., 1998). Biomimetic and nanobacterial mineralization in general contributes to Lowenstam's concept of induced mineralization and changes our interpretation of the origin and early stages of this process. Both types of induced biomineralization are largely physical–chemical, rather than biological processes. Living organisms supply organic molecules and/or supramolecular structures as final products of their metabolism, rather than parts of organisms. During biological-induced mineralization, the resulting minerals and their crystalline aggregates are equivalent in their habitus and properties to minerals formed through inorganic processes. During the bacteria-induced mineralization a wide range of minerals may be formed, including metal ores, depending on the medium in which mineralization occurs. Induced biomineralization was the first stage of the evolutionary development of this process; it is widely represented in Prokaryota and is also present in Eukaryota. In Eukaryota, a second, more advanced type of mineralization matrix-mediated biomineralization of Lowenstam (1981) (currently called controlled biomineralization) prevails. Controlled mineralization results in crystals that have specific morphology and shape. Their polycrystalline structure is genetically controlled and shows a high regularity of their axes and crystalline structure. These biominerals are different from minerals formed inorganically. They usually have multileveled hierarchical inner structure, which is not observed in abiogenic minerals. This suggests the presence of several fractions of biomolecules participating at different stages of mineralization from nucleation to the arrest of crystalline growth. In addition, some minerals formed by biologically controlled mineralization do not precipitate in the modern hydrosphere through abiogenic synthesis (gypsum, fluorite, celestine, magnetite, etc.). During controlled mineralization a restricted number of minerals containing calcium, phosphorus, iron, and silicon, i.e. elements playing an important role in the cell's metabolism, the content of which is rigidly controlled in the cytoplasm. Although matrix-mediated biomineralization is known in several groups of bacteria, its mass occurrence is related to the appearance of Eukaryota and the development of mineralized skeletons. It is still not known whether Eukaryota inherited the mechanism of matrix-mediated biomineralization, or in various groups of eukaryotes, including metazoans, matrix-mediated biomineralization developed independently (Lowenstam, 1984).

3.2 *Evolution of the Composition, Structure, and Functions of Biominerals*

The major aspects of the relationships between the biological evolution and biomineralization were considered earlier (Barskov, 1982, 1984; Lowenstam, 1984; Lowenstam and Weiner, 1989). It is very likely that some inorganic minerals worked as templates and matrices for the formation of the first self-replicating organic compounds, predecessors of the cellular organization of life. This may be an explanation of their chiral selectivity, which would be unlikely in the case of the non-matrix-mediated abiogenic synthesis. Biomineralization is essentially a reverse process, when biological molecules and supramolecular structures work as mediators of mineralization. The composition and diversity of bacterially induced minerals largely depend on the environment and suggest the diversity of biogeochemical functions in prokaryotes, rather than differences in their evolutionary level. Mineralogical composition of the products of the skeletal matrix biomineralization is restricted to four minerals: calcite, aragonite, silica and calcium phosphate. Siliceous mineralization is mainly observed in Protista (radiolarians, diatoms, and closely related groups) and lower metazoans (sponges). Phosphatic mineralization is observed in advanced phyla of invertebrates (brachiopods and arthropods) and chordates. Phosphatic biomineralization of the early Cambrian small shelly fossils, the "first" skeletal organisms in the Phanerozoic, is likely to result from the diagenetic biomimetic mineralization against a background of spotted carbonate mineralization (Lowenstam, 1984). Calcite and aragonite biomineralization are universally found in almost all eukaryotic groups (from Protista to Vertebrata). In some phyla, organisms have only one mineral (either calcite or aragonite); in some phyla there are both types. In the evolution of corals, calcitic skeletons precede aragonitic. In bivalves and gastropod molluscs, quite the reverse occurs, taxa with calcitic shells appeared in the geological record later than those with aragonitic shells. The structure, i.e. spatial organization and size hierarchy of crystalline elements, is the ultimate morphological demonstration of skeletal matrix-mediated biomineralization. The supposed stages of the gradual transition from the induced to controlled mineralization considered by Lowenstam (1984) are observed in the organization of the size and shape of crystals and of their arrangement in polycrystalline units and their integration in skeleton formations. This process occurred in the Late Precambrian and first resulted in patchy mineralization of the external skeleton, composed of isolated sclerites, and later massive skeletons (Knoll, 2004; Lowenstam, 1984). This transition is usually thought to be connected with the Cambrian explosion. This crisis resulted in the skeleton acquiring a morphogenetic function and subsequently the acceleration of morphological evolution in the animal world and the diversity of skeleton structures, which became as specific as morphological characters of the soft body.

Bacterially induced biomineralization, which prevailed in the Precambrian, played a large role in the development of the composition of sediments and geochemical circulation of many elements and in the conditioning of organisms' habitats. Controlled matrix-mediated biomineralization became an innate, genetically fixed property of metabolism, whereas its products became parts of the organism, essentially a new organ responsible for various functions. As mentioned above, the primary function of biomineralization was to detoxify the organism. The integration of mineralization into the external tissues and internal structures of organisms facilitated the acquisition of various functions by minerals. The function of orientation in the gravity field in mobile animals is performed by statocysts and statocones of various mineral compositions (gypsum, celestine, calcite, aragonite, phosphate). The formation of magnetite crystals in different groups (from magnetotactic bacteria to birds and man) provides extrasensory orientation in the magnetic field. Piezoelectric crystals of apatite may work as extrasensory receptors. Mineral formations work as depots for metabolically important elements (Ca, P, Fe) and energy depots providing homeostasis and relative independence of functions from the changes in the environment. The inner skeleton of vertebrates performs supporting functions and encloses blood-producing organs.

Thus the interaction between the mineral and biological worlds can be traced throughout the entire period of their existence that is available to study. For the world of minerals, this coevolution enabled the existence of some minerals in the thermodynamically unbalanced conditions of the biosphere and allowed minerals to acquire functions, other than the substrate for life. For biological world, biomineralization was a powerful means for conditioning its environment, facilitated the acceleration of the biological evolution and growth of diversity, and allowed the acquisition of new organs responsible for various biological functions.

Acknowledgment I would like to thank Dr T.B. Leonova for her time and efforts, Dr S.V. Nikolaeva for help in translation and contribution to this manuscript.

References

Barskov, I.S. (1982) Biomineralization and evolution. Paleontol. Zh. 4, 5–13 (in Russ.).
Barskov, I.S. (1984) Paleontological aspects of biomineralization. Reports of 27th International Geological Congress, Moscow, USSR, Paleontology Sect. C.02, vol. 2, 61–66 (in Russ.).
Benzerara, K., Menguy, N., Guyot, F., Dominici, C. and Gillet, P. (2003) Nanobacteria-like calcite single crystals at the surface of the Tataouine meteorite. Proc. Natl Acad. Sci. USA 100(13), 7438–7442.
Chafetz, H.S., Akdim, B., Julia, R. and Reid, A. (1998) Mn- and Fe-rich black travertine shrubs: bacterially (and nanobacterially) induced precipitates. J. Sediment. Res. 68, 404–412.
Cisar, J.O., Xu, D.-Q., Thompson, J., Swaim, W., Hu, L. and Kopecko, D.J. (2000) An alternative interpretation of nanobacteria-induced biomineralization. Proc. Natl Acad. Sci. USA 97(21), 11511–11515.
Danilov-Danilian, V.I., Losev, K.S. and Reif, I.E. (2005) To the main defiance of civilization. View from Russia. Infra-M, Moscow (in Russ.).
Dove, P.M., DeYoreo, J.J. and Weiner, S. (Eds) (2004) Biomineralization. Mineralogical Society of America.
Ehrlich, P.A. and Raven, P.R. (1964) Butterflies and plants. Evolution 18(4), 586–608.
Folk, R.L. (1992) Bacteria and nannobacteria revealed in hardgrounds, calcite cements, native sulfur, sulfide materials, and travertines (abstract). Geol. Soc. Am. Annu. Prog. Abstr., p. 104.
Folk, R.L. (1993) SEM imaging of bacteria and nannobacteria in carbonate sediments and rocks. J. Sediment. Petrol. 63, 990–999.
Folk, R.L. and Lynch, F.L. (1998) Carbonaceous objects resembling nannobacteria in the Allende meteorite. Proceedings of the International Symposium on Optical Science, Engineering, and Instrumentation (SPIE), vol. 3441, pp. 112–122.

Hoover, R.B., Rozanov, A.Yu., Jerman, G.A. and Coston, J. (2004) Microfossils in CI and CO carbonaceous meteorites. In: R.B. Hoover, G.V. Levin and A.Yu. Rozanov (Eds), Instruments, Methods and Missions for Astrobiology VII, Proc. SPIE, vol. 5163, 7–22.

Kajander, E. and Ciftçioglu, N. (1998) Nanobacteria: an alternative mechanism for pathogenic intra- and extracellular calcification and stone formation. Proc. Natl Acad. Sci. USA 95(14), 8274–8279.

Kirkland, B.L., Lynch, F.L., Rahnis, M.A, Folk, R.L., Molineux, I.J. and McLean, R.J.C. (1999) Alternative origins for nannobacteria-like objects in calcite. Geology 27(4), 347–350.

Knoll, A.H. (2004) Biomineralization and evolutionary history. In: P.M. Dove, J.J. DeYoreo and S. Weiner (Eds), Reviews in Mineralogy and Geochemistry, 54(1), 329–356.

Lowenstam, H.A. (1981) Minerals formed by organisms. Science 221, 1126–1131.

Lowenstam, H.A. (1984) Processes and products of biomineralization. Evolution of biomineralization. Reports of 27th International Geological Congress, Moscow, USSR, Paleontology Sect. C.02, vol. 2, 51–56 (in Russ.).

Lowenstam, H. and Weiner, S. (1989) On Biomineralization. Oxford University Press, Oxford.

Mann, S. (2001) Biomineralization. Oxford University Press, Oxford.

McKay, D.S., Gibson, E.K., Jr, Thomas-Keprta, K.L., Vali, H., Romanek, C.S., Clemett, S.J., Chiller, X.D., Maechling, C.R. and Zare, R.N. (1996) Search for past life on Mars: possible relic of biogenic activity in Martian meteorite ALH84001. Science 273, 924–930.

Moissejev, N.N. (1997) Human and Biosphere. Jung Quardian, Moscow (in Russ.).

Sarikaya, M. (1994) An introduction to biomimetics: a structural viewpoint. Microsc. Res. Technol. 27, 360–375.

Zhegallo, E.A., Rozanov, A.Yu., Ushatinskaya, G.T., et al. (2000) Atlas of Microorganisms from Ancient Phosphorites of Khubsugul (Mongolia). NASA, Huntsville, AL, 167 pp.

Visualization of the Silicon Biomineralization in Cyanobacteria, Sponges and Diatoms

Ye. V. Likhoshway, E. G. Sorokovikova, O. I. Belykh, OL. V. Kaluzhnaya, S. I. Belikov, Ye. D. Bedoshvili, OK. V. Kaluzhnaya, Ju. A. Masyukova, and T. A. Sherbakova

Abstract Organisms of three kingdoms – cyanobacteria, sponges and diatoms – played a key role in the global cycle of silicon at certain moments of formation of the biosphere. At present, only diatoms retain this leading position. Using microscopy techniques, we studied mineralization of Si by these organisms. Analysis of silicateins-proteins, which take part in condensation of silica in sponges – helped to establish phylogeny of sponges of Lake Baikal. The presence of the gene of Silicic Acid Transport protein in chrysophycean algae suggests that this protein was "invented" long before diatoms appeared. Data on biomineralization of Si, analysis of silicic acid transport and of silica-condensing proteins suggest that the biotic pathway of the global Si cycle appeared at an early stage of the evolution and involved cyanobacteria, sponges and some algae.

1 Introduction

Silicon is the second highest abundant element in the Earth's crust. The cycling of silica, along with geological processes, involves the work of live organisms (Ragueneau et al., 2000). Diatom algae which give a significant contribution to the global primary production sink to the bottom of oceans, seas and freshwater bodies forming siliceous sediments up to several kilometer thickness. Ancient fossil sponges formed large reefs (Conway et al., 2006). Cyanobacteria are believed to have changed the Earth atmosphere to oxygenic at an early stage of the development of life, about 2 billion years BP (Rozanov, 2006; Sergeev et al., 2002). Cyanobacteria, sponges and diatoms belong to three different kingdoms of living beings; their origin is separated by many hundred million years (Table 1).

Ye.V. Likhoshway
Limnological Institute of the Siberian Branch of the Russian Academy of Sciences, Irkutsk

Table 1 Time of nascence on Earth of silicifying organisms

Organisms	Time of appearance	Reference
Cyanobacteria	Silicified microfossils found in deposits of an age beginning with 3.5 billion years BP	Gerasimenko and Ushatinskaya (2002)
Sponges	The most ancient siliceous sponge spicules were found in sediments belonging to Late Pre-Cambrian (650–543 My BP)	Gehling and Rigby (1996)
Diatom algae	The most ancient diatoms are no older than 240 My according to molecular clock calculations based on sequences of four genes	Medlin and Kaczmarska (2004, p. 245)

Cyanobacteria from thermal springs, siliceous sponges and diatoms share a common property of being able to process biomineralization of silica—uptake of silicic acid and its reworking into sediments. The structures which they form are diverse, such as silicified trichomes, spicules and bivalve frustules. Visualization of the processes of mineralization of Si in different organisms by methods of light, scanning and transmission electron microscopies (LM, SEM and TEM) can disclose common features and help us to approach understanding of the mechanisms given below.

2 Silicification of Cyanobacteria

Cyanobacteria, among other environments (seas and fresh waters), inhabit thermal springs where they form bacterial mats. Siliceous deposits formed by cyanobacteria are found in tuffs of many thermal springs (Canet et al., 2005; Jones et al., 2005; Konhauser et al., 2001; Renaut et al., 2002).

Live cyanobacteria form siliceous deposits outside their cells, but does not involve the cytoplasm (Konhauser et al., 2001). Experiments have shown that deposits of silica occur as amorphous sediments on the outer surfaces of polysaccharide sheaths sometimes built in part of nanospheres (Benning et al., 2004; Phoenix et al., 2000); due to silicification, the diameter of *Calothrix* sp. sheaths increased two to three times.

Our experiments on cultivation of cyanobacterial strains at a high concentration of Si (20 mM) isolated from thermal springs of the Baikal rift zone have shown that already in 24 h the density of sheaths increased; layers of the sheaths become more distinct. There was no silica between the sheaths and the cell wall, as well as in the cytoplasm. There were no abnormalities (Fig. 1a–c). SEM revealed that mineralization of Si by cyanobacteria occurs in three ways: thickening and increase of density of sheaths (Fig. 1d and e); deposition of amorphous silica around the trichome

Silicon Biomineralization in Cyanobacteria

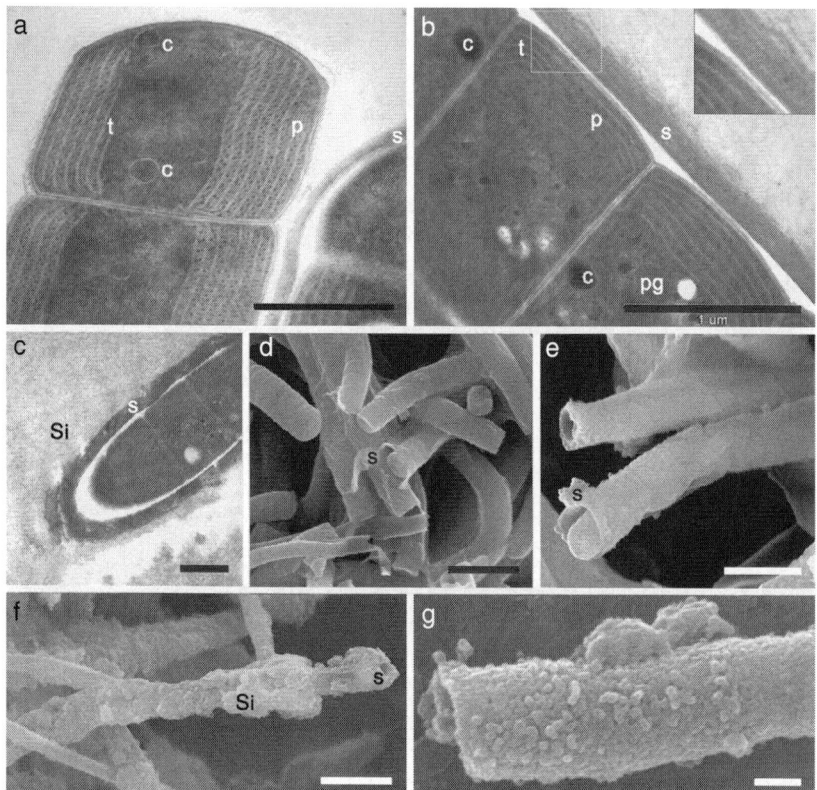

Fig. 1 Biosilicification of strain *Pseudanabaena* sp. 0411, isolated from Kotelnikovsky spring. (**a–c**) TEM: (**a**) without Si, (**b**) cells on the sixth day of the experiment, (**c**) silica deposition on the 24th day. (**d**) without Si; (**e–g**) different types of Si biomineralization, SEM: (**e**) sheath silicification; (**f**) deposition of amorphous silica around the trichomes; (**g**) nanospheres (treatment with 30% H_2O_2). *c* cyanophycin granules, *p* phycobilisomes in the interthylakoidal space, *pg* polyphosphate granules, *s* sheath, *t* thylakoid. Scale bars: **a–c, g** 1 μm; **d–f** 5 μm

(Fig. 1f); and deposition as spheres of submicrometer sizes (Fig. 1g) (Likhoshway et al., 2006a).

It is possible that cyanobacteria have certain adaptations which help to precipitate silica. The silicified cover can provide protection from UV light (Phoenix et al., 2000), from heat and from excess of silicic acid and other detrimental substances. On the other hand, precipitation of silicic acid and accompanying salts favorably changes the environment for the benefit of cyanobacteria themselves and of other organisms.

The structure of the siliceous covers of cyanobacteria does not have species-specific peculiarities, unlike that of siliceous structures of sponges and diatoms which form species-specific siliceous parts of their exoskeletons—spicules and frustules. The latter are built within cells in specialized subcellular particles, silica deposition vesicles (SDVs) encircled by special membranes—silicalemmas (Drum and Pankratz, 1964; Reimann, 1964)—under control of genomes. The first stages of the synthesis of spicules also occur within SDVs of special cells, scleroblasts, but mature spicules are formed extracellularly due to assistance of cells of other types (Uriz et al., 2003).

3 Sponges

Sponges are the only extant multicellular organisms which build their skeletons of silica. Spicules of sponge skeletons have a different form and architecture of surface. In the middle of spicules, there are channels which have different forms in different species (Uriz et al., 2000; Weaver and Morse, 2003). Spicules of a sponge from the Lake Baikal *Lubomirskia baicalensis* (Color Plate 3, see p. 395) contain a round axial channel (Color Plate 3c, see p. 395). Side branches of the central channel determine the outer form of spicules (spines) (Belikov et al., 2005; Kaluzhnaya et al., 2005a; Müller et al., 2006). The channel hosts an axial filament which can be stained blue with Coumassee after dissolution of spicule with ammonium fluoride (Color Plate 3d–f, see p. 395) (Kaluzhnaya et al., 2005b; Müller et al., 2006).

The central filaments of marine sponges contain special proteins, silicateins, belonging to the class of cysteine proteinases cathepsins-L. Electrophoresis in a gradient of polyacrylamide (4–20%) revealed three fractions having molecular masses about 30 kDa which were named silicateins-α, -β and -γ (Shimizu et al., 1998). A broader diversity of proteins was found in spicules of the freshwater

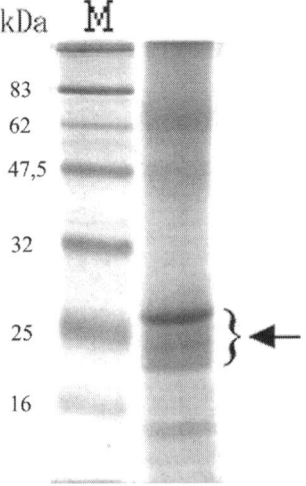

Fig. 2 SDS-PAGE of *L. baicalensis* spicule proteins. An *arrow* shows bands corresponding to silicateins by molecular weights (*M* molecular weight marker)

sponge *L. baicalensis*: along with proteins of the size of silicateins, two more proteins of masses 45 and 62 kDa were found (Kaluzhnaya et al., 2007) (Fig. 2).

We were the first who determined nucleotide sequences of silicatein gene for Baikal species (Kaluzhnaya et al., 2005b) and demonstrated the multiplicity of silicateins in freshwater sponges sequencing four silicatein cDNAs. Comparison of silicatein sequences for freshwater *L. baicalensis*, *Ephydatia fluviatilis* and marine species shows the significant variation at the N-terminus, which are cleaved during sequence maturation. Mature proteins are considerably similar and also conserved at active center sites (Fig. 3).

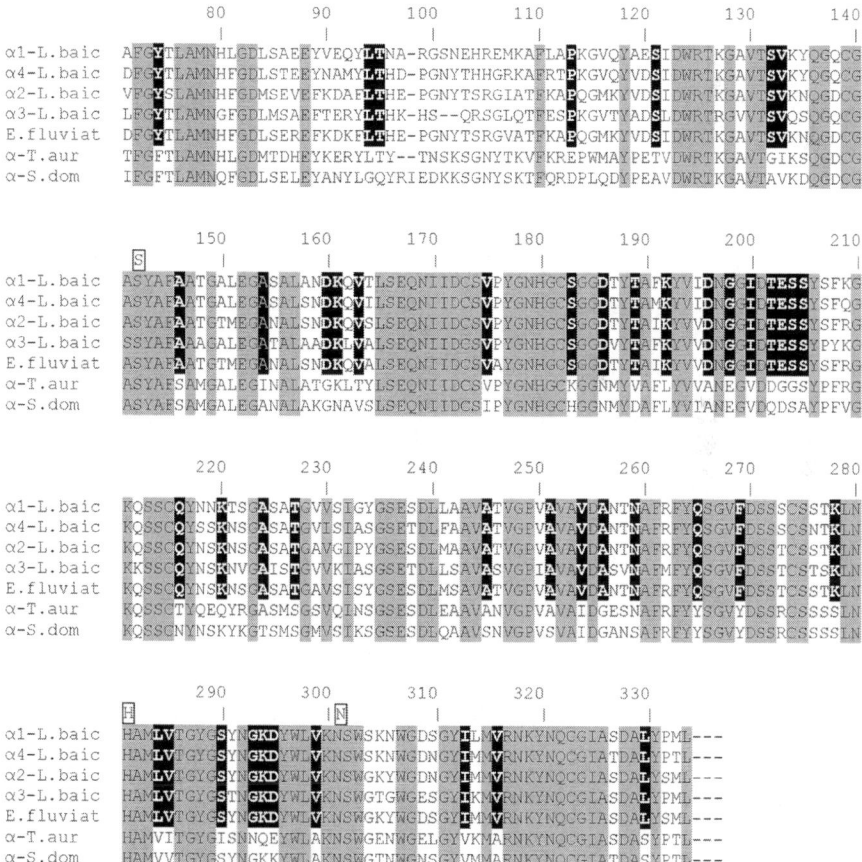

Fig. 3 Comparison of silicatein amino acid sequences for freshwater and marine sponges. Conserved amino acids are in *gray*; sites conserved among freshwater sponges sequences are in *black*. The amino acid residues involved in the silicatein active center are *boxed* above sequences. Denotation of species names: *L. baicalensis* α1-L.baic–α4-L.baic; freshwater *E. fluviatilis* E.fluviat; marine species *Thetya aurantia* and *Suberites domuncula* α-T.aur and α-S.dom, respectively

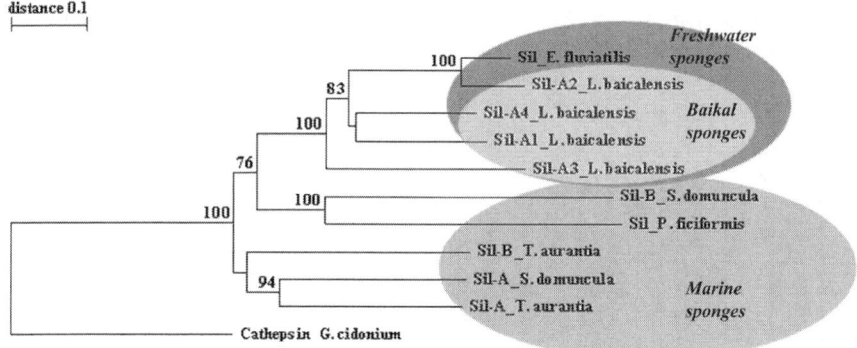

Fig. 4 Neighbor-joining tree based on the nucleotide sequences of silicatein α genes of freshwater *L. baicalensis* and *E. fluviatilis* and marine species

As has been shown, genes coding freshwater sponge silicateins, significantly vary from those of marine sponges (Kaluzhnaya et al., 2005b). Multiplication of freshwater sponge silicatein-α genes has also been demonstrated. With phylogenetic approach we found the evidence that silicatein genes during the evolution originated from a common ancestor via two duplication events (Kaluzhnaya et al., 2007; Wiens et al., 2006) (Fig. 4).

Experiments in vitro have shown that silicateins are able to hydrolyze esters of silicic acid forming a precipitate of silica on isolated axial filaments (Cha et al., 1999; Krasko et al., 2000). Recombinant silicatein-β of marine sponges *Tethya aurantia* induced the formation of siliceous 120–300 nm nanospheres (Weaver and Morse, 2003).

Siliceous nanospheres whose sizes, depending on the conditions, vary between <50 and 1000 nm are formed in silicic acid solutions in the presence of polyamines and silaffins isolated from diatom valves (Kröger et al., 2000).

4 Diatoms

Biosilification in the process of synthesis of diatom valves is covered by a few excellent reviews, the most recent by Hildebrand and Wetherbee (2003). Of many approaches to understanding the mechanisms of formation of siliceous structures of diatoms, we shall consider the following two.

4.1 Studies of Morphogenesis of Early Stages of Diatom Valves

An efficient way to study forming valves is by staining with fluorescent dyes which bind with silica deposited within the SDV. Using such dyes it is possible to observe optically the formation of valves of diverse diatoms, including tiny valves (2–3 μm, *Synedra acus*) (Color Plate 5a–f, see p. 397). Staining with Rhodamine 123 (R 123) allowed to study the development of valves and girdle bands of a centric diatom *Ditylum brightwellii* and to see daughter valves while they are still within the mother cell (Li et al., 1989). R 123 was also used to study the deposition of silica during the cell cycle of a centric marine diatom *Thalassiosira weissflogii* with flow cytometry (Brzesinski and Conley, 1994). By measuring the decrease of silicic acid concentration in the growth medium and the accumulation of R 123 in cells during a few cell division cycles, it was found that the dye applied at a concentration of 2 μg/L does not affect the growth, and that 1 molecule of R 123 is absorbed per 17 million molecules of SiO_2.

Shimizu et al. (2001) recommended to use another fluorescent PDMPO which produces a more intense fluorescence compared to R 123 and allows to study the development of valves in more detail. Using PDMPO Hazelaar et al. (2005) visualized the very earliest stages of morphogenesis of a marine pennate diatom *Navicula salinarum*. It was found that at the initial stage the valve grows in two dimensions and grows very fast—less than 15 min. Subsequent growth in third dimension takes a much longer time, a significant part of the time of valve synthesis which equals 240 min.

Study by means of SEM (some examples in Color Plate 5g and h, see p. 397) and TEM (some examples in Color Plate 5i–k, see p. 397) of valves after decomposition of organic matter with acid revealed stages of morphogenesis in diatoms of all known classes: for review see Pickett-Heaps et al. (1990) and Hildebrand and Wetherbee (2003). These studies determined stages of morphogenesis of all siliceous structures which are important for the identification of species and served as a basis for the proposed scheme of their evolution (Cox, 1999; Cox and Kennaway, 2004; Round et al., 1990).

TEM of cross-sections reveals electron-dense matter within SDVs of valves and SDVs of girdle bands (Color Plate 5l–n, see p. 397). TEM is used to study the role of different subcellular particles in synthesis of diatom cell walls for more than 40 years (Drum and Pankratz, 1964; Pickett-Heaps et al., 1990; Reiman, 1964; Schmid and Schultz, 1979), but its key issues have not been understood yet and remain topics for continuous discussions (Hildebrand and Wetherbee, 2003).

4.2 Discovery of Silicic Acid Transporters

Long-term studies of processes of silicification and morphogenesis in diatoms performed in the laboratory of B.E. Volcani (University of California, San Diego) have finally led to the discovery of a gene of a silicic acid transporter protein (SIT) in a marine diatom *Cylindrotheca fusiformis* (Hildebrand et al., 1997). A homologous gene was subsequently found in a freshwater diatom *S. acus* (Grachev et al., 2002). Up to date, complete and partial sequences of

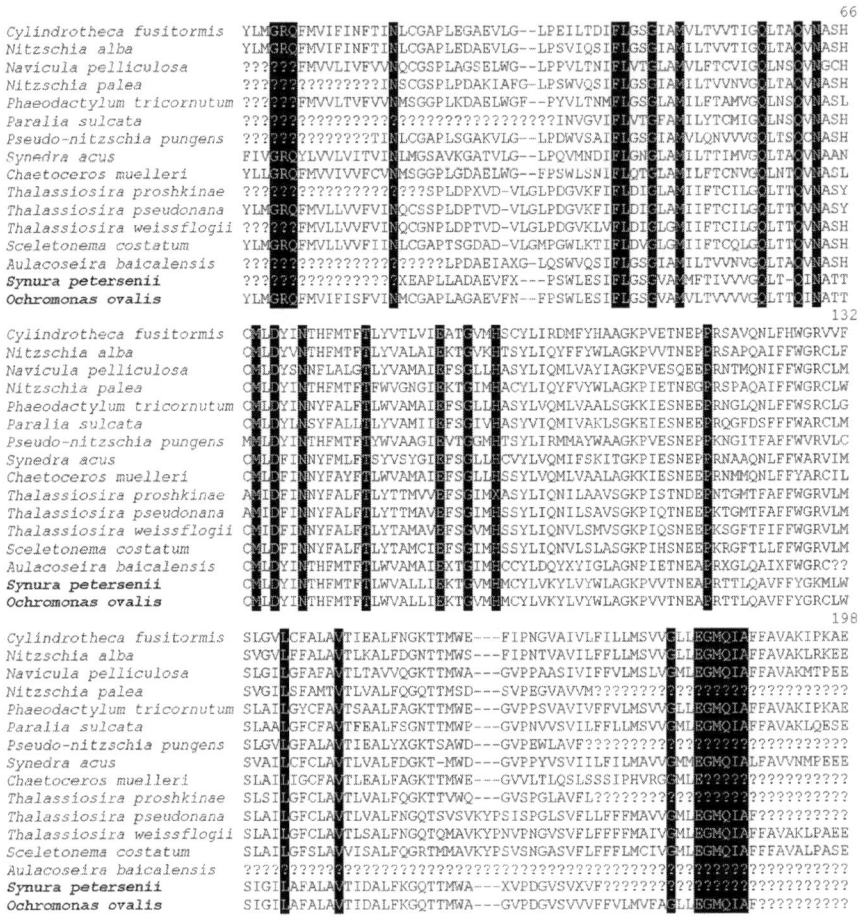

Fig. 6 Fragments of predicted SIT sequences of chrysophyte algae and diatoms. Identical amino acid residues are printed on *black* background. Names of chrysophytes are in *bold*

sit genes are published for more than 100 diatom species of all known classes (Sherbakova et al., 2005; Thamatrakoln and Hildebrand, 2005; Thamatrakoln et al., 2006; Alverson et al., 2007 and our unpubl. data). Analysis of the sequences revealed that SITs diverge rather fast. During million years of evolution, a few motifs have remained intact (Fig. 6); in spite of numerous substitutions the physicochemical properties of these membrane proteins have kept that is confirmed by significant similarity of hydropathy profiles (Sherbakova et al., 2005; Likhoshway et al., 2006b). All the SIT sequences are predicted to contain sites of post-translational modification, i.e., glycosilation, myristylation, phosphorylation. A motif YxxØ, a potential target of clathrin adaptors, is also present (Sherbakova et al., 2005). Examination of deciphered SIT sequences of an araphid freshwater diatom *S. acus* and a raphid marine species

Chaetoceros muelleri allowed to propose a mechanism of interaction of SIT proteins with silicic acid (Grachev et al., 2002). According to this hypothesis, silicic acid is coordinated via a zinc ion with a motif CML(I)D (Grachev et al., 2005; Sherbakova et al., 2005). Another model assumes that conservative repeats GXQ join silicic acid by means of hydrogen bonds, inducing changes in the protein conformation which helps silicic acid to move inside the cell (Thamatrakoln et al., 2006).

The origin of diatoms and their ability to use silica for cell walls construction have been discussed in many publications (for review see Sims et al., 2006). Modern phylogeny based on sequences of the nuclear gene of 18S rRNA places diatoms close to Bolidophyceae, planktonic algae without any signs of silicification (Guillou et al., 1999; Medlin and Kaczmarska, 2004). Nikolaev et al. (2001) based on paleontological data believe that a precursor of diatoms might have been silicified bivalved spores of Xanthophyta. According to other hypothesis, diatoms arose from a scaled ancestor (Round and Crawford, 1981, 1984) or from a cyst-like form similar to the extant Parmales in the chrysophyte algae (Mann and Marchant, 1989).

Vegetative cells of some chrysophyte algae are covered by scales, and at certain stages of their development form siliceous spherical cysts, or stomatocysts. Both for chrysophyte algae and diatoms silicon is an obligatory element of their metabolism. In accord with this dependence, chrysophyte species *Ochromonas ovalis* and *Synura petersenii* contain *sit* genes (Likhoshway et al., 2006b). SIT proteins of diatoms and chrysophytes have evident features of homology, the differences between them are similar to those between SITs of diatoms belonging to evolutionary distant classes. Figure 5 shows that all SITs have identical amino acid residues separated by non-identical sites. Discovery of *sit* genes in distant relatives of diatoms—chrysophyte algae— suggests that these genes are ancient (Likhoshway et al., 2006b) because scales of chrysophytes were found in deposits of the Late Precambrian (Blackwell and Powell, 2000; Knoll, 1992). If the identification of scales is correct, SIT was invented much earlier than the first diatoms appeared. Nevertheless, no *sit* genes have been found in sponges so far. Hence, the question put by Franz Brümmer (2003, p. 8): "And what about a silicon transporter in sponges?" still remains an open one. However, it is noteworthy that siliceous sponges also appeared in the Late Precambrian (Table 1)—the time of expansion of aerobic eukaryotes into the ocean.

According to one of the most recent hypotheses, life in the hot, salty and silica-saturated ocean was limited to bacteria, including cyanobacteria. Global glaciations cooled the water and increased the concentration of dissolved oxygen making the ocean more favorable for diverse aerobic eukaryotic organisms (Knauth, 2005). Appearance of silicon-reworking molecular machinery was timely because it gave a possibility to use a "cheap" raw material for the construction of functional inorganic exoskeletons, a matter of subsequent improvement in the course of evolution.

Acknowledgment The present study was supported by the Program of the Presidium of RAS "Origin and evolution of the biosphere" N 18.4.

References

Alverson A.J., Jansen R.K., Theriot E.C. (2007) Bridging the Rubicon: Phylogenetic analysis reveals repeated colonizations of marine and fresh waters by thalassiosiroid diatoms. Molecular Phylogenetics and Evolution 4(1), 193–210.

Belikov, S.I., Kaluzhnaya, O.V., Schröder, H.C., Krasko, A., Müller, I.M. and Müller, W.E.G. (2005) Expression of silicatein in spicules from the Baikalian sponge *Lubomirskia baicalensis*. Cell Biol. Int. 29, 943–951.

Benning, L.G., Phoenix, V.R., Yee, N. and Konhauser, K.O. (2004) The dynamics of cyanobacterial silicification: an infrared micro-spectroscopic investigation. Geochim. Cosmochim. Acta 68, 743–757.

Blackwell, W.H. and Powell, M.J. (2000) A review of group filiation of the strameopiles, additional approaches to the question. Evol. Theory 12, 49–88.

Brümmer, F. (2003) Living inside a glass box—silica in diatoms. In: W.E.G. Müller (Ed.), Silicon Biomineralization. Springer, Berlin, pp. 3–10.

Brzesinski, M.A. and Conley, D.J. (1994) Silicon deposition during the cell cycle of *T. weissflogii* using dual Rhodamine 123 and propidium iodide staining. J. Phycol. 30, 45–55.

Canet, C., Prol-Ledesma, R.M., Torres-Alvarado, I., Gilg, H.A., Villanueva, R.E. and Lozano-Santa Cruz, R. (2005) Silica-carbonate stromatolites related to coastal hydrothermal venting in Bahia Concepcion, Baja California Sur, Mexico. Sediment. Geol. 174, 97–113.

Cha, J.N., Shimizu, K., Zhou, Y., Christianssen, S.C., Chmelka, B.F., Stucky, G.D. and Morse, D.E. (1999) Silicatein filaments and subunits from a marine sponge direct the polymerization of silica and silicones in vitro. Proc. Natl Acad. Sci. USA 96, 361–365.

Conway, K.W., Krautter, M., Barrie, J.V., Whitney, F., Thomson, R.E., Reiswig, H., Lehnert, H., Mungov, G. and Bertram, M. (2006) Sponge reefs in the Queen Charlotte Basin, Canada: controls on distribution, growth and development. In: A. Freiwald and J.M. Roberts (Eds), Cold-Water Corals and Ecosystems. Springer, Berlin, pp. 605–621.

Cox, E.J. (1999) Variation in patterns of valve morphogenesis between representatives of six biraphid diatom genera. J. Phycol. 35, 1297–1312.

Cox, E.J. and Kennaway, G.M. (2004) Studies of valve morphogenesis in pennate diatoms: investigating aspects of cell biology in a systematic context. In: M. Poulin (Ed.), Proceedings of the 17th International Diatom symposium. Biopress Ltd, Bristol, pp. 35–48.

Drum, R.W. and Pankratz, H.S. (1964) Post mitotic fine structure of *Gomphonema parvulum*. J. Ultrastruct. Res. 10, 217–223.

Gehling, J.G. and Rigby, J.K. (1996) Long expected sponges from the Neoproterozoic Ediacara fauna of South Australia. J. Paleontol. 70(2), 185–195.

Gerasimenko, L.M. and Ushatinskaya, G.T. (2002) Cyanobacteria, cyanobacteria/bacteria associations, mats, biofilms. In: A.Yu. Rozanov (Ed.), Bacterial Paleontology. Paleontological Institute RAS, Moscow, pp. 36–46.

Grachev, M.A., Denikina, N.N., Belikov, S.I., Likhoshwai, E.V. (Likhoshway, Ye.V.), Usoltseva, M.V., Tikhonova, I.V., Adelshin, R.V., Kler, S.A. and Shcherbakova (Sherbakova), T.A. (2002) Elements of the active center of silicon transporters in diatoms. Mol. Biol. 36, 535–536.

Grachev, M., Sherbakova, T., Masyukova, Yu. and Likhoshway, Ye. (2005) A potential Zink-binding motif in silicic acid transport proteins of diatoms. Diatom Res. 20(2), 409–441.

Guillou, L., Chretiennot-Dinet, M.-J., Medlin, L.K., Claustre, H., Loiseaux-de Goër, S. and Vaulot, D. (1999) *Bolidomonas*: a new genus with two species belonging to a new algal class, the Bolidophyceae (Heterokonta). J. Phycol. 35, 368–381.
Hazelaar, S., Strate, H.J., Gieskes, W.C. and Vrieling, E.G. (2005) Monitoring rapid valve formation in the pennate diatom *Navicula salinarum* (Bacillariophyceae). J. Phycol. 41, 354–358.
Hildebrand, M. and Wetherbee, R. (2003) Components and control of silicification in Diatoms. In: W.E.G. Müller (Ed.), Silicon Biomineralization. Springer, Berlin, pp. 11–58.
Hildebrand, M., Volcani, B.E., Gassmann, W. and Schröder, J.I. (1997) A gene family of silicon transporters. Nature 385, 688–689.
Jones, B., Renaut, R.W. and Konhauser, K.O. (2005) Genesis of large siliceous stromatolites at Frying Pan Lake, Waimangu geothermal field, North Island, New Zealand. Sedimentology 52, 1229–1252.
Kaluzhnaya, O.V., Belikov, S.I., Schröder, H.C., Rothenberger, M., Zapf, S., Kaandorp, J.A., Borejko, A., Müller, I.M. and Müller, W.E.G. (2005a) Dynamics of skeleton formation in the Lake Baikal sponge *Lubomirskia baicalensis*. Part I. Biological and biochemical studies. Naturwissenschaften 92, 128–133.
Kaluzhnaya, O.V., Belikov, S. I., Schröder, H.C., Wiens, M., Giovine, M., Krasko, A., Müller, I.M. and Müller, W.E.G. (2005b) Dynamics of skeleton formation in the Lake Baikal sponge *Lubomirskia baicalensis*. Part II. Molecular biological studies. Naturwissenschaften 92, 134–138.
Kaluzhnaya, O.V., Belikova, A.S., Podolskaya, E.P., Krasko, A., Müller, W.E.G. and Belikov, S.I. (2007) Silicatein identification of the freshwater sponge *Lubomirskia baicalensis*. Mol. Biol. 4,554–561.
Knauth, L.P. (2005) Temperature and salinity history of the Precambrian ocean: implication for the course of microbial evolution. Palaeogeogr. Palaeoclimatol. Palaeoecol. 219, 53–69.
Knoll, A.H. (1992) The early evolution of eukaryotes: a geological perspective. Science 256, 622–627.
Konhauser, K.O., Phoenix, V.R., Bottrell, S.H., Adams, D.G. and Head, I.M. (2001) Microbial–silica interactions in Icelandic hot spring sinter: possible analogues for some Precambrian siliceous stromatolites. Sedimentology 48, 415–433.
Krasko, A., Batel, R., Schröder, H.C., Müller, I.M. and Müller, W.E.G. (2000) Expression of silicatein and collagen genes in the marine sponge *Suberites domuncula* is controlled by silicate and myotrophin. Eur. J. Biochem. 267, 4878–4887.
Kröger, N., Deutzmann, R., Bergsdorf, C. and Sumper, M. (2000) Species-specific polyamines from diatoms control silica morphology. Proc. Natl Acad. Sci. USA 97, 14133–14138.
Li, C.-W., Chu, S. and Lee, M. (1989) Characterizing the silica deposition vesicle of diatoms. Protoplasma 151, 158–163.
Likhoshway, Ye.V., Sorokovikova, E.G., Belkova, N.L., Belykh, O.I., Titov, A.T., Sakirko, M.V. and Parfenova, V.V. (2006a) Silicon mineralization in the culture of cyanobacteria from hot springs. Dokl. Biol. Sci. 407, 201–205.
Likhoshway, Ye.V., Masyukova, Yu.A., Sherbakova, T.A., Petrova, D.P. and Grachev, M.A. (2006b) Detection of the gene responsible for silicic acid transport in Chrysophycean algae. Dokl. Biol. Sci. 408, 256–260.
Mann, D.G. and Marchant, H.J. (1989) The origins of the diatom and its life cycle In: J.C. Green, B.S.C. Leadbeater and W.L. Diver (Eds) The Chromphyte Algae: Problems and Perspectives. Clarendon Press, Oxford, pp. 307–323.
Medlin, L.K. and Kaczmarska, I. (2004) Evolution of the diatoms: V. Morphological and cytological support for the major clades and a taxonomic revision. Phycologia 43, 1–29.
Müller, W.E.G., Kaluzhnaya, O.V., Belikov, S.I., Rothenberger, M., Schröder, H.C., Reiber, A., Kaandorp, J.A., Manz, B., Mietchen, D. and Volke, F. (2006) Magnetic resonance imaging of the siliceous skeleton of the demosponge *Lubomirskia baicalensis*. J. Struct. Biol. 153, 31–41.

Nikolaev, V.A., Harwood, D.M. and Samsonov, N.I. (2001) Early Cretaceons Diatoms. Nauka, St Petersburg.
Phoenix, V.R., Konhauser, K.O. and Adams, D.G. (2000) Cyanobacterial viability during hydrothermal biomineralization. Chem. Geol. 169, 329–338.
Pickett-Heaps, J.D., Schmid, A.-M.M. and Edgar, L. (1990) The cell biology of diatom wall morphogenesis. In: F.E. Round and D.J. Chapman (Eds.) Progress in Phycological Research. Biopress, Bristol, pp. 2–168.
Ragueneau, O., Tréguer, P., Leynaert, A., Anderson, R.F., Brzezinski, M.A., DeMaster, D.J., Dugdale, R.C., Dymond, J., Fischer, G., François, R., Heinze, C., Maier-Reimer, E., Martin-Jézéquel, V., Nelson, D.M. and Quéguiner, B. (2000) A review of the Si cycle in the modern ocean: recent progress and missing gaps in the application of biogenic opal as a paleoproductivity proxy. Global Planet. Change 26, 317–365.
Reimann, B.E.F. (1964) Deposition of silica inside a diatom cell. Exp. Cell. Res. 34, 605–608.
Renaut, R.W., Jones, B., Tiercelin, J.J. and Tarits, C. (2002) Sublacustrine precipitation of hydrothermal silica in rift lakes: evidence from Lake Baringo, central Kenya Rift Valley. Sediment. Geol. 148, 235–257.
Round, F.E. and Crawford, R.M. (1981) The lines of evolution of Bacillariophyta. I. Origin. Proc. R. Soc. Lond. B 211, 237–260.
Round, F.E. and Crawford, R.M. (1984) The lines of evolution of Bacillariophyta. II. Origin. Proc. R. Soc. Lond. B 221, 169–188.
Round, F.E., Crawford, R.M. and Mann D.G. (1990) The Diatoms Biology and Morphology of the Genera. Cambridge University Press, Cambridge.
Rozanov, A.Yu. (2006) Precambrian geobiology. Paleontol. J. 40(4), 434–443.
Schmid, A.-M.M. and Schulz, D. (1979) Wall morphogenesis in diatoms: depositions of silica by cytoplasmic vesicles. Protoplasma 100, 268–288.
Sergeev, V.N., Gerasimenko, L.M. and Zavarzin, G.A. (2002) The proterozoic history and present state of cyanobacteria. Microbiology 71, 623–637.
Sherbakova, T.A, Masyukova, Yu.A., Safonova, T.A., Petrova, D.P., Vereshagin, A.L., Minaeva, T.V., Adelshin, R.A., Triboy, T.I., Stonik, I.I., Aizdaitcher, N.A., Kozlov, M.V., Likhoshway, E.(Ye.)V. and Grachev, M.A. (2005) Conservative motif CMLD in silicic acid transport proteins of diatom algae. Mol. Biol. 39, 269–280.
Shimizu, K., Cha, J, Stucky, G.D. and Morse, D.E. (1998) Silicatein alpha: cathepsin L-like protein in sponge biosilica. Proc. Natl Acad. Sci. USA 95, 6234–6238.
Shimizu, K., Del Amo, Y., Brzezinski, M.A., Stucky, G.D. and Morse, D.E. (2001) A novel silica tracer for biological silification studies. Chem. Biol. 8, 1051–1060.
Sims, P.A., Mann, D.G. and Medlin L.K. (2006) Evolution of the diatoms: insights from fossil, biological and molecular data. Phycologia 45(4), 361–402.
Thamatrakoln, K. and Hildebrand, M. (2005) Approaches for functional characterization of diatom silicic acid transporters. J. Nanosci. Nanotechnol. 5, 1–9.
Thamatrakoln, K., Alverson, A.J. and Hildebrand, M. (2006). Comparative sequence analysis of diatom silicon transporters: toward a mechanistic model of silicon transport. J. Phycol. 42, 822–834.
Uriz, M.J., Turon, X. and Becerro, M.A. (2000) Silica deposition in Demospongiae: spiculogenesis in *Crambe crambe*. Cell. Tissue Res. 301, 299–309.
Uriz, M.J., Turon, X., Becerro, M.A. and Agell, G. (2003) Siliceous spicules and skeleton frameworks in sponges: origin, diversity, ultrastructural patterns, and biological functions. Microsc. Res. Technol. 62, 279–299.
Weaver, J.C. and Morse, D.E. (2003) Molecular biology of Demosponge axial filaments and their roles in biosilicification. Microsc. Res. Technol. 62, 356–367.
Wiens, M., Belikov, S.I., Kaluzhnaya, O.V., Krasko, A., Schröder H.C., Perovic-Ottstadt, S. and Müller, W.E.G. (2006) Molecular control of serial module formation along the apical–basal axis in the sponge *Lubomirskia baicalensis*: silicateins, mannose-binding lectin and mago nashi. Dev. Genes Evol. 216(5), 229–242.

Transformational Changes in Argillaceous Minerals due to Cyanobacteria

T. V. Alekseeva, L. M. Gerasimenko, E. V. Sapova, and A. O. Alekseev

Abstract The aim of the study was to investigate the transformations of bentonite, illite, kaolin and smectite–zeolite clay at the laboratory experiments under the growth and fossilization of alkaline cyanobacteria *Microcoleus Chthonoplastes*. Cyanobacteria influenced the chemical properties of studied clays. It had no visible influence on the mineralogy of bentonite and kaolin. Whereas the development of more smectitic layers within the illite matrices in case of illite clay and defect "island" layer in the interlayer space of montmorillonite in smectite–zeolite clay were found. Cyanobacteria affected the properties of iron compounds, which are present in all studied clays as impurities. Both dissolution and precipitation processes of iron compounds were observed. In the experiments with smectite–zeolite and kaolin an increase in magnetic susceptibility and magnetization values connected with the precipitation of Fe in the form of metastable ferrihydrite followed by the formation of goethite were found. This process correlated with the mineralization of organic matter, which plays the role of inhibitor and prevents goethite crystallization during the growth of cyanobacteria. In the case of bentonite a decrease in both magnetic susceptibility and magnetization values connected with oxidation of Fe^{2+} in the magnetite (maghemite) structure and precipitation of goethite took place. Ferrihydrite is a key mineral in the biogenic cycle of Fe.

1 Introduction

Cyanobacteria or blue–green algae are oxygenic photoautotrophic organisms. Silicified microfossils of cyanobacteria (stromatolites) are found in deposits from 1.9–2.0 billion years ago. At present they create relict communities in extreme ecological conditions—regions with hydrothermal activity, hyper

T. V. Alekseeva
Institute Physical, Chemical and Biological Problems of Soil Science RAS, Pushchino
e-mail: alekseeva@issp.serpukhov.su

saline reservoirs and alkaline lakes (Dubinin et al., 1992; Gerasimenko et al., 1996; Phoenix et al., 2000; Schultze-Lam et al., 1995; Tazaki, 1998; Tazaki et al., 2003; Zavarzin, 1993). Besides organic matter, bacterial mats contain mineral components—carbonate, silica, Fe and Mn oxides /hydroxides, clay minerals, phosphates, sulfides of biogenic origin (Tazaki, 1998; Zavarzin, 1993). Biomineralization is a widespread phenomenon which occurs in different geochemical environments in the presence of liquid. In the most cited paper on biominerals, Lowenstam (1981) has shown that 31 biominerals have been found and more than 30% are formed with the participation of microorganisms. The most frequent biominerals are carbonate, opal and Fe oxides/hydroxides (ferrihydrite and magnetite).

Recent findings show that microorganisms also participate in the synthesis of layer silicates—smectite–celadonite (Geptner et al., 1997), glauconite–nontronite (Geptner and Ivanovskaya, 1998; Ueshima and Tazaki, 2001), imogolite–allophane–chamosite (Kawano and Tomita, 2002), saponite (Geptner et al., 2002). Based on electron microscopy observations of fresh water microorganism communities, Konhauser and Urrutia (1999) concluded that these communities are mineralized and are associated with amorphous (Fe, Al)-silicates with a composition close to chamosite/bertherine. These authors concluded that this phenomenon has global distribution.

Mineralization processes in the presence of cyanobacteria are extensively studied for hot spring communities (Likhoshway et al. 2006; Phoenix et al., 2000), but there are few publications on mineralization in the presence of alkaline species (Ushatinskaya et al., 2006).

The aim of the study was to investigate the possible transformations of kaolinite, smectite–zeolite clay, illite and bentonite SWy-2 in laboratory experiments under the growth and fossilization of alkaline cyanobacteria *Microcoleus Chthonoplastes*.

2 Experimental

2.1 Objectives

In our experiments we used a culture of alcalophilic filamentous cyanobacteria *Microcoleus chthonoplastes*, a unique member of laminated benthic community, isolated from soda lake Khilganta (Buryatiya, Russia).

The stimulating growing of cells was done using "S" medium (g/l): KCl 1.0; K_2SO_4 1.0; $NaHCO_3$ 16.8; K_2HPO_4 0.5; $NaNO_3$ 2.5; $NaCl$ 30.0; $MgSO_4.7H_2O$ 0.2; $CaCl_2$ 0.04; $FeSO_4$ 0.01; A5 1ml (Gerasimenko et al., 1996). pH of the medium was 8.5. The culture was grown autotrophically under continuous illumination (white fluorescent light) at approximately 3000 lx (10,000E S m^{-2} s^{-1}) and 28°C with CO_2 provided by bubbling air. After 4 days stimulating growth 4 ml of suspension containing *Microcoleus chthonoplastes* was added to bottles contained: clay 0.5 g, NaCl 2.0 g, NaHCO$_3$ 1.2 g and sterile distilled water 100 ml. Cyanobacteria were incubated for 7, 14, 28 and 60 days at 28°C in the light. Control bottles without cyanobacteria were incubated under the same conditions. All experiments have been done in duplicate. It is well known that the intensity of weathering and their trends are determined by the mineral's structure and composition. We studied the influence of *Microcoleus chthonoplastes* on the properties of different clays: 1:1 (kaolinite), 2:1 non-swelling (illite), 2:1 swelling (bentonite) and zeolite. The information about clays and some of their properties are given in Table 1.

2.2 *Methods*

The content of chlorophyll *a* was detected by spectrophotometric analysis at 665 nm wavelength after extraction in ethanol (Gerasimenko et al., 1989). The velocity of oxygen flow was determined by polyarographic method using a Clark electrode and 1 mm cell (Trebst, 1980) Organic matter content in clays before and after incubation with cyanobacteria was measured by the dichromate method. Exchangeable base cations were extracted with 1.0 mol L^{-1} ammonium acetate and Ca^{2+} and Mg^{2+} in solution were determined by atom absorption spectroscopy (AAS) and Na$^+$ and K$^+$ by flame photometry. Element concentrations in solids were determined with a "SPECTROSCAN MAKC-GV" XRF crystal diffraction scanning spectrometer. Mineralogical composition of clays was analyzed by X-ray diffraction (XRD, CuKα radiation). Oriented specimens were prepared by sedimentation on glass-slides and step-scanned in steps 0.1°2θ and 10 s/step counting time. Mg- and

Table 1

Sample	Mineralogy	Content (%)		CEC, Cmol/kg	χ (10^{-8}m^3 kg^{-1})	OC (%)
		Fe$_2$O$_3$	K$_2$O			
Kaolin (Vimianzo, Spain)	Kaolinite, muscovite, quartz	0.69	0.92	2.25	0.9	0.24
Smectite–zeolite (Russia)	Montmorillonite, clinoptilolite, quartz, muscovite, calcite	1.53	1.40	34.6	2.73	0.04
Illite (Hungary)	Illite, quartz, microcline	0.60	6.20	12.84	0	0.15
Bentonite SWy-2	Montmorillonite, quartz, feldspars, calcite	3.10	0.50	94.53	10.47	0.86

K-saturated clays were examined at room temperature, after ethylene glycol saturation and after heating to 300°C and 550°C.

The magnetic characteristics of sediments reflect the amount and quality of ferruginous minerals they contain and are connected with their content, mineralogy and the grain size. To investigate the influence of cyanobacteria on the properties of iron compounds present in all studied clays as impurities, solids were investigated by a complex of magnetic measurements, including: low- and high-frequency magnetic susceptibility (MS), anhysteretic remanence (ARM) and incremental acquisition of magnetic remanence (IRM) and demagnetization, and a high-field IRM (HIRM). These magnetic techniques are non-destructive and sensitive to trace amounts of magnetic minerals (Maher and Thompson, 1999; Thompson and Oldfield, 1986). Magnetic susceptibility (χ) reflects the total concentration of ferrimagnetic or total concentration of paramagnetic minerals and antiferromagnetic with low content of ferromagnetics. IRMs were imparted in pulsed fields of 10, 20, 50, 100 and 300 mT (Molspin pulse magnetizer) and a dc field of 1000 mT (Newport 4" Electromagnet). All magnetic remanences were measured using a fluxgate magnetometer (Molspin Ltd, sensitivity $\sim 10^{-7}$ A m^2) and JR-6 magnetometer (AGICO Ltd). As the majority of ferrimagnetic is fully saturated in fields of 100 mT, IRM$_{100\ mT}$/SIRM allows evaluating the content of ferrimagnets (magnetite, maghaemite). In our experiment a remanence acquired in a field of 1T is referred to as SIRM (SIRM = IRM1000 mT). HIRM = SIRM-IRM$_{300\ mT}$ can be used to approximate the total concentration of high coercitivity minerals (haematite + goethite). Magnetic hysteresis was measured using vibrating sample magnetometer (VSM Molspin Ltd) in the magnetic field between +1 and –1 Tesla.

Concentrations of Fe, Si and Al in solutions were measured by AAS and K by flame photometry. For the observation of dynamics of *Microcoleus* growth in the presence of clays, SEM-EDAX (CamScan with Link-860) was used.

3 Results and Discussion

3.1 Biological Activity of Cyanobacteria in the Presence of Clays

SEM micrographs show that the behavior of cyanobacteria in the presence of studied clays is similar: on 7–14th days the trichomes are well developed; on 28th day the formation of glycocalyx in a form of extracellular polysaccharide sheath was found. On 60th day trichomes become lithified and hardly recognized on the surface of clays (not shown).

The pH values for all experiments were 9.2–9.5 and similar for the systems with cyanobacteria and controls. Figure 1 shows the dynamics of chlorophyll *a* concentration and the oxygen flow for *Microcoleus chthonoplastes* in the presence of clays. The peaks of both parameters are observed for the first 7

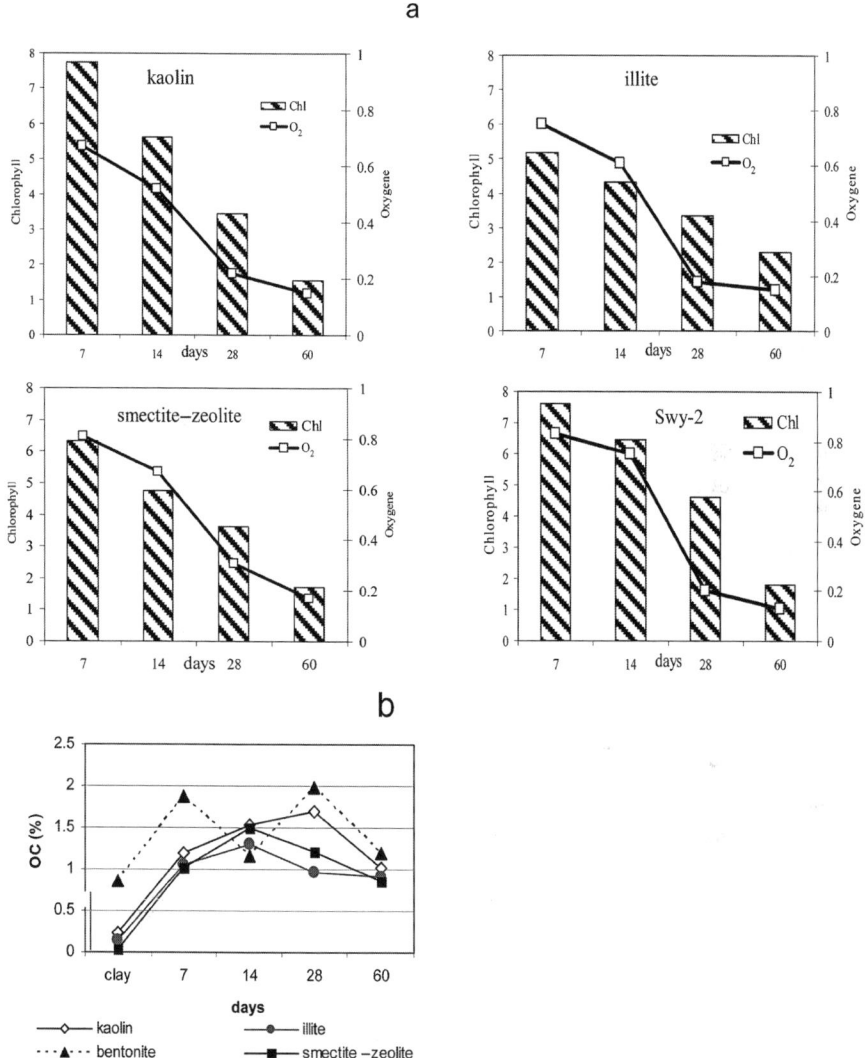

Fig. 1 Biological activity of *Microcoleus chthonoplastes*: (**a**) dynamics of chlorophyll á concentration (mg/ml) and oxygene (mmol/ml^h); (**b**) dynamics of organic carbon content

days of the experiments. Afterwards a gradual decrease in the biological activity takes place. The largest chlorophyll concentration is observed for the experiments with bentonite and kaolin and the smallest—with illite.

Parent clays have different concentrations of organic C (Table 1). For the experiments with illite and smectite–zeolite clay, 14 days of incubation with cyanobacteria lead to the accumulation of OC in spite of the decline of biological activity. For the experiment with kaolin an increase in the OC content took place till 28th day, after which the destruction of OC prevailed. For bentonite 2 steps were observed with a decline of the OC content after 14 days followed by an increase on 28th day.

The intermediate decline is most probably connected with the destruction of OC presented in the parent bentonite. The obtained results allow the period of culture growth (7–14 days) and destruction of organic substances (after 14–28 days) to be distinguished.

3.2 Changes in Argillaceous Minerals After Incubation with Cyanobacteria

Both the content of dissolved Si, Al, Fe and K and the composition of solids after incubation with *Microcoleus* show that kaolin and bentonite clays have the greatest stability, for which no visible changes have been observed but only the formation of Mg-calcite in the presence of kaolin after 28 days of incubation (not shown).

The main component of the illite clay (Füzèrradvany, Hungary) is an irregularly interstratified illite—smectite mineral which contains 17–26% of swelling smectite layers. Incubation of cyanobacteria with illite resulted in a depletion of K^+ and the development of more smectitic layers within the illite matrices which is observed after 7 and 14 days (Fig. 2). This conclusion is based upon the appearance of a new peak at 17.3 Å after ethylene glycol treatment of the specimen. K-saturation resulted in the collapse of newly formed smectite phase.

Formation of traces of defect "island" layer in the interlayer space of montmorillonite from smectite–zeolite clay was observed after 28 days of incubation where the collapse of smectite structure on heating is not complete (Fig. 3b and c). The latter most probably connected with the exchange of K^+ for Al (or other)—hydroxyls or their complexes with organic anions. Ransom et al. (1999) showed that in biological systems swelling smectite type minerals play the role of "catch pit" where the accumulation of ammonia and other products of metabolism could occur. The described transformations are accompanied by the increase in concentrations of dissolved cations in supernatants after incubation with cyanobacteria—K, Si and Fe for illite; K, Si, Fe and Al for smectite–zeolite.

Clay minerals usually possess paramagnetic properties. Magnetic properties of clays are caused by impurities of iron oxide minerals which have ferromagnetic properties (dispersed forms of magnetite or maghaemite, hematite and iron hydroxides).

Iron is involved in the variety of the physiological and biochemical functions of organisms and being the element of variable valence is exclusively sensitive to the changes of surrounding conditions (pH, oxidation–reduction etc.). The participation of iron-reducing bacteria in the biogeochemical cycle of Fe is

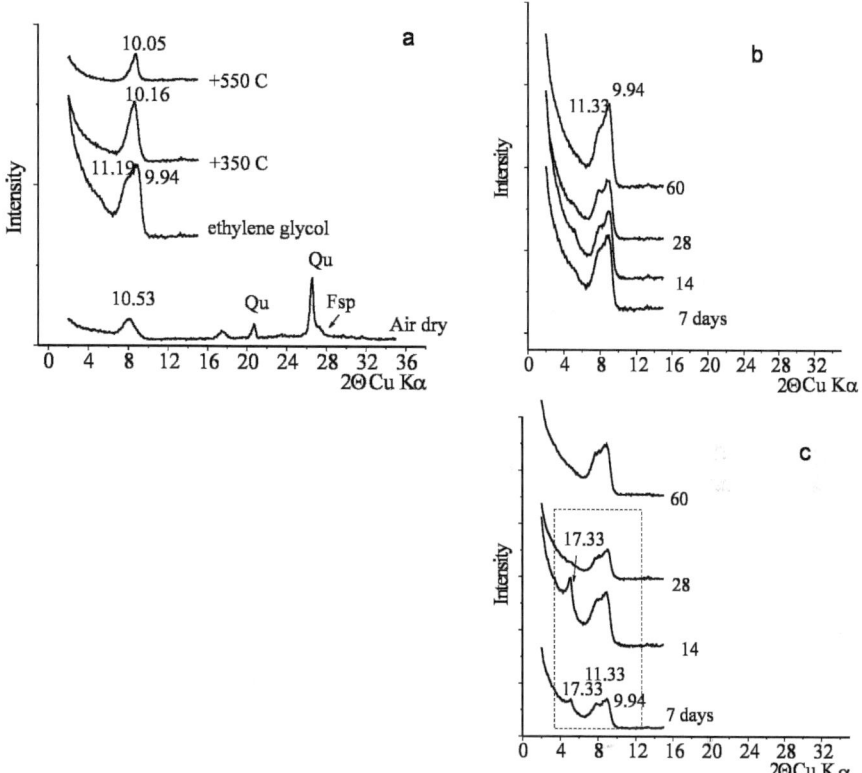

Fig. 2 XRD patterns of illite: (**a**) parent illite air dry, ethylene glycol solvated and heated to 350 and 550°C; (**b**) control illite, samples ethylene glycol solvated after 7, 14, 28 and 60 days incubation without cyanobacteria; (**c**) the same after incubation with cyanobacteria. Qu, quartz; fsp, feldspar. Values are given in Å

well known (Lovley et al., 1998; Murad and Fischer, 1988; Zavarzina et al., 2003). Recently the participation of phototrophic microorganisms in a cycle of oxidation–reduction of iron has also been shown (Konhauser et al., 2002).

Microcoleus chthonoplastes visibly influenced the properties of iron compounds. Both dissolution and precipitation processes were observed. Kaolin and smectite–zeolite demonstrate the increase of magnetic susceptibility and magnetization values (Fig. 4a) which connect with the precipitation of Fe in the form of metastable ferrihydrite ($5Fe_2O_3 \cdot 9H_2O$) followed by the formation of more stable goethite (α-FeOOH). Robert and Berthelin (1986) underlined that weathering in biochemical systems promotes the prevalence of components with a low degree of the order, such as allophone, ferrihydrite et al.

In our experiments the formation of poor ordered ferrihydrate is caused by the properties of the system—large concentration of oxygen which promotes the fast oxidation of Fe, the presence of organic matter which, probably, plays

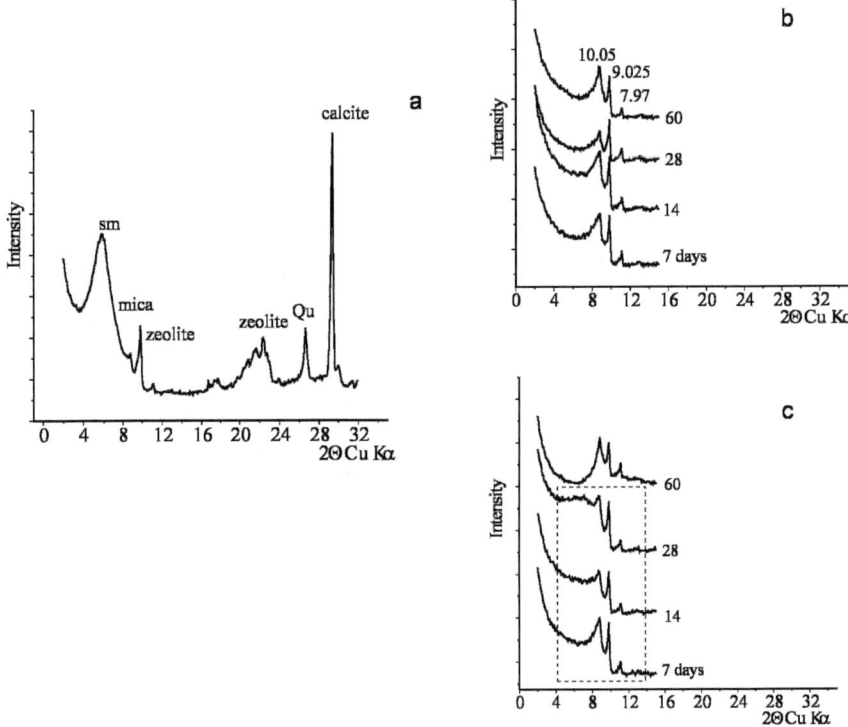

Fig. 3 XRD patterns of smectite–zeolite clay: (**a**) parent clay, air dry; (**b**) control samples heated to 350°C after 7, 14, 28 and 60 days incubation without cyanobacteria; (**c**) the same after incubation with cyanobacteria. Sm, smectite; m, mica; ka, kaolinite; Qu, quartz. Values are given in Å

the role of inhibitor and prevents crystallization of goethite. The appearance of goethite was observed at the end of the experiment (28–60th days) and correlated with the mineralization of organic matter. Our experimental parameters of the metastable mineral, which are similar to the low coercivity minerals (magnetite) with full saturation at low fields of ~200 mT, low magnetic susceptibility value (comparable with goethite or hematite) allow us to confirm the ferrihydrate composition of this mineral (Pannalal et al., 2005).

The Fe_2O_3 content in SWy-2 bentonite is 3.10% (Table 1), its magnetic susceptibility is 10.5 units of SI 10^{-8} m^3 kg^{-1} and magnetizations of saturation (SIRM)—90 to 130×10^{-5} Am kg^{-1}. The analysis of hysteresis loops demonstrates the presence of the mixture of ferrimagnetic and paramagnetic parts. The relatively high magnetic values are caused by impurities of iron oxide minerals (superparamagnetic hematite, goethite and traces of magnetite). The behavior of magnetic properties correlates with the biological activity of cyanobacteria. Destruction of organic mater on 14th day of the experiment is accompanied by a jump of a paramagnetic signal and a corresponding decrease

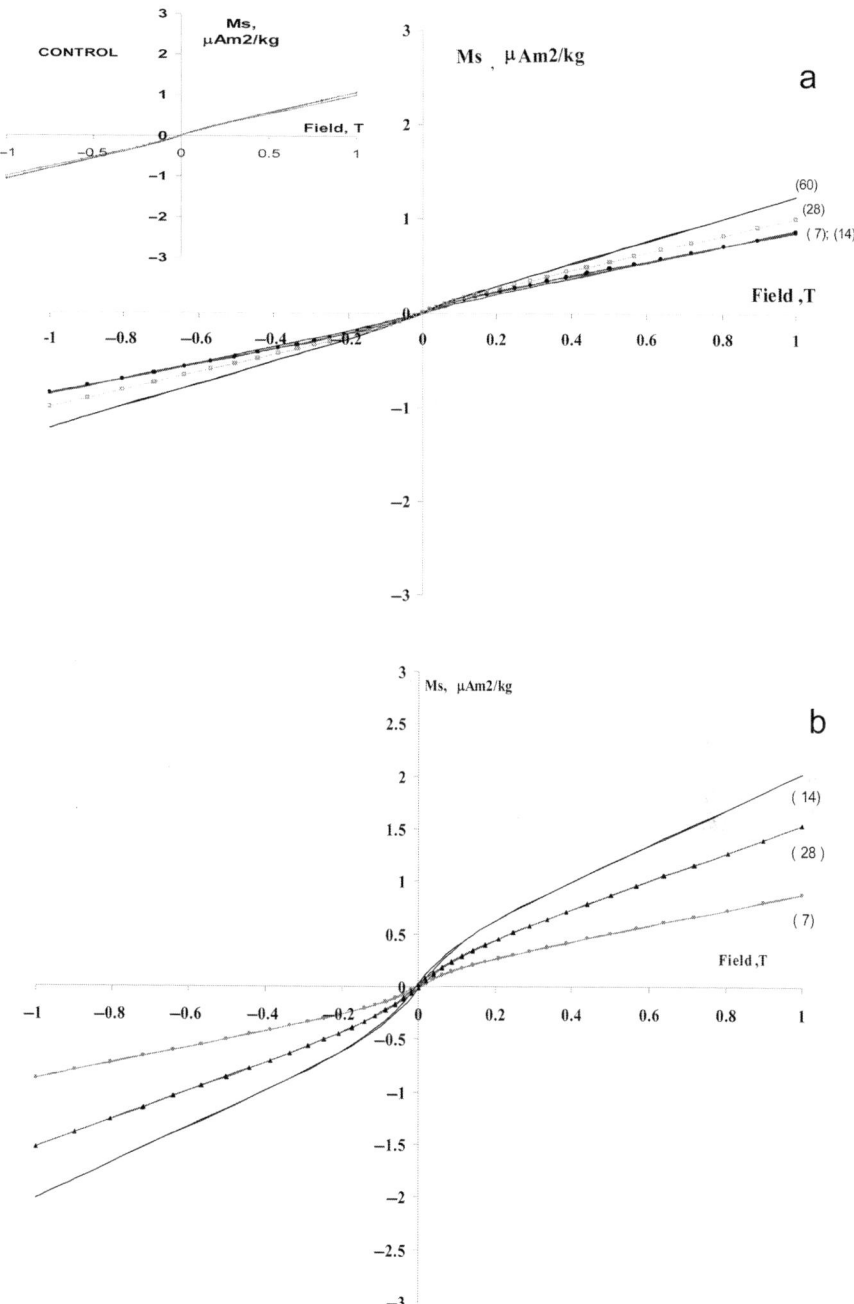

Fig. 4 Hysteresis loops for clays after incubation with cyanobacteria: (**a**) smectite–zeolite; (**b**) bentonite (in brackets duration of experiment in days)

in the ferromagnetic signal (Fig. 4b). A decrease in both magnetic susceptibility and magnetization took place and most probably are connected with oxidation of Fe^{2+} in the magnetite (maghemite) structure and precipitation of ferrihydrite and goethite.

4 Conclusions

Cyanobacteria *Microcoleus chthonoplastes* affects the properties of studied clays. It has no visible influence on the mineralogy of bentonite and kaolin. Whereas the development of more smectitic layers within the illite matrices in the case of illite clay and defect "island" layer in the interlayer space of montmorillonite in smectite–zeolite clay were found. Cyanobacteria influences the properties of iron compounds, which are present in all studied clays as impurities. Both dissolution and precipitation processes of iron compounds were observed. We suppose that ferrihydrite is the key mineral in the biogenic cycle of Fe and it can, probably, be synthesized via "green rust" minerals. "Green rusts" are natural hydrotalcites with the formula: $Fe^{2+}_4 Fe^{3+}_2(OH)_n A$, where A is SO_4, Cl, CO_3 or other. They are unstable in the presence of oxygen and hardly detectable. We suppose that these minerals could be one of the intermediate metastable phases in the biochemical cycle of Fe in the studied systems which contain Fe^{2+}, Fe^{3+}, HCO_3^-, Cl^-. Fast oxidation of Fe in the structure of "green rust" in the presence of organic C most probably results in ferrihydrite synthesis (Hansen, 2001). Obtained data show that the biochemical transformations we observed have the multi-step nature. The initial precipitates differ from the posterior form and one mineral can substitute to another. These processes connect with the conditions of the medium (temperature, pH, redox conditions, rate of Fe oxidation, presence of inhibitors—organic matter, Si and etc). This conclusion applies to transformations of layer silicates as well as Fe oxides (hydroxides).

Acknowledgment The investigation was supported by Program 25 of the Russian Academy of Sciences "Biosphere Origin and Evolution." We are very grateful to Prof. B. Maher (Centre for Environmental Magnetism and Paleomagnetism, Lancaster University, UK) for the possibility to use the equipment.

References

Dubinin, A.V., Gerasimenko, L.M. and Gusev, M.V. (1992) Physiological characteristics of *Microcoleus chthonoplastes* from hyper saline lake. Microbiology 61(1), 63–69 (in Russian).
Geptner, A.R. and Ivanovskaya, T. A. (1998) Biochemogenic genesis of the glauconite–nontronite series minerals in present-day sediments of the Pacific Ocean. Lithol. Miner. Resour. 33(6), 503–517.

Geptner, A.R., Petrova, V.V., Sokolova, A.L. and Gor'kova, N.V. (1997) Biochemogenic formation of phyllosilicates during hydrothermal alteration of basalts in Iceland. Lithol. Miner. Resour. 32(3), 216–225.

Geptner, A., Kristmannsdottir, H., Kristjansson, J. and Marteinsson, V. (2002) Biogenic saponite from an active submarine hot spring, Iceland. Clays Clay Miner. 50(2), 174–185.

Gerasimenko, L.M., Nekrasova, V.K., Orleansky, V.K., Venetskaya, S.L. and Zavarzin, G.A. (1989) The primary production of halophilic *Cyanobacterium* cenoses. Microbiology 58(3), 507–514 (in Russian).

Gerasimenko, L.M., Dubinin, A.V. and Zavarzin, G.A. (1996) Alkaliphilic cyanobacteria from soda lakes of Tuva and their ecophysiology. Microbiology 65(6), 736–740.

Hansen, H.C.B. (2001) Environmental Chemistry of iron(II)–iron (III) LDHs (green rusts). In: V. Rivers (Ed.), Layered Double Hydroxides: Present and Future. Nova Science Publishers, New York, pp. 413–434.

Kawano, M. and Tomita, K. (2002) Microbiotic formation of silicate minerals in the weathering environment of a pyroclastic deposit. Clays Clay Miner. 50(1), 99–110.

Konhauser, K.O. and Urrutia, M.M. (1999) Bacterial clay authigenesis: a common biogeochemical process. Chem. Geol. 161, 399–413.

Konhauser, K.O., Hamade, T., Raiswell R., Morris, R.C, Ferris, G, Southam, G. and Canfield D.E. (2002) Could bacteria have formed the Precambrian banded iron formations? Geology 30, 1079–1082.

Likhoshway, E.V., Sorokovikova, E.G., Bel'kova, N.L., Belykh, O.I., Titov, A.T., Sakirko, M.V. and Parfenova V.V. (2006) Silicon Mineralization in the Culture of Cyanobacteria from Hot Springs. Doklady Biol. Sci. 407, 201–205.

Lovley, D.R., Fraga, J.L., Blunt-Harris, E.L., Hayes, L.A., Phillips, E.J.P. and Coates J.D. (1998) Humic substances as a mediator for microbially catalyzed metal reduction. Acta Hydrochim. Hydrobiol. 26, 152–157.

Lowenstam, H.A. (1981) Minerals formed by organisms. Science 211, 1126–1130.

Maher, B.A. and Thompson R. (1999) Quaternary Climates, Environments and Magnetism. Cambridge University Press, Cambridge.

Murad, E. and Fischer, W.R. (1988) The geobiochemical cycle of iron. In: J.W. Stucki, B.A. Goodman and U. Schwertmann (Eds), Iron in Soils and Clay Minerals. NATO ASI series, Vol. 217, pp. 1–18.

Pannalal, S.J., Crowe, S.A., Cioppa, M.T., Symons, D.T.A., Sturm, A. and Fowle, D.A. (2005) Room-temperature magnetic properties of ferrihydrite: a potential magnetic remanence carrier? Earth Planet. Sci. Lett. 236, 856–870.

Phoenix, V.R., Adams, D.G. and Konhauser, K.O. (2000) Cyanonacterial viability during hydrothermal biomineralisation. Chem. Geol. 169, 329–338.

Ransom, B., Bennet, R.H., Baerwald, R., Hulbert, M.H. and Burkett P-J. (1999) *In situ* conditions and interactions between microbes and minerals in fine-grained marine sediments: A TEM microfabric perspective. Am. Miner. 84, 183–192.

Robert, M. and Berthelin, J. (1986) Role of biological and biochemical factors in soil mineral weathering. In: Interactions of Soil Minerals with Natural Organics and Microbes. SSSA special publication 17. Soil Science Society of America, Madison, pp. 453–495.

Schultze-Lam, S., Forris, F.G., Konhauser, K.O. and Wiese, R.G. (1995) *In situ* silicification of an Icelandic hot spring microbial mat: implications for microfossil formation. Can. J. Earth Sci. 32, 2021–2026.

Tazaki, K. (1998) A New world in the science of biomineralization—environmental biomineralization in microbial mats in Japan. The Science Reports of Kanazawa University, Japan. XLII (No. 1 and 2).

Tazaki, K., Okrugin, V., Okuno, M., Belkova, N., Islam, A.R., Chaerun, S.K., Wakimoto, R., Sato, K. and Moriichi S. (2003) Heavy metallic concentration in microbial mats found at hydrothermal systems, Kamchatka, Russia. Science reports of Kanazawa University, Japan, 47 (No. 1–2).

Thompson, R. and Oldfield, F. (1986) Environmental Magnetism. Goerge Allen and Unwin, London.

Trebst, A. (1980) Inhibitors in electron flow: tools for the functional and structural localization of carriers and energy conversation sites. Meth. Enzymol. 69, 675–715.

Ueshima, M. and Tazaki, K. (2001) Possible role of microbial polysaccharides in nontronite formation. Clays Clay Miner. 49(4), 292–299.

Ushatinskaya, G.T., Zaitseva, L.V., Orleansky, V.K. and Gerasimenko, L.M. (2006) Significance of bacteria in natural and experimental sedimentation of carbonates, phosphates and silicates. Paleontol. J. 40(Suppl. 4), 524–531.

Zavarzin, G.A. (1993) Development of microbial community in the Earth's history. In: Problems of Before Anthropogenic Evolution of Biosphere. Nauka, Moscow, pp. 212–222 (in Russian).

Zavarzina, D.G., Alekseev, A.O. and Alekseeva, T.V. (2003). Iron-reducing bacteria and formation of magnetic properties of steppe soils. Eurasian Soil Sci. 36(10), 1085–1094.

Part IV
Coevolution of Geological and Biological Events in Phanerozoe

Ecological Revolution Through Ordovician Biosphere (495 to 435 Ma ages): Start of the Coherent Life Evolution

A. V. Kanygin

Evolution of the biosphere can be presented as the following processes: (1) emergence of new ecologically specialized groups (guilds), providing a more efficient use, transfer, and transformation of matter and energy in ecosystems; (2) spatial expansion of life throughout the Earth (gradual transition from a discrete to continual exploration of new bionomic zones and biotopes); (3) complication of the trophic structure of ecosystems (from simple Archean autotrophic–heterotrophic prokaryotic systems to modern global trophic pyramid); (4) variations in the spatial and dynamic parameters of biogeochemical cycles.

The main evolutionary strategy at the early stages of biosphere development (including the Cambrian) was involved with the formation of optimal physiological mechanisms for interaction of living organisms with the physicochemical environment (widening of the tolerance range) with minimization of the dependence of the biotic environmental factors (Kanygin, 2004). To follow Krasilov (1977), who coined the terms "coherent ecosystems" (ecologically saturated) and "incoherent ecosystems" (ecologically unsaturated), we may refer to this evolution as "incoherent evolution" to distinguish it from the coherent evolution, in which the biotic interaction of organisms, together with physicochemical factors, plays a decisive role in the evolution (Kanygin, 2001). The Ordovician period is the early stage of transformation of noncoherent sea ecosystems to coherent ones, when they acquired the "Phanerozoic" habit, according to Sepkoski (1982). Starting from the Ordovician, as the taxonomic diversity increased to the upper limit with successive emergence of new biomes in marine areas, fresh-water reservoirs, and on the land, the rate of evolution increased at each stage of the spatial expansion of the biosphere.

Early Paleozoic (Cambrian, Ordovician) is a transitional stage between the primordial, locally distributed Precambrian-type ecosystems and the mature, tiered Phanerozoic-type ecosystems. The Cambrian and Ordovician periods

A. V. Kanygin
Trofimuk Institute of Petroleum Geology and Geophysics, Novosibirsk, Russia
e-mail: kanyginav@ipgg.nsc.ru

mark the two fundamental revolutionary events that determined the principal structural-functioning and spatial parameters of marine ecosystems: (1) the appearance of major skeletal hydrobionts and the explosion of biotic diversity in benthic communities dominated by grazing heterotrophs (trilobites and soft-bodied worm-type organisms) in the Early Cambrian (Fig. 1); (2) in the middle Ordovician—the rapid diversification of benthic groups of filter-feeding fauna (cnidarians, bryozoans, crinoids, brachiopods, etc.) and small-sized trophic generalists (ostracodes), the rise of a new group of autotrophs (chitinozoans), specialized groups of zooplankton (graptolites, radiolarians) and nektonic predators (conodontophores, nautiloids, early agnostids), the flourishing of colony-grade organisms, being an effective means of ecospace conquering and food resource utilization (Fig. 2). The principal chorological and trophic restructuring in marine ecosystems in Cambrian and Ordovician was controlled by (1) the changes in the biological productivity and spatial (tiered) distribution of autotrophic components in food chains (benthic cyanobacteria being replaced with phytoplankton), the formation of a vertical vector and a system network in trophic conveyers; and (2) the emergence and rapid diversification of

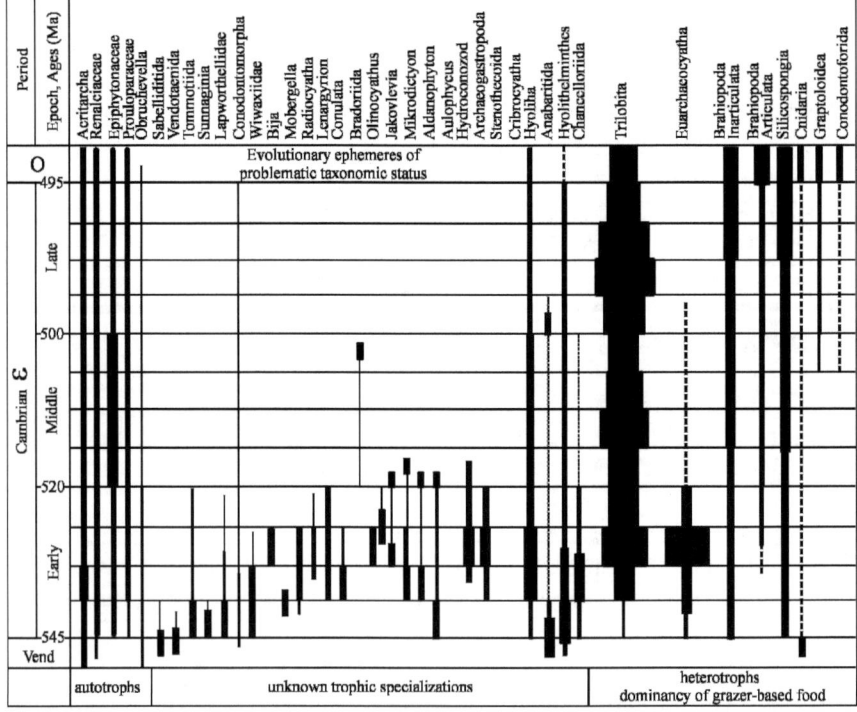

Fig. 1 Taxonomic and trophic structure of Cambrian biota: the model of Siberian platform (after Phanerozoic of Siberian, 1984, with additions)

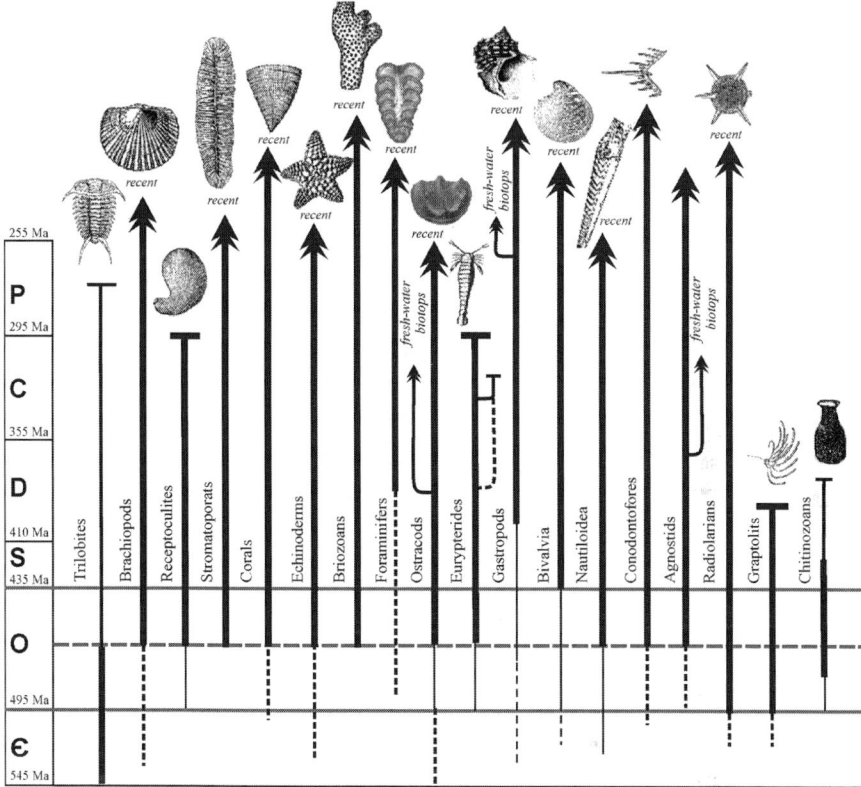

Fig. 2 Main ecological quilds of the Ordovician period. Dominancy oxyphil groups of long-time philolines

new ecological guilds in the benthial and pelagial, thereby facilitating the diversity of trophic specializations, the amplification of the system of depositing, transportation, and spacial distribution of food resources, and the dominance of detritivore-based food webs over grazer-based food webs.

The Ordovician was marked by a unique combination of geological and biological events (Kanygin, 2001). Here are the most important Ordovician biological events:

1. The most powerful explosion of biological divergence in the Phanerozoic (after the sudden emergence of main marine animal taxa in the early Cambrian). This resulted in a threefold increase in taxonomic diversity.
2. Rapid (geologically instantaneous in the case of some taxa) global colonization of the bottoms and water columns of seas by new ecological guilds, which was accompanied by the formation of a multistage trophic structure of ecosystems.

3. Spatial rearrangement and taxonomic rotation of producers (photosynthesizing organisms) at the Early–Middle Ordovician boundary (Llanvirn), which changed dramatically the spatial and biological structure of food resources in the benthic and pelagic zones of seas.
4. Emergence of a constant zoopelagic zone (instead of the former facultative one) owing to the emergence and efflorescence of specialized planktonic and nektonic organisms.
5. The prosperity of oxyphil colonial (cnidaria, bryozoans), aggregated (crinoids) filter-feeding organismus with carcass skeletons and predators (conodontophores, nautiloids, asters, early agnostids).
6. The decrease of the diversity and the loss of dominant part of grazing heterotrophs (trilobites and soft-bodied worm-type organisms) among the bottom communities.
7. Dramatic decrease in taxonomic diversity of marine biotas in the late Ordovician. It was not, however, accompanied by complete extinction of any ecological guilds, unlike later major biological crises (e.g., at the Permian–Triassic and Cretaceous–Paleogene boundaries).

Global geological events also occurred in the Ordovician. They caused dramatic changes in abiotic environmental settings in the euphotic zones of seas and a wide expansion of the ecological space:

1. Contrasting climatic changes: traces of profound coolings and aridization in South America and on the Siberian Platform, global warming in the Middle Ordovician, and Late Ordovician continental glaciations in Gondwana.
2. The widest occurrence of Phanerozoic epicontinental seas in the Middle Ordovician and their dramatic reduction during the Late Ordovician continental glaciation.
3. Abrupt change in the Mg:Ca ratio in marine sediments at the Early–Middle Ordovician boundary. The evidence is a wide occurrence of dolomites and dolomitic limestones in the Lower Ordovician of many epicontinental basins (particularly on the Siberian Platform) and their nearly complete absence from the Middle and Upper Ordovician, dominated by limestones and marls.
4. The greatest peaks in the content of CO_2 in the atmosphere, volumes of volcanic rocks, and fossilized in carbonate rocks (similar large-scale changes in the balance of these indices occurred in the Carboniferous and Cretaceous); CO_2 is the main biophil gas for the biosphere autotrophic level.
5. Dramatic increase in the content of oxygen in the atmosphere (to about 10% of the present-day value) and, correspondingly, its partial concentrations in the hydrosphere and formation of the ozone screen (Berkner–Marshall point), which protected life from lethal fluxes of the short-wave solar radiation on the Earth's surface and in the euphotic zones of water areas.

O_2 is the main biophil gas for the biosphere heterotrophic level. Taken together, these profound changes in biotas and environmental settings greatly affected further development of the biosphere.

Presently available paleontological, paleoecological, and geological data (Budyko et al., 1985; Kraft and Fatka, 1999; Li and Droser, 1999; Morrow et al., 1996; Phanerozoic of Siberian, 1984; Ronov, 1993; Rozhnov, 2005; Severtsev, 1939; Walliser, 1996) allow outlining the ecological regularities of ecosystem evolution in the Ordovician.

Three stages are recognized in the evolution of Ordovician biotas: Early Ordovician (495 to 455 Ma ages), Middle Ordovician (465 to 455 Ma ages), and Late Ordovician (455 to 435 Ma ages). The most dramatic biota changes occurred at the Early–Middle Ordovician boundary. The Early and Middle Ordovician biotas differed not only in the range of taxa but also in abundance (population density) and clearly discrete bionomic distribution, both lateral and stratigraphic (Kraft and Fatka, 1999; Morrow et al., 1996; Walliser, 1996). The Early Ordovician fauna was much closer to the Cambrian than the Middle Ordovician fauna in taxonomy and structure (range of species, short food chains, and discrete location of biotopes). According to Sepkoski (1982), the boundary between the "Cambrian" and "Phanerozoic" faunas should be drawn at the base of the Middle Ordovician rather than at the base of the Lower Ordovician. In life density, the bottoms of Early Ordovician seas can be compared with modern desert landscapes, with rare oases and scarce traces of life in the rest of the area. Early Ordovician seas, like Cambrian ones, were dominated by mud-eating trilobites and soft-bodied worm-type organisms (identified due to varions ichnofossils), often forming dense accumulations at the muddy bottom, rich in organic matter. Less frequent were brachiopods, gastropods, monoplacophorans, and sponges.

Exotic living forms of uncertain systematics appeared almost everywhere in the shallow areas of the Siberian Platform before the Middle Ordovician taxonomic explosion. They were Clyptolichinaria (morphologically close to tabulates), Tolmachovia (presumably either crustaceans or mollusks), Soanites (having features of archaeocyaths, sponges, and corals), and Angarella (having features of cap-shelled gastropods and monoplacophorans). Some of these taxa occur in other regions of the world at approximately the same stratigraphic levels. All of them existed for a short time, much shorter than any other species; therefore, they can be called evolutionary ephemeres. They did not leave direct evolutionary descendants, but the morphoanatomical variants they tested appeared, in a modified form, in related taxa during saltations.

The appearance and rapid extinction of the evolutionary ephemeres on the eve of the global taxonomic radiation of evolutionary long-livers is a good example of the role of inadaptive taxa in search for adaptational compromise between the changing environment and living organisms. Mass appearance of evolutionary ephemeres also preceded the Early Cambrian large-scale biological radiation (Fig. 1).

In the Middle Ordovician, new filter feeders of attached benthos (bryozoans, stromatoporates, tabulates, rugoses, and crinoids) began large-scale colonization of the sea bottom (Fig. 2). The appearance and rapid dispersal of these carcass organisms abruptly changed the stage and trophic structure of benthic biocenoses. All of them are sessile filter feeders capable of consuming nutrient particles varying in size not only from bottom water flows but also from vertical gravitational flows. The rapid radiation of colonial organisms and their mass dispersal in the shelf zones of seas point to a rapid increase in food resources near the bottom and their stepwise distribution. The colonial forms proved to be efficient for occupation of living space by rapid propagation of polyps, continual colony growth, and more complete food utilization by cooperative action of tentacles, which ensures constant water circulation around the colonies and catch of nutrient particles.

The beginning of the Middle Ordovician was also marked by rapid radiation of echinoderms, particularly crinoids, which formed large monotaxic associations (Rozhnov, 2005). They occupied space no less efficiently than it could be done by colonial cooperation of organisms. Ordovician crinoids had a new feeding tool absent from their Cambrian ancestors: mobile tentacles with enlarged food-conducting grooves. This allowed them to rise to a higher stage (above 1 m, whereas eucrinoids existed no more than 10–15 cm above the bottom), widen the zone of food catch from bottom currents and gravitational flows, and utilize food particles of a broader size range. Eucrinoids, as deduced from examination of their present-day descendants, had a less diverse diet consisting of dissolved organic matter and bacteria, which were supplied to the body by brachioles.

The high diversification rate of another guild, bryozoans, may also be related to their capability for occupying and efficiently using food space by combining two types of cooperation: colonial habit and aggregation. They could also have formed monopoly settlements in different biotopes owing to varying colony shapes (from mechanically rigid spherical to mobile bushy) and topographic distribution.

The aggregational habit of settlements, which increases the adaptational potential owing to cooperation, characterized early stages of ecosystem evolution. It is also typical of migrational successions related to changes in physicochemical indices of the environment. Modern analogs of various aggregation types can be observed in extreme landscapes, such as tundra. According to paleontological evidence, aggregations were the main type of the topographic structure of associations in the Cambrian and Early Ordovician seas. The fundamental difference between the topographic structures of the Cambrian–Early Ordovician and Middle Ordovician associations was that in the Middle Ordovician, the local discrete ecological space gave way to a continual-mosaic one. Three reasons can be suggested for the rearrangement of the topographic structure of associations: (1) change in the spatial structure of the primary producer link, i.e., the autotroph stage; (2) appearance of new guilds and increase in biodiversity at the heterotroph stage; (3) appearance of ecological generalists. All these factors led to a broad extension of ecological space (range of available food resources) and heterogeneity in the distribution of producers and consumers accompanied by an increase in the interdependence of associations related to increasing intricacy of food webs.

Geological data indicate that at the Early–Middle Ordovician boundary, sea basins became less rich in stromatoliths (traces of bacterial–algal associations). Probably, they provided the bulk of food resources for benthic biota by photosynthesis. It is difficult to estimate the role of phytoplankton in the total food balance from paleontological evidence, though there is much proof for its early (Precambrian) appearance. According to indirect data, significant amounts of specialized forms of zooplankton appeared only in the Ordovician. They could ensure vertical trophic transition of primary products and expand its horizontal transition within the euphotic zone of seas. Important arguments for the dramatic change in the balance between the bottom and pelagic phytoproduction, in addition to the abrupt disappearance of abundant stromatoliths, are a dramatic increase in the taxonomic diversity of pelagic (graptolites and radiolarians) and nektonic (conodonts and nautiloids) organisms, a principal change in the ratio of filter-feeding and mud-eating (crawling and burrowing) animals, and formation of a multistage trophic structure in benthic associations. Moreover, with regard to the Middle Ordovician large-scale occurrence of benthic colonial organisms propagating through the planktonic larva stage, meroplankton acquired a great significance in the food webs of the bottom parts of seas.

Ostracodes became an important link in the global food cycle of Middle Ordovician sea ecosystems. Their spatial expansion and diversification occurred most rapidly. From the outset, ostracodes had the widest range of adaptive tools of all aquatic organisms: the ability to crawl, burrow, and drift with water currents owing to small sizes. In the Middle Ordovician, ostracodes acquired tools for passive hovering and active swimming in the water and were among the first aquatic animals colonizing the pelagic zone. Nearly all alimentation modes characteristic of aquatic invertebrates are known in ostracodes: sestonic, filtering, detritivorous, carrion feeding, parasitic, etc. Moreover, ostracodes, like all short-lived small animals capable of rapid propagation, could have filled the smallest ecological niches under the conditions of dense occupation of the ecological space. In the Ordovician, being universally eurybiotic, ostracodes were predominant at the lower trophic level and were then dispersed throughout the Earth's hydrosphere, including various fresh-water biotopes.

The abrupt restructuring of food resources in Middle Ordovician seas is also confirmed by significant changes in the proportions of former predominant groups: trilobites and articulate brachiopods. The Early Ordovician seas, as well as the Cambrian ones, were clearly dominated by mud-eating trilobites, and the role of filter-feeding brachiopods was of little significance, whereas in the Middle Ordovician, the range of brachiopod species reached its acme, and trilobites began to dwindle. The Middle Ordovician is marked by the maximum rate of emergence of new adaptive tools in trilobites. For example, the length of ocular stems in some species of the family Asaphidae increased dramatically. This allowed them to burrow deep into mud, maintaining visual monitoring of the situation. Some trilobites acquired tools for active and passive swimming above the bottom. This enhanced their ability to obtain food under the conditions of rapidly increasing competition for food resources.

The distribution of skeletal over Cambrian and Ordovician biostratons demonstrates that the total bioproductivity of sea ecosystems increased by an order of magnitude on the average (Kanygin, 2001; Li and Droser, 1999). The bioproductivity difference between Early and Middle Ordovician ecosystems is still more pronounced if the increasing role of living organisms in trophic decomposition of dead ones (particularly, mollusks) and reduction of skeletal elements by bioturbation (particularly, graptolites) is taken into account.

The main feature of the Middle Ordovician saltation, in comparison with the Early Cambrian one, is that it was not accompanied by fundamental anatomical changes. Biodiversification of all new ecological guilds involved all construction elements which appeared earlier. However, the structures of locomotory and food-catching organs of new taxa underwent numerous modifications, advantageous for colonization of living space and utilization of food resources. These modifications determined new types of adaptation to the biotic rather than abiotic environment. Thus, their scale and evolutionary consequences allow them to be assigned to aromorphs, i.e., changes which substantially affect the energy potential and adaptability of organisms and determine appearance of new modes of interaction with other organisms in ecosystems (Severtsev, 1939). Taxonomically, the new aromorphs manifested themselves in rapid divergence and broad expansion of new high-rank groups: corals, brachiopods, bryozoans, ostracodes, stromatoporates, echinoderms, graptolites, radiolarians, conodonts, and nautiloids. All these groups were oxyphil and long-lived. Many of them still exist.

Thus, in the Middle Ordovician, the evolutionary strategy of marine biotas changed, first in benthic associations and then in pelagic ones. The general evolutionary trend was regulated mainly by ecological interactions between organisms (i.e., adaptation to the biotic environment) rather than by physicochemical adaptation, which was the main driving and stabilizing force in earlier ecosystems. In the Middle Ordovician, evolution of benthic ecosystems became coherent. As a result, the total range of species in benthic ecosystems stabilized and very slowly increased thereafter (Fig. 3).

The Ordovician was the time of formation of constantly functioning pelagic zones inhabited by the first specialized groups of marine skeletal zooplankton: graptolites and radiolarians. This time was also marked by large-scale development of nektonic organisms (nautiloids and conodonts) and the advent of agnathans. The first ostracodes capable of broad dispersal and facultative dwelling in the pelagic zone are also dated to the Middle Ordovician.

The global ecological event was accompanied by the greatest (in the Phanerozoic) burst of the diversity of Ordovician marine biotas followed by rapid stabilization. Later, the stability was maintained by a phylogenetic succession of ecologically equivalent taxa, with some ecological guilds replaced at critical borderlines. Thus, in the Ordovician, sea ecosystems became multistage, their trophic structure became more complex, and a global closed biogeochemical cycle formed for the first time throughout the sea area. The Ordovician global biotic events matched large-scale geological events (abrupt

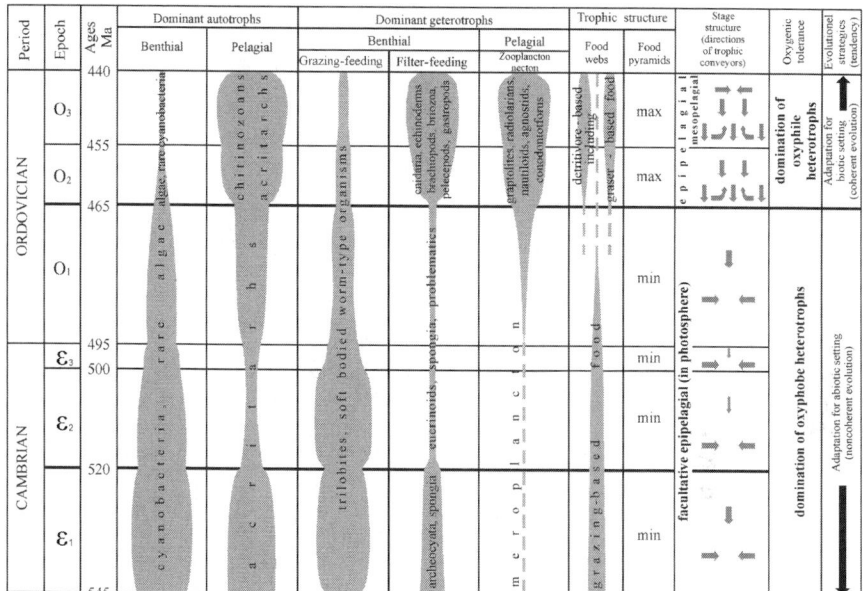

Fig. 3 Evolution of taxonomic, trophic, and space structure of marine ecosystems through early Paleozoic

climatic changes, maximum range of transgressions and regressions of epicontinental seas, changes in Mg and Ca balance in marine sediments, increase in the content of oxygen in the Earth's atmosphere and hydrosphere, and formation of the ozone screen; Budyko et al., 1985; Morrow et al., 1996; Ronov, 1993). It is supposed that the increase in the content of oxygen in seawater had a crucial impact on the settling of heterotrophs in the pelagic zone and formation of coherent (ecologically complete) benthic ecosystems. At the initial metastable stage of the development of the ozone screen, eustatic fluctuations of the world ocean level caused dramatic biodiversity fluctuations in bottom and pelagic associations determined by profound changes in spatial parameters of sea shelves, the main habitat of biota. The Late Ordovician extinction of marine biotas resulted from an abrupt shrinkage of the shelf habitat caused by a lowering of the world ocean, which, in turn, resulted from the fixation of huge volumes of water in continental glaciers after the Ordovician transgression maximum.

References

Budyko, M.I., Ronov, A.B. and Yanshin, A.L. (1985) History of the Atmosphere. Gidrometeoisdat, Leningrad, 208 pp. [in Russ.].
Kanygin, A.V. (2001) The Ordovician explosive divergence of the Earth's organic realm: causes and effects of the biosphere evolution. Russ. Geol. Geophys. 42(4), 599–633.

Kanygin, A.V. (2004) Geological Setting of Life Evolution. SB RAS, Novosibirsk, pp. 23–32 [in Russ.].
Kraft, P. and Fatka, O. (Eds) (1999) Qua vadis Ordovician? Short papers of the International Symposium on Ordovician System, Universitatus Carolinae. Geologica 43(1/2), 533 pp.
Krasilov, V.A. (1977) Evolution and Biostratigraphy. Nauka, Moscow, 256 pp. [in Russ.].
Li, X. and Droser, M. (1999) Lower and Middle Ordovician shell beds from the basin and range province of the Western United States (California, Nevada and Utah). Palaios 14, 2i5, 215–233.
Morrow, J., Schindler, E. and Walliser, O. (1996) Phanerozoic development of selected global environmental features. Global Events and Event Stratigraphy in the Phanerozoic. Springer, Berlin Heidelberg New York, pp. 53–61.
Phanerozoic of Siberian (1984) T. 1. Vend, Paleozoic. Nauka, Novosibirsk, 192 pp. [in Russ.].
Ronov, A.B. (1993) Stratisphere, or the Sedimentary Shell of the Earth. Nauka, Moscow, 144 pp. [in Russ.].
Rozhnov, S.V. (2005) Morphological Regularity of Echidermes Higher Taxons Establishment and Evolution. KMK, Moscow, pp. 156–170 [in Russ.].
Sepkoski, J.J. (1982) Spec. Pap. Geol. Soc. Am. 247, 283–290.
Severtsev, A.N. (1939) Morphological Regularity of Evolution. AS USSR, Moscow, 610 pp. [in Russ.].
Walliser, O. (Ed.) (1996) Global Events and Event Stratigraphy in the Phanerozoic. Springer, Berlin Heidelberg New York, 333 pp.

Part V
Ecosystems and Molecular Genetic Factors of Organism Evolution

Evolution by Gene Duplications: from the Origin of the Genetic Code to the Human Genome

S. N. Rodin and A. S. Rodin

Abstract Evolution *ab simplecioribus ad complexiora* is based on duplications. In a working live system novelties almost never emerge by chance. New genes, exons, or even smaller functional units typically originate with minute changes in the duplicate(s) of preexisting sequences. Extant gene and protein sequences often harbor the periodicity that unambiguously points to their duplication-based origins (Ohno, 1987, 1988). Here we consider two fundamental paradoxes associated with duplications. The first paradox dates back to the origin of encoded protein synthesis, and it can be explained away by what arguably was the single most important duplication event in the history of life, the duplication of a presumable short precursor of a transfer RNA that shaped it into a major adaptor of the genetic code. The second paradox belongs to the realm of already quite complex and advanced life; it concerns survival of gene duplicates per se and appears to be most pronounced in the human genome. In a sense, the entire history of life (as we see it presently) can be said to lie in between the above two series of events.

1 Origin of the Genetic Code: Duplication in tRNA

1.1 Domains of tRNAs and the Paradox of Two Codes

The assignment of nucleotide triplets (codons) to amino acids (aa) represents the universal code (Table 1) for synthesizing gene-instructed proteins.

Although other codes have arisen during evolution (Weiss and Buchanan, 2005), the universal genetic code is accepted as the basic (most fundamental) one. Any hypothesis of its origin must somehow evade the "chicken-or-egg" conundrum. This brings us to a paradox that concerns the nature of tRNA, a

---

S.N. Rodin
Beckman Research Institute of the City of Hope, Duarte, CA, USA
e-mail: srodin@coh.org

Table 1 The conventional representation of the genetic code with indicated assignment of aaRSs to class I (White) and class II (gray)

1	2				3
	U	C	A	G	
U	UUU Phe	UCU Ser	UAU Tyr	UGU Cys	U
U	UUC Phe	UCC Ser	UAC Tyr	UGC Cys	C
U	UUA Leu	UCA Ser	UAA stop	UGA stop	A
U	UUG Leu	UCG Ser	UAG stop	UGG Trp	G
C	CUU Leu	CCU Pro	CAU His	CGU Arg	U
C	CUC Leu	CCC Pro	CAC His	CGC Arg	C
C	CUA Leu	CCA Pro	CAA Gln	CGA Arg	A
C	CUG Leu	CCG Pro	CAG Gln	CGG Arg	G
A	AUU Ile	ACU Thr	AAU Asn	AGU Ser	U
A	AUC Ile	ACC Thr	AAC Asn	AGC Ser	C
A	AUA Ile	ACA Thr	AAA Lys	AGA Arg	A
A	AUG Met	ACG Thr	AAG Lys	AGG Arg	G
G	GUU Val	GCU Ala	GAU Asp	GGU Gly	U
G	GUC Val	GCC Ala	GAC Asp	GGC Gly	C
G	GUA Val	GCA Ala	GAA Glu	GGA Gly	A
G	GUG Val	GCG Ala	GAG Glu	GGG Gly	G

I class AARS
II class AARS

necessary bifunctional code-translation adaptor. It is as follows: transfer RNAs bring the code into action by using the codon's precise complementary copy, the anticodon. The problem with the anticodon is that it is located at the *opposite* end of the tRNA molecules to the corresponding aa-binding site (Fig. 1). This makes the anticodon-specific self-aminoacylation of tRNAs very unlikely. Indeed, even the simplest forms of life always have 20 enzymes, the aminoacyl-tRNA synthetases (aaRS), each of which "charges" its tRNAs with a cognate amino acid. Obviously, to perform this substrate-specific aminoacylation, the aaRSs must recognize their tRNAs as selectively as possible, by interacting either directly with the anticodon triplet or with some other site(s) within the same tRNA molecule. In the latter case, the anticodon-to-aa assignment is established indirectly, due to the physical linkage of the anticodon with the aa-specific site. In reality, both mechanisms seem to have been used (Schimmel and Beebe, 2006). And yet, when tRNAs are truncated to the mini-helices (or even smaller fragments of the acceptor stem) they can still be selectively charged with the correct amino acid (Schimmel and Beebe, 2006; Schimmel et al., 1993). Such anticodon-independent aminoacylation of tRNAs was detected for half of amino acids (Schimmel and Beebe, 2006; Schimmel et al., 1993). Moreover, in reciprocal experiments with truncated aaRSs (including extreme cases in which a truncated enzyme was unable to reach the anticodon) the specificity of aminoacylation also remained unchanged (Schimmel and Beebe, 2006; Schimmel et al., 1993).

This remarkable autonomy of aa-binding domains in tRNAs, as well as aaRSs, led to the concept of an *operational* code that is localized mainly in the

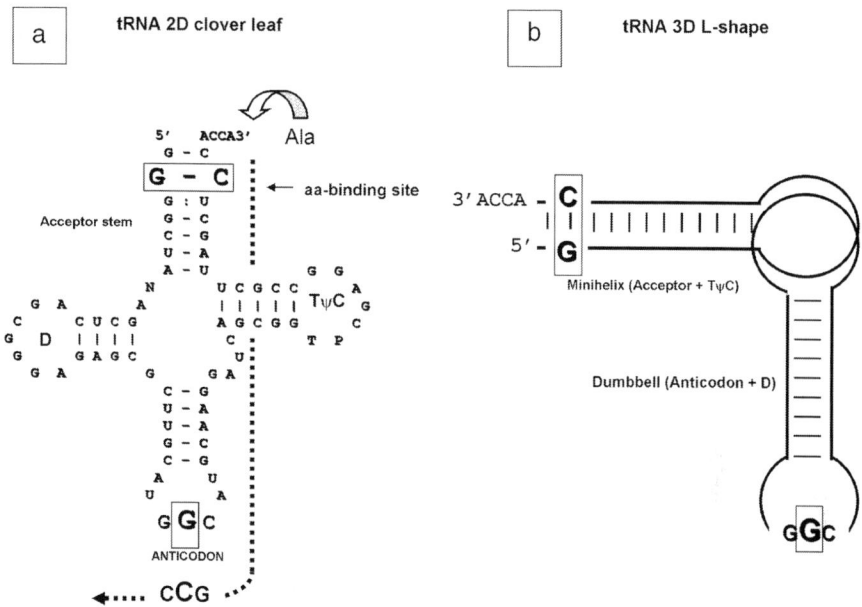

Fig. 1 tRNA cloverleaf (**a**) and L-shaped 3D structure (**b**) consisting of two halves, *top* minihelix (acceptor stem plus TψC arm) and *bottom* dumbbell (anticodon arm plus D arm). One of *E. coli* tRNAs^{Ala} (with GGC anticodon) is shown, with the characteristic 3G:U72 "wobbling" complementary base pair that determines the identity of the Ala tRNAs across all species (Hou and Schimmel, 1988). The second base of the anticodon and the second base pair of the acceptor (showing dual complementarity—see text) are enlarged and boxed. The hypothetical aminoacylating ribozyme is schematically shown with a *dotted line* (to the right, **a**)

acceptor stem of tRNAs (Schimmel et al., 1993; see also De Duve, 1988) as opposed to the *classic* genetic code that is associated with the anticodons.

Importantly, it seems logical to postulate the similar modularity for the ribozymic precursors of aaRS (Fig. 1). The obvious advantage would be that such ribozymes can easily recognize the anticodon by means of direct complementary base pairing. However, in order to attach appropriate aas to the tRNAs, this ribozyme must, in turn, have had the aa-specific binding site, and this site would again be localized very far from the "anti-anticodon" site—in the vicinity of the tRNA 3' terminus, to be exact (Fig. 1). It thus appears that the ribozymic precursors of aaRSs had exactly the same problem with two (operational and classic) codes: to aminoacylate the cognate tRNA with high specificity they would need the catalysts of their own, i.e., "ribozymes of the second order"…which, in turn, would require their own catalysts again, and so on (Rodin and Rodin, 2006a).

It is the operational code, rather than the classic code, which is directly responsible for the tRNA's correct aminoacylation at its 3'' end by the cognate aaRS (Schimmel et al., 1993). And it is the operational code that makes the classic code possible—it was suggested, therefore (Schimmel et al., 1993) that the top halves of tRNAs, which contain the embedded operational code for specific aa attachments, might be older than the bottom halves, which are associated with the anticodon–codon recognition (Fig. 1b). Consistent with this hypothesis are the minihelix-like tags that are found at the ends of RNA genomes—these short tRNA precursors might have been used originally as the replication initiation sites (Weiner and Maizels, 1987, 1999).

Furthermore, the very existence of the operational code revived the idea of the classic code (Table 1) being a "frozen accident" of sorts (Crick, 1968; Schimmel et al., 1993). An alternative to the "frozen accident" scenario is the possibility of a direct stereochemical affinity between amino acids and cognate triplets (anticodons and/or codons) (Szathmary, 1999; Woese, 1965; Yarus, 1998; Yarus et al., 2005). Theoretical modeling (Shimizu, 1982) and RNA aptamers that were selected in evolutionary experiments with random RNA pools (Caporaso et al., 2005; Yarus, 1998; Yarus et al., 2005) clearly demonstrate that this affinity, weak though it might be, does exist.

Such stereochemical preference necessarily generates the "similar triplets for similar aas" coding pattern. However, the frozen accident-based evolution of the classic code does not necessarily preclude its optimization along the "similar triplets for similar aas" trajectory (Crick, 1968). Indeed, the close association (within the same tRNA molecule) of the operational and classic codes allows them to evolve concordantly. For such co-evolutionary processes to initiate successfully, and to continue proceeding smoothly (and with lower selection costs), it would be beneficial if the two (presently very different) codes initially had the same ancestor. This implies that the dumbbell module of tRNA might have originated from the minihelix module, or vice versa (Fig. 1b), by duplication (Di Giulio, 1992; Rodin et al., 1996).

1.2 Dual Complementarity in tRNAs Points to Ancient Duplication

A direct search for anticodon- or codon-like duplicates within the acceptor stem, in the vicinity of the base-determinator and 3′ terminal site of aa attachment, yielded nothing. However, traces of this crucial duplication event do show up—not directly in individual tRNAs, but rather in pairs of consensus/ ancestral tRNAs with complementary anticodons. Specifically, such pairs have complementary bases at the second positions of their acceptor stems as well (Fig. 1) (Rodin et al., 1996). We interpreted this dual complementarity as a signature of the common ancestry of two codes (classic and operational). Moreover, the anticodon triplet and the first three bases of the acceptor stem originally could have been the same entity (Rodin and Ohno, 1997; Rodin et al., 1996). However, since this concerted complementarity is observed only for the second base pair in the acceptor, even a triplet structure of the primordial operational code remains difficult to explain away. How, in general, could the two codes co-evolve during the expansion of the amino acid repertoire? We have recently addressed this question by testing the updated collection of 8246 tRNA gene sequences (Sprinzl and Vassilenko, 2005) with respect to dual complementarity. The sequences cover three main kingdoms, eubacteria, archaebacteria and eukaryotes. We have manually aligned the sequences of concatenated tRNA genes and built the phylogenetic trees for each of these kingdoms. We subsequently reconstructed the ancestral tRNA sequences (supposedly representing the respective latest universal common ancestors) separately for each anticodon, for each phylogenetic tree and for the complete set of tRNAs.

Two groups of ancestral tRNA pairs have been tested: the pairs with fully complementary anticodons and the pairs in which only the second bases of the anticodons are complementary. Only the first group exhibited the dual complementarity thus suggesting (Rodin and Rodin, 2006a) that:

1. Ancestral tRNAs gained the dual complementarity when the three-letter translation frame was already in use.
2. Although the double-stranded anticodon (codon)-like trinucleotides could indeed have represented the ancestral operational code, the latter was highly ambiguous. In fact, only the second base pair was established at the time when the first protein aaRSs replaced their ribozymic precursors.
3. It was most likely the ancient simultaneous translation (in the same frame) of both sense and antisense strands (under which the new tRNAs entered primitive translation in pairs with complementary anticodons) that preserved the dual complementarity.

1.3 Simultaneous Sense–Antisense Coding Might Predetermine Two Complementary Modes of tRNA Aminoacylation

The main premise of our ongoing study is that, in contrast to the advanced "bilingual" life of nucleic acids and proteins with only one of the DNA strands (sense strand) used for encoding proteins, in the hypothetical preceding "monolingual" (and thus arguably more strand-symmetric) RNA world both complementary replicas of a gene could have been used not only as catalysts (Kuhns and Joyce, 2003) but also, later, as first templates for encoded protein synthesis (Carter and Duax, 2002; Pham et al., 2007; Rodin and Ohno, 1995, 1997; Rodin and Rodin, 2006a). Accordingly, we expected to find the fundamental "fingerprints" of this primordial strand symmetry not only in tRNAs but also in aaRSs and in the genetic code organization itself.

The division of aaRSs into two unrelated classes, I and II (Eriani et al., 1990), is remarkable in this regard. Their modes of tRNA recognition appear to be complementary to each other: class I and II aaRSs approach the acceptor stem from the opposite sides, minor versus major groove, and attach aas to the 2'-OH and 3'-OH of the terminal A76, respectively (ibid). Accordingly, we rearranged the classic genetic code table by putting the complementary codons vis-à-vis each other (Table 2, A). This rearrangement revealed the striking mirror

Table 2 A subcode for two modes of tRNA recognition reveated by complementary transformation of the genetic code (Table 1)

A

1	2				3
	U	C	A	G	
U	UUU Phe	UCU Ser	UAU Tyr	UGU Cys	U
U	UUC Phe	UCC Ser	UAC Tyr	UGC Cys	C
U	UUA Leu	UCA Ser	UAA stop	UGA stop	A
U	UUG Leu	UCG Ser	UAG stop	UGG Trp	G
C	CUU Leu	CCU Pro	CAU His	CGU Arg	U
C	CUC Leu	CCC Pro	CAC His	CGC Arg	C
C	CUA Leu	CCA Pro	CAA Gln	CGA Arg	A
C	CUG Leu	CCG Pro	CAG Gln	CGG Arg	G
A	AUU Ile	ACU Thr	AAU Asn	AGU Ser	U
A	AUC Ile	ACC Thr	AAC Asn	AGC Ser	C
A	AUA Ile	ACA Thr	AAA Lys	AGA Ser/Gly	A
A	AUG Met	ACG Thr	AAG Lys	AGG Ser/Gly	G
G	GUU Val	GCU Ala	GAU Asp	GGU Gly	U
G	GUC Val	GCC Ala	GAC Asp	GGC Gly	C
G	GUA Val	GCA Ala	GAA Glu	GGA Gly	A
G	GUG Val	GCG Ala	GAG Glu	GGG Gly	G

Minor groove side

Major groove side

B

⇒ YUN ▷   ⇒ YGИ ▷
◁ RAИ ⇐   ◁ RCN ⇐

1	2	3	1	2	3	1	2	3	1	2	3
Y	U	N	И	A	R	Y	G	И	N	C	R
R	U	N	И	A	Y	R	G	И	N	C	Y

⇒ RUN ▷   ⇒ RGИ ▷
◁ YAИ ⇐   ◁ YCN ⇐

(A) The genetic code representation in which white and gray mark two modes of tRNA recognition (instead of class I and II of AARSs shown in Table 1)—from the minor and major groove sides of the acceptor stem, respectively. A lighter shade of gray (Lys) reflects the minor groove side recognition of its tRNA by a class I synthetase in some archaebacteria (Ibba et al., 1997). The "nonsense" UGA is assigned to Trp, as in many nuclear and mitochondrial codes (Knight et al., 2001). Two other stop codons are shown in white because the known cases of their "capture" by amino acids are mostly of the minor groove side variety (Knight et al., 2001). Also, codons AGR are assigned to gray Ser or Gly, as they are in mitochondria (Knight et al., 2001). Alternatively, one can assume a Arg↔Lys swap between codons AGR and AAR (Szathmary, 1991). This swap is consistent with AAA triplets found in the Arg-binding site of selected RNA aptamers (Caporaso et al., 2005). (B) The genetic code representation from (A) is further rearranged to put complementary codons *vis-à-vis* each other. The code is "compressed," with N (and complementary И) standing for all four nucleotides; R, purine (G or A); Y, pyrimidine (C or U). The opposite-directed arrows poiint at four pairs of codons on the complementary stands that might have determined subcode for two modes of tRNA aminoacylation in early RNP world. See Rodin and Rodin (2006b) for details.

symmetry of the two modes of tRNA recognition (Table 2, B) (Rodin and Rodin, 2006b). Specifically, two types of complementary codons (and, symmetrically, anticodons) immediately attract attention. We will call them "type 1" and "type 2." Type 1 complementary codons contain YY versus RR at the second and either first or third positions. They are in the same recognition mode, meaning that their aaRSs approach the tRNA acceptor from the same side (either minor, shown in white in Table 2B, or major, shown in gray in Table 2B, groove side). In contrast, type 2 complementary codons contain RY versus YR at the aforementioned positions. They are in the different recognition modes, their aaRSs approaching from the opposite groove sides (one from the major groove side and the other from the minor groove side).

It is important to note here that the dinucleotides in type 2 complementary codons, namely YR and RY, include palindromes CG, GC, UA and AU (of course, RR and YY dinucleotides in type 1 complementary codons cannot possibly be palindromes). Palindromes, by definition, are identical to their complements. Therefore, a single one-base shift in recognition of corresponding tRNAs would dramatically increase the recognition confusion risk. Such errors were likely quite frequent in the primordial RNA life catalyzed by ribozymes. We see the only way to avoid this confusion—the recognition of tRNAs by the earliest two (most likely ribozymic) precursors of aaRSs should be spread from their anticodons in the opposite directions (Rodin and Rodin, 2006b). In the most parsimonious scenario, this suggests complementarity of the two ribozymes themselves. Accordingly, they spread tRNA recognition and reached amino acid attachment site from the opposite sides (minor versus major groove) of the acceptor stem.

Let us assume also that the ancestral genes of two protein aaRSs were physically linked with their ribozymic precursors. If we take into account the general principle of evolutionary continuity, then all of the above suggest that the two protein aaRSs were originally encoded by the complementary strands of the same ancestral gene. Interestingly, we put forward this hypothesis much earlier (Rodin and Ohno, 1995), from a different inquiry and on the basis of the surprising complementarity shown in "head-to-tail" alignments of the signature motifs from the catalytic domains of class I and II aaRSs (Fig. 2). The real precedent for sense–antisense coding of class I and II homologs (Carter and Duax, 2002), as well as the recent artificial creation of a "minimalist" TrpRS catalytic domain (Pham et al., 2007), strongly supports this hypothesis.

To summarize, the early genetic code might have had its operational duplicate in the acceptor helix. Moreover, the code might have begun its expansion under the constraints of archaic double strand translation. This double-stranded beginning is likely the reason why the code is not mutationally optimized for the subsequent differentiation of DNA strands into the coding (sense, non-transcribed) and non-coding (anti-sense, transcribed) strands (Rodin and Ohno, 1997). Using this strand asymmetry, we proposed a

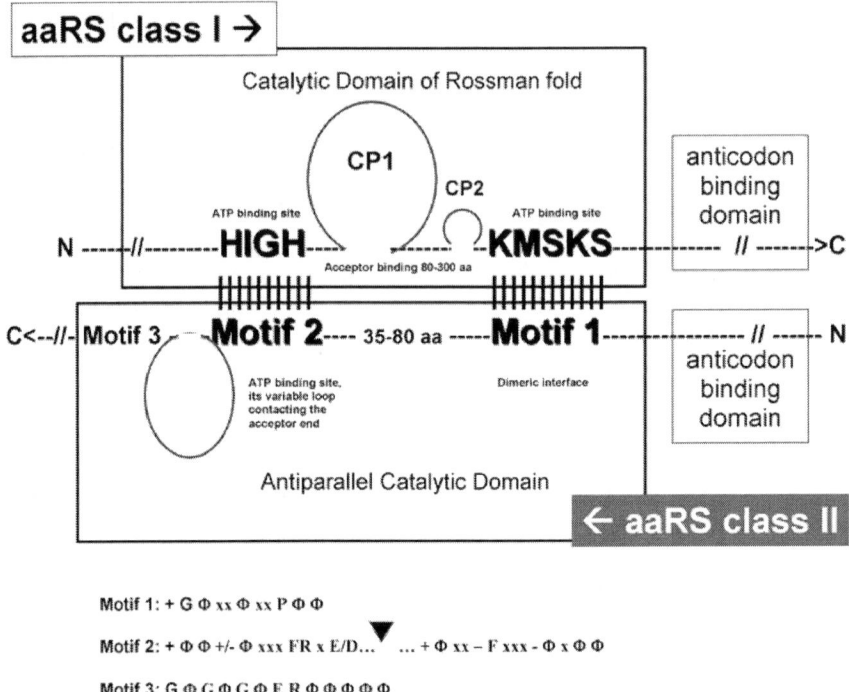

Motif 1: + G Φ xx Φ xx P Φ Φ

Motif 2: + Φ Φ +/- Φ xxx FR x E/D... ▼ ... + Φ xx − F xxx - Φ x Φ Φ

Motif 3: G Φ G Φ G Φ E R Φ Φ Φ Φ Φ

Fig. 2 Schematic diagram representing class I versus class II aaRSs complementarily ("head-to-tail") mapped to each other (adopted from Rodin and Ohno, 1995). Class-defining signature motifs are HIGH (class I) against motif 2 (class II) and KMSKS (class I) against motif 1 (class II). CP1 and CP2 are the hypervariable (in length) connective polypeptides. Three signature motifs of class II aaRSs are shown at the *bottom*. Amino acid groups are designated as follows: x—any residue, (±)—polar residue, (+)—positively charged residue, (−)—negatively charged residue, (−)—hydrophobic residue. In this representation, signature motifs from the opposite classes appear to be highly complementary, suggesting that the two aaRSs might have originated as products of in-frame translation of complementary sense and antisense strands of the same primordial gene (Rodin and Ohno, 1995). Recently, Pham et al. have successfully deleted the anticodon-binding domain from TrpRS, have reduced CP1 and have fused the discontinuous segments comprising its active site (Pham et al., 2007). The resulting 130-aa-long minimal catalytic domain maintained the original TrpRS's ability to activate tryptophan. This proves in vitro that the conserved class I signature motifs HIGH and KMSKS could have indeed originated simultaneously with the class II signature motifs 2 and 1 from complementary strands, as proposed in Rodin and Ohno (1995)

novel indicator of selection via calculation of the ratio of non-synonymous to synonymous base substitutions at CpG palindromes (Rodin et al., 1998, 2005a). This indicator is insensitive to the local site-specific mutability; at the same time, it distinguishes the selection- and drift-driven trajectories of molecular evolution. Therefore, it is particularly suitable for the study of duplicate genes.

2 Epigenetic Regulation of Expression, Repositioning and Evolutionary Fate of New, Redundant Gene Copies

2.1 Loss-or-Gain Dilemma, Effective Population Size and the Duplication–Degeneration–Complementation Model

An intriguing paradox lies at the very heart of evolution *ab simplecioribus ad complexiora* (Rodin and Riggs, 2003; Rodin et al., 2005a,b). All other things being equal, more complex organisms have to be more complex genetically. An increase in genetic complexity necessarily suggests the emergence of new genes and/or new, non-coding regulatory elements. Both of these novelties, in turn, necessarily suggest the presence of DNA duplications. In the classic model of evolution to a new function (neo-functionalization) by gene duplication, one gene copy retains the original function and remains under strong control by negative (purifying) selection, whereas its duplicate gene is free of selective constraints, so spontaneous mutations and positive selection may gradually shape it into a novel gene (Ohno, 1970). However, any mutation, either deleterious or advantageous, in the redundant gene copy becomes neutral and may be spread in a population by random drift. Deleterious mutations originate far more frequently than advantageous ones. Therefore, one would predict that instead of neo-functionalization, deleterious mutations will almost always cause the deterioration of the redundant gene into a functionless pseudogene and, eventually, into junk DNA. Population genetic models (Ohta, 1980) and genome-wide studies of duplicate genes support this prediction (Lynch and Conery, 2000; Lynch et al., 2001; Mounsey et al., 2002).

Moreover, the smaller the effective population size N_e, the more likely that pseudogenization will occur, relative to the selective fixation of advantageous mutations (Lynch and Conery, 2000; Lynch et al., 2001; Mounsey et al., 2002). This means that the "loss-or-gain" alternative is especially biased toward "loss" in higher organisms with complex, slow development and relatively small N_e. This is a striking paradox (Rodin and Riggs, 2003; Rodin et al., 2005a,b), because the complexity is originally defined, and subsequently increased, by the new genes and/or regulatory elements that are generated by duplication. Ironically, in order to detect positive selection at work on gene duplicates, one usually turns to relatively simple organisms with large N_e. In other words, the more complex the organisms are, the more difficult neo-functionalization should be. At some point, presumably long before achieving, say, human genome levels of complexity, neo-functionalization becomes prohibitively difficult. And yet, here we are!

Members of a typical multigene family are descendents of duplication events, but they do not degenerate into pseudogenes because each member of the family has a particular developmental period and/or tissue-specific expression pattern, when and/or where it is exposed to selection. That is, the individual expression patterns do not overlap, but rather complement each other temporally and spatially. This complementation is employed in the duplication–degeneration–complementation (DDC) model of duplicate genes survival (Force et al., 1999; Lynch et al., 2001). This model assumes that, although degenerative mutations occur in gene duplicates they may affect different, relatively independent regulatory elements that are responsible for stage/tissue-specific expression of the gene. In this case, duplicate genes come back under the control of selection, thus preserving themselves and opening new vistas toward further adaptive divergence. The earlier a duplicate gene comes under selective pressure, the more likely the duplicate will escape from mutational degradation. Keeping this in mind, it is important to consider the role of epigenetic silencing, which seems to occur much faster than mutational silencing.

2.2 The Epigenetic Complementation Model

Indeed, the newly created duplicates can easily be inactivated epigenetically—via methylation, heterochromatinization, homologous RNAi-mediated silencing or other processes that involve heritable chromatin structure (Rodin and Riggs, 2003; Rodin et al., 2005a,b). In this epigenetic complementation (EC) model, degenerative mutations appear as an effect rather than a cause of the complementary silencing of duplicate genes.

A quantitative analysis of the model (Rodin and Riggs, 2003; Rodin et al., 2005a,b) showed that:

- The complementary silencing of duplicates notably accelerates their functional divergence in relatively small populations (in the range $1 < N_e s < 100$, where s is the selection factor).
- This accelerative advantage is inversely dependent on the gene-specific adaptive/degenerative mutation ratio: the rarer the adaptive mutation is (compared to the degenerative mutations), the larger the beneficial effect of the epigenetic stage/tissue-complementary silencing of duplicates will be (thus clearly pointing to the protective role of this mechanism against pseudogenization).

Notably, the EC model predicts more efficient "escape" from pseudogenization and more efficient functional expansion of multigene families in organisms with small N_e, and methylated genomes. Our genome-wide study confirmed the prediction (Rodin and Riggs, 2003). Another attribute of the EC-based survival of extra gene copies is their repositioning in a genome.

2.3 Repositioning Favors Survival of Duplicate Genes

The local chromatin environment (whether on the same or different chromosomes) has a strong effect on gene expression (Brown et al., 2001; Cockell and Gasser, 1999). Therefore, translocations may, and do, change gene expression (Aladjem et al., 1998; Simon et al., 2001). Accordingly, we hypothesized that repositioning might play an important role in the evolution of gene duplicates. Repositioning is of different importance for the DDC and EC models. It is essentially unnecessary for the DDC model, because two independent strongly deleterious mutations in different regulatory sites are all that is needed for the extra gene copy to escape pseudogenization. In the EC model, survival is greater for those duplicates that move to a different chromatin domain in the genome (Rodin and Riggs, 2003). As a baseline analysis of the repositioning effect, we compared syntenic (located on the same chromosome) and non-syntenic (located on different chromosomes) groups of gene pairs (Rodin and Parkhomchuk, 2004). Their evolutionary dynamics contrast strongly (Fig. 3b). About 90% of very young duplicates (synonymous substitutions per site, sps < 0.05) are syntenic. However, as the number of synonymous substitutions per site increases (≥0.05 sps), a steep decline in the number of duplicates is observed. Non-syntenic duplicates show the opposite trend; they rapidly grow in number with time, apparently at the expense of syntenic counterparts. Remarkably, in the initial period (0 < sps < 0.025) the "syntenic decrease" noticeably exceeds the "non-syntenic increase," suggesting that the large fraction of newly created (most likely tandem) duplicates perish if they stay at the same chromosome location, whereas translocation to ectopic sites, often on a different chromosome, favors their survival.

Computer simulation of random base substitutions in protein-encoding genes showed that at ~ 0.1 sps the majority of human gene duplicates that are destined to degenerate (pseudogenes-to-be) have already become identifiable pseudogenes, and accordingly have been removed from the database of intact genes (Rodin and Parkhomchuk, 2004). By superposing the two genome-wide plots (Fig. 3a and b) we noticed that exactly at the point where the number of pseudogenes-to-be runs short, the group of syntenic duplicate genes drops to its minimum, whereas the group of non-syntenic duplicate genes reaches its maximum, and then both groups, presumably largely purged of pseudogenes, decline slowly (Fig. 3b) (Rodin et al., 2005a,b). Since such pseudogene-free pairs are predominantly non-syntenic (Fig. 3b, sps > 0.1), we conclude that repositioning to ectopic sites often saves duplicate genes from degradation and promotes their functional divergence, probably through epigenetic developmental stage- and tissue-complementary silencing of their expression (Rodin and Riggs, 2003). Pairs of non-syntenic duplicates certainly belong to this category, whereas the majority of pairs with a degrading "pseudogene-to-be" are from the syntenic group.

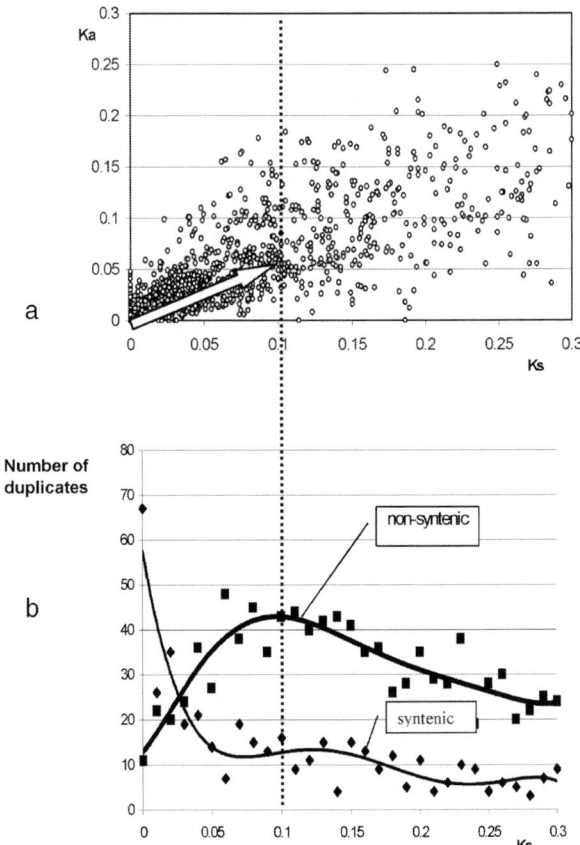

Fig. 3 Survival of human duplicates depends on their repositioning. (**a**) The standard K_a versus K_s plot of duplicates' divergence in the human genome. The plot depicts the number of non-synonymous mutations (resulting in aa replacement, Y axis) per site versus the number of synonymous (no aa replacement, X axis) mutations per site. Each point represents a pair of homologs ($\geq 60\%$ amino acid identity) (Rodin et al., 2005a). (**b**) "Syntenic versus non-syntenic" comparison of human gene duplicates. To minimize overrepresentation of genes from large families, we only used the families with less than six members. In addition, the analysis used the "serial" pairing of homologous genes so that the total number of gene pairs is equal to that of duplication events minus one (see Rodin and Parkhomchuk, 2004 for details). The number of duplicates is shown as function of the number of silent substitutions per site (sps). The "same chromosome" (syntenic) group is comprised of the pairs in which both homologous genes are located on the same chromosome; the pairs of homologous genes located on different chromosomes constitute the "different chromosomes" (non-syntenic) group. The *arrow* denotes the evolutionary "neutral" stream with one of two genes in each pair being functionally redundant and destined to become a pseudogene. One can see that even some of the rather old gene pairs (sps $\gg 0.1$) still show high K_a/K_s ratio. However, a closer inspection reveals that majority of such cases are associated with the immune/inflammatory response genes that were directly or indirectly involved in prolonged co-evolutionary process (such as antigen–antibody co-evolution) driven by the strong positive selection

2.4 Repositioning Increases Mutational Asymmetry

The new chromatin environment may also influence local mutability, at least in the genomes that consist of isochores, long (>300 kb) segments of DNA that differ in GC content (Bernardi, 2000). A relocated duplicate may accumulate mutations that shift its GC content toward the GC content surrounding its new location (Jabbari et al., 2003; Rodin and Parkhomchuk, 2004). Our genome-wide studies revealed such changes for different eukaryotic species.

Specifically (Fig. 4):

- Duplicate genes show significant asymmetry in their GC content.
- For human and other mammalian genomes, this GC asymmetry correlates with the relative GC content of the isochores in which the GC-asymmetric duplicates are located (Rodin and Parkhomchuk, 2004).
- Apparently, the GC asymmetry results from significant differences between gene duplicates in the rate and direction of mutations (Fig. 4b).

Shown in Fig. 4 are mutation data for the third codon position, where nucleotide substitutions are mostly silent and therefore reflect the duplicate-asymmetric neutral mutational noise. We observed the same, but less pronounced, asymmetries of gene duplicates for the first and second codon positions. Increased mutation pressure on the re-positioned gene copy could be a great help in that copy's functional divergence from its parental twin: it does not change the ratio of detrimental to advantageous mutations, but essentially shortens the "waiting time" required for the acceptable non-silent (quasi-neutral or, rarely, advantageous) mutations to occur. Very telling in this regard is the difference between mammalian α- and β-globin gene clusters (Rodin et al., 2005a):

- In the mammalian β-globin cluster, adult and fetal genes have been accumulating non-silent mutations significantly faster than embryonic genes, whereas in the mammalian, avian and amphibian α-globin clusters adult genes and their embryonic paralogs evolved at approximately the same speed.
- Also, silent mutations have been accumulating significantly faster in β- than in α-globin loci of mammals, thus pointing to a higher background mutability of the β genes "neighborhood."

These differences correlate with the fact that mammalian α- and β-globin gene clusters are non-syntenic, and have contrasting subnuclear environments (GC content, CpG islands, methylation levels, replication timing and interphase positioning). Accordingly, the α locus is expressed in many tissues (like housekeeping genes) whereas the β locus is expressed solely in erythroid cells (Brown et al., 2001).

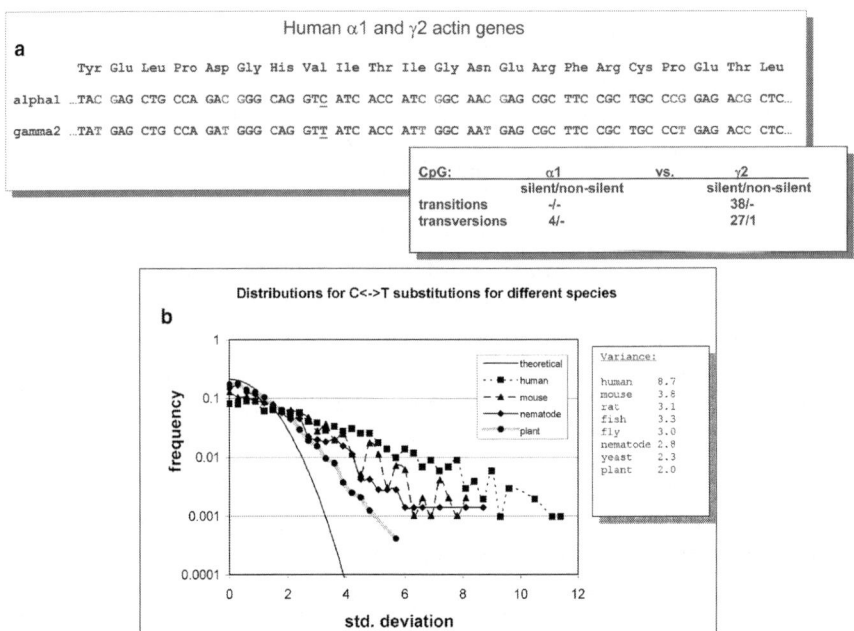

Fig. 4 Mutational asymmetry of duplicate genes. (**a**) A fragment of aligned human α1 and γ2 actin genes with seven "silent" nucleotide differences, six of which occur at CpG dinucleotides (shown in *bold*). Below are the "gene-total" ratios of silent to non-silent transitions and transversions that presumably occurred in the α1 and γ2 lineages. (**b**) The asymmetry of gene duplicates in C→T transitions in human, mouse, nematode and mustard weed genomes. Shown are frequency distributions of deviations from expected mean values of C↔T transitions. Each point represents a frequency (*Y*-axis) of gene pairs with the corresponding "oddness" of pair-wise asymmetry measured in unit variance (*X*-axis). The duplicates with nucleotide identity > 80% were excluded since they might contain many still unrecognized pseudogenes ("pseudogenes-to-be") (see Fig. 3a). The *solid line* shows the normal (Gaussian) distribution that is expected under the assumption of equal substitution rates in gene duplicates. If a value is higher than expected, the mutation rates are duplicate-asymmetric in the corresponding pairs. The *inset table* shows the variance of this distribution in different eukaryotic species. From (Rodin and Parkhomchuk, 2004)

2.5 Adaptive Evolution of Young Duplicates

It is very difficult to directly observe positive selection at work on the extra gene copy. First, the period when the copy evolves in an adaptive manner (when $K_a/K_s \gg 1$) is very short (Fay and Wu, 2001, 2003). Second, advantageous mutations are rare—they emerge only in very few key sites, whereas silent mutations originate across the entire duplicated gene, almost in each codon (Conant and Wagner, 2003; Ophir et al., 1999). As a consequence, gene pairs with $K_a/K_s > 1$ are very rare and are statistically indistinguishable from

$K_a/K_s \approx 1$ cases of neutral pairs. Third, the K_a and K_s components are usually calculated for different sites that may vary greatly in mutability, even within the same gene or its particular region. Finally, gene duplicates themselves may differ in background mutation rate, with the difference being position-dependent, especially when one of the duplicates translocates and experiences a mutational pressure for alternative base composition. The fact that evolutionary time measured in synonymous mutations runs differently in different genomic locations calls for adopting the position-independent and more selection-sensitive ratio of missense to same-sense base substitutions as an unbiased indicator of natural selection.

We have proposed an unbiased indicator that employs transitions (C→T and G→A) and transversions (G→T and C→A) at CpG dinucleotides (Rodin and Rodin, 2005; Rodin et al., 1998, 2002): these account for 30% of base substitutions in mammals. The putative primary mutational events for both CpG →TpG and CpG→CpA transitions are most likely the same (specifically, deamination of methylated Cs), but occur on opposite DNA strands (Rideout et al., 1990; Yang et al., 1996). CpG is a strand-symmetric palindrome. However, if mC deamination occurs on the sense (non-transcribed) strand, it generates a CpG→TpG transition, whereas the same deamination event on the antisense (transcribed) strand requires replication to appear in the sense strand as a complementary CpG→CpA transition (Fig. 5a).

Similarly, two transversions at CpG dinucleotides, G→T and C→A, seem to have the same primary lesion at the guanines residues, suggesting that the G→T transversion originates directly in the coding (non-transcribed) strand, while C→A is the strand-complementary conversion of the G→T transversion that originated in the non-coding (transcribed) strand (Fig. 5b) (Rodin and Rodin, 2005; Rodin et al., 2002).

Importantly, in both cases the complementary substitutions originate on the opposite strands with equal probabilities, i.e., the substitutions themselves are strand-symmetric, but, due to the genetic code, have extremely asymmetric functional consequences. This asymmetry for transitions and transversions in codons CCG (Pro) and CGG (Arg), respectively, which shows silent versus strong non-silent mutations with major physical–chemical changes in charge, size and hydrophobicity, is illustrated in Fig. 5.

When tested for *p53* tumor suppressor and some other disease-associated genes, this indicator of selection provided the expected results. For example, the ratio of C→T transitions over total transitions at CpG dinucleotides in codons NCG and CGN (i.e., the ratio $N_{C \to T}/(N_{C \to T} + N_{G \to A})$) was 0.5 for *p53* pseudogenes, in full agreement with the strand-independent transition rate in the absence of selection. When selection acted, this ratio was strand-asymmetric, 0.06–0.2 for purifying selection in evolution of functional p53 genes and >>0.5 for the strong tumorigenic selection that favored non-transcribed (coding) strand-biased loss-of-function mutations in *p53* genes from human cancers (Rodin et al., 1998, 2002).

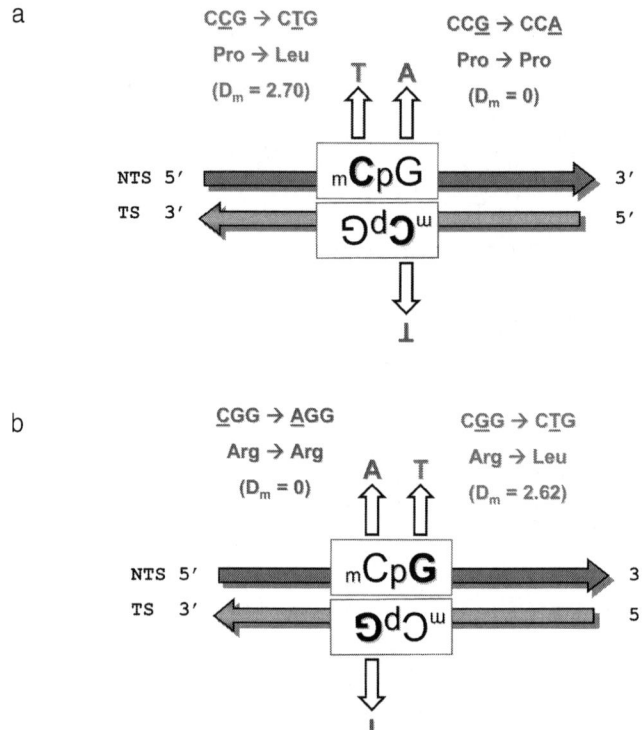

Fig. 5 Strand-specific origin of transitions (**a**) and transversions (**b**) at methylated CpG mutagenic sites. Shown are two strand-mirror CpG dinucleotides located in the coding ("sense") non-transcribed strand (NTS) and non-coding ("antisense") transcribed strand (TS) of DNA, respectively. The mutagenic sites are shown in *bold*. C→T transitions result from deamination of methylated cytosines in the NTS; same changes in TS result in G→A transitions. G→T transitions are assumed to result from oxidation, polycyclic aromatic hydrocarbons adducts and other changes of guanines in the NTS; same changes in TS result in C→A transversions. Two CG-containing codons, CCG (Pro) and CGG (Arg), exemplify the genetic code-based functional strand asymmetry of these substitutions. Non-silent mutations are shown in *bold*, and "quantified" by D_m values (based on changes in charge, size and hydrophobicity of amino acids (Miyata et al., 1979)). Silent mutations have $D_m = 0$. For strand-asymmetric D_m values in all other CG-containing codons, see Rodin et al. (2005a)

As a natural control, one can use common CpG sites that cover two neighboring codons (i.e., NNCpGNN) and that produce the non-silent G→A versus the silent C→T, thus mirroring the non-silent C→T and silent G→A in codons NCG. Four putative silent C→T transitions at such CpG sites in the line leading to the human γ2 actin gene serve a good example (Fig. 4a).

The rationale for using the CpG-focused metric of selection in genomic analyses of gene duplicates is that a translocated duplicate gene might be under relatively strong mutational pressure in its new location (Fig. 4). At the same time, according to the EC model, in order to escape pseudogenization the duplicate needs to change its expression pattern and come under the control of selection as soon as possible. Therefore, the repositioning of a gene duplicate affects both its mutability and survival (Figs. 3 and 4). Additional study is required to estimate the relative contributions of the primary mutagenesis and subsequent selection to the dynamics (such as shown in Fig. 3); mutationally strand-symmetric/selectionally strand-asymmetric CpG-based criteria will be helpful in achieving this goal.

Adaptation to a new genomic environment might require advantageous mutations. However, compared to silent substitutions, advantageous substitutions are rare and, consequently, the repositioned duplicate might show a significantly smaller K_a/K_s value than its parental gene at the old location, thus obscuring any positive selection activity. When we calculate K_a and K_s for only the NCG and neighboring NNC and GNN codons, instead of calculating them for entire gene sequences, we are guaranteed the same initial mutability for the silent and non-silent mutations that are under comparison. This leads to the more reliable interpretation of the K_a/K_s ratio in terms of the dominating evolutionary mode, i.e., whether the corresponding stage of the duplicate's divergence was driven by selection (positive or negative) or neutral random drift.

Finally, an additional code-based opportunity is associated with the pair of complementary CpG transversions G→T and C→A in Arg codons (CGA and CGG); C→A is silent, whereas its complementary counterpart G→T produces CTA and CTG that encode Leu, which is functionally very different from Arg. We expect these two specific codons to be particularly useful in genome-wide studies of duplicate genes *encoding transcription factors* since the most important domains of transcription factors directly interact with DNA targets and, accordingly, are rich in positively charged amino acids such as Arg and Lys. Five mutationally "hot" Arg codons in the DNA-binding domain of the *p53* tumor suppressor gene serve as a representative, and practically very important, example (Rodin and Rodin, 2005).

2.6 EC Model and G-Value Paradox

Gene duplicates in the human genome appear to be more mutationally asymmetric than those in any other genomes tested so far (Fig. 4), undoubtedly due to the depletion of methylated CpG dinucleotides. Methylation is also of growing interest as a marker for the epigenetic processes that likely assist with the neo-functionalization of extra gene copies.

However, the human genome has only about 30,000 genes in total. This seems unimpressive compared, for example, with the 17,000 genes of the nematode genome. This lack of linear proportionality between the organism complexity and the total number of genes is known as a G-value paradox (Hahn and Wray, 2002). Organism complexity does correlate with the tissue- and stage-associated diversity of expression patterns, which in turn depends on the number of genes encoding transcription factors and co-factors, and the number of their DNA targets—promoters, enhancers, silencers, etc. (Levine and Tjian, 2003). However, this does not resolve the paradox; instead, it simply re-targets it from the genes per se to their regulatory regions. We posit that during the initial phase local epigenetically contrasting chromatin domains could have provided a "launching platform" of sorts for evolution by gene duplication (Rodin et al., 2005b). To create genes with novel functions, evolution, perhaps more often than not, experiments with already existing genes and regulatory elements by moving their duplicates (or sometimes even single genes themselves) to new locations in the genome and changing their tissues and stages of expression, instead of creating a truly novel gene through the gradual accumulation of rare advantageous mutations that change the old gene's function (Rodin et al., 2005b). If the repositioned gene or element has a positive effect, the repositioned duplicate will be further modified by perhaps no more than a very few adaptive mutations, to better suit the functional demands of its new environment. Further genome-wide analyses are needed to examine this hypothesis in detail.

Acknowledgment: The authors thank Paul Schimmel, Eors Szathmary, Arthur Riggs, Gerald Holmquist and Dmitry Parkhomchuk for valuable suggestions.

References

Aladjem, M.I., Rodewald, L.W., Kolman, J.L. and Wahl, G.M. (1998) Genetic dissection of a mammalian replicator in the human beta-globin locus. Science 281, 1005–1009.
Bernardi, G. (2000) Isochores and the evolutionary genomics of vertebrates. Gene 241, 3–17.
Brown, K.E., Amoils, S., Horn, J.M., Buckle, V.J. and Higgs, D.R. (2001) Expression of α- and β-globin genes occurs within different nuclear domains in haemopoetic cells. Nature Cell Biol. 3, 602–606.
Caporaso, J.G., Yarus, M. and Knight, R. (2005) Error minimization and coding triplet/binding site associations are independent features of the canonical genetic code. J. Mol. Evol. 61, 597–607.
Carter, C.W., Jr. and Duax, W.L. (2002) Did tRNA synthetase classes arise on opposite strands of the same gene? Mol. Cell 10, 705–708.
Cockell, M. and Gasser, S.M. (1999) Nuclear compartments and gene regulation. Curr. Opin. Genet. Dev. 9, 199–205.
Conant, G.C. and Wagner, A. (2003) Asymmetric sequence divergence of duplicate genes. Genome Res. 13, 2052–2058.
Crick, F.H.C. (1968) The origin of the genetic code. J. Mol. Biol. 38, 367–380.
De Duve, C. (1988) The second genetic code. Nature 333, 117–118.
Di Giulio, M. (1992) On the origin of the transfer RNA molecule. J. Theor. Biol. 159, 199–214.
Eriani, G., Delarue, M., Poch, O., Gangloff, J. and Moras, D., (1990) Partition of aminoacyl-tRNA synthetases into two classes based on mutually exclusive sets of conserved motifs. Nature 347, 203–206.
Fay, J.C. and Wu, C-I. (2001) The neutral theory in the genomic era. Curr. Opin. Genet. Dev. 11, 642–646.

Fay, J.C. and Wu, C-I. (2003) Sequence divergence, functional constraint, and selection in protein evolution. Annu. Rev. Genomics Hum. Genet. 4, 213–235.

Force, A., Lynch, M., Pickett, B., Amores, A., Yan, Y-l. and Postlethwait, J. (1999) Preservation of duplicate genes by complementary, degenerative mutations. Genetics 151, 1531–1545.

Hahn, M.V. and Wray, G.A. (2002) The G-value paradox. Evol. Dev. 4, 73–75.

Hou, Y.-M. and Schimmel, P. (1988) A simple structural feature is a major determinant of the identity of a transfer RNA. Nature 333, 140–145.

Jabbari, K., Rayko, E. and Bernardi, G. (2003) The major shifts of human duplicated genes. Gene 317, 203–208.

Knight R.D., Freeland S.J. and Landweber L.F. (2001) Rewriting the keyboard: evolvability of the genetic code. Nat. Rev. Genet. 2, 49–58.

Kuhns, S.T. and Joyce, G.F. (2003) Perfectly complementary nucleic acid enzymes. J. Mol. Evol. 56, 711–717.

Levine, M. and Tjian, R. (2003) Transcription regulation and animal diversity. Nature 424, 147–151.

Lynch, M. and Conery, J.C. (2000) The evolutionary fate and consequences of duplicate genes. Science 290, 1151–1155.

Lynch, M., O'Hely, M., Walsh, B. and Force, A. (2001) The probability of preservation of a newly arisen gene duplicate. Genetics 159, 1789–1804.

Miller, S.L. (1987) Which organic compounds could have occurred on the prebiotic earth. Cold Spring Harbor Symp. Quant. Biol. 52, 17–27.

Miyata, T., Miyazawa, S. and Yasunaga, T. (1979) Two types of amino acid substitutions in protein evolution. J. Mol. Evol. 12, 219–236.

Mounsey, A., Bauer, P. and Hope, I.A. (2002) Evidence suggesting that a fifth of annotated *Caenorhabditis elegans* genes may be pseudogenes. Genome Res. 12, 770–775.

Ohno, S. (1970) Evolution by Gene Duplication. Springer, Berlin.

Ohno, S. (1987) Early genes that were oligomeric repeats generated a number of divergent domains on their own. Proc. Natl Acad. Sci. USA 84, 6486–6490.

Ohno, S. (1988) On periodicities governing the construction of genes and proteins. Anim. Genet. 19, 305–316.

Ohta, T. (1980) Evolution and Variation of Multigene Families. Springer, Berlin.

Ophir, R., Itoh, T., Graur, D. and Gojobori, T. (1999) A simple method for estimating the intensity of purifying selection in protein-coding genes. Mol. Biol. Evol. 16, 49–53.

Pham, Y., Li, L., Kim, A., Erdogan, O., Weinreb, V., Butterfoss, G.L., Kuhlman, B. and Carter, C.W. Jr. (2007) A minimal Trp RS catalytic domain supports sense/antisense ancestry of class I and II aminoacyl-tRNA synthetases. Mol. Cell 25, 851–862.

Rideout, W.M., III, Coetzee, G.A., Olumi, A.F. and Jones, P.A. (1990) 5-Methylcytosine as an endogenous mutagen in the human LDL receptor and p53 genes. Science 249, 1288–1290.

Rodin, S. and Ohno, S. (1995) Two types of aminoacyl-tRNA synthetases could be originally encoded by complementary strands of the same nucleic acid. Orig. Life Evol. Biosph. 25, 565–589.

Rodin, S.N. and Ohno, S. (1997) Four primordial modes of tRNA-synthetase recognition, determined by the (G,C) operational code. Proc. Natl Acad. Sci. USA 94, 5183–5188.

Rodin, S.N. and Parkhomchuk, D.V. (2004) Position-associated GC asymmetry of gene duplicates. J. Mol. Evol. 59, 372–384.

Rodin, S.N. and Riggs, A.D. (2003) Epigenetic silencing may aid evolution by gene duplication. J. Mol. Evol. 56, 718–729.

Rodin, S.N. and Rodin, A.S. (1998) Strand asymmetry of CpG transitions as indicator of G1 phase-dependent origin of multiple tumorigenic p53 mutations in stem cells. Proc. Natl Acad. Sci. USA 95, 11927–11932.

Rodin, S.N. and Rodin, A.S. (2005) Origins and selection of p53 mutations in lung carcinogenesis. Semin. Cancer Biol. 15, 103–112.

Rodin, S.N. and Rodin, A.S. (2006a) Origin of the genetic code: first aminoacyl-tRNA synthetases could replace isofunctional ribozymes when only the second base of codons was established. DNA Cell Biol. 25, 365–375.

Rodin, S.N. and Rodin, A.S. (2006b) Partitioning of aminoacyl-tRNA synthetases in two classes could have been encoded in a strand-symmetric RNA world. DNA Cell Biol. 25, 617–626.

Rodin, S., Rodin, A. and Ohno, S. (1996) The presence of codon–anticodon pairs in the acceptor stem of tRNAs. Proc. Natl Acad. Sci. USA 93, 4537–3542.

Rodin, S.N., Holmquist, G.P. and Rodin, A.S. (1998) CpG transition strand asymmetry and hitch-hiking mutations as measures of tumorigenic selection in shaping the p53 mutation spectrum. Int. J. Mol. Med. 1, 191–199.

Rodin, S.N., Rodin, A.S., Juhasz, A. and Holmquist, G.P. (2002) Cancerous hyper-mutagenesis in p53 genes is possibly associated with transcriptional bypass of DNA lesions. Mutat. Res. 510, 153–168.

Rodin, S.N., Parkhomchuk, D.V., Rodin, A.S., Holmquist, G.P. and Riggs, A.D. (2005a) Repositioning-dependent fate of duplicate genes. DNA Cell Biol. 24, 529–542.

Rodin, S.N., Parkhomchuk, D.V. and Riggs, A. D. (2005b) Epigenetic changes and repositioning determine the evolutionary fate of duplicated genes. Biochemistry (Moscow) 70, 559–567.

Schimmel, P. and Beebe, K. (2006) Aminoacyl tRNA synthetases: from the RNA world to the theater of proteins. In: R.F. Gesteland, T.R. Cech and J.F. Atkins (Eds) The RNA World. Cold Spring Harbor Laboratory Press, New York, pp. 227–255.

Schimmel, P., Giege, R., Moras, D. and Yokoyama S. (1993) An operational RNA code for amino acids and possible relation to genetic code. Proc. Natl Acad. Sci. USA 90, 8763–8768.

Shimizu, M. (1982) Molecular basis for the genetic code. J. Mol. Evol. 18, 297–303.

Simon, I., Tenzen, T., Mostoslavsky, R., Fibach, E., Lande, L., Milot, E., Gribnau, J., Grosveld, F., Fraser, P. and Cedar, H. (2001) Developmental regulation of DNA replication timing at the human beta globin locus. EMBO J., 20, 6150–6157.

Sprinzl, M. and Vassilenko, K.S. (2005) Compilation of tRNA sequences and sequences of tRNA genes. Nucl. Acids Res. 1(33), D139–D140.

Szathmary, E. (1991) Codon swapping as a possible evolutionary mechanism. J. Mol. Evol. 32, 178–182.

Szathmary, E. (1999) The origin of the genetic code: amino acids as cofactors in an RNA world. Trends Genet. 15, 223–229.

Weiner, A.M. and Maizels, N. (1987) TRNA-like structures tag the 3' ends of genomic RNA molecules for replication: implications for the origin of protein synthesis. Proc. Natl Acad. Sci. USA 84, 7383–7387.

Weiner, A.M. and Maizels, N. (1999) The genomic tag hypothesis: modern viruses as molecular fossils of ancient strategies for genomic replication and clues regarding the origin of protein synthesis. Biol. Bull. 196, 327–330.

Weiss, K.M. and Buchanan, A.V. (2005) "The" genetic code? Evol. Anthropol. 14, 6–11.

Woese, C.R. (1965) On the evolution of the genetic code. Proc. Natl. Acad Sci. USA 54, 1546–1552.

Yang, A.S., Jones, P.A., Shibata, A. (1996) The mutational burden of 5-methylcytosine. In: V.E.A. Russo, R.A. Martienssen, A.D. Riggs (Eds) Epigenetic Mechanisms of Gene Regulation. Cold Spring Harbor Laboratory Press, New York, pp. 77–94.

Yarus, M. (1998) Amino acids as RNA ligands: a direct-RNA-template theory for the code's origin. J. Mol. Evol. 47, 109–117.

Yarus, M., Caporaso, J.G. and Knight, R. (2005) Origins of the genetic code: the escaped triplet theory. Annu. Rev. Biochem. 74, 125–151.

Evolution of the Translation Termination System in Eukaryotes

G. A. Zhouravleva, O. V. Tarasov, V. V. Schepachev, S. E. Moskalenko, N. I. Abramson and S. G. Inge-Vechtomov

Abstract Proteins eRF1 and eRF3 are key components of translation termination in eukaryotes. The highly conserved translation termination factor eRF1 decodes stop codons, while another eukaryotic release factor (RF) eRF3 stimulates eRF1 in GTP-dependent manner. Functional C-terminal domain of eRF3 is necessary for cell viability and reveals high degree of similarity between all known eRF3 and elongation factor eEF1A. Unlike the C-terminal part, the N-terminal region of eRF3 proteins is not conserved and contains "prion forming domain" (PFD). In mammals, eRF3 homologous proteins can be divided into two subfamilies based on the sequence of their N termini, GSPT1 (eRF3a) and GSPT2 (eRF3b). In our work we hypothesize that *GSPT2* gene originated through retrotransposition of processed *GSPT1* transcript after divergence between placental and marsupial mammals. Data obtained on the order *Rodentia* indicate that nucleotide sequence encoding N-terminal part of *GSPT2* maybe used as a new marker for philogenetic analysis to distinguish between families.

1 The Evolutionary Origin of Termination Factors

Protein synthesis is an essential process, highly conserved among different organisms such as Eubacteria, Archaea and Eucarya. Termination of translation (recognition of stop codons and hydrolysis of peptidyl-tRNA) is also well conserved among eukaryotes. Two types of translation termination factors,

---

G. A. Zhouravleva
Department of Genetics and Breeding, St. Petersburg State University, 199034, St. Petersburg, Russia
e-mail: zhouravleva@rambler.ru

class-I and class-2 polypeptide RF, participate in the termination of protein synthesis both in prokaryotes and eukaryotes (Kisselev et al., 2003). In eukaryotes, a single factor, eRF1, decodes all three stop codons, while eRF3 stimulates termination through a GTP-dependent mechanism by forming a complex with eRF1. Homologues of the eRF1 have been identified in different species. In all examined cases eRF1 proteins from higher eukaryotes were functional when tested in yeast *S. cerevisiae* (Urbero et al., 1997). Comparisons between eRF1 homologues from animals, fungi, plants revealed high degree of similarity (Kisselev et al., 2003). Contrary to eRF1, eRF3 is much more divergent, especially in its N-terminal domain. In all known eRF3 proteins only the C-terminal domain (highly conserved with strong similarity to the elongation factor eEF1A) is required for the translation termination and is indispensable for cell viability (Le Goff et al., 2002). Release factors eRF1 and eRF3 are thought to have an independent origin (Fig. 1). However, according to phylogenetic analysis, eRF1 has the common origin with the elongation factor eEF-2 and has arisen from the prokaryotic elongation factor EF-G (Inagaki and Ford, 2000); while eRF3 has originated from the bacterial elongation factor EF-Tu (Nakamura and Ito, 1998; Inagaki and Ford, 2000). Thus, eRF3 has the same ancestor protein as eEF1-A (a eukaryotic homolog of bacterial elongation factor EF-Tu) and Hbs1. Hbs1 protein possesses sequence similarly with both eEF-1A and eRF3 (Wallrapp et al., 1998) and participates in "no-go decay" (NGD), a special type of mRNA degradation (Doma and Parker, 2006). Possibly eRF3 arose at the early step of eukaryotic evolution because neither bacterial nor archaebacterial genomes contain any homologs of eRF3.

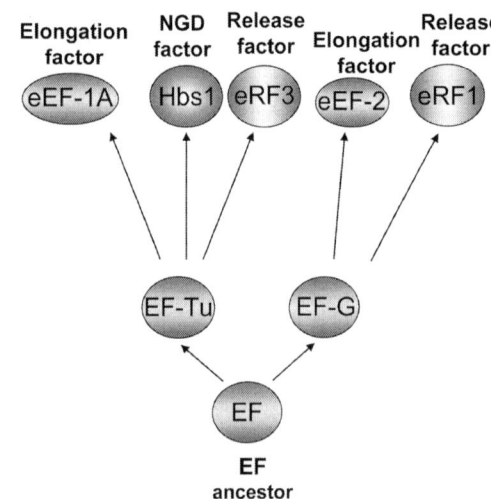

Fig. 1 A model proposing evolutionary origin of eRFs proteins. It was proposed that the progenitors of prokaryotic EF-G and EF-Tu proteins first diverged from a common ancestral GTPase, and then each of them gave rise to two protein families corresponding to the elongation and the termination factors (Nakamura and Ito, 1998; Inagaki and Ford, 2000). EF – elongation factor; RF – release factor; e – eukaryotic; NGD - "no-go decay"

2 The Structural Organization of Termination Factor eRF3

In mammals eRF3s are represented by two related proteins, GSPT1 (eRF3a) and GSPT2 (eRF3b), each encoded by a distinct gene, *GSPT1* and *GSPT2*, respectively (Fig. 2). Homologous genes were found in rat, dog, elephant and cow genomes. Only one eRF3-homologous gene was discovered in each of full sequenced genomes of *Gallus gallus*, *Caenorhabditis elegans* and *Drosophila melanogaster*. As in fungi, the eRF3-C region remains greatly conserved from yeast to human, while the eRF3-N regions exhibit high degree of divergence.

All eukaryotic eRF3 proteins studied so far (except *G. lamblia*) (Inagaki et al., 2002) have long N-proximal extension with unknown functions. In several species of fungi, closely related to *S. cerevisiae*, this domain is responsible for [PSI$^+$] prion induction and propagation, however it seems that prionization ability of eRF3 is highly restricted to budding yeast (reviewed in Zhouravleva

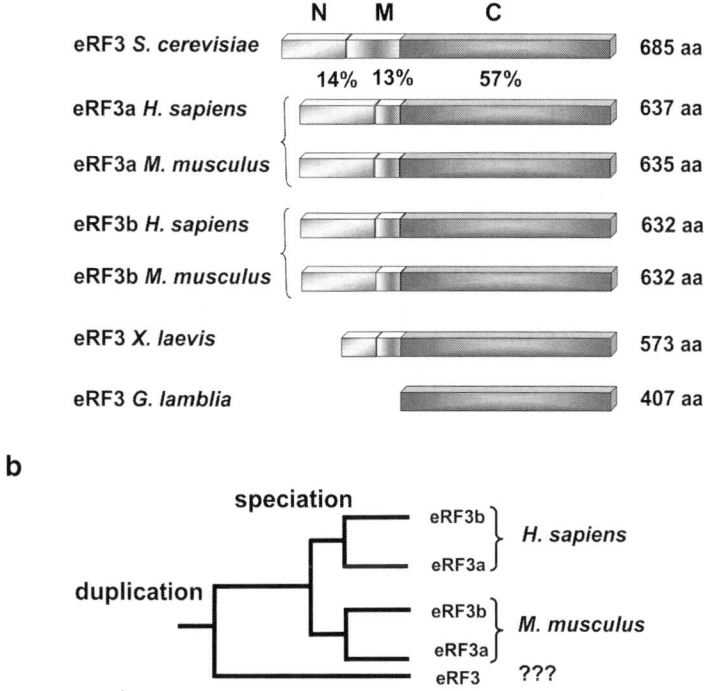

Fig. 2 Organization and evolutionary relationships of eRF3s of different species. **a.** Translation termination factor eRF3 consists of three domains, N, M and C. Percentage of identity between different domains of eRF3a *Homo sapiens* and eRF3 of *S. cerevisiae* is shown. Numbers on the right correspond to the size of eRF3 proteins. **b.** Duplication of eRF3-encoding genes, *GSPT1* and *GSPT2*, during evolution. (???)—time of duplication is unknown

et al., 2002). It was also shown that some sub-regions of N-terminal domain participate in interaction with such proteins as Pab1, Sla1, IAPs (reviewed in Inge-Vechtomov et al., 2003), nevertheless the role of most part of the N-domain remains unknown.

The most intriguing property of eRF3 N-terminal domain is its unusual amino acid composition. N-terminal part of *S. cerevisiae* eRF3 contains a so called "prion forming domain" (PFD) with two sub-regions which influence the prion propagation: QN-rich stretch (aa 6-33), and OR, region of oligopeptide repeats (aa 41-97) (Fig. 3). The presence of G-rich oligopeptide repeats (consensus sequence PQGGYQQ-YN), which are similar to those found at approximately the same position in mammalian PrP (consensus sequence PHGGGWGQ), is one of the most puzzling features of the eRF3 PFD. It is rich in Q and N residues (45%, compared to 10% for the average yeast protein), and also in G and Y residues (33%, compared to 8% average in the yeast proteome). Capability of forming prions has not been proven as yet for eRF3 homologues from species other than budding yeast. The eRF3-N region of a distant relative of budding yeast, the fission yeast *Shizosaccharomyces pombe*, does not contain QN and OR regions and exhibits very low amino acid sequence homology (18% identity) with the corresponding domain of *S. cerevisiae*, while the eRF3-C region remains highly conserved (64% identity) However, another distantly related species, *Podospora anserina*, contains the N-proximal region

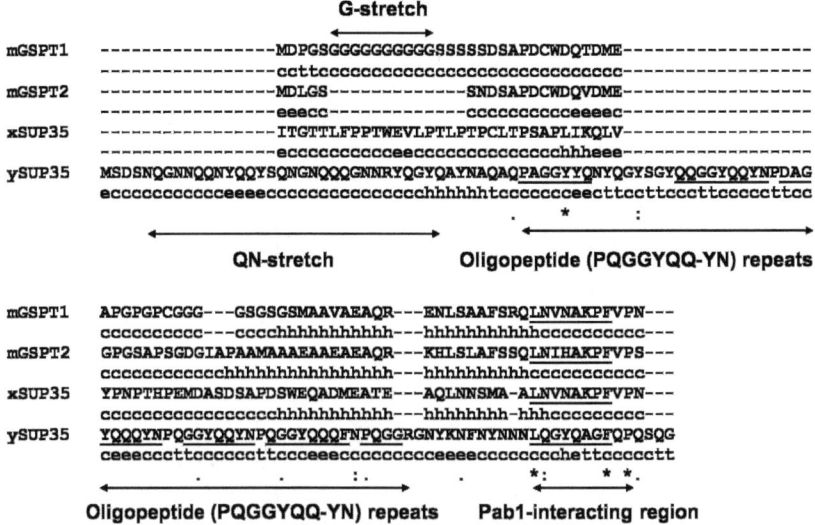

Fig. 3 N-terminal domain of eRF3 is not conserved in evolution. SOPM method was used for the secondary structure prediction (Geourjon and Deleage, 1994, see http://npsa-pbil.ibcp.fr/cgi-bin). Alpha helix: h; extended strand: e; random coil: c; beta turn: t. mGSPT1 and mGSPT2: eRF3a and eRF3b from *Mus musculus*, respectively; xSup35 and ySup35: eRF3 from *X. laevis* and *S. cerevisiae*, respectively

resembling that of budding yeast by a high content of Q/N residues (40%) and the presence of at least four repeats with a consensus sequence QQGQGYG (Zhouravleva et al., 2002).

While the N-terminal domains of higher eukaryotes eRF3 proteins show no obvious sequence similarity to the corresponding domains of lower eukaryotes, they still exhibit an unusual aa composition. For example, N-domain of mGSPT1 (mouse eRF3a) contains high percentage of P, S and G residues (10, 15 and 20%, respectively). Instead of the QN-stretch and oligopeptide repeats found in yeast eRF3-N regions, mGSPT1 and hGSPT1 (human eRF3a) proteins contain long poly-G tracts, similar to those observed in the so-called "homopeptide proteins" (Karlin and Burge, 1996). Homopeptides, or SSR (single sequence repeats) are regions within proteins that comprise a single homopolymeric tract of a particular amino acid (Faux et al., 2005). Most eukaryote RCPs (repeat-containing proteins) are involved in transcription/translation or interact directly with DNA, RNA or chromatin (Faux et al., 2005). It was shown that uncontrolled genetic expansions of SSR regions lead to the development of some neurodegenerative disorders, such as, for example, Huntington's disease in the case of expanded poly-Q tract. It was also demonstrated that many SSRs are toxic to cells and/or lead to protein aggregation or misfolding (Fandrich and Dobson, 2002). Interestingly, both human, rat and mouse GSPT1s contain homopeptide sequences of poly-G and of poly-S tracts (Fig. 4). Moreover, G_{11} in hGSPT1 is encoded from $(GGC)_{10} (GGG)_1$ showing no variation in codon usage. The peptide S_7 of hGSPT1 protein immediately downstream of G_{11} is encoded from $(AGC)_3 (GGC) (AGC)_4$. Thus, the difference in the length of homopeptide sequences between hGSPT1 and mGSPT1 can be explained by expansion of trinucleotides GGC and AGC.

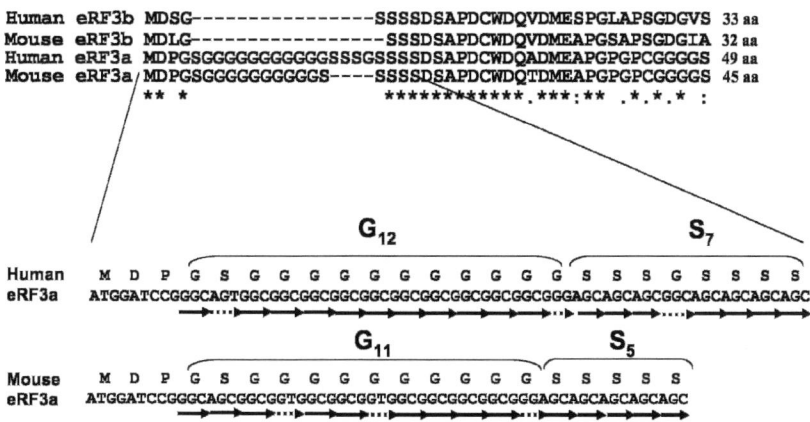

Fig. 4 Localization of GGC- and AGC-repeats in sequences encoding N-terminal parts of *H. sapiens* and *Mus musculus* eRF3 proteins. G_{11}, G_{12}, S_7 and S_5—homopeptide repeats. *Continuous arrows*—repeats of the same codon GGC or AGC, *dotted arrows*—codons different from GGC or AGC

Regardless of the low level of amino acid identity among N-domains of mGSPT1 with mGSPT2 and Xenopus eRF3 proteins (49 and 11%, respectively) they show similar secondary structure with α-helices in the same positions (Fig. 3). This indicates that the mammalian mGSPT1 and hGSPT1 N-proximal regions may also exist in highly flexible or partially unfolded conformations. Taken together, these data show that all eRF3 homologues exhibit an unusual aa composition of their N-proximal regions, pointing to a possibility of their high structural flexibility, the property potentially related to the capability to undergo prion-like switches.

3 Evolution of Genes Encoding for eRF3

Only one of mammalian paralogues, *GSPT2* is intronless and expressed only in mouse brain (Hoshino et al., 1998). Its expression is not detected in any of human cell lines (Chauvin et al., 2005). In contrast, all *GSPT1* genes are intron-containing and ubiquitously expressed. We have proposed that the *GSPT2* gene is a functional retrogene, or intronless paralogue of *GSPT1* (which contains 15 exons) that has arisen as a result of retrotransposition of processed *GSPT1*'s transcript into the genome (Zhouravleva et al., 2006) (Fig. 5). CAGE analysis (http://fantom.gsc.riken.jp) suggests dual promoter activity of *GSPT1* gene that in its turn indicates the presence of cryptic promoter in 5' region of mRNA *GSPT1*.

We have estimated the divergence time for GSPT1 and GSPT2 genes of human and mouse using QDate algorithm (Rambaut and Bromham, 1998). The obtained time is 119 ± 15 million years. We hypothesize that the duplication event took place after the divergence between placental and marsupial mammals, which is proposed to have happened from 160 to 135 Mya (Arnason

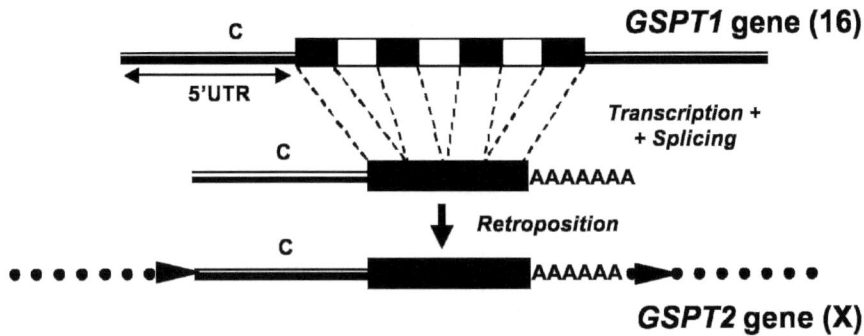

Fig. 5 *GSPT2* is a retrogene, which arose by retroposition of mRNA *GSPT1*. *GSPT2* arises from mRNA transcript of *GSPT* that is reverse transcribed and re-integrated in chromosome X. Only four exons of 15 exons in *GSPT1* are shown for simplicity. 5'UTR—5' untranslated region, CP—cryptic promoter, AAAAA—polyadenine sequence

```
              (512)
M. musculus   GAGATGCAGGGCCCCCAGAAGAAAGTGTCAAGGAAGTGATGGAGGAGAAAGAGGAAGTAAGGAAAT
E. europaeus  GAGATTCAGGGCAACCAGAAGAAAATGGTCAGGAAATGATGGAGGAGAAAGAGGAAGTG-AAAAAC
M. lucifugus  GAGATGCAGGGCCCCCAGAAGAAAGTGGCCAGGAAATGATGGAGAA---AGAGGAAAGA-GAAAAT
T. belangeri  GAGATTCAGGGCCTCCAGAAGAAAGTGGCCAGGAA-TGATGGAGGA---AAAAGAGACATGAGAAA

              (578)
M. musculus   CAAAATCTGTGTCCATACCATCAGGTGCACCTAAGAAAGAACACGTAAATGTGGTCTTCATTGGGC
E. europaeus  TTAAATCTGTGGCCGTACCCTCAGGTACTCCTAAAAAAGAACACGTAAATGTGGTATTCATTGGGC
M. lucifugus  CTAAATCTGTGGTCCTACCCTCAGGTGCTCCTAAAAAAGAACACGTAAATGTTGTATTCACTGGGC
T. belangeri  TCAAATCTGTGGTCGTGCCTTC-GGTGCTC-TAAAAAGGAACATGTAA-TGTGGTCTTCATCGGGC
```

Fig. 6 Nucleotide sequence alignment of three proposed pseudogenized *GSPT2* genes with *GSPT2* from *Mus musculus*. Numbers designate nucleotide position in *GSPT2* gene from *M. musculus*. Underlined triplet is an in-frame nonsense codon. Note single point deletions in the proposed pseudogenes compared to mouse *GSPT2*

et al., 2000; Hasegawa et al., 2003). The recent release of the complete gene build for the grey short-tailed opossum (*Monodelphis domestica*), a representative of methatherian lineage, reveals three sequences homologous to eRF3 genes (http://www.ebi.ac.uk/ensembl/). One of them contains 14 introns and encodes protein, carrying a poly-G stretch. The other two are intronless pseudogenes, which do not contain sequence, corresponding to N-domain of the protein. In contrast to methatheria genomes all mammals tested have intact GSPT2 loci.

Pseudogenes were also found in some but not all eutherian species tested. Thus at least in some mammalian genomes there are three loci homologous to GSPT (*Tupaia belangeri, Macaca mulatta, Pan troglodytes, Echinops telfairi Dasypus novemcinctus, Myotis lucifugus, Erinaceus europaeus and Bos taurus*).

Surprisingly, it appears that in some recently sequenced eutherian species, open reading frames of *GSPT2* genes are interrupted with frame shifts and nonsense codons, thus resembling the methatherian genome (one intact and two pseudogenised loci). This status of *GSPT2* gene was observed in case of *E. europaeus, T. belangeri* and *Myotis lucifugus* (see Fig. 6). Some uncertainty remains to be resolved though whether these pseudogenes arose due to mistakes in sequencing or they really exist in genomes. If the former case is true we speculate that *GSPT2* gene is dispensable for living and performs inessential functions though it is present in vast majority of eutherian genomes.

According to molecular phylogeny of mammals these three species fall into two different superorders and thus cannot be considered to be closely related (Springer et al., 1997). Thus one might propose that pseudogenisation of *GSPT2* gene took place independently in two lineages. This in its turn puts under question the estimate of divergence between *GSPT1* and *GSPT2* by highlighting the possibility of pseudogenisation of *GSPT2* in opossum.

Functional significance of pseudogenes found in methatherian and eutherian genomes remain to be elucidated. There are two types of eRF3 pseudogenes in mammals. Long variants correspond to full length *GSPT2* gene whereas short variants correspond to 5′ part of *GSPT2* encoding C-domain only. In cases of bovine genome (*Bos taurus*) pseudogene retains intact sequence corresponding to the C-domain. In yeast truncated variants of eRF3 are capable of supporting viability. Shortened variants of eRF3 in mammals can still be implicated in some molecular processes. The appearance of pseudogene of *GSPT2* in cow seems to have occurred relatively recently because it shares 97.9% identity in nucleotide sequence with *GSPT2*.

4 GSPT2 as a New Phylogenetic Marker

Since 1980s, the comparative analysis of nucleotide sequences of different parts of genome (coding and noncoding regions of nuclear and mitochondrial DNA), in a broad sense—molecular markers, has become the most popular method in the studies of phylogeny and systematics of various organisms. Today it is practically impossible to imagine the further development in these fields without the application of molecular markers. Meanwhile, up to the present time, for most groups of animals only few molecular markers have been used for the study of phylogeny (most common were 18S RNA and different mtDNA genes). Mammals are the group where the widest range of molecular markers was applied for the study of phylogeny. But even within this well-studied group the markers for the phylogenetic study are selected most frequently in a random manner, without any preliminary study of their evolution or variation. The adequate choice of molecular marker in correspondence with taxonomic level of the group under study is a very problematic point to which attention is rarely paid. The algorithms of a phylogenetic analysis independent of quality and number of data will anyway create some trees. The necessity of thorough selection of a character or set of characters adequate to the taxonomic level of the group under study is well known to morphologists and classical systematicians: it is impossible to use one and the same characters to the study of phylogeny of taxa at different levels. There are only some general assumptions from empirical data on the rate of variation of this or that marker proceeding from which they could be more reasonably used at the species, generic or other level. In practice it is quite often that one and the same marker is used for the analysis of several different taxonomic ranks. As a rule the choice of the molecular marker for the phylogenetic study is highly influenced by the data on the rate of its evolution in other groups, by the convenience of their usage and by the presence of comparative material in the Genebank. However, if the rate of evolution of molecular marker is slower than the times of speciation in the group under study or a too small fragment is analyzed, then it is highly probable that the number of unique mutations (synapomorhies) will be insufficient for the statistically robust tree. If the observed small variability is restricted to the small number of sites then the probability of parallel and reciprocal changes (homoplasies) will highly increase. On the other hand if molecular marker evolves too fast in relation to the time of divergence of the group then the high level of homoplasies will be achieved due to the mutational saturation and will result in severe bias on the obtained phylogenetic trees. Thus the evolution and variation of the new molecular markers should be studied. To

Evolution of the Translation Termination System in Eukaryotes

do that one needs to perform the validation of a new marker using a group of animals with a phylogeny well established by means of classical systematics.

All mentioned above is the reason why we tried to estimate whether *GSPT1* and *GSPT2* genes could be used as molecular phylogenetic markers. We used the order *Rodentia* as a model group because among mammals the phylogeny of this order is one of the best investigated.

The C-terminal part of eRF3s does not carry any significant phylogenetic signal because of high conservativeness. Thus, we only aimed to explore N-terminal parts of eRF3a and eRF3b. In eRF3a, there are several exons encoding N-terminal part, among which the longest (about 350 bp) is the first one. Most positions which differ in eRF3a and eRF3b, are situated in this part. *GSPT2* which encodes eRF3b is an intronless gene, thus being easier for investigation. For the first step, we attempted to amplify the first exon of *GSPT1*. We failed to design primers specific enough to obtain the necessary PCR product. Supplementary approaches also did not lead to optimization of PCR reaction. This is the reason why for further work only *GSPT2* was used.

We succeeded in amplifying and sequencing the part of *GSPT2* genes encoding the N-terminal part of eRF3b (about 730 bp) obtained from 11 rodent species of three families *(Muridae, Gerbillidae* and *Cricetidae)*. We also used Genebank data for *Mus musculus* and *Rattus norvegicus* genes. In the family

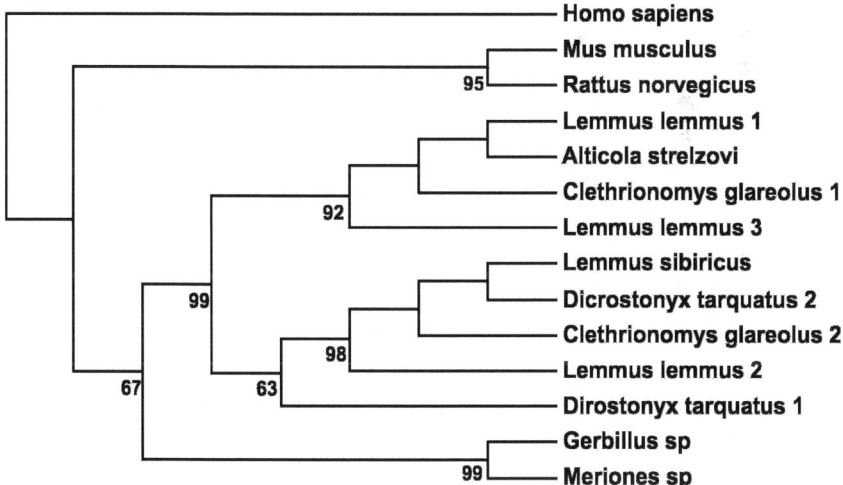

Fig. 7 The phylogenetic tree inferred from *GSPT2* 5′-part obtained from 10 rodent species. Phylogenetic relationships were analyzed by maximum parsimony (MP) and neighbor-joining (NJ) methods using MEGA3 program (http://www.megasoftware.net/; Kumar et al., 2004). NJ analysis was performed with distance matrices calculated with the Kimura two-parameter model. For each method bootstrap analysis (1000 repetitions) was performed. *H. sapiens* (Genbank data) was used as an outgroup. Branching of MP and NJ trees was the same so only MP tree is presented. Bootstrap indexes above 60 are shown. Species for which several different alleles were obtained are indicated with numerals

Cricetidae (hamsters and voles), we had several sample species for some genera and even several geographically distant samples for three species. Using these data, we estimated the variability of those sequences at the level of species and genera. We found out that there was no correlation between gene divergence and species divergence in this family. Almost all the variability observed was a result of a "switch" between the set of same substitutions in same positions. This may be a case of "gene sorting" of alleles or paralogues inherited from the common ancestor of all the family.

Further analysis of the data indicates that the variability of the gene of our interest may be enough to distinguish between families inside one order, though this result is to be proven by enlarged set of data. In maximum parsimony and neighbor joining trees four nodes with high bootstrap indexes can be found. Three of them correspond to family divisions (families *Muridae, Gerbillidae, Cricetidae*), while the fourth one is a subdivision inside *Cricetidae* described above (Fig. 7). Thus the part of *GSPT2* gene encoding the N-terminal part of eRF3b is potentially useful molecular phylogenetic marker at family level.

Acknowledgment This work was supported by grant from the Russian Foundation for Basic Research (07-04-00605), Lider Scientific Schools (7623.2006.4) and the program Origin and Evolution of the Biosphere of the Presidium of the Russian Academy of Sciences.

References

Arnason, U., Gullberg, A., Burguete, A.S. and Janke, A. (2000) Molecular estimates of primate divergences and new hypotheses for primate dispersal and the origin of modern humans. Hereditas 133, 217–228.

Chauvin, C., Salhi, S., Le Goff, C., Viranaicken, W., Diop, D. and Jean-Jean, O. (2005) Involvement of human release factors eRF3a and eRF3b in translation termination and regulation of the termination complex formation. Mol. Cell Biol. 25, 5801–5811.

Doma, M.K. and Parker, R. (2006) Endonucleolytic cleavage of eukaryotic mRNAs with stalls in translation elongation. Nature 440, 561–564.

Fandrich, M. and Dobson, C. (2002) The behaviour of polyamino acids reveals an inverse side chain effect in amyloid structure formation. EMBO J. 21, 5682–5690.

Faux, N., Bottomley, S., Lesk, A., Irving, J., Morrison, J., de la Banda, M. and Whisstock, J. (2005) Functional insights from the distribution and role of homopeptide repeat-containing proteins. Genome Res. 15, 537–551.

Geourjon, C. and Deleage, G. (1994) SOPM: a self-optimized method for protein secondary structure prediction. Protein Eng. 7, 157–164.

Hasegawa, M., Thorne, J.L. and Kishino, H. (2003) Time scale of eutherian evolution estimated without assuming a constant rate of molecular evolution. Genes Genet. Syst. 78, 267–283.

Hoshino, S., Imai, M., Mizutani, M., Kikuchi, Y., Hanaoka, F., Ui, M. and Katada, T. (1998) Molecular cloning of a novel member of the eukaryotic polypeptide chain-releasing factors (eRF). Its identification as eRF3 interacting with eRF1. J. Biol. Chem. 273, 22254–22259.

Inagaki, Y. and Ford, D.W. (2000) Evolution of the eukaryotic translation terminationsystem: origins of release factors. Mol. Biol. Evol. 17, 882–889.

Inagaki, Y., Blouin, C., Doolittle, W. and Roger, A. (2002) Convergence and constraint in eukaryotic release factor 1 (eRF1) domain 1: the evolution of stop codon specificity. Nucleic Acids Res. 30, 532–544.

Inge-Vechtomov, S., Zhouravleva, G. and Philippe, M. (2003) Eukaryotic release factors (eRFs) history. Biol. Cell 95, 195–209.

Karlin, S. and Burge, C. (1996) Trinucleotide repeats and long homopeptides in genes and proteins associated with nervous system disease and development. Proc. Natl. Acad. Sci. USA 93, 1560–1565.

Kisselev, L., Ehrenberg, M. and Frolova, L. (2003) Termination of translation: interplay of mRNA, rRNAs and release factors? EMBO J. 22, 175–182.

Kumar, S., Tamura, K. and Nei, M. (2004) MEGA3: Integrated software for molecular evolutionary genetics analysis and sequence alignment. Brief. Bioinf. 5, 150–163

Le Goff, C., Zemlyanko, O., Moskalenko, S., Berkova, N., Inge-Vechtomov, S., Philippe, M. and Zhouravleva, G. (2002) Mouse *GSPT2*, but not *GSPT1*, can substitute for yeast eRF3 *in vivo*. Genes Cells 7, 1043–1057.

Nakamura,Y. and Ito,K. (1998) How protein reads the stop codon and terminates translation. Genes Cells 3, 265–278.

Rambaut, A. and Bromham, L. (1998) Estimating divergence dates from molecular sequences. Mol. Biol. Evol. 15, 442–448.

Springer, M.S., Cleven, G.C., Madsen, O., de Jong, W.W., Waddell, V.G., Amrine, H.M. and Stanhope, M.J. (1997) Endemic African mammals shake the phylogenetic tree. Nature 388, 61–64.

Urbero, B., Eurwilaichitr, L., Stansfield, I., Tassan J-P., Le Goff, X., Kress, M. and Tuite, M. (1997) Expression of the release factor eRF1 (Sup45p) gene of higher eukaryotes in yeast and mammalian tissues. Biochimie 79, 27–36.

Wallrapp, C., Verrier, S., Zhouravleva, G., Philippe, H., Philippe, M., Gress, T. and Jean-Jean, O. (1998) The product of the mammalian orthologue of the *Saccharomyces cerevisiae HBS1* gene is phylogenetically related to eukaryotic release factor 3 (eRF3) but does not carry eRF3-like activity. FEBS Lett. 440, 387–392.

Zhouravleva, G., Alenin, V., Inge-Vechtomov, S. and Chernoff, Y. (2002) To stick or not to stick: Prion domains from yeast to mammals. Recent Res. Develop. Mol. Cell. Biol. (3), 185–219.

Zhouravleva, G., Schepachev, V., Petrova, A., Tarasov, O. and Inge-Vechtomov, S. (2006) Evolution of translation termination factor eRF3: is *GSPT2* generated by retrotransposition of *GSPT1*'s mRNA? IUBMB. Life 58, 199–202.

The Hedgehog Signaling Cascade System: Evolution and Functional Dynamics

K. V. Gunbin, D. A. Afonnikov, L. V. Omelyanchuk, and N. A. Kolchanov

Abstract Here, the results we obtained in the analysis of the parametric robustness of the Hh signaling cascade and molecular gene evolution are compared. Emphasis is on the molecular evolution events that match the corresponding divergence of the major Bilateria taxonomic types. It is demonstrated that positive selection is characteristic of the genes that encode proteins of the Hh-cascade whose function is related to the molecular morphogenesis mechanisms and matches with the divergence events of the Bilateria types. It was found that the gene products of the Hh-cascade that are subject to positive selection define the kinetic parameters whose change produces the greatest shift in the dynamics of the Hh-cascade. The evolutionary implications for the phenomenon are discussed.

1 Introduction

The molecular-genetic mechanisms that control morphogenesis in the higher eukaryotes pose thought-provoking problems. A large body of evidence has been accumulated to treat their multifaceted aspects (Held, 2002; Rossant and Tam, 2002). Breakthroughs came with increasing knowledge about genes controlling cell differentiation, the patterning of developmental processes by some of them and also about the precise functions of the genes. Furthermore, the transduction pathways have been determined for the Hh, Wnt, TGF-β, RTK, JAK/STAT, Notch signals and also for the nuclear receptors (Pires-daSilva and Sommer, 2003). There loomed an impasse, with the experimental data seemingly incompatible and the systems appearing so intractable (Held, 2002; Rossant and Tam, 2002). Clearly, recourse had to be taken to novel high-throughput computer technologies and approaches to bring the data together

---

K. V. Gunbin
Institute of Cytology and Genetics SB RAS, Novosibirsk, Russia
e-mail: genkvg@bionet.nsc.ru

and to simulate computational events (Eppig et al., 2005; Grumbling et al., 2006).

The theoretical studies of eukaryote morphogenesis currently tend to (1) build and analyze numerical models of morphogenesis, and (2) make feasible comparative and/or evolutionary analyses of genes controlling morphogenesis. Mathematical modeling is usually of three steps, including building of a model, identification of its parameters and analysis of its behavior in terms of the model robustness. When intricate molecular-genetic systems with their multiple kinetic parameters are simulated, insights are gained into how the systems function so that, importantly, the system responses to changes in kinetic parameters can be estimated and the classification of kinetic parameters can be made (de Jong, 2002). Researchers are usually interested in those molecular-genetic system parameters whose small changes cause a strong response in the system's function (the hyper-responsive parameters) and, conversely, parameters whose significant changes cause no significant changes in the molecular-genetic system response (the inert parameters; Dillon et al., 2003; Eldar et al., 2002; Lai et al., 2004). Genes that define the hyper-responsive parameters are of biological interest. The morphogenetic process may prove to be very sensitive to impairment or weak mutational changes in the function of these genes. This makes their research timely and appropriate, particularly with reference to cancerogenesis and/or formation of developmental abnormalities.

It is pertinent to recall that analyses of the modes of molecular evolution of these genes have disclosed their evolutionary features. As is well known, the analyses were based on Kimura's (1983) neutral evolution theory, according to which genes or their regions accomplishing important functions due to negative selection accumulated nonsynonymous substitutions slower than those under smaller functional load. A part of the genes could most rapidly accumulate nonsynonymous substitutions, thereby evolving in the positive selection mode (Kimura, 1983). Thus, evolutionary analysis of the accumulation rates of different types of nucleotide substitutions (synonymous and nonsynonymous) made it possible to rank genes according to the degrees of their variability and functional importance.

Here we study the Hh signaling cascade (Lum and Beachy, 2004) and compare the results we obtained using the model of the insect Hh-cascade (Gunbin et al., 2007a) with those yielded by evolutionary analysis of the genes involved in the functioning of the cascade (Gunbin et al., 2007b). The Hh-cascade is involved in numerous morphogenetic processes, being one of the most amply studied (Lum and Beachy, 2004; Nybakken and Perrimon, 2002). The structure of the gene network that controls the cascade has been described (Gunbin et al., 2004). The set of families which participate in the cascade work is the same in all Bilateria. It includes the Hh morphogene; the Hh morphogene secretion mechanism (the Disp proteins); the cellular receptors Ptc and Smo; the Cos2, PKA, Slmb, Su(Fu) and Fu proteins; the components of the high molecular complex that transduces the Hh signal into the cell nucleus, the transcription factors Ci (Gli) that regulate the *ptc* gene expression (in vertebrates and invertebrates) and the *Gli* expression (in vertebrates only).

Taking into consideration the differences in the Hh-cascade for signal transduction between vertebrates and invertebrates, we compared the parametric robustness of the Hh-cascade model in invertebrates (Gunbin et al., 2007a) and vertebrates (Lai et al., 2004). As a result, we identified a parameter set whose small changes cause a strong response in the model's normal behavior and, in this way, we identified a set of hyper-responsive genes of the Hh-cascade. The analysis of the molecular evolution of genes of the Hh-cascade was successful in that it allowed us to (1) identify the genes that at the divergence step of the major Bilateria types evolved in the positive selection mode and (2) demonstrate that the genes are predominantly those we identified as hyper-responsive. Based on the obtained results, we hypothesize that the hyper-responsive genes might have served as evolutionary "internal reserves" for the compensatory changes during the structural reorganization of the Hh signaling cascade.

2 Modeling of the Hedgehog Signaling Cascade

In the current study, we performed a comparative analysis of parametric robustness of the vertebrate (Fig. 1a) and invertebrate (Fig. 1b) Hh-cascade models. These models take into account the kinetic parameters obtained in experimental research (Gunbin et al., 2007a; Lai et al., 2004). Lai and coauthors have analyzed the robustness of the switchlike behavior of the vertebrate Hh-cascade (Lai et al., 2004); their simulation has demonstrated that bistable behavior of the Hh-system is consistent with the experimental data; furthermore, they have described the necessary and sufficient conditions for the molecular Hh-system to adopt one of the alternate states.

The dynamics of the invertebrate Hh-cascade has been studied in Gunbin et al. (2007a). It has been found that the model of invertebrate Hh-cascade behaves in a manner consistent with the experimental data (Gunbin et al., 2007a) and also the consequences of mutational changes in the various system components were predicted (Gunbin et al., 2007a), but the invertebrate Hh-cascade did not behave switchlike, consistent with the previous studies (von Dassow and Odell, 2002; von Dassow et al., 2000).

It should be noted that the above models of the Hh signaling cascade in invertebrates (Gunbin et al., 2007a) and vertebrates (Lai et al., 2004) are complementary in their common subnet for the Hh-cascade gene network (Fig. 1). The model for the Hh-cascade in invertebrates offers a more detailed description of the processes that mediate Hh signal transduction within the cell. This enables to study the features of the kinetics of the processes. Also, the model of vertebrate Hh-cascade contains a subnet of the Hh-cascade gene network specific to vertebrates and allows to take into account the crucial

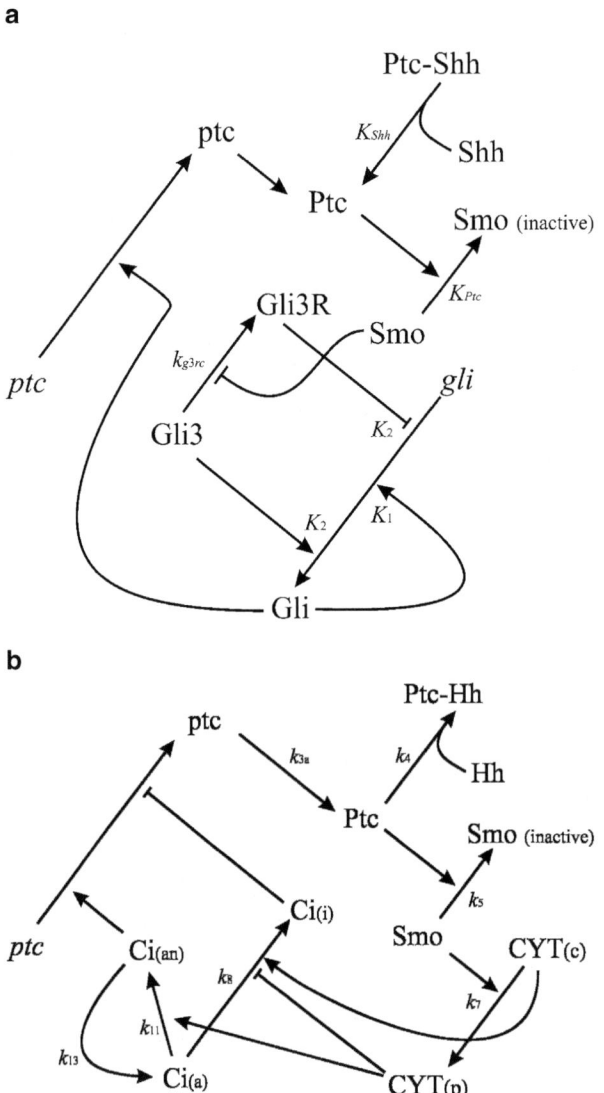

Fig. 1 Models for Hh signaling cascade: **a** vertebrates (Lai et al., 2004), **b** invertebrates (Gunbin et al., 2007a). *Sharp arrows* point to activation, *blunt arrows* to inhibition

differences in the molecular mechanisms of the Hh-cascade between vertebrates and invertebrates (Huangfu and Anderson, 2006).

From the analysis of the parametric robustness of the vertebrate and invertebrate Hh-cascade models, it followed that there exist regions highly sensitive to changes in the gene network of the Hh signaling system (Table 1). The following regions were distinguished: (1) the formation and/or spreading of

Table 1 Hyper-responsive parameters of the models for vertebrate and invertebrate Hh signaling cascade

Parameter	Parameter description	Proteins (roman) or/and genes (italic) defining parameter
D_H^{**}	Hh diffusion coefficient	Disp, Hh
K_{Shh}^{*}	Dissociation constant for Ptc-Hh protein complex	Ptc, Hh
M^{**}	Constant of Ptc inactivation rate	Ptc, Hh
K_{Ptc}^{*}	Half-maximum concentration of Ptc protein required for Smo protein inhibition	Ptc, Smo
k_5^{**}	Half-maximum concentration of Ptc protein required for Smo protein inhibition	Ptc, Smo
k_{3a}^{**}	Synthesis rate constant of the membrane Ptc protein form	Ptc
kd_1^{**}	Constant of the rate of Ptc protein degradation	Ptc
I^{**}	Background activity of the *ptc* gene	*ptc*
k_3^{**}	Constant of the rate of *ptc* mRNA degradation	*ptc*
k_0^{*}	Maximum transcription rate of ptc gene	*ptc*, Ci
$v_{max,G}$	Maximum transcription rate of the *Gli* genes	*Gli*, Gli3
K_2^{*}	Dissociation constant of the Gli3 protein and the *ptc* gene promoter	Gli3
kd_3^{**}	Constant of the rates of degradation/inactivation of proteins and protein complex that mediate Hh signal transduction within the cell	Fu, Su(Fu)

Designations: models for * – the Hh-cascade in vertebrates (Lai et al., 2004); ** – the Hh-cascade in invertebrates (Gunbin et al., 2007a).

the Hh morphogene and its interaction with the Ptc receptor, (2) Ptc–Smo protein interaction and Ptc synthesis, (3) the relatedness degree between the Ci (Gli) transcription factors and gene enhancers and the interaction with the Fu and Su(Fu) proteins (Table 1).

In comparison of the dynamics of the vertebrate and invertebrate Hh-cascade, one must be keenly aware that the difference in the gene networks of vertebrate and invertebrate Hh-systems in the regulatory mechanism of *Gli/Ci* genes expression is the key factor that enables to make the system monostable dynamics bistable (Lai et al., 2004; von Dassow and Odell, 2002; von Dassow et al., 2000). It may be suggested that stable functioning of Hh-cascade gene network can be ensured in the case of such significant structural changes by coordinated compensatory shifts in the system kinetic parameters in the widely variable stretches of the Hh-cascade gene network, i.e., in the mechanisms of interaction between the Ptc and Smo proteins, between the transcription factors Ci/Gli and Ci/Gli genes enhancers and also in the formation/spread of the Hh morphogene and its interaction with the Ptc receptor.

3 Molecular Evolution of Hedgehog Signaling Cascade

The above-described comparative structural–functional analysis of the vertebrate and invertebrate Hh-cascade demonstrated that their topology underwent significant changes at the time when vertebrates diverged from invertebrates. We analyzed single nucleotide substitutions in the families of genes encoding proteins, the participants of Hh-cascade (*Hh, Ptc, Smo, Disp, PKA, Slmb, Su(Fu), Fu* and *Ci*), to define the evolutionary mode of the genes at a particular step of evolution. We proceeded on the assumption that during the evolution of the gene network, topology reorganized under selection impact to stabilize its function, a part of the genes incorporated into the morphogenetic network could accumulate nonsynonymous substitutions that compensated structural changes to provide function (Gunbin et al., 2007b).

A result of the performed analysis was identification of the gene (protein) regions in the Hh-cascade genes subjected to positive selection (Table 2). When the location of the gene regions subjected to positive selection and the functional role of the encoding protein domains were related, it proved that positive selection mainly acts on proteins and their domains that are involved in protein–protein interaction within the Hh signaling cascade and that these proteins and their domains define the hyper-responsive parameters (Tables 1 and 2). To illustrate, changes in the intein domain of the Hh protein and also the N-end of the Disp protein can affect the rate at which Hh morphogene forms and/or spreads (Table 1, the D_H and K_{Shh} parameters). Regarding the *ptc* gene, selection was found to be positive for the regions that determine interaction features between Hh and Ptc proteins (Table 1, the M, kd_1 and K_{Shh} parameters). Selection in the *smo* gene is positive in those regions that define the interaction with the high molecular weight complex Fu/Cos2/Ci (Table 1, the k_5 parameter). The regions of the *ptc* and *smo* genes defining the interaction between the Ptc and Smo proteins also underwent positive selection (Table 1, the k_5 and K_{Ptc} parameters). In the *su(fu)* and *fu* genes, the regions define the interaction between the Su(Fu) and Fu proteins and also between them and the Ci protein. These regions are subjected to positive selection, possibly due to the evolutionary changes in the Hh signal transduction rate into the cell nucleus (Table 1, the kd_3 parameter). Finally, in the *ci* gene and its *Gli* homologs, positive selection was observed for the coding domain regions responsible for the retention of Ci (Gli) proteins in the cytoplasm and for binding to the transcription cofactors (Table 1, the k_0, K_2 and $v_{max,G}$ parameters)

Analysis of the phylogenetic trees did not reveal positive selection on the branches for the *Ptc, PKA* and *Slmb* genes (Gunbin et al., 2007b). The stages throughout which positive selection acted in the *Hh, Disp, Smo, Su(Fu), Fu* and *Ci(Gli)* genes corresponded to the stages of the divergence between invertebrates and vertebrates (Gunbin et al., 2007b). Thus, it may be reasonably inferred that the timescales when positive selection acted in the evolving Hh-cascade matched with the divergence steps of the major Bilateria groups: Ecdysozoa (Insecta,

Table 2 Positive selection in the Hh-cascade gene regions encoding protein domains (Gunbin et al., 2007b)

Protein name	Domain protein composition	Domain function	Positive selection
Hh	N-terminal	Binding with Ptc receptor	−
	C-terminal, intein	Self-excision	+
Ptc	12 transmembrane loops		
	Sterol-sensitive	Intracellular vesicular traffic	−
	Exporters of RND superfamily (predicted)	Ion transport	−
	Extracellular domain between the 1st and 2nd transmembrane loops	Contact with Hh morphogene	+
	Extracellular domain between the 7th and 8th transmembrane loops	Contact with Hh morphogene	+
Disp	12 transmembrane loops		
	Sterol-sensitive	Intracellular vesicular traffic	−
	Exporters of RND superfamily (predicted)	Ion transport	−
	Extracellular domain between the 1st and 2nd transmembrane loops	Possibly, secretion of the Hh morphogene by cells	+
	Extracellular domain between the 7th and 8th transmembrane loops		−
Smo	7 transmembrane loops		
	N-terminal, extracellular	Possibly, binding to Ptc	−
	G-receptors	G-protein bindings	+
	C-terminal	Binding to high molecular complex Fu/Cos2/Ci	+
PKA	N-terminal	Unknown	+
	Serine/threonine kinases, catalytic	Protein phosphorylation	−
	C-terminal	Unknown	−
Slmb	WD-domain	Protein–protein interactions,	−
	F-box	Binding to the SCF ubiquitin-ligase complex	−
Su(Fu)	SU(FU)-domain	Interaction with Ci	+
Fu	Protein kinase catalytic domain	Protein phosphorylation	−
	C-terminal	Inhibition of the Hh signaling, interaction with Su(Fu) and Cos2	+
Ci	Cytoplasm anchoring (at C-end and N-end from zinc finger)	Retaining the protein in the cytoplasm (only in insects)	+
	Zinc finger	Binding to DNA	−
	Transcription cofactors binding	Binding to transcription cofactors	+

Crustacea), Lophotrochozoa (Mollusca, Annelida), Deuterostomia (Chordata, Echinodermata). This means that the evolutionary steps subjected to positive selection and the emergence of large Bilateria taxa matched.

4 Comparison of the Evolution Mode of the Hh Signaling Cascade Genes with Their Functional Load and Response Type

Summing up, based on analyses of (1) the parametric robustness of mathematical models for Hh-cascade in vertebrates and invertebrates, the hyper-responsive parameters of the models and genes related to these parameters were identified; (2) the molecular evolution of the Hh-cascade genes, genes that underwent positive selection soon, in evolutionary terms, after vertebrates and invertebrates had diverged, were identified.

Table 3 compares the evolutionary mode of the Hh-cascade genes (proteins), their functional load and response type. From the tabulated data it follows that, of the nine genes under study, five can be assigned to the hyper-responsive genes and two to the potentially hyper-responsive genes (Table 3). It is of interest that all these genes belong to the developmental according to their functions. Rank correlation analysis demonstrates that parametric robustness correlates with gene function ($r = 1$; $p < 0.05$). Another interesting feature, also shown in Table 3, is the consistency between the response type of a gene (hyper-response present) and the positive selection mode of the gene at the divergence time of the major Bilateria groups. Of the seven hyper-responsive genes, the evolutionary mode under positive selection was identified for six, and the positive selection mode was not identified for two not hyper-responsive genes. Rank correlation analysis also demonstrates that gene evolutionary mode correlates with gene function ($r = 1$; $p < 0.05$) and with parametric robustness ($r = 0.75$; $p < 0.05$).

Heterogeneity of parametric robustness and evolutionary liability of various Hh-cascade components may be, in the first approximation, related to protein–protein interaction number and protein (gene) expression level. However, rank correlation analysis demonstrates that the estimates of gene evolutionary mode do not correlate with those of the number of protein–protein associations ($r = 0.07$) and with the protein (gene) expression level ($r = 0.07$). Rank correlation analysis also demonstrates that parametric robustness estimates do not correlate with the number of protein–protein associations ($r = 0.45$) and with gene expression level ($r = 0.45$).

5 Discussion

The performed analysis concerned the parametric robustness of the models for the Hh-cascade system and evolution of the system genes. Based on the analysis of parametric robustness, the hyper-responsive genes were identified in the Hh-cascade.

Table 3 Relation between gene evolution modes, divergence of Bilateria taxonomic types, and hyper-responsive kinetic parameters

Protein (gene) name	Kinetic parameters of the models corresponding to protein function	Network response corresponding to change in kinetic parameters	Functional protein group (D–Developmental; H–Housekeeping)	Events of positive selections related with divergence of taxonomic Bilateria types	Number of protein–protein associations (L – ≥ 20; S – ≤ 15)	Gene (protein) expression level (U – ubiquitous; L –local)
Hh	D_H^{**}; K_{Shh}^{*}	$+^{***}$	D	+	L	L
Ptc	kd_1, M, k_5^{**}; K_{Ptc}^{*}	$+^{***}$	D	–	L	L
Smo	k_5^{**}; K_{Ptc}^{*}	$+^{***}$	D	+	S	U
Disp	D_H^{*}; K_{Shh}^{*}	$+^{***}$	D	–	–	U
PKA	k_8, k_9, kd_2^{**}; k_{g3rc}, K_{g3rc}^{*}	$-^{***}$	H	–	S	U
Slmb	k_8, k_9, kd_2^{**}; k_{g3rc}, K_{g3rc}^{*}	$-^{***}$	H	–	S	U
Su(Fu)	k_{11}, k_{12}, kd_3^{**}	$+/-^{**}$	D	+	S	U
Fu	k_{11}, k_{12}, kd_3^{**}	$+/-^{**}$	D	+	S	U
Ci	k_0^{**}; $v_{max,G}$, K_2^{*}	$+^{***}$	D	+	L	U

Number of protein–protein associations extracted from BioGRID (Stark et al., 2006). Gene (protein) expression level extracted from FlyBase (Crosby et al., 2007). Designations: models for * – the Hh-cascade in vertebrates (Lai et al., 2004); ** – the Hh-cascade in invertebrates (Gunbin et al., 2007a); + hyper-response (positive selection), – inertness (negative selection), +– intermediate effect

For the Hh-cascade developmental genes under positive selection, there were time spans that matched with the divergence time between the major groups of Bilateria—Ecdysozoa and Deuterostomia. This was taken to mean that amino acid substitutions accumulated rapidly during periods when the structure of Hh-cascade changed drastically. In contrast, positive selection is not a characteristic feature of the *PKA* and *Slmb* genes that encode proteins involved in interactions with the other signaling pathways or the catalytic domains of the proteins under study. Possibly, these components can manage and be tasked to perform a continuous function to provide the vital activities of the cell; this explains why their evolution was in the negative selection mode.

The question is "How structural changes possibly occurred in the Hh signaling cascade?" The greatest shifts in the dynamics of a functioning Hh-cascade were brought about by change in the input portion of the Hh-cascade, the Hh morphogene production by the cells (the Disp, Hh genes under study) and its diffusion (the D_H, M (Gunbin et al., 2007a) and K_{Shh} (Lai et al., 2004) parameters) and also change in the molecular-genetic system that senses the Hh morphogene, i.e., change in its Ptc receptor (the M, kd_1 (Gunbin et al., 2007a) and K_{Shh} (Lai et al., 2004) parameters). Thus, the positive selection in the Disp, Hh and Ptc proteins would provide profound reorganization of the Hh-cascade functioning by appropriate changes in the kinetics of the formation and diffusion of the Hh morphogene.

Changes in the proteins of the middle portion of the Hh-cascade might have provided important changes of rates and forces of cell responsiveness to the Hh morphogene. As known, the main Bilateria groups (Ecdysozoa and Deuterostomia) differ in the composition of proteins involved in the work of the middle portion of the Hh-cascade (the k_5, kd_3 (Gunbin et al., 2007a) and K_{Ptc} (Lai et al., 2004) parameters). For example, it is known that in Drosophila the N-end of the Smo protein is required for Hh signal transduction, while a deletion in the region is without significant effect on the vertebrate Hh-cascade (Nakano et al., 2004). Moreover, in invertebrates the Cos2 and Fu proteins form an important complex that controls the penetration of the transcription factor Ci into the nucleus (Ascano and Robbins, 2004; Ruel et al., 2003), but the experimental data on the Cos2 necessity in vertebrates are not straightforward (Tay et al., 2005).

Changes in the proteins of the output portion of the Hh-cascade could provide drastic changes in the cell response type (the passage from the continuous to switchlike response) to the Hh morphogene by modifying the type for transcription regulation of the *ci* (*gli*) and *ptc* genes. Ecdysozoa and Deuterostomia differ fundamentally by the composition and function of the Ci(Gli) transcription factors (Huangfu and Anderson, 2006). In contrast to invertebrates, vertebrates have three Gli (Gli1, Gli2 and Gli3) paralogs (Huangfu and Anderson, 2006). Moreover, Gli transcription factors activate the expression of the *gli1* and *gli2* genes, thereby forming a regulatory positive feedback loop missing in invertebrates (the k_0 (Gunbin et al., 2007a), $v_{max,G}$ and K_2 (Lai et al., 2004) parameters; Dai et al., 1999; Huangfu and Anderson, 2006). Thus, vertebrates possess additional feedback circuits that set a definite expression level for the effector genes of Hh-cascade. The presence of paralog groups of the transcription factors Gli in vertebrates broadens the repertoire of genes whose expression can regulate the Hh-cascade, thereby giving more opportunities for getting specific response of various tissues to the same morphogene (Huangfu and Anderson, 2006).

The results of the current comparative analysis of the evolution of proteins and their domains in the Hh signaling cascade in the Bilateria, in conjunction with that of the dynamics of the Hh-cascade function, are highly suggestive. The drastic changes in the Hh-cascade occurred at the time when the main taxonomic groups of the Bilateria Ecdysozoa (Insecta, Crustacea) and Deuterostomia (Chordata, Echinodermata) might have diverged.

6 Conclusions

The results of the current analysis are dual. To begin with, the two mathematical models for the Hh signaling cascade and their numerical representation revealed that among the participant genes in the cascade a set of hyper-responsive genes that define the model parameters is distinguishable. Small changes in hyper-responsible parameters produce significant changes in the dynamics of the function of the molecular-genetic network. Genes of this type may be of particular importance for stabilization of the network function. In fact, very small derangements of hyper-response gene function by small spontaneous weakly damaging mutations can make the network malfunctioning. It is of interest that all these genes belong to the developmental according to their functions.

Second, analysis of nucleotide substitutions demonstrated that evolution under the pressure of positive selection is a feature of the hyper-responsive genes. Positive selection began to operate immediately after the major groups of the Bilateria taxonomic types Ecdysozoa and Deuterostomia had diverged, i.e., very soon after the molecular events associated with the rearrangements in the Hh-cascade topology. It may be assumed that the hyper-responsive genes had an important role in evolution. Even small changes in their function might have resulted in great changes in the function of the entire network and, hence, these genes could serve as good candidate genes for "the within" sources of compensatory shift produced by mere point mutations. In summary, this is the putative evolutionary role we assign to the hyper-response genes we identified here.

Acknowledgment The work is supported by the Ministry of Education of the Russian Federation grant "Development of the Higher School Scientific Potential" 2.1.1.4935, Russian Foundation of the Basic Research (05-04-49141-a, 05-07-98012-p, 03-04-48506-a), SB RAS integration projects 49, N10104-34/П-18/155-270/1105-06-001/28/2006-1. The computation was performed in part at the High Performance Computing Center, SB RAS. The authors are grateful to A.N. Fadeeva for translating the manuscript from Russian to English.

References

Ascano, M., Jr and Robbins, D.J. (2004) An intramolecular association between two domains of the protein kinase Fused is necessary for hedgehog signaling. Mol. Cell. Biol. 24, 10397–10405.

Crosby, M.A., Goodman, J.L., Strelets, V.B., Zhang, P., Gelbart, W.M. and the FlyBase Consortium (2007) FlyBase: genomes by the dozen. Nucleic Acids Res. 35, D486–D491. Version FB2006_01, December 8, 2006.

Dai, P., Akimaru, H., Tanaka, Y., Maekawa, T., Nakafuku, M. and Ishii, S. (1999) Sonic hedgehog-induced activation of the Gli1 promoter is mediated by GLI3. J. Biol. Chem. 274, 8143–8152.

de Jong, H. (2002) Modeling and simulation of genetic regulatory systems: a literature review. J. Comput. Biol. 9, 69–105.

Dillon, R., Gadgil, C. and Othmer, H.G. (2003) Short- and long-range effects of sonic hedgehog in limb development. Proc. Natl Acad. Sci. USA 100, 10152–10157.

Eldar, A., Dorfman, R., Weiss, D., Ashe, H., Shilo, B.Z. and Barkai, N. (2002) Robustness of the BMP morphogen gradient in Drosophila embryonic patterning. Nature 419, 304–308.

Eppig, J.T., Bult, C.J., Kadin, J.A., Richardson, J.E., Blake, J.A. and Mouse Genome Database Group (2005) The Mouse Genome Database (MGD): from genes to mice—a community resource for mouse biology. Nucleic Acids Res. 33, D471–D475.

Grumbling, G., Strelets, V. and The FlyBase Consortium (2006) FlyBase: anatomical data, images and queries. Nucleic Acids Res. 34, D484–D488.

Gunbin, K.V., Omelyanchuk, L.V. and Ananko, E.A. (2004) Two gene networks underlying the formation of the anterior–posterior and dorso-ventral wing imaginal disc compartment boundaries in *Drosophila melanogaster*. Proceedings of the Forth International Conference on Bioinformatics of Genome Regulation and Structure, BGRS'2004, vol. 2, 56–59.

Gunbin, K.V., Omelyanchuk, L.V., Kogai, V.V., Fadeev, S.I. and Kolchanov, N.A. (2007a) Model of the reception of hedgehog morphogen concentration gradient: comparison with an extended range of experimental data. J. Bioinform. Comput. Biol. 5, 491–506.

Gunbin, K.V., Afonnikov, D.A. and Kolchanov, N.A. (2007b) The evolution of the Hh-signaling pathway genes: a computer-assisted study. In Silico Biol. 7, 0047.

Held, L.I. (2002) Imaginal Discs: The Genetic and Cellular Logic of Pattern Formation. Cambridge University Press, Cambridge.

Huangfu, D. and Anderson, K.V. (2006) Signaling from Smo to Ci/Gli: conservation and divergence of hedgehog pathways from Drosophila to vertebrates. Development 133, 3–14.

Kimura, M. (1983) The Neutral Theory of Molecular Evolution. Cambridge University Press, Cambridge.

Lai, K., Robertson, M.J. and Schaffer, D.V. (2004) The sonic hedgehog signaling system as a bistable genetic switch. Biophys. J. 86, 2748–2757.

Lum, L. and Beachy, P.A. (2004) The hedgehog response network: sensors, switches, and routers. Science 304, 1755–1759.

Nakano, Y., Nystedt, S., Shivdasani, A.A., Strutt, H., Thomas, C. and Ingham, P.W. (2004) Functional domains and sub-cellular distribution of the hedgehog transducing protein Smoothened in Drosophila. Mech. Dev. 121, 507–518.

Nybakken, K. and Perrimon, N. (2002) Hedgehog signal transduction: recent findings. Curr. Opin. Genet. Dev. 12, 503–511.

Pires-daSilva, A. and Sommer, R.J. (2003) The evolution of signalling pathways in animal development. Nat. Rev. Genet. 4, 39–49.

Rossant, J. and Tam, P. (2002) Mouse Development: Patterning, Morphogenesis, and Organogenesis. Academic Press, San Diego.

Ruel, L., Rodriguez, R., Gallet, A., Lavenant-Staccini, L. and Therond, P.P. (2003) Stability and association of Smoothened, Costal2 and Fused with Cubitus interruptus are regulated by hedgehog. Nat. Cell Biol. 5, 907–913.

Stark, C., Breitkreutz, B.J., Reguly, T., Boucher, L., Breitkreutz, A. and Tyers, M. (2006) BioGRID: a general repository for interaction datasets. Nucleic Acids Res. 34, D535–D539. Version 2.0.27, May 1, 2007.

Tay, S.Y., Ingham, P.W. and Roy, S. (2005) A homologue of the Drosophila kinesin-like protein Costal2 regulates hedgehog signal transduction in the vertebrate embryo. Development 132, 625–634.

von Dassow, G., Meir, E., Munro, E.M. and Odell, G.M. (2000) The segment polarity network is a robust developmental module. Nature 406, 188–192.

von Dassow, G. and Odell, G.M. (2002) Design and constraints of the drosophila segment polarity module: robust spatial patterning emerges from intertwined cell state switches. J. Exp. Zool. B Mol. Dev. Evol. 294, 179–215.

Approaches to the Resolution of Contradictions Between Phylogenetic Systems Based on Paleontological and Molecular Data

G. S. Rautian, A. S. Rautian, and N. N. Kalandadze

Abstract Essential differences between molecular and morphofunctional characteristics as sources of information about evolutionary development are discussed. Global historical geography of terrestrial vertebrates is considered as a source of information on the basic events in mammalian phylogeny and on the dates of emergence of certain lineages. It is emphasized that some aspects of zoogeographical reconstructions are supported by new data of the fossil record and comparative molecular studies.

1 Introduction

Phylogenetic reconstruction based on paleontological and molecular data has much in common, but, at the same time, shows certain essential differences. It is reasonable to suppose that, as contradictions between final phylogenetic reconstructions are resolved, new insights into the patterns of evolution and the correlation of evolutionary changes at different levels of biological organization will be achieved. An important obstacle to this is the profound differences in generally accepted methods in different fields of biology, and their mostly independent development. An alarming sign of this discordance is the recently emerging practice to introduce new taxonomic names based on the results of molecular studies, regardless of the absence of any analogs in taxonomic systems based on classical approaches. Taking into account the fact that molecular studies mostly involve similarities and dissimilarities, the biological significance of which is poorly understood, such newly created molecular taxa cannot be diagnosed in a biologically meaningful way. Reasonable diagnoses are presently possible only at the morphophysiological level, although even in this case the degree of biological understanding is still a long way from perfect.

G. S. Rautian
Paleontological Institute, Russian Academy of Sciences, Moscow, 117997 Russia
e-mail: gsrautrian@mtu-net.ru

In other words, a taxonomic group recognized based on molecular traits should necessarily be comprehended in terms of the morphophysiological properties of organisms. For the moment, molecular taxa remain "a thing to themselves." They add little to the understanding of the nature of organisms, or to ordering our knowledge of biodiversity and, hence, cannot provide the basis for further biological studies. At the same time, the search for a constructive dialog aimed to reveal and overcome the contradictions between phylogenetic reconstructions at molecular and morphophysiological levels is the most topical problem of modern phylogenetics.

The major distinctive feature of phenotypic evolution is the fact that it is directly controlled by natural selection, whereas the dependence of genetic characteristics on selective pressure is always indirect and is determined by the extent to which they have a selectively significant phenotypic manifestation.[1] As early as 1938, Schmalhausen (1938, 1949) formulated that phenotypic characters are potentially much more stable than the genetic traits providing the basis for their development. This concept agrees with the empirical generalization that Timofeev-Resovsky (1958) named the Chetverikov's principle (Chetverikov, 1926, 1983). It states that uniform phenotypes of the *wild type* (Schmalhausen named it the *adaptive norm*) are underlain by a vast genetic diversity, which can be discovered in laboratory conditions through inbreeding but usually has no phenotypic manifestation in nature. Thus, according to Schmalhausen, the stability of the adaptive norm (= wild type) is maintained in spite of genotypic diversity rather than due to genotypic uniformity.

Stabilizing (canalizing) selection produces and constantly maintains structural variants that are capable of persisting over potentially infinite periods of geological history (irrespective of the accumulation of genetic change). Therefore, the rates of morphological evolution not only vary within a very wide range but, more importantly, can be extremely low, supporting successful variants of organization. This is corroborated, in particular, by the existence of living fossils and long-lived groups at any time in the geological history of Earth. Thus, the organismal level of biological organization has the potential (not necessarily realized) for self-preservation over an arbitrarily long time period.

Molecular genetic characteristics are only indirectly influenced by natural selection; in particular, they are only indirectly subjected to stabilization (Schmalhausen, 1938, 1949). Therefore, in contrast to morphofunctional characteristics, they are incapable of remaining constant over arbitrary long periods and *inevitably change* in the course of evolution. In particular, even under conditions of long stabilizing selection of the phenotypes (for example, in living fossils), more or less neutral genetic changes are constantly accumulated and, hence, result in genetic divergence between species even when they belong to

---

[1] The more complex is the morphofunctional organization, the longer is the ontogenetic path from a nucleotide sequence to phenotype. Therefore, this statement is particularly important with reference to higher metazoans.

closely related taxa. A prominent example is provided by the Siberian newt (*Salamandrella keyserlingii*), which was examined in regard to the nucleotide sequence of the mitochondrial gene for cytochrome *b* (Berman et al., 2005). It was shown that the Magadan, Sakhalin, Chukchi and Ural populations are rather similar both morphologically and genetically (genetic variability was estimated as 0.38%), while the population from the Primorye region combined small morphological differences from other populations with high differences in mtDNA (9.8-11.6%). The researchers emphasized that similar genetic distances between the nucleotide sequences of the cytochrome *b* gene are observed in different species of the same genus of the family Hynobiidae. Thus, the interspecific distances within the genus *Batrachuperus* (with six species included in the comparison) range from 4.4 to 9.3%; those in the genera *Pseudohynobius* (two species) and *Hynobius* (three species) are 15.5 and 10.7-12.7%, respectively.[2] This example distinctly illustrates the theoretically derived statement that genetic differences are accumulated even when morphological evolution is slowed down.

The more neutral the genetic change that occurs in a certain lineage, the more reliable the information on the sequence of branching and the time of divergence they provide, but the less suitable they are for the recognition of boundaries between taxa, because they cannot provide a measure of the biological significance of changes. Thus, the tasks of splitting phylogenetic branches into particular taxa and the creation of hierarchical phylogenetic classification are still solvable only on the basis of morphofunctional characteristics. In other words, genetic and morphological approaches mostly supplement each other rather than impose rigid restrictions on the interpretation of results.

A number of recent studies have considered the divergence times between various groups and species. This is important for assessing evolutionary rates of morphological and molecular changes as well as for estimating the geological and ecological context of the diversification events and for understanding the speciation mechanisms. However, the calibration points based on the fossil record always give younger dates than true points of divergence between taxa examined because split between particular lineages occurs deep in the ancestral taxon where derived characters have not yet developed. Therefore, even if it were possible to find an earliest representative of a lineage in the fossil record it would not be placed in the daughter taxon. In addition, the fossil record is rather incomplete, particularly in regard to small mammals, since the probability to find an extinct taxon directly depends on the biological progress sensu Severtsov, implying high taxonomic diversity and abundance of subordinate taxa, and on the body size. In particular, mammals were suppressed by dinosaurs over a large part of their early history and remained cryptic until they blossomed forth in abundance in the Cenozoic. At the same time they

---

[2] For comparison, intergeneric distances (estimated based on the nucleotide sequences of the cytochrome *b* gene) between the elephantids *Elephas, Loxodonta* and *Mammuthus* are only 3–4% (Rautian and Dubrovo 2001, 2003; Rautian et al., 2006).

dominated in the small-sized class beginning from the later part of the Late Jurassic as a result of the greatest crisis in the history of land tetrapods (Kalandadze and Rautian, 1993a,b).

It is possible to obtain additional calibration points for dating certain divergence events in mammalian phylogenetic tree based on the results of global historical zoogeography. Regarding the estimation of divergence times of particular lineages zoogeographical and genetic approaches have much in common, since both reflect the loss of links between initial populations irrespective of their morphofunctional and ecological characters. Thus, in contrast to the morphological approach, they provide time estimates for the divergence of the stem groups.

2 Historical Vertebrate Zoogeography and Molecular Phylogeny of Mammals

2.1 Global Historical Geography of Land Vertebrates as a Source of Information on Divergence Events in Mammalian Phylogeny

The history of intercontinental faunal contacts was established using all accessible data (mainly published) on geographical and stratigraphic distribution of fossil and modern tetrapods, excluding marine or flying taxa. Each taxon under study was considered to be monophyletic, monotopic and monochronous in origin (Skarlato and Starobogatov, 1974). The faunas were compared by the degree of their taxonomic similarity at the generic and family levels. To avoid subjectivity of paleogeographical judgments, faunas were compared with each other in pairs, regardless of their present or past position on the Earth's surface. The reconstruction of faunal connections and zoogeographical division were built for each stage of the stratigraphic scale.

Discarding the false (inherited from the previous epochs) and mediated faunal links from each reconstruction was made through the combination of two operations.

Operation of subtraction. If the direct link between two faunas shown by a certain taxon can be explained by its actually known distribution in the previous stages, such a taxon was eliminated from consideration in the zoogeographical reconstruction for the given stage.

Operation of reduction. If the direct link between faunas of two regions was weaker, or could be fully reduced to two (or a greater number) mediating links (i.e., all or overwhelming majority of taxa of direct link corresponded to the same taxa of indirect links), such a link was considered to be false, produced by the connection between two faunas through a third, or several faunas.

The valid faunal links between paleogeographical regions were interpreted as indication of direct land contact between them during the whole or a part of the given geological stage, while the absence of links was evidence of marine isolation of these regions during the whole stage. It is noteworthy that only one paleogeographical reconstruction was in agreement with the strict zoogeographical requirements imposed.

In addition, the principles of irreversibility of evolution, monophyletic, monotopic and monochronic origin of taxa of phylogenetic system allowed the reconstruction of a number of missing links based on the following assumptions (Rasnitsyn, 1988, 2002, 2006): (1) a disjunctive range of a taxon is evidence for the existence of its representatives in the intermediate area; and (2) the presence of members of a taxon in isolated faunas is evidence to its existence in the epoch when the faunas of the territories were in direct or indirect contacts. These assumptions of geographical and stratigraphical distribution of taxa allowed the improvement of the general scenario of faunagenesis, in particular, a suitable paleogeographical reconstruction for the Early and Middle Jurassic (which were impossible to obtain based on the fossil record of that time because of extremely scanty data).

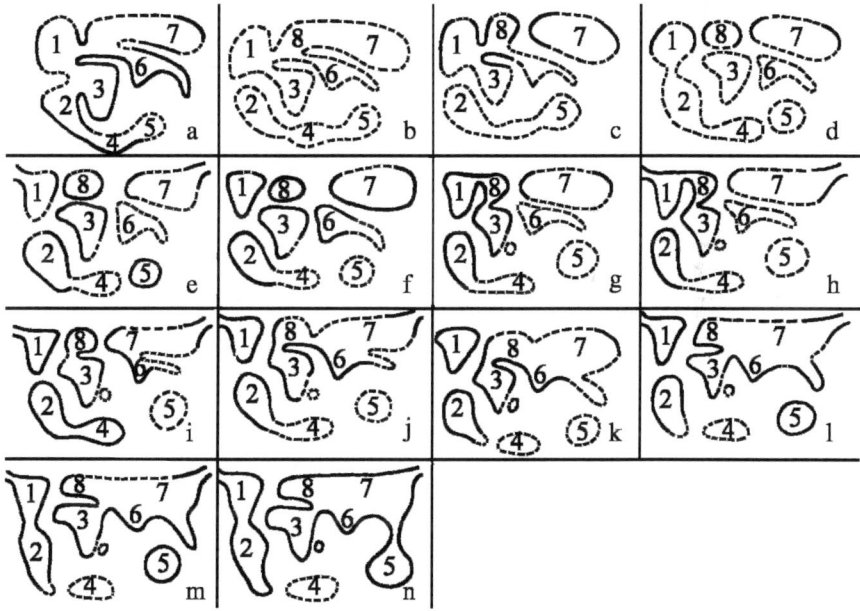

Fig. 1 Global paleogeographical reconstructions established based on the data on historical zoogeography of terrestrial tetrapods: (**a**) Late Triassic–early Early Jurassic, 227–195 Ma; (**b**) late Early Jurassic–early Middle Jurassic, 195–159 Ma; (**c**) early Middle Jurassic–terminal Jurassic, 159–142 Ma; (**d**) early Early Cretaceous, 142–127 Ma; (**e**) late Early Cretaceous–early Late Cretaceous, 127–85 Ma; (**f**) late Late Cretaceous, 85–65 Ma; (**g**) Early Paleocene, 65–61 Ma; (**h**) Middle Paleocene–Early Eocene, 61–49 Ma; (**i**) Middle–Late Eocene, 49–34 Ma; (**j**) Early Oligocene, 34–28 Ma; (**k**) Late Oligocene, 28–24 Ma; (**l**) Early–Middle Miocene, 24–11 Ma; (**m**) Late Miocene–Middle Pliocene, 11–3 Ma; (**n**) Late Pliocene–Recent, 3–0 Ma. Designations of faunas of: (*1*) North America, (*2*) South America, (*3*) Africa, (*4*) Antarctic, (*5*) Australia, (*6*) India, (*7*) Asia, (*8*) Europe

Based on the paleogeographic reconstruction obtained (Fig. 1), it is possible to date certain important events in the history of mammals.

2.2 Basic Events in the History of Mammalian Faunas

Mammals compose a terminal group of an extensive trunk of Theromorpha, which includes not less than two-thirds of fossil land vertebrates (Synapsida). The mammal-like reptile branch differentiated not later than in the middle of the Carboniferous (Westphalian 310–320 Ma), i.e., before the Carboniferous separation of the Euramerican and Gondwanian zoogeographical regions. However, they probably emerged even earlier. The distribution of Eothyrididae and Edaphosauria is confined to the Euramerican Zoogeographical Realm established by Kalandadze and Rautian (1980, 1981, 1983), which expanded in the Early Permian (283–270 Ma) at least to western China.

This reconstruction is supported by the high level of specialization of the earliest known Synapsida and the presence of primitive pelycosaurs (Varanopseidae, seu "Anningiomorpha") in the Late Permian (265–270 Ma) of South Africa (Romer and Price, 1940) and northern Eastern Europe (Caseidae and Varanopseidae: Mesenosaurinae) (Carroll, 1988). In addition, Therapsida, a sister group of pelycosaurs, was widespread in the Gondwanan Zoogeographical Realm. This group included Cynodontia, direct ancestors of mammals. The Upper Triassic (227–205 Ma) of either zoogeographical realm yielded such taxa as Diplocaulinae, Herpetospondyli, Dissorophoidea, Captorhynomorpha and Bolosauria (Kalandadze and Rautian, 1980, 1981, 1982, 1983, 1991, 1992, 1995a,b, 1997, 1998a,b).

Thus, the development of both remote (Therapsida) and direct ancestors of mammals (Cynodontia) was primarily connected with Gondwana (South America–Africa–Antarctic–Australia–India). Mammals (both Prototheria and Theria) achieved the global distribution at the very beginning of their history, in the Late Triassic (about 227–205 Ma). At that time the greatest consolidation of all tetrapod faunas was observed, all continents came in direct or indirect contact (Fig. 1a), which has never occurred later in the Earth's history (Kalandadze and Rautian, 1980, 1981, 1983, 1991). Zoogeographical Pangea of the Late Triassic destroyed all previous zoogeographical divisions.

The separation of the southern block of continents (including South America–Antarctica–Australia) occurred in the middle of the Early Jurassic (200–190 Ma, Fig. 1b). This event marks the split of Prototheria into Monotremata and Multituberculata (extinct groups) as well as Theria into Metatheria and Eutheria, the signs of which are clearly seen even in the modern mammal zoogeography. The origin of Monotremata and Metatheria was connected with the southern block, while Multituberculata and Eutheria emerged in the northern block (including Africa).

The next geological event marked by paleozoogeography was the isolation of Asia from the other continents of the northern block (Fig. 1c), which occurred in the late Middle Jurassic (169–159 Ma); at that time, the Turgai Sea (in place of modern Western Siberia) and epeiric seas in the Fore-Urals isolated Asia from Europe. A number of placental lineages, including Cimolodonta (Kielantheria, Deltatheridia, which were ancestral to Creodonta), Scandentia (including Anagalida), Dermoptera and their relatives (Plesiadapiformes and Apatotheria) and Lagomorpha and their relatives (Mixotheridia, Mixodontia and Taeniodontia) appeared in Asia.

In the western center (Africa–Europe–North America), Plagiaulacida, Feliformia, Lipotyphla, Xenarthra, Tubulidentata, Pholidota, ancestors of Primates (Strepsirhini + Haplorhini) and Rodentia emerged. The absence of Lipotyphla from the Asian center and the absence of Menotyphla (sensu Butler, 1956) from the western center suggest that the basal placentals were carnivorous animals closer to predators (Carnivora sensu lato, including Aegialodontia, Deltatheridia, Creodonta) rather than to insectivores, as is usually thought. Similarly, in the parallel Metatheria trunk, insectivorous groups (Peramelida, Myrmecobiidae, Notoryctidae, Caenolestidae, Necrolestidae) evolved from the predator trunk Marsupicarnivora (Dasyuroidea + Didelphoidea).

At the Jurassic–Cretaceous boundary (about 142 Ma), the western block of continents split into isolated Africa, Europe and North America (Fig. 1d). The fauna of Madagascar, which was isolated for at least the whole of the Cenozoic, shows essential elements of the autochthonous African fauna, with which the origin of Feliformia, Tenrecoidea, Strepsirhini, Miomorpha (Cricetidae) and Tubulidentata was connected.

Not later than at the middle of the Early Cretaceous (132–121 Ma), South America came in contact with North America (Fig. 1d); this was the only possibility for some placental groups to enter South America. As a result of the faunal exchange between these continents, Didelphoidea penetrated from South America into North America, while placentals, such as Tarsiiformes (descendants of which gave rise to New World monkeys, Platyrrhini), ancestors of Caviomorpha, Xenarthra, Notoungulata (Notioprogonia) (probable ancestors of Astrapotheria and Pyrotheria) and Condylarthra (Mioclaenidae) (ancestors of Litopterna) penetrated from North America into South America. The absence from South America of Artiodactyla and Perissodactyla strongly suggests that they differentiated in North America later, when the bridge between South and North Americas was broken. The autochthonous (pre-Pliocene) fauna of South America lacks placentals of Asian origin (descendants of Menotyphla, Scandentia, Dermoptera or Lagomorpha). This supports that the earliest Bering land (bridge between Asia and North America in place of the Bering Strait) that provided intense faunal exchange between these continents occurred at the end of the Early Cretaceous (121–99 Ma) when South America had become completely isolated from the other continents (Fig. 1e). The next faunal exchange between South and North Americas occurred as late as the Late Miocene (see Simpson, 1980).

In the autochthonous fauna of South America, carnivorous mammals are represented by marsupials (Didelphoidea, Borhyaenoidea and Coenolestidae), while phytophagans are almost exclusively placentals (except for Polydolopidae, Groeberiidae and Argyrolagidae): rodents (Caviomorpha), Xenarthra and ungulates (Notoungulata, Astrapotheria, Pyrotheria, Didolodontidae and Litopterna). North America is the native land of ungulates (Notoungulata and Condylarthra) and the unique continent free from autochthonous predatory mammals. Apparently, at the moment of contact between the marsupial and placental faunas in the Early Cretaceous, marsupials were represented by more specialized and competitive predators, while at least some placentals possessed pronounced pre-adaptations to phytophagy.

Marsupials (Didelphoidea) probably stimulated specialization of ungulate ancestors in North and South Americas. Predatory specialization of condylarthrs, such as Mesonychia and Arctocyonia, was probably secondary. The absence of a Bering bridge from the end of the Cretaceous (Campanian–Maastrichtian) to the Early Paleocene (83–58 Ma; Fig. 1e–h) combined with the origin of Artiodactyla in North America and Cetacea on the coast of the Tethys suggest the Late Cretaceous age (85–83 Ma) of their common ancestor (predatory condylarths Eparctocyonia).

The absence of marsupial phytophagans of Australian origin (Diprotodonta) in America and the absence of placentals in the Miocene (24 Ma) and earlier faunas of Australia indicate that Australia became isolated before the middle of the Early Cretaceous (> 130–140 Ma; Fig. 1d and e). The finds of South American tetrapod groups (Polydolopidae, Ancodonta, Astrapotheria, Litopterna, Phorusrhacidae and Sebecidae) in Antarctic Continent (Hooker, 1992) corroborated the reconstruction where the faunal link between Antarctic and South America existed longer than a bridge between Antarctic and Australia (Kalandadze and Rautian, 1982) (Fig. 1d–k).

2.3 Zoogeographical Dating Compared with the Phylogenetic Reconstruction Based on Molecular Data

When the global historical tetrapod zoogeography was first developed and associated with contacts and isolation between faunas inhabiting particular landmasses, it provided surprisingly early dates of divergence of a number of therian lineages (Kalandadze and Rautian, 1980, 1981, 1983, 1992), so that many researchers could not accept these results. However, recent studies have provided new fossil data on Early Cretaceous and Jurassic therian mammals (Flynn et al., 1999; Ji et al., 2002; Rauhut et al., 2002; Rich et al., 1999), which reduce the gap between zoogeographical estimates and data of the fossil record. In addition, molecular evidence for the early divergence of mammals, in particular, placentals have been reported (Arnason and Janke, 1996; Easteal et al., 1995; Hasegawa et al., 2003; Huchon et al., 2002; Kumar and Hedges, 1998; Madsen et al., 2001; Murphy et al., 2001; Penny et al., 1999; Scally et al., 2002; Springer et al., 2003; etc.). The dates for particular nodes of phylogenetic trees range depending on the material, technique and calibration points used; however, they are much closer to the dates obtained on the basis of tetrapod zoogeography than the dates derived directly from the fossil record. For example, the divergence between placentals and marsupials is dated from 120 to 180 Ma.

At the same time, these molecular-based estimates are still younger than what is predicted by zoogeography (190–200 Ma). This is not surprising, because they are also based on particular calibration dates that mark the appearance of taxa in the fossil record, i.e., mostly the state of biological progress of these taxa. In actuality, divergences between lineages always occurred somewhat earlier. In this respect, it is interesting that the molecular date that is closest to zoogeographical estimate was obtained when the divergence between synapsids and diapsids (ca. 310 Ma) was taken for calibration (Kumar and Hedges, 1998). This is additional evidence for the fact that therian mammals reached biological progress rather late in historical development and were less completely represented in the fossil record than other tetrapod groups. In other words, many amphibian and reptile groups reached biological progress soon after their emergence and, hence, they appeared in the fossil record relatively early. In contrast, all therian groups played a subordinate role in communities throughout most of their history. This raises the question of the reason for the acquisition of the key position.

Mammals mastered the majority of adaptive zones only in the Cenozoic when they became free from large reptiles. It is evident that mammals could not supersede adult dinosaurs through competition. This replacement probably resulted from the development of Multituberculata after the Great Middle Jurassic Crisis in the tetrapod community (Kalandadze and Rautian, 1992, 1993a). This group comprised the earliest small efficient tetrapod phytophagans, which gave rise to a special food chain within the subdominant community sensu Olson (1966) of small tetrapods (Kalandadze and Rautian, 1995c). Since they were progressive phytophagous animals, they produced abundant biomass and were prerequisite for the development of predators of a new type adapted for feeding on these and similar tetrapod animals. As such predators developed, they could consume juvenile dinosaurs as a supplementary food resource. At the same time, dinosaurs were formed under conditions of the absence of this type of predators, in particular, they were not adapted for taking care of their posterity. For them it appeared impossible to develop these mechanisms. Thus, one could say dinosaurs became extinct because of vulnerability of their juvenile stage. On the contrary, mammals showed pronounced juvenile care and, hence, were pre-adapted for specialized predators feeding on them. This distinction anticipated the Cenozoic progress of mammals, including the development of large-sized taxa.

It is important to keep in mind the subordinate position of early mammals, when taking calibration points from the fossil record. Additional restrictions on the presence in the fossil record are imposed by the biotopes. Inhabitants of forest landscapes, especially, small arboreal animals infrequently occur in the fossil record (Simpson, 1980). This point is particularly important with reference to primates and rodents. They are recorded in South America beginning from the Oligocene. However, zoogeographical data suggest that both groups entered South America not later than 120 Ma; large differences in molecular sequences between these and closest primate and rodent groups, respectively, from other continents are in general agreement with these results. Thus, calibration points based on primate and rodent fossils should be used with caution.

Thus, recent studies show agreement between the results of tetrapod zoogeographical reconstructions and molecular phylogeny of mammals, which support each other, although certain disagreements still remain and require the development of study in this field.

Acknowledgment The study was supported by the programs of the Presidium of the Russian Academy of Sciences 'Origin and Evolution of the Biosphere,' "Biological Resources of Russia: Fundamental Bases of Efficient Use" and the Russian Foundation for Basic Research, project no. 05-04-48493a.

References

Arnason, U. and Janke, A. (1996) Mitogenomic analyses of eutherian relationships. Cytogenet. Gen. Res. 96, 20–32.
Berman, D.I., Derenko, M.V., Malyarchuk, B.A., Grzybowski, T., Kryukov, A.P. and Miscicka-Sliwka, D. (2005) Intraspecific genetic differentiation of the Siberian Newt (@*Salamandrella keyserlingii*,@ Amphibia, Caudata) and the cryptic species @*S. schrenckii*@ from southeastern Russia. Zool. Zh., 84(11), 1–15.
Butler, P.M. (1956a) The skull of *Ictops* and the classification of the Insectivora. Proc. Zool. Soc. London, Ser. B 126, 453–481.
Carroll, R.L. (1988) Vertebrate Paleontology and Evolution. Freeman, New York.
Chetverikov, S.S. (1926) On certain points of the evolutionary process from the point of view of modern genetics. Zh. Eksp. Biol., Ser. A. 2, 3–54.
Chetverikov, S.S. (1983) Problems of General Biology and Genetics (Reminiscences, Research Works, and Lectures.Nauka, Novosibirsk.
Easteal, S., Collet, C.C. and Betty, D.J. (1995) The Mammalian Molecular Clock. Springer, Texas.
Flynn, J.J., Parrish, M., Rakotosamimanana, B., Simpson, W.F. and Wyss, A.R. (1999) A Middle Jurassic mammal from Madagascar. Nature 401, 57–60.
Hasegawa, M., Thorne, J.L. and Kishino H. (2003) Time scale of eutherian evolution estimated without assuming a constant rate of molecular evolution. Genes Genet. Syst. 78, 267–283.
Hooker, J.J. (1992) An additional record of a placental mammals (Order Astrapotheria) from the Eocene of West Antarctica. Antarct. Sci. 4(1), 107–108.
Huchon, D., Madsen, O., Sibbald, M., Ament, K., Stanhope, M.J., Catzeflis, F., Jong, W. and Douzery, E.J.P. (2002) Rodent phylogeny and a timescale for the evolution of Glires: evidence from an extensive taxon sampling using three nuclear genes. Mol. Biol. Evol. 19(7), 1053–1065.
Ji, Q., Luo, Z.-X., Yuan, C.-X., Wible, J.R., Zhang, J.-P. and Georgi, J.A. (2002) The earliest known eutherian mammal. Nature 416, 816–822.

Kalandadze, N.N. and Rautian, A.S. (1980) On historical zoogeography of terrestrial tetrapods of the terminal Paleozoic and Early Mesozoic. In: Paleontology and Stratigraphy. Nauka, Moscow, pp. 93–102.

Kalandadze, N.N. and Rautian, A.S. (1981) Intercontinental contacts of terrestrial tetrapods and resolution of the problem of the Scottish Elgin Fauna. In: Life on Ancient Continents: Formation and Development. Nauka, Leningrad, pp. 124–133.

Kalandadze, N.N. and Rautian, A.S. (1982) Historical zoogeography of mammals. In: Mammals of the USSR: III Congress of the All-Union Theriological Society, Moscow, Vol. 1, p. 82.

Kalandadze, N.N. and Rautian, A.S. (1983) The position of Central Asia in the zoogeographical history of the Mesozoic. In: Extinct Reptiles of Mongolia. Nauka, Moscow, pp. 6–44.

Kalandadze N.N. and Rautian A.S. (1991) Late Triassic zoogeography and reconstruction of the terrestrial tetrapod fauna of North Africa. Paleontol. J. 25(1), 1–12.

Kalandadze, N.N. and Rautian, A.S. (1992) Mammal system and historical zoogeography. In: Phylogenetics of Mammals. Mosk. Gos. Univ., Moscow, pp. 44–152.

Kalandadze, N.N. and Rautian, A.S. (1993a) Jurassic ecological crisis in terrestrial tetrapod communities and the heuristic model for the conjugated evolution of the biota and community. In: Problems of Pre-Anthropogene Evolution of the Biosphere. Nauka, Moscow, pp. 60–95.

Kalandadze, N.N. and Rautian, A.S. (1993b) Symptoms of Ecological Crises. Stratigr. Geol. Correlation 1(5), 473–478.

Kalandadze, N.N. and Rautian, A.S. (1995a) Interpreted geochronological schedule ("calendar") of the major events in the phylocenogenesis of community (taxocene) of terrestrial vertebrates: Part 1. In: Ecosystem Rearrangements and Evolution of the Biosphere. Paleontol. Inst. RAS, Moscow, Vol. 2, pp. 8–11.

Kalandadze, N.N. and Rautian, A.S. (1995b) Interpreted geochronological schedule ("calendar") of the major events in the phylocenogenesis of community (taxocene) of terrestrial vertebrates: Part 2. In: Ecosystem Rearrangements and Evolution of the Biosphere. Paleontol. Inst. RAS, Moscow, Vol. 2, pp. 12–15.

Kalandadze, N.N. and Rautian, A.S. (1995c) Physiology prerequisites for the utilization of plant resource by land vertebrates. Paleontol. J. 29(4), 179–185.

Kalandadze, N.N. and Rautian, A.S. (1997) Historical zoogeography of terrestrial tetrapods and a new method of global paleogeographical reconstruction. In: Evolution of the Biosphere. Queen Victoria Museum and Art Gallery Publ., Launceston, pp. 95–98 (Records of the Queen Victoria Museum and Art Gallery, Launceston. Vol. 104).

Kalandadze, N.N. and Rautian, A.S. (1998a) Interpreted geochronological schedule ("calendar") of the major events in the phylocenogenesis of community (taxocene) of terrestrial vertebrates: Part 3. In: Ecosystem Rearrangements and Evolution of the Biosphere. Paleontol. Inst. RAS, Moscow, Vol. 3, pp. 38–41.

Kalandadze, N.N. and Rautian, A.S. (1998b) Interpreted geochronological schedule ("calendar") of the major events in the phylocenogenesis of community (taxocene) of terrestrial vertebrates: Part 4. In: Ecosystem Rearrangements and Evolution of the Biosphere. Paleontol. Inst. RAS, Moscow, Vol. 3, pp. 42–46.

Kumar, S. and Hedges, S.B. (1998) A molecular timescale for vertebrate evolution. Nature 392, 917–920.

Madsen, O., Scally, M., Douady, C.J., Kao, D.J., Debry, R.W., Adkins, R., Amrine, H., Stanhope, M.J., de Jong, W.W. and Springer M.S. (2001) Parallel adaptive radiations in two major clades of placental mammals. Nature 409, 610–614.

Murphy, W.J., Eizirik, E., O'Brien, S.J., Madsen, O., Scally, M., Douady, C., Teeling, E., Ryder, O.A., Stanhope, M.J., de Jong, W.W. and Springer, M.S. (2001) Resolution of the early placental mammal radiation using Bayesian phylogenetics. Science 294, 2348–2351.

Olson, E. (1966) Community evolution and the origin of mammals. Ecology 47(2), 291–302.

Penny, D., Haseqawa, M., Waddell, P.J. and Hendy, M.D. (1999) Mammalian evolution: timing and implications from using log determinant transform for proteins of differing amino acid composition. Syst. Biol. 48, 76–93.

Rasnitsyn, A.P. (1988) Phylogenetics. In: Modern Paleontology. Nedra, Moscow, Vol. 1, pp. 480–497.

Rasnitsyn, A.P. (2002) Evolutionary process and methodology of systematics. Tr. Ross. Entomol. Ob-va 73, 1–107.

Rasnitsyn, A.P., (2006) Ontology of evolution and methodology of taxonomy. Paleontol. J. 40(6 Suppl.), 679–737.

Rautian, G.S. and Dubrovo, I.A. (2001) The study of mammoth DNA. In: Mammoth and Its Environment. Geos, Moscow, pp. 112–123.

Rautian, G.S. and Dubrovo, I.A. (2003) Data on DNA give evidence for parallel development in mammoths and elephants. Deinsea 9, 381–394.

Rauhut, O.W.M., Martin, T., Ortiz-Jaureguizar, E. and Puerta, P. (2002) A Jurassic mammal from South America. Nature 416, 165–168.

Rautian G.S., Rossina V.V. and Rautian A.S. (2006) Approaches to the resolution of contradictions between phylogenetic reconstructions based on morphofunctional and genetic data. Paleontol. J. 40, 508–523.

Rich, T.H., Vickers-Rich, P., Constantine, A., Flannery, T.F., Kool, L., van Klaveren, N. (1999) Early Cretaceous mammals from Flat Rocks, Victoria, Australia. Rec. Queen Victoria Mus. 106, 1–35.

Romer, A.Sh. and Price, L.I. (1940) Review of the Pelycosauria. Geol. Soc. Am. Spec. Pap. 28, 1–538.

Scally, M., Madsen, O., Douady, C.J., de Jong, W.W., Stanhope, M.J. and Springer M.S. (2002) Molecular evidence for the major clades of placental mammals. J. Mammal. Evol. 8(4), 239–277.

Schmalhausen, I.I. (1938) Organism As the Whole in Individual and Historical Development. Akad. Nauk SSSR, Moscow.

Schmalhausen, I.I. (1949) Factors of Evolution: Theory of Stabilizing Selection. Blakiston, Toronto.

Simpson, G.G. (1980) Splendid Isolation. Yale Univ. Press, New Haven.

Skarlato, O.A. and Starobogatov, Ya.I. (1974) Phylogenetics and principles of construction of natural system. In: Theoretical Questions of Taxonomy and Phylogeny of Animals. Nauka, Leningrad, pp. 30–46.

Springer, M.S., Murphy, W.J., Eizirik, E., and O'Brien, S.J. (2003) Placental mammal diversification and the Cretaceous–Tertiary boundary. PNAS 100(3), 1056–1061.

Timofeev-Resovsky, N.V. (1958) Microevolution: elementary evolutionary events, materials, and factors of the evolutionary process. Bot. Zh. 43(3), 317–336.

Chromosomes and Speciation

P. M. Borodin

Abstract We analyzed a relative contribution of chromosomal and genetic divergence into reproductive isolation and speciation in three different taxa of mammals. We demonstrated that male sterility in hybrids between different subspecies of South American rodent *Thrichomys* and hybrids between geographically isolated population of the house musk shrew *Suncus murinus* was determined solely by genetic incompatibility, and chromosomal changes played minor, if any, role in the speciation. The genetic incompatibility in these two taxa was controlled by a very small number of loci affecting the early stages of meiosis of the hybrids, while the viability of the hybrids was not affected. The analysis of the hybrid zone between two chromosome races of the common shrew *Sorex araneus* revealed more complex picture. Their genetic divergence for habitat preferences led to formation of ecological premating isolation. Divergence for the genes controlling morphological traits affected the development of the hybrids and, apparently, their viability, thus contributing into postmating isolation. Meiosis in the hybrid males was also affected. In the hybrids we observed a high frequency of synaptic aberrations, which were apparently determined by chromosomal incompatibility. Thus, we may conclude that the fixation of different selectively neutral chromosomal rearrangements in geographically isolated populations does not drive these populations to speciation; however in some cases chromosomal differences may reinforce postmating isolation already established by the divergence of the genes controlling development and chromosome pairing and recombination at meiosis.

P. M. Borodin
Institute of Cytology and Genetics, Siberian Department of Russian Academy of Science, Novosibirsk 630090, Russia
e-mail: borodin@bionet.nsc.ru

1 Introduction

Many closely related species of animals show considerable differences in their karyotypes. There is a large body of evidences that chromosomal rearrangements reduce the fertility of heterozygous hybrids. These facts led to suggestion that chromosomal changes play a causative role in speciation (King, 1993; White, 1978). An alternative view is that the accumulation of chromosomal differences between populations is largely incidental to speciation (Coyne and Orr, 1998, 2004; Rieseberg, 2001).

In this chapter I review our studies in relative contribution of chromosomal and genetic changes to reproductive isolation in three mammalian models:

1. A group of sibling species of South American caviomorph rodent *Thrichomys* (Rodentia, Echimyidae) (Borodin et al., 2006);
2. Geographically isolated populations of the house musk shrew (*Suncus murinus*, Eulipotyphla, Soricidae) (Aulchenko et al., 1998; Axenovich et al., 1998; Borodin et al., 1998; Rogatcheva et al., 1997, 1998, 2000a,b);
3. Parapatric chromosome races of the common shrew (*Sorex araneus*, Eulipotyphla, Soricidae) (Pack et al., 1993; Polyakov et al., 1996, 1997a,b, 2000a,b, 2001, 2002, 2003).

These models differ for the time of intra-group divergence and complexity of chromosomal changes thus allowing us to assess the role of chromosomal rearrangements and genetic divergence in formation of genetic incompatibility, reproductive isolation and finally in speciation.

2 Sibling Species of South American Caviomorph Rodent *Thrichomys*

All taxa of genus *Thrichomys* show very low morphological variation and the genus has been for long considered as monospecific. Recently, this view has been challenged on the basis of morphological and cytogenetic evidences, and several sibling species and subspecies have been suggested: *T. pachyurus, T. apereoides apereoides, T. apereoides laurentius* and others (Bonvicino et al., 2002). They are geographically isolated and occupy rather different habitats. Cytogenetic analyses revealed karyotypic divergence between these taxa. *T. pachyurus* has diploid chromosome number $2n = 34$ and differs from *T. a. laurentius* ($2n = 30$) and *T. a. apereoides* ($2n = 28$) for a series of chromosomal rearrangements. *T. a. apereoides* differs from *T. a. laurentius* for a tandem fusion of a large acrocentric chromosome with a small metacentric chromosome.

When we crossed specimens of three taxa of genus *Thrichomys*: *T. pachyurus, T. a. laurentius* and *T. a. apereoides* we found that they were able to mate in captivity and produce viable progeny. We did not detect underdominance effects of interspecies hybridization on litter size and body mass of the hybrids at birth and weaning. The reproductive success of the majority of between-species crosses was similar to that of within-species crosses.

Some female hybrids produced viable offspring in the crosses with the males of the parental species. All male hybrids were sterile. Histological examination revealed meiotic arrest at the primary spermatocyte stage. No sperm was detected (Fig. 1).

Electron microscopic analysis of surface spread synaptonemal complexes revealed a complete failure of chromosome pairing in F1 hybrids of *T. pachyurus* with *T. a. laurentius* and *T. a. apereoides*.In the male hybrids between *T. a. apereoides* and *T. a. laurentius* all chromosomes including the heteromorphic ones paired orderly, however, meiosis did not proceed beyond diplotene.

Results of our analysis indicated that in the hybrids between *T. a. laurentius* and *T. a. apereoides* the heterozygosity for the tandem fusion was not the main cause of the male hybrid sterility.

We found that the chromosomes involved in the rearrangement paired orderly in the majority of the F1 hybrid male pachytene cells. Chromosomal cause of male sterility in the hybrids of *T. pachyurus* with *T. a. laurentius* and *T. a. apereoides* also appears to be unlikely. Indeed, spermatogenesis in these hybrids was already affected at premeiotic stage: the number of spermatogonia was greatly reduced. In a few of them that entered meiotic prophase, we observed complete pairing failure in all chromosomes, both heteromorphic and homomorphic.

Spermatogenesis abnormalities detected in the hybrids of *Thrichomys* were rather different from those described in mice heterozygous for chromosomal rearrangements. The main cause of low fertility in mouse heterozygotes was the non-disjunction of rearranged chromosomes and formation of unbalanced gametes (Capanna and Castiglia, 2004; Castiglia and Capanna, 2000; Hauffe and Searle, 1998), while the hybrids of *Thrichomys* showed a decreased number

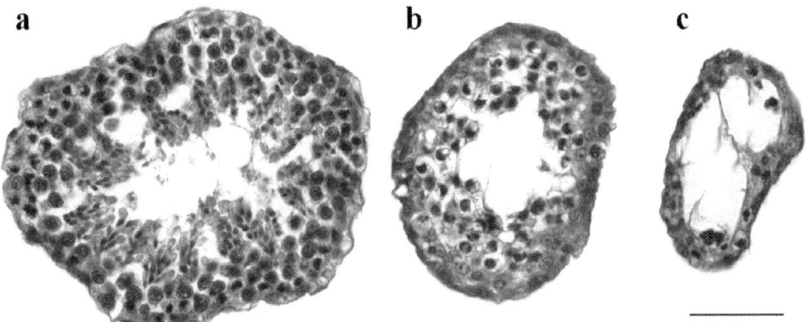

Fig. 1 Sections of seminiferous tubules of pure breed *T. a. apereoides* (**a**), and its hybrids with *T. a. laurentius* (**b**) and *T. pachyurus* (**c**). Bar 50 μm

of spermatogonia and primary spermatocytes due to arrested meiosis before and during prophase (Fig. 1).

The stage of meiotic arrest in the hybrids depended on the phylogenetic distance between the parental species. Both cytogenetic analysis and the analysis of DNA sequence variation within cytochrome *b* among species of the genus *Thrichomys* (Braggio and Bonvicino, 2004) indicate earlier branching *T. pachyurus* and rather recent divergence between *T. a. laurentius* and *T. a. apereoides*. The hybrids with *T. pachyurus* showed a very early arrest of germ cell development. Meiosis in the hybrids of more closely related *T. a. laurentius* and *T. a. apereoides* was less affected.

Another strong argument against the chromosomal cause of the sterility of the hybrids between *T. a. laurentius* and *T. a. apereoides* comes from the result of the analysis of backcross males. The types of sterility observed among them fit results expected under models of genetic incompatibility arising from epistatic interactions between two parental loci that have diverged in allopatry (Dobzhansky, 1936, Muller, 1940). Under such models, two types of incompatibilities should be observed among backcross progeny: between two heterozygous loci derived from different species (e.g. A^1A^2; B^1B^2) and a heterozygous and a homozygous loci (e.g. A^1A^1; B^1B^2) (Turelli and Orr, 2000).

Although our sample of backcross males did not permit detailed statistical analysis of segregation, both classes predicted by the Turelli–Orr incompatibility model were observed. One sterile male, which showed meiotic arrest after pachytene (similar to F1 males), could be explained as heterozygous for two incompatibility loci (A^lA^a; B^lB^a, where *l* and *a* refer to *laurentius* and *aperoides* alleles respectively) while two males with very early meiotic arrest could be assumed homozygous for one locus of *T. a. laurentius* origin and heterozygous for another (A^lA^l; B^lB^a). The fertile male was apparently homozygous for loci of *T. a. laurentius* origin (A^lA^l; B^lB^l).

Thus, all these data indicate that the male sterility of the hybrids between the species of genus *Thrichomys* is due to genetic but not chromosomal incompatibility of the parental taxa.

3 Chromosome Races of the House Musk Shrew

The house musk shrew is a widespread species which occur from East Africa to East Asia. The most common karyotype of *S. murinus* consists of 40 chromosomes (Rogatcheva et al., 1997; Yosida, 1982). Variation in the chromosome number, attributed to fixation of novel metacentrics formed by Robertsonian fusion has been reported from Malaysia ($2n = 40-35$), Sri Lanka ($2n = 30$), and southern India ($2n = 30-32, 37, 40$). Sex chromosome variation has been recorded in the *S. murinus*, with Y-chromosome polymorphism found to be very extensive (Yosida, 1982). Analysis of variation in mtDNA, allozymes and morphological traits revealed a high genetic diversity of local populations of

S. murinus, comparable with genetic distances between different subspecies of the house mouse (Yamagata et al., 1995).

In the house musk shrew we were able to estimate a relative contribution of chromosomal and genic divergence in formation of reproductive isolation between geographically isolated populations. We crossed in the laboratory two strains of the shrew: KAT and SRI. The KAT strain was derived from a wild population in Katmandu, Nepal and has been shown to have 2 = 40. The SRI strain was derived from the animals captured on the West Coast of Sri Lanka. SRI strain differed from KAT strain by five Robertsonian fusions and morphology of chromosomes 7, X and Y (Rogatcheva et al., 1997, 2000a,b).

Being so diverged in karyotype SRI and KAT shrews were able to hybridize in the laboratory and produce fertile offspring. A hybrid stock SK was set up. SK shrews had various combinations of variable chromosomes, having them either heterozygous or homozygous. In general, they demonstrated a sufficient reproductive performance. However, many sterile males were found among them. Meiosis in sterile SK males did not progress beyond pachytene. Electron-microscopic analysis of synaptonemal complexes (SC) in fertile hybrids demonstrated an orderly pairing of all chromosomes, including heteromorphic combinations. In the sterile males chromosome pairing was distorted not only in heterozygous pairs, but also in homozygous ones (Borodin et al., 1998).

Heterozygotes for each of the rearrangements were found among both sterile and fertile males. There was no correlation between the number of heterozygous rearrangements per individual and its fertility (Fig. 2).

Sterile males had all possible combinations of the sex chromosomes; hence compatibility of the sex chromosomes does not seem to play any role in the

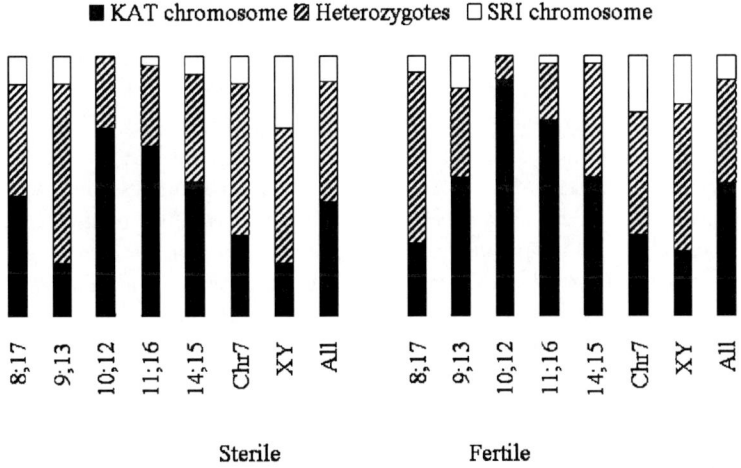

Fig. 2 Frequency of different cytotypes in sterile and fertile male hybrids between *Suncus murinus* from Nepal (KAT) and Sri Lanka (SRI)

control of fertility. Therefore we concluded that meiotic arrest and sterility of interracial hybrids was not determined by structural heterozygosity.

Using segregation analysis we tested several hypotheses of autosomal genetic control of male sterility. The results of this analysis of pedigree data indicated that it can be described within a framework of the triallele model, assuming that the KAT strain was A2A2 homozygous, the SRI was polymorphic for allele A1 and A3, and the hybrids (A2A3) were sterile (Aulchenko et al., 1998; Axenovich et al., 1998).

Thus, our data on *Suncus murinus* clearly demonstrated that chromosome divergence played a negligible (if any) role in fertility problems of the hybrids. A single gene controlling early stages of male meiosis was responsible for male hybrid sterility. It is especially interesting that the hybrid sterility in mammals (in contrast to Drosophila, for example) is usually determined by relatively simple oligogenic systems (one locus in *Suncus murinus* and two loci in *Thrichomys*).

4 Chromosome Races of the Common Shrew

The common shrew (*Sorex araneus*) populates a huge area from British Islands on the west to the Lake Baikal on the east. It shows one of the most remarkable chromosomal polymorphism in mammals. To date more than 60 chromosome races were described (Wojcik et al., 2003) and the actual number of distinct races goes probably far beyond 100. This chromosomal variation is determined by occurrence and fixation of various Robertsonian and whole-arm reciprocal translocations. It is postulated that the ancestral karyotype of the common shrew was characterized by acrocentrics and the karyotype evolution has proceeded in the direction from high to low $2n$ numbers (Polyakov et al., 2001, Volobouev, 1989). Each of the ancestral acrocentric chromosomes in the common shrew is labeled by an italicized letter of the alphabet (e.g. *g, h, i, k*), and so the different karyotypic races can be characterized by their specific arm combination (e.g. race 1 may have metacentrics *gh* and *ik*, race 2 may have *gi* and *hk* and race 3 may have metacentric *gi* and acrocentrics *h* and *k*).

We carried out a detailed analysis of geographic distribution, morphological differences and meiotic properties of the Tomsk (*gk, hi, mn, o, p, q, r*) and Novosibirsk races (*go, hn, ik, mp, qr*) and their hybrids. The Tomsk race is distributed mainly on Altai Mts. and Kuznetzky Alatau highlands while the Novosibirsk race was found on a huge area of the West Siberian Plain (Polyakov et al., 1996, 1997a, 2001).

To interpret the characteristics and distributions of the Novosibirsk and Tomsk races, the events of the last glaciation and the transition to the Holocene, are likely to be of particular significance (Searle, 1986). During the last glaciation, much of the western Siberia is thought to have been flooded by a glacial lake (Frenzel et al., 1992). Currently, over much of the area of West Siberian

Plain common shrews have the Novosibirsk race karyotype. This suggests that Novosibirsk race shrews were able to colonize and spread over this territory before other races had a chance to. It can be presumed that the Novosibirsk race colonized from a glacial refugium in northern parts of the Ural Mts. The similarity in karyotype between the Serov race, an undoubted Ural form, and the Novosibirsk race (Polyakov et al., 1997b) supports this model. Likewise, the Altai Mts. could have been a refugial area for the Tomsk and more easterly races that show karyotypic affinities (Polyakov et al., 2000a,b, 2001). The current distribution of the Tomsk race is certainly consistent with south–north colonization from a glacial refugium in the Altai Mts. into low-lying areas to the east of the River Yenisey. When the Novosibirsk and Tomsk races spread from their glacial refugia, it appears that they made contact near Novosibirsk such that there was introgression of metacentric qr chromosome into the Tomsk race and the acrocentrics q, r into the Novosibirsk race.

To check this hypothesis we carried out a detailed study of the hybrid zone between the Novosibirsk and Tomsk races (Polyakov et al., 2002, 2003). We found that the Novosibirsk–Tomsk F1 hybrids occurred in very low frequency and in a very narrow segment of the hybrid zone (no more than 500 m wide), while the recombinant individuals were rather frequent and found as far as 50 km from the center of the hybrid zone. This difference in frequency and dispersal of the F1 hybrids and recombinants may indicate a low viability of the hybrids or/and rather strong pre-mating isolation between the parental races. A strict ecological affinity of the parental races gives evidence in favor of the latter suggestion.

We have demonstrated altitudinal partition of the races (Polyakov et al., 2003). The Tomsk race occurs in the highlands while the Novosibirsk race is found in the lowlands. A relatively narrow contact zone between the races runs at about 200 m above sea level. The 200 m isocline separates two types of vegetation in West Siberia: forest-steppe habitats at lower altitudes, typical for most of the West Siberian Plain, and taiga habitats at higher altitudes, widely distributed in the Siberian highlands. There is no reason to believe that the particular chromosomal combinations that characterize these races somehow predisposed their adaptations to highland and lowland habitats, respectively. These adaptations have come through selective exposure of the races to different environments.

The chromosomal differences between the parental races contribute to their reproductive isolation at postmating stage. Immunofluorescent and electron microscopy of the synaptonemal complexes of the male hybrids demonstrated that in some cells the rearranged chromosomes were able to pair orderly (Fig. 3).

However, a majority of their meiotic cells contained univalents and other synaptic aberrations (from 40 to 85% in different hybrids) while in the pure-race and recombinant males the frequency of such pairing aberration did not exceed 5–10%. We also found a significant decrease of recombination frequency both in multivalents and normal bivalents of the hybrid.

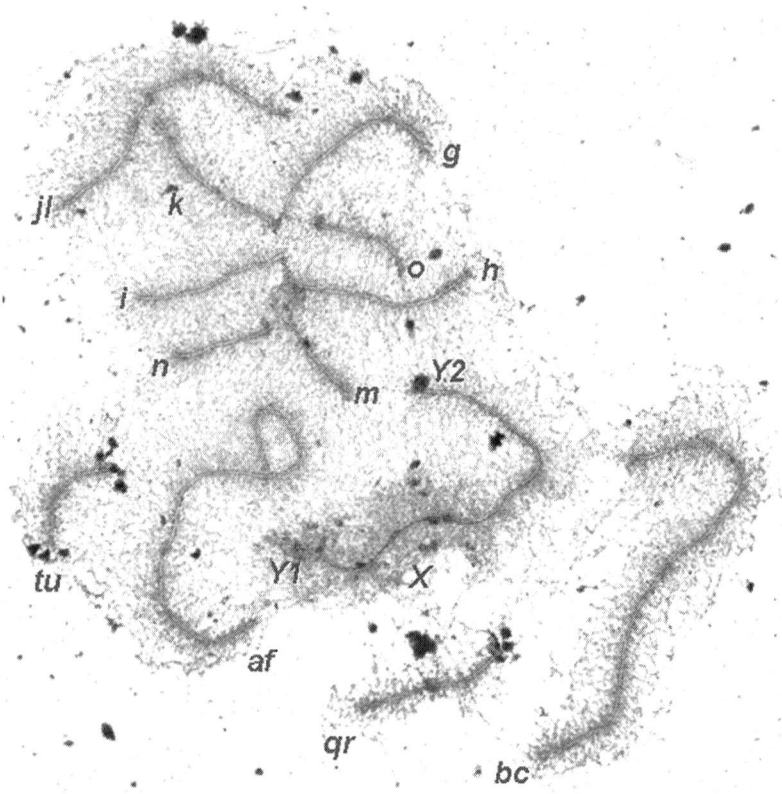

Fig. 3 Chromosome pairing in the male hybrid between the Tomsk and Novosibirsk races of the common shrew

Random segregation of such non-crossover chromosomes may result in the high frequency (up to 50%) of aneuploid genetically unbalanced gametes. These data indicate that the fertility of the male hybrids should be severally reduced

Our analysis of variation in 22 traits of cranial and postcranial skeleton of the shrews trapped in the center of the hybrid zone three distinct groups of individuals, which corresponded to the three karyotypic categories involved in the analysis: Tomsk, Novosibirsk and hybrids. The first discriminant function reflected the differences in the size of skeletal elements. The Novosibirsk shrews and the hybrids were significantly smaller than the Tomsk shrews. However, it was not the variation in the separate morphometric features that allowed a reliable discrimination of the hybrids from the pure races, but the covariation between them. In the hybrids the pattern covariation was apparently disrupted (Polyakov et al., 2002).

This may indicate that the divergence between the races was not only concerned with the set of genes controlling skeletal morphology, but also with the coordination of those genes. Lack of coordination in the development of the hybrids may decrease their viability and thus contribute to postmating isolation.

Thus, both pre-mating and postmating mechanisms of reproductive isolation apparently restrict the gene flow between the races in the hybrid zone. We may consider the Tomsk and Novosibirsk chromosome races as biological species at the early stage of speciation. Due to random genetic processes (occurrence and accumulation of chromosome rearrangements and gene mutations) and natural selection in different ecological conditions they diverged both chromosomally and genetically up to the point of partial chromosomal and genetic incompatibility.

5 Conclusions

We may conclude that the fixation of different selectively neutral chromosomal rearrangements in geographically isolated populations does not drive these populations to speciation; however, in some cases, chromosomal differences may reinforce postmating isolation already established by the divergence of the genes controlling development and chromosome pairing and recombination at meiosis.

Acknowledgment This work was supported by research grants from the Programs of RAS "Biosphere Origin and Evolution" and "Biodiversity," INTAS, RFBR. We thank the Microscopic Center of the Siberian Branch of the Russian Academy of Sciences for granting access to microscopic equipment.

References

Aulchenko, Yu.S., Oda, S.-I., Rogatcheva, M.B., Borodin, P.M. and Axenovich, T.I. (1998) Inheritance of litter size at birth in the house musk shrew (Suncus murinus, Insectivora, Soricidae). Genet. Res. 71, 65–72.

Axenovich, T.I., Rogatcheva, M.B., Oda, S.-I. and Borodin, P.M. (1998) Inheritance of male hybrid sterility in the house musk shrew (Suncus murinus, Insectivora, Soricidae). Genome 41, 825–831.

Bonvicino, C.R., Otazu, I.B. and D'Andrea, P.S. (2002) Karyologic evidence of divesification of the genus Thrichomys (Rodentia, Echimyidae). Cytogenet. Genome Res. 97, 200–204.

Borodin, P.M., Rogatcheva, M.B., Zhelezova, A.I. and Oda S. (1998) Chromosome pairing in inter-racial hybrids of the house musk shrew (Suncus murinus, Insectivora, Soricidae). Genome 41, 79–90.

Borodin, P.M., Barreiros-Gomez, S.C., Zhelezova, A.I., Bonvicino, C.R. and D'Andrea P.S. (2006) Reproductive isolation due to the genetic incompatibilities between Thrichomys pachyurus and two subspecies of Thrichomys apereoides (Rodentia, Echimyidae). Genome 49, 159–167.

Braggio, E. and Bonvicino, C.R. (2004) Molecular divergence in the genus Thrichomys (Rodentia, Echimyidae). J. Mammal. 85, 316–320.

Capanna, E. and Castiglia, R. (2004) Chromosomes and speciation in Mus musculus domesticus. Cytogenet. Genome Res. 105, 375–384.

Castiglia, R. and Capanna, E. (2000) Contact zone between chromosomal races of Mus musculus domesticus. 2. Fertility and segregation in laboratory-reared and wild mice heterozygous for multiple robertsonian rearrangements. Heredity 85, 147–156.

Coyne, J.A. and Orr H.A. (1998) The evolutionary genetics of speciation. Philos. Trans. R. Soc. Lond. B. Biol. Sci. 28, 287–305.

Coyne, J.A. and Orr, H.A. (2004) Speciation. Sinauer Associates, Sunderland, Massachusetts.

Dobzhansky, Th. (1936) Studies on hybrid sterility. II. Localization of sterility factors in Drozophila pseudoobscura hybrids. Genetics 21, 113–135.

Frenzel, B., Pecsi, M. and Velichko, A.A. (1992) Atlas of Palaeoclimates and Palaeoenvironments of the Northern Hemisphere. G. Fischer, Stuttgart.

Hauffe, H.C. and Searle, J.B. (1998) Chromosomal heterozygosity and fertility in house mice (Mus musculus domesticus) from Northern Italy. Genetics 150, 1143–1154.

King, M. (1993) Species Evolution: the Role of Chromosome Change. Cambridge University Press, Cambridge.

Muller H.J. (1940) Bearings of the Drosophila work on systematics. In: J. Huxley (Ed.), The New Systematics. Clarendon Press, Oxford, pp. 185–268.

Pack, S.D., Borodin, P.M., Serov, O.L. and Searle, J.B. (1993) The X-autosome translocation in the common shrew (Sorex araneus L.): late replication in female somatic cells and pairing in male meiosis. Chromosoma 102, 355–360.

Polyakov, A.V., Volobouev, V.T., Borodin, P.M. and Searle J.B. (1996) Karyotypic races of the common shrew (Sorex araneus) with exceptionally large ranges: the Novosibirsk and Tomsk races of Siberia. Hereditas 125, 109–115.

Polyakov, A.V., Borodin, P.M., Lukanova, L., Searle, J.B. and Zima, J. (1997a) The hypothetical Old-Northern chromosome race of Sorex araneus found in the Ural Mts. Ann. Zool. Fin. 34, 139–142.

Polyakov, A.V., Chadova, N.B., Rodionova, M.I., Panov, V.V., Dobrotvorsky, A.K., Searle, J.B. and Borodin, P.M. (1997b) Novosibirsk revisited 24 years on: chromosome polymorphism in the Novosibirsk population of the common shrew Sorex araneus L. Heredity. 79, 172–177.

Polyakov, A.V., Zima, J., Banaszek, A., Searle, J.B. and Borodin, P.M. (2000a) New chromosome races of the common shrew Sorex araneus from Eastern Siberia. Acta Theriol. 45, 11–18.

Polyakov, A.V., Zima, J., Searle, J.B., Borodin P.M. and Ladygina, T.Y. (2000b) Chromosome races of the common shrew Sorex araneus in the Ural Mts: a link between Siberia and Scandinavia? Acta Theriol. 45, 19–26.

Polykov, A.V., Panov, V.V., Ladygina, T.Yu., Bochkarev, M.N., Rodionova, M.I. and Borodin, P.M. (2001) Chromosomal Evolution of the Common Shrew Sorex araneus L. from the Southern Urals and Siberia in the Postglacial Period. Rus. J. Genet. 37, 351–357.

Polyakov, A.V., Onischenko, S.S., Ilyashenko, V.B., Searle, J.B. and Borodin, P.M. (2002) Morphometric difference between the Novosibirsk and Tomsk chromosome races of Sorex araneus in a zone of parapatry. Acta Theriol. 47, 381–387.

Polyakov, A.V., Volobouev, V.T., Aniskin, V.M. and Borodin, P.M. (2003) Altitudinal partitioning of two chromosome races of the common shrew (Sorex araneus) in West Siberia. Mammalia 67, 201–207.

Rieseberg, L.H. (2001) Chromosomal rearrangements and speciation. Trends Ecol. Evol. 16, 351–358.

Rogatcheva, M.B., Borodin, P.M., Oda, S.-I. and Searle J.B. (1997) Robertsonian chromosomal variation in the house musk shrew (Suncus murinus, Insectivora: Soricidae) and the colonisation history of the species. Genome 40, 18–24.

Rogatcheva, M.B., Oda, S.-I., Aulchenko, Yu.S., Axenovich, T.I., Searle, J.B. and Borodin, P.M. (1998) Chromosomal segregation and fertility in Robertsonian chromosomal heterozygotes of the house musk shrew (Suncus murinus, Insectivora, Soricidae). Heredity 81, 335–341.

Rogatcheva, M.B., Aulchenko, Yu.S., Oda, S.-I., Zhelezova, A.I., Serova, I.A., Axenovich, T.I. and Borodin, P.M. (2000a) Chromosomal and genic mechanisms of reproductive isolation: the case of Suncus murinus. Acta Theriol. 45, 147–161

Rogatcheva, M.B., Ono, T., Sonta, S.-I, Oda, S.-I. and Borodin P.M. (2000b) Robertsonian metacentrics of the house musk shrew (Suncus murinus, Insectivora, Soricidae) lose the telomeric sequences in the centromeric area. Genes Genet. Syst. 75, 155–158.

Searle, J.B. (1986) Factors responsible for a karyotypic polymorphism in the common shrew, Sorex araneus. Proc. R. Soc. Lond. B. 229, 277–298.

Turelli, M. and Orr, H.A. (2000) Dominance, epistasis and the genetics of postzygotic isolation. Genetics 154, 1663–1679.

Volobouev, V.T. (1989) Phylogenetic relationships of the Sorex araneus-arcticus species complex (Insectivora, Soricidae), based on high-resolution chromosome analysis. J. Hered. 80, 284–290.

White, M.J.D. (1978) Modes of Speciation. Freeman, San Francisco.

Wojcik, J.M., Borodin, P.M., Fedyk, S., Fredga, K., Hausser, J., Mishta, A., Orlov, V.N., Searle, J.B., Volobouev, V. and Zima, J. (2003) The list of the chromosome races of the common shrew Sorex araneus (updated 2002). Mammalia 67, 169–178.

Yamagata, T., Ohishi, K., Faruque, M.O., Masangkay, J.S., Ba-Loc, C., Vu-Binh, D., Mansjoer, S.S., Ikeda, H. and Namikawa, T. (1995) Genetic variation and geographic distribution on the mitochondrial DNA in local populations of the musk shrew, Suncus murinus. Jap. J. Genet. 70, 321–337.

Yosida, T.H. (1982) Cytogenetical studies on Insectivora. II. Geographical variation of chromosomes in the house shrew, Suncus murinus (Soricidae), in East, Southeast and Southwest Asia, with a note on the karyotype evolution and distribution. Jap. J. Genet. 57. 101–111.

Biotic Turnover in Superorganism Systems: Several Principles of Establishment and Sustenance (Theoretical Analysis, Debatable Issues)

V. G. Gubanov and A. G. Degermendzhy

Abstract Biosphere in first approximation (taking into consideration the most general characteristics) can be characterized by the inflow of light energy, the presence of differentiated alive and inert matter and high degree of thermodynamic closedness. The material cycle based on closure, driven by energy flow based on functional distribution of organisms, is formed to become the basis for long-term existence and evolution of the biosphere. The problem of formation and maintenance of the cycle on all biologically important chemical elements in terms of the biosphere seems urgent because the cycle is the most important attribute of biosphere existence. The biosystem closed on biotic matter cycle is the system with substantially short biocycle of its existence. In the planetary scale the possibility of biocycles destruction due to the formation of biologically non-degradable matters ("deadlocks") can be absolutely real, besides other mechanisms of biosphere degradation (toxicity, warming, ozone holes, etc.). The same problem exists in constructing artificial closed ecosystems, e.g. with the aim of life support for crew of spacecrafts, underwater and arctic settlements and the possible model prototype of noosphere. The aim of the work is to describe and make primary mathematical analysis of two principles of biological closure: "ecological" and "evolutionary." The ratio of any ith biogenic element flow on the producer link to the sum of the same flow and the flow of element coming into the deadlock sediments is the closure measure of any ith element (coefficient Cli).

1 Introduction

The problem of formation and maintaining steady state of the matter cycle on all necessary chemical elements in natural ecosystems (Odum, 1975), on the biosphere level (Kamshilov, 1966, 1979; Pechurkin, 1982, 1988; Vernadsky,

---

V. G. Gubanov
Institute of Biophysics SB RAS, Krasnoyarsk, Russia
e-mail: guban@ibp.ru

1940, 1967), in life support systems (Gitelson et al., 1975, 2003; Nelson et al., 1993), laboratory microcosms (Beyers, 1964; Burdin, 1978; Carpenter, 1996; Folsome and Hanson, 1986; Gubanov et al., 1984; Kovrov and Fishtein, 1978; Nixon, 1969; Sugiura, 1998; Taub, 1969a,b, 1974) and other similar superorganism systems seems extremely urgent. In this case, the intensity and the closure degree (or closure) are the most important characteristics. This work will consider the latter characteristic.

For the Earth's biosphere on the whole the closure biotic of matter cycle, calculated based on biologo-geographical data, is approximately 99.99%.

Apparently, the high matter cycle closure in biosphere does not mean the same high closure of natural ecosystems (ES) in the biosphere. The matter, which is taken out from the cycle (deadlock matter) of one ES, can be involved in the cycle of a neighbour and a farther (in distance and time) ES. Instead of the matter fallen out of the cycle, almost the same amount of the matter from either abiogenic supply or another ES is involved in the cycle. ES in its existence depends upon the introduction of matters from the outside.

Probably, primary ES, due to constant mutational pressure, occasionally but necessarily forms species which initiate the matter cycle and increase its closure degree. Thanks to this property obtained in the process of evolution, organism communities resist the pressure of lack of nutrition supply.

Since the nutrient resource is limited, the time of biotic systems with small closure (at that, complete and required nutrient resource does not recycle and thus stops sooner or later) existence is highly limited in comparison with more closed systems. Therefore, additional opportunities for further evolutionary improvement occur for populations inhabiting these systems (Yablokov and Yusufov, 1989), which leads to an increase in the system life-time and favours further selection, etc. where the biotic system itself as a whole will be the regulator and the test object. Probably, ES should simultaneously be selected according to matter cycle closure. Supposedly, systems with higher matter cycle closure should be more reliable or, at least, more competitive than those with lower closure—they do not need "thinking" about the sources required for living under conditions of limited matter resource.

Therefore, ES in the process of evolution, probably, should increase the matter cycle closure and, at least, be selected in this direction on primary stages of evolution. It is noteworthy that the mentioned sign (closedness) favours the sustainment of the integrity of the properties of biotic system.

The urgency of understanding and analysing the situation in this direction can be justified by the observed reduction and destruction of many natural biocycles by human impact. To estimate the methods of sustaining the cycling diversity in nature, one needs to study the mechanisms of formation and maintenance of various types of biocycles. In this case, important questions appear: what species (with what kinetic characteristics) are capable of forming and maintaining highly closed stable biotic matter cycle in different systems? How can the cycle closure in quantity and change (increase) be estimated? What are the mechanisms (methods) of maintaining highly closed biocycle?

This work is dedicated to the analysis of possible ways and some principles of formation and sustainment of high closure of biotic material cycle. It would be convenient to start solving this problem by taking into consideration all possible (in structure and functional relations) systems capable of long-term autonomous functioning.

2 The Measure of Closure of Biotic Matter Cycle for Systems Based on Matter Supply

Biotic matter cycle (BMC) is the movement of matter through *production* of organic matter (*start* of matter moving–ascending branch) and through its *destruction* with restoration of *initial elements required and used in biosynthesis* (descending branch). Notably, the descending branch is directed to *producers*, i.e. *to the start* of the matter movements, which is *principal*.

In this sense, we will not consider here populations, communities, associations, artificial systems on flow channels (like chemostat), biosystems with matter recycling *only* on heterotrophic level. There is no biotic matter cycle in them in certain sense, as in that case there would be no matter return *to the start*—*to producers*. We consider the *ecosystem* level and higher.

Notably and of *principle*, when talking of matter cycle formation in superorganism systems, we will consider only one forming force – *biological* transformation of matters (compounds), *biological* mechanism of cycle formation, without taking into consideration any physical or artificial forces. Therefore, we speak *particularly* about *biological* i.e. *biotic* (i.e. formed only by biotic components) matter cycle.

Formalize the concept of degree of closure (closure coefficient Cl) of the matter biocycle in superorganism systems—the measure of *biotic* closure. In this case we require the correlation $0 \leq Cl \leq 1$ to hold always. With $Cl = 1$ the system is completely closed biotically (*no deadlocks formation*) and is capable of existing for indefinitely long time. With $Cl = 0$ the system has no biotic cycle (in the sense defined above) elements and, in this case, its existence is completely dependent on the matter supply from outside. The biotic closure coefficient allows, to a certain extent, ways of enhancing the matter cycle closedness to be evaluated.

It is noteworthy that the parameter which characterizes *the internal potential of the biotic system to organization of some biocycle*, though is not responsible for system openness in relation to external matter flows, should be taken into consideration. For example, the completely closed biocycle system ($Cl = 1$, *no deadlocks formation and this has a principal meaning*) can be set into the flow conditions. In this case, if the system structure does not change, the species potentially are still able to form closed biocycle (in this case $Cl = 1$ for such a system), though there exists the flow through the system which shows its incomplete closure on biocycle. In reality, in this case the system is open on matter flow through it (*thermodynamically open*) but, according to species' characteristics, can exist for indefinitely long time even if the matter flow is disrupted (i.e. *biotically closed*), since as before there is no deadlock formation. In nature the partially closed lakes are among these objects. It is noteworthy that all known methods of closure coefficient formalization (Bartsev, 2002; Bartsev et al., 1996; Finn, 1976, 1978; Gitelson et al., 1975; Lisovsky and Tikhomirov, 2002; Morozov, 1977; Voronin and Polivoda, 1967) suffer from not taking *the most important* BMC property into account.

Let the system on matter supply be the initial point. The system boundaries will be set so that the deadlock matter, which falls out of the cycle, would be left as system component. In this case such system is closed from the thermodynamic point of view (i.e. does not exchange matter with the external environment, only the energy).

The degree (coefficient) of the biotic material cycle closure (Gubanov and Degermendzhy, 2002, 2003; Tikhomirov et al., 2003) for supply systems is the ratio of the flow rate of the material coming from heterotrophic organisms to producers [$\dot{\Omega}$] to the sum of the flow rates of the material coming from heterotrophs to producers ($\dot{\Omega}$) and the material going to the dead end [$\dot{U}$]:

$$\text{Cl}_i = \frac{\sum_k \dot{\Omega}_{ik}}{\sum_k \dot{\Omega}_{ik} + \sum_l \dot{U}_{il}}, \tag{1a}$$

$$\text{Cl} = \frac{\sum_i \sum_k \dot{\Omega}_{ik}}{\sum_i \sum_k \dot{\Omega}_{ik} + \sum_i \sum_l \dot{U}_{il}}. \tag{1b}$$

where Cl_i, Cl are coefficients of the cycle closure for the ith biogenic element and the matter as a whole, k and l are all possible channels through which the substances move from heterotrophic organisms to producers and to the dead end. Obviously $0 \leq \text{Cl}i, \text{Cl} \leq 1$.

Systems *on matter supply* include artificial (microcosms), complex technical systems (e.g. biological life support systems, BLSS) and other natural systems, e.g. those without access to closed lakes, etc. (including biosphere).

A priori, we can assert that the degrees of closure on various elements are different. They become equal only in the case of complete closure, i.e. when all closure coefficients are equal to 1. From this assertion and determination of the cycle degree of closure (Eqs. 1) it follows that

$$\text{Cl}_j, \min < \text{Cl} < \text{Cl}_k, \max, \tag{2}$$

where Cl_j, min, Cl_k, max are the minimum and the maximum closure on some j and k elements, respectively. To all appearances, the correlation (2) is a general property.

From the formulas (Eqs. 1), we can see that the change of Cl goes in the same direction with change of Cl_i for any ith chemical element. But it is obvious that Cl change happens not necessarily unidirectionally with Cl_i change for any ith chemical element. It is obvious that, if on any account, the degree of closure of matter biotic cycle as a whole (Cl) stays invariable at the change of degree of closure on some ith element (Cl_i), then there should be the change of degree of closure (Cl_m) *at least on one element* m with the coefficient opposite to i element. Thus, saying in the form of the statement: *a change in the biotic cycle degree of closure on some element with the same coefficient of closure on the matter as a whole having the same sign results in the change in the closure degree of the cycle with opposite sign on other element (or elements).*

Obviously, this statement is the corollary of the law of matter conservation in the system, i.e. assignable.

3 Mathematical Model of Simple Homogenous Closed Ecosystem on Matter Supply

3.1 Base Model Description

Consider a mathematical model of a simplest homogenous biotic system with constant and limited amount of the matter turning over with a certain degree of closedness (large store system) (see Fig. 1):

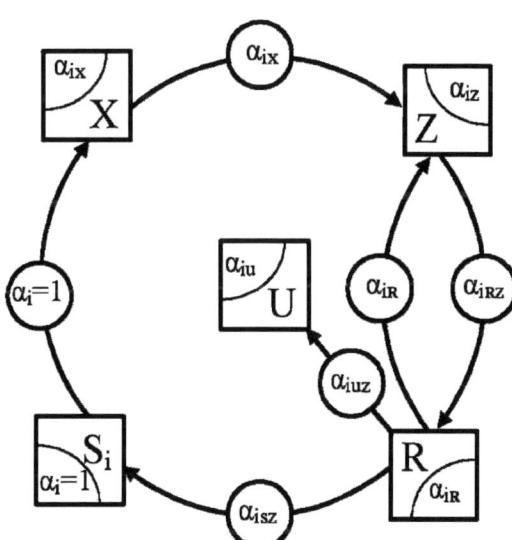

Fig. 1 Flows of substance in the model. *Arrows with circles* denote direction of flows of substance in the system, X producer, R reducer, Z dead organic matter, S_i the ith biogenous element in the environment, α_{iA} specific content of the ith element in A component or respective flow of the substance

$$\dot{X} = \mu_x X - \gamma_x X, \quad \dot{R} = \mu_R R - \gamma_R R,$$
$$\dot{Z} = \gamma_x X + \gamma_R R - \frac{\mu_R}{Y_{RZ}} R,$$
$$\dot{Si} = -\frac{\mu_x X}{Y_{Xi}} + \alpha_{iSZ}\varphi\mu_R R, \quad \dot{U} = \mu_R R\left(\frac{1}{Y_{RZ}} - 1 - \varphi\right), \qquad (3)$$

where μ are specific growth rates of respective organisms; γ are specific die-off (metabolism) rates; Y are the economic yield coefficients for organisms on the respective substrate; α iSZ is the specific content of the ith biogene in the flow of mineralized substance into the background; φ is the mineralization coefficient indicating the amount of dead organic matter mineralized by the reducer in increment rate unit.

It should be noted that for the sake of simplicity of analysis all biogenous elements feature identical properties (typical with respect to their involvement in the turnover). Special properties of such elements as C, O, H, Si and some other are given no consideration.

We should also note that deadlock U has no effect on establishment and sustenance of the cycle; to take it into account explicitly is important for this model from the viewpoint of control of balance of substances in the system, for the analysis of the biotic cycle formation mechanism and for the introduction of the biotic cycle closure measure—its closure degree (coefficient). The presence of the natural deadlocks for introduction of the biotic cycle closure term and measure has *the principal* meaning.

It is common knowledge that the element composition of organisms (specific content of elements in the biomass) is not constant and varies with food, conditions of existence, physiological condition of organisms, etc. For the sake of simplicity and without loss of generalization of reasoning these parameters (specific content) in the models are assumed to be constant.

Consider a more general case: *variability* of these parameter values to evaluate the effect of variability of the element composition of the biotic system components on the closure degree of matter cycle and, thus, on the feasibility of implementing closure of "ecological" type (see below).

If A is the biomass of a certain component, and α iA is the specific (relative) content of the ith component in A, then

$$\overset{\bullet}{(\alpha i_A A)} = \alpha \dot{i}_A A + \dot{\alpha} i_A A, \qquad (4)$$

where $\overset{\bullet}{(\alpha i_A A)}$ is the variation rate of the ith element in component A equal to the difference between the rate of arrival (involvement) of the ith element into A and its removal (elimination) from A. For example,

$$(\alpha_{iZ}Z)^{\bullet} = \alpha_{iX}\gamma_x X + \alpha_{iR}\gamma_R R - \alpha_{iZ}\frac{\mu_R}{Y_{RZ}}R,$$

and the rest by analogy.

From Eq.4,

$$\dot{\alpha}iA = \frac{(\alpha i_A A)^{\bullet} - \alpha i_A \dot{A}}{A} \qquad (5)$$

Then

$$\dot{\alpha}_{iX} = \mu_x\left(\frac{1}{YXi} - \alpha iX\right),$$

$$\dot{\alpha}_{iR} = \mu_R(\alpha_{iRZ} - \alpha_{iR}),$$

$$\dot{\alpha}_{iZ} = \frac{(\alpha_{ix} - \alpha_{iz})\gamma_x X + (\alpha_{iR} - \alpha_{iz})\gamma_R R}{Z}, \qquad (6)$$

$$\dot{\alpha}_{iU} = \frac{\mu_R R}{U}(\alpha_{iUZ} - \alpha_{iU}).$$

where $\alpha\ iRZ$ is the specific content of the ith element in the flow of the organic matter onto the reducer; $\alpha\ iuz$ is the specific content of the ith element in the flow of the matter into the deadlock.

Even if, in this case, α_{iR} and α_{iX} are assumed to be constant (then $\dot{\alpha}_{iX} = \dot{\alpha}_{iR} = 0$, $\alpha_{iRZ} = \alpha_{iR}$, $\alpha_{iX} = 1/Y_{Xi}$), then α_{iZ} and α_{iU} are variable when $\alpha_{iX} \neq \alpha_{iR}$.

As the components consist only of elements comprising the system only, and no other, then

$$\sum_i \alpha_{iX} = \sum_i \alpha_{iR} = \sum_i \alpha_{iZ} = \sum_i \alpha_{iRZ} = \sum_i \alpha_{iSZ} = \sum_i \alpha_{iUZ} = \sum_i \alpha_{iU} = 1 \qquad (7)$$

3.2 Conditions for Establishment of Biotic Matter Cycle

The system to be considered is thermodynamically closed, i.e. without exchange of the matter with the outside.

Then, since in this case the amount of the matter within the system is constant,

$$\alpha_{iX}X + \alpha_{iR}R + \alpha_{iZ}Z + Si + \alpha_{iU}U = Mi = \text{const}_i \qquad (8)$$

where M_i is the total amount of the ith element in the system for all components.

This equality is the corollary of the requirement of the thermodynamic "closedness" of the system:

$$(\alpha_{iX}\dot{X}) + (\alpha_{iR}\dot{R}) + (\alpha_{iZ}\dot{Z}) + \dot{S}i + (\alpha_{iU}\dot{U}) = 0, \qquad (9)$$

and the ratio (8) is the first integral of the system (Eqs. 3) on the ith element.

Substituting the right sides of Eqs. 3 into Eq. 9 and expanding Eq. 9 for each i, derive the "closedness condition" for the system, i.e. the conditions for the constancy of amount of each ith element in the system:

$$\alpha_{iU}Z\left(\frac{1}{Y_{RZ}} - 1 - \varphi\right) = \frac{\alpha_{iZ}}{Y_{RZ}} - \alpha_{iRZ} - \alpha_{iSZ}\varphi. \qquad (10)$$

These conditions only indicate that the system does not spontaneously generate or eliminate the matter, they do not demonstrate the nature of conditions sustaining prolonged operation of the system based on stable cycle of the matter.

What species (with what properties) are capable of establishing and sustaining biotic cycle of the matter? Find for the system (Eqs. 3) conditions required to maintain closure of biotic cycle and call them "closure conditions." If a system is capable of sustaining stable turnover of the matter for indefinitely long time (which means that *the time of matter turnover within a system is considerably smaller than the time of its existence, while in the limit this ratio equals zero*), it is the case of completely closed biotic cycle, i.e. *no deadlock formation*. Therefore, isolating deadlock U from the system (Eqs. 3) and imposing requirements analogous to Eq. 9:

$$((\alpha_{iX}\dot{X}) + (\alpha_{iR}\dot{R}) + (\alpha_{iZ}\dot{Z}) + \dot{S}i + (\alpha_{iU}\dot{U}))_{(\alpha_{iU}U) \to 0} = (\alpha_{iX}\dot{X}) + (\alpha_{iR}\dot{R}) + (\alpha_{iZ}\dot{Z}) + \dot{S}i = 0 \qquad (11)$$

have sought for conditions for the given element i—closure conditions (CC):

$$\frac{\alpha_{iZ}}{Y_{RZ}} = \alpha_{iRZ} + \alpha_{iSZ}\varphi \qquad (12)$$

Condition (11) is the method of increasing the biocycle closure in some system under study, here for the system (Eqs. 3) it is the left part of the equivalence (12).

When conditions (12) hold for all elements the system is "closed" for the entire matter on the whole; at the same time the "closedness conditions" (10) are met for every element i, whence in this case for the matter on the whole follows

$$\frac{1}{Y_{RZ}} = 1 + \varphi, \qquad (13)$$

which actually means total absence of deadlock formation (build-up).
For the system (Eqs. 3) it is easy to show that

$$\text{Cl}i = \frac{Y_{RZ}\varphi \alpha_{iSZ}}{\alpha_{iZ} - \alpha_{iRZ}Y_{RZ}}, \text{Cl} = \frac{\varphi Y_{RZ}}{1 - Y_{RZ}}. \tag{14}$$

If $\text{Cl}_i = 1$, the closure conditions (12) are fulfilled, and vice versa. For $\text{Cl} = 1$ condition (13) is fulfilled analogically. This indicates the *immanent* relation of closure conditions and the introduced closure measure, which implies that the closure coefficient on any element or matter on the whole can reach *1*, if the corresponding CC are realized, and vice versa.

Even though conditions (12) are fairly trivial (because of the simplicity of the system (Eqs. 3)), they mean that only species with such properties are capable of establishing a completely closed cycle of the matter. These conditions resemble a *stoichiometric* relation, but in the general case it is not so, because they depend on the method of deadlock formation and can involve kinetic characteristics of the species. Call such a type of closure of a biotic cycle—permanent maintenance of these conditions in invariable state— "ecological." Conditions close to the ratio (12) when sign "=" is replaced by sign "≈" are the conditions of quasiclosure and, accordingly, the cycling realized is quasiclosed.

Conditions sufficient to sustain the cycle result from the requirement of positiveness of cycle intensity. If the system structure is invariable this means that all living components are positive, here: $X, R > 0$. As it was demonstrated (Abrosov et al., 1976; Alekseyev, 1973; Svirezhev, 1978) for X and R to be positive for systems of this type it is necessary to fulfil condition $M_i > M_i^*$, where M_i^* is a certain critical value of the total (for all components) content of the *i*th biogenous element in the system. Then, the stability requirement is imposed.

Writing the variability rate of the deadlock substance in the system (Eqs. 3) element by element, we have

$$\dot{U}_i = \mu_R R \left(\frac{\alpha_{iz}}{Y_{Rz}} - \alpha_{iRZ} - \alpha_{iSZ}\varphi \right) \tag{15}$$

whence we can see that for element *i* not to form a deadlock it is required to fulfil condition (12), i.e. the same "closure condition" for the *i*th element, and vice versa, the absence of a deadlock leads to the fulfilment of CC.

Systems with completely closed biotic cycle are non-existent in nature. Yet, systems with the matter cycle are ubiquitous and the case in point is evaluation of completeness of this cycle, methods to increase its closure and intensity. In our case system of this kind can be formed (as follows from Eqs. 3) by the species for which the following conditions are fulfilled.

for the whole matter

$$1 + \varphi \leq 1/Y_{RZ,} \quad (16a)$$

for every element

$$Y_{RZ}(\alpha_{iRZ} + \alpha_{iSZ}\varphi) \leq \alpha i z. \quad (16b)$$

It is possible to demonstrate that when $\alpha_{iSZ} > \alpha_{iUZ}$, then

$$B_i > \alpha_{iZ} > C_{i,} \quad (17)$$

where

$$B_i = \alpha_{iSZ}(1 - Y_{RZ}) + \alpha_{iRZ} Y_{RZ}, C_i = Y_{RZ}(\alpha i_{RZ} + \varphi \alpha_{iSZ}). \quad (18)$$

if $\alpha i_{UZ} > \alpha_{iSZ}$, then

$$A_i > \alpha i_Z > B_i, \quad (19)$$

where

$$A_i = 1 - Y_{RZ}(1 + \varphi) + Y_{RZ}(\alpha i_{RZ} + \varphi \alpha i_{SZ}) \quad (20)$$

These correlations (17–20) result from the law of conservation of the quantity of matter in the closed system and add to correlations (10) and (16) for the requirement of formation and sustenance of matter cycle in a natural manner in the model—biotic cycle.

Particularly, the choice of coefficients for the model of initial conditions should meet correlations (10), (16), (17) and (19). Otherwise, various "exotic" regimes of model system functioning can occur in the *model* and be very similar and hardly differentiated from the real regimes. However, they would still be "exotic" going, so to say, with the conservation law disruption.

Yet, as $\sum_i \alpha_{iSZ} = \sum_i \alpha_{iUZ} = \sum_i \alpha_{iZ} = 1$, for certain elements j fulfilled are correlations (17) and at the same time for other elements k realized are correlations (19). This fact means that the larger is the part of some element to form the deadlock (removed from the turnover) the larger is to be its part in the dead organic matter to maintain the turnover of the matter and vice versa. Noteworthy is the buffer role of detritus (Odum, 1975).

4 Examples of Closure Coefficients Calculation

Closure coefficients can be calculated for different systems. Experimental investigations with the microecosystems presenting bacterial communities with *Spirulina platensis* as a producer were conducted Bolsunovsky et al.,

1997; Bolsunovsky, 1999; Bolsunovsky and Kosinenko, 2000. The aim of the experiments was to find stable regimes of cultivation of spirulina at different conditions and obtain appropriate kinetic dependencies and characteristics. The culture *Sp. platensis* used in the experiments was grown on a Zarruk nutrient solution (Pinevich et al., 1970) in the absence of phosphorus and subsequently diluting the culture with the phosphorus-free Zarruk nutrient in the culture filtrate. In the definite periods of the experiments additional portions of mineral or organic phosphorus were added into flasks with *Sp. platensis* culture. Thus in the experiments the systems in which intracellular poll of spirulina was filled rather completely (well-formed), moderately (filled) and poorly (unfilled) were investigated. The effect of different forms of nitrogen on the spirulina growth was studied in two variants of the experiment. In the first variant ammonium chloride (NH_4Cl) was used as a source of nitrogen, in the second $(NH_2)_2CO$. Development of spirulina was controlled daily by registering changes in the optical density and dry weight of the culture.

On the basis of the model (Eqs. 3) mathematical models were constructed where *Sp. platensis* was taken as a producer and its growth rate was limited by the intracellular concentration of phosphorus or carbamide nitrogen. It was shown that the mathematical models were in good agreement with the experimental data.

Figures 2–4 show the calculated dynamics of closure coefficients for different variants of experimental systems limitation. Notably, the variant for the system with non-filled phosphorus pool and without compensating it by adding phosphorus (012410, Fig. 2) is rather exotic—closure coefficient is more than 1. In this case the required correlation of primary conditions for the law of matter conservation in the system was disrupted specially for demonstration, conditions (16–20). The figures indicate that the closure coefficient can be the marker of correct choice of model coefficients and their primary values.

Analysis of Fig. 4 shows that the dynamics of biocycle closure coefficient on nitrogen for studied systems, in which *Sp. platensis* development is limited by

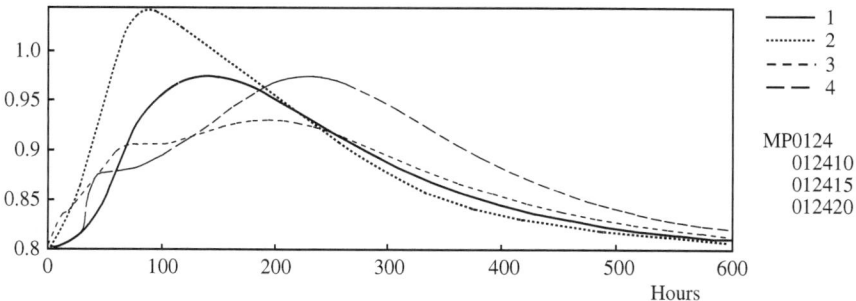

Fig. 2 Dynamics of coefficient of system closure in terms of phosphorus (calculated) for well-formed pool (MPO124) and for the system with unfilled pool (012410, 012415, 012420). It is obvious that the version without compensation is exotic: closure is higher than 1

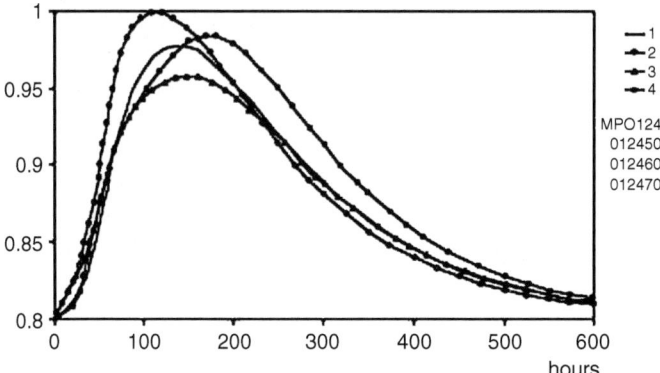

Fig. 3 Calculated dynamics of coefficients of cycle closure for systems with filled phosphorus pool. All coefficients are correct

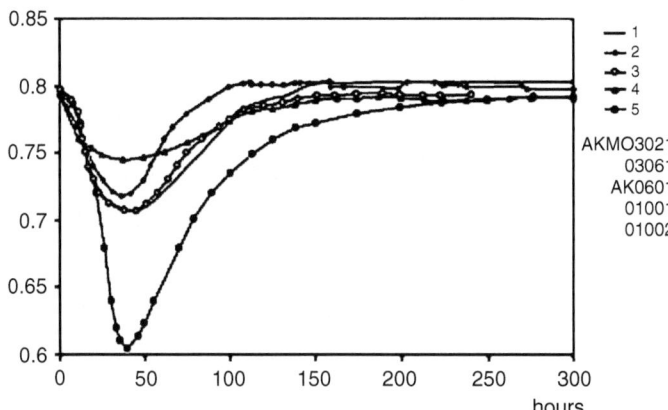

Fig. 4 Calculated dynamics of closure coefficient for systems with *Spirulina* growth limited by carbamide nitrogen in the medium

carbamide nitrogen, differs from the dynamics of the similar closure coefficient for phosphorus for systems limited by phosphorus (Figs. 2 and 3). The example of this difference is principal and important. For BLSS with crew, in case of emergency leading to matter cycle disruption, return of the biocycle closure degree back to primary state (required) is preferable according to the curve in Figs. 2 and 3 than according to Fig. 4, because the state of the crew will be determined by the depth of closure degree fall and its restoration rate. This is the advantage of the suggested dynamic coefficient of cycle closure, which allows estimating non-standard situations in dynamics, which is sometimes useful.

It is noteworthy that the calculated closure coefficients on phosphorus were higher than on nitrogen.

Biotic Turnover in Superorganism Systems

5 On Realizability of Some Mechanisms of Sustaining Highly Closed Cycle of the Matter

5.1 Ecological Principle of Biological Cycle Closure

Consider now the feasibility of the "ecological" principle (type) of closure.

As Cl_i for the system (Eqs. 3) is a monotonically decreasing hyperbolic function of α_{iz}, passing all its values from $\breve{\alpha}_{iz}$ to 1, it achieves its maximum ($=1$) and minimum ($=\breve{\alpha}_{iz}$) values at the boundary of the interval on which the function is defined, i.e. by conditions (12), (18) and (19) with α_{iz} equal to C_i or A_i, respectively. In other words, the matter cycle cannot be closed for some element so that the specific content of this element in the forming non-living organic matter could naturally be adjusted by the organisms in the course of functioning to completely close in terms of this element, i.e. to have a certain degree of freedom = to be realized in terms of α_{iz} within the interval from C_i to A_i, and not at the boundary of the interval on which the function is defined with $\alpha_{iz} = C_i$. The requirement is quite rigorous, but with respect to $Cl_i = 1$ it seems to have generality. Though if this can be realized, it would be almost impossible to sustain it for a long period at the Cl_i definition (= at the point) boundary.

The complete closure can be most "simply settled" under the requirement for the specific content of each biogene both in the producer and the reducer to be constant and equal to each other. Hence it is clear why the biochemical composition maintained by nature in different organism species is on the whole approximate. At the same time, as organisms with absolutely identical qualitative and quantitative element composition are non-existent in nature, biotic systems truly closed in matter cycle can hardly be realized, i.e. speaking about "closure" as a desirable property we should mean "quasiclosure."

Thus, in biotic systems the "ecological" type of the matter cycle can be realized under very rigorous requirements and is highly susceptible to external disturbances. This means that with large times it is difficult for ecosystems with poor species composition to sustain high closure of the cycle, provided that it is sustained "ecologically" only. Besides, the degree of cycle closure being fixed in the characteristics of species that comprise the system hides from the view possible avenues to increase it.

5.2 "Evolutionary" Type of Closure

Given the poor feasibility of "ecological" type of closure, what are the other ways and opportunities to enhance closure of matter cycle in biotic systems? How can a cycle with high closure form? For example, can the Darwinian evolution (in this case it is selection by growth rates) affect the matter cycle closure coefficient? The answer seems to be the following: If the evolution (selection) is going by individual signs unconnected (not correlating) with the signs responsible for the closure of the matter cycle the Darwinian evolution does change the degree of closure of the matter cycle. So, the way to enhance biotic closure evolutionally must be different.

Natural existence of ES generates deadlocks—substances which are not transformed by the existing set of species. Assuming that a new species emerges in the process of evolution and involves these matters in cycling but generates new deadlock, i.e. the trophic chain length increases = "vertical" species formation occurs (Fig. 5).

Figure 6 shows the combined model dynamics of the producer and three sequentially occurring reducers for the model of type (Eqs. 3) constructed

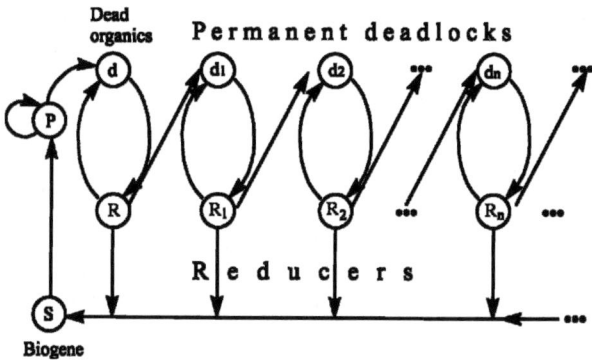

Fig. 5 General scheme of cycle closure evolutionary principle

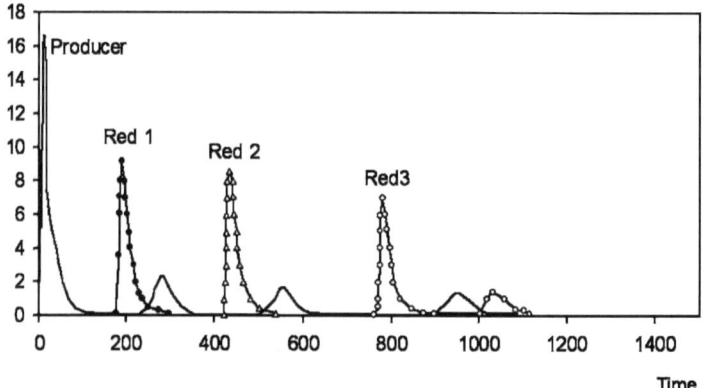

Fig. 6 Combined dynamics of producer and three sequentially occurring reducers for scheme in Fig. 2

according to the scheme of Fig. 5 (we will mark this model 3* and will not display it here).

Feasibility of such an "evolution" mechanism increasing the cycle closure is based on the assumption of convergence of series of biomass values of emerging and co-existent reducer species and deadlocks generated by them. Analysis of such a diagram in Fig. 5 with the model of 3* type disproves this assumption—the diagram is realized for the infinitely large value of the total mass of the system. Besides, the feasibility of such a diagram is limited by dissipation of energy along the trophic chain (Svirezhev, 1978).

5.3 Interaction of Microevolutionary Population Parameters

Many problems of population microbiology (especially problems of microevolution) are solved based on description of population dynamics and DDGF equations (equations in terms of density-dependent and growth-controlling factors; Adamovich et al., 1987; Degermendzhy et al., 1979), which involve various kinetic microparameters of populations characterizing different relations of organisms to environment. These parameters include maximum specific growth rates, coefficients of substrate consumption (yield), Michaelis–Menten coefficients, coefficients of various released materials, etc. By "colliding" simple populations with varied microparameters in the models, we study the consequences of such collision (removal, co-existence, dominance). In this case, it is considered that these microparameters are *independent* and, correspondingly, we can get the result when the predominant population grows quicker and has higher substrate application coefficient and other advantages.

Here a principal question arises about the degree of population microparameters relation, when one of them varies due to, e.g. mutations. If adequate model would exist and describe the inter-related change of these microparameters, these questions could be answered. However, such models do not exist. Even in regard to a frequently used microparameter—maximum specific growth rate—there are several different schemes for description of specific growth rate dependence on DDGF: enzymatic, populational, etc. However, it follows from the enzymatic scheme that the hypothesis on *independence* of microparameters is beneath all criticism. Let us consider this in detail. Let the mechanism of specific growth rate dependence upon the limiting substrate described by real enzymatic "bottleneck" reaction (Fig. 7) be

$$g = \mu S / (K_S + S) \tag{21}$$

where g is the specific growth rate, μ is the maximum reaction rate ($\mu = K_2 E$), K_S is Michaelis–Menten constant ($K_S = (K_2 + K_{-1})/K_1$).

Obviously, the mutations change the microparameters (K_1, K_2, K_{-1} and E). If, for example, K_2 increases, both μ and K_S will increase as well, i.e. the curve of

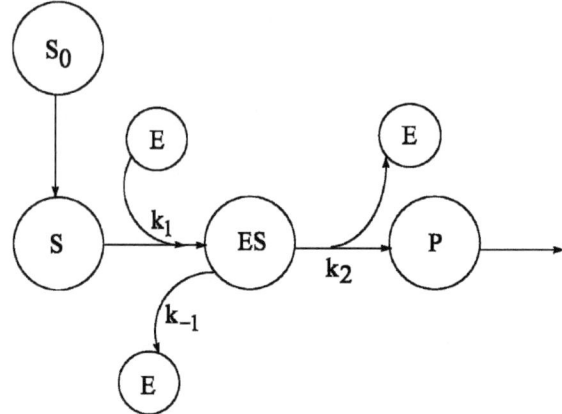

Fig. 7 Block diagram of "bottleneck" enzymatic reaction. S_0 input concentration of substrate, S background concentration of substrate, E general concentration of enzyme, ES enzyme–substrate complex, P final product concentration (biomass), k_1 rate of enzyme–substrate complex formation, k_{-1} back reaction rate, k_2 rate of product, P formation (biomass)

specific growth rate dependence will become more plain, but will rise at high S (Fig. 8). Change of K_1 will influence only K_S, whereas μ will not change (Fig. 9). The number of such examples can be continued. Consequently, even the simplest models show that microparameters are *independent* of one type of mutation (on K_1, K_{-1}, E) and *dependent* on the other type (K_2).

The "direct" way of solving this problem depends upon the type of chosen cellular or populational level model describing the microparameters similar to the above. However, when the knowledge about the whole model of cellular biosynthesis and its regulation is not enough, this way cannot be used. There exists an alternative method of phenomenological type without analysis of the relation mechanisms between the microparameters yet, though based on statistical analysis. Imagine we have a collection of one strain mutants with calculated microparameters. Then, the multi-dimensional regression parametrical

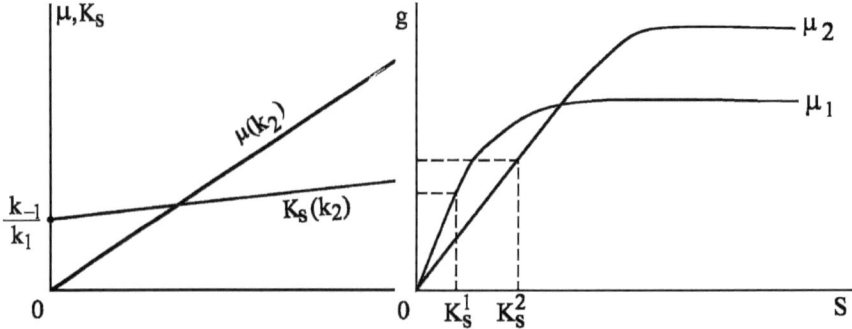

Fig. 8 "Mutation" increase in microparameter k_2 (constant of product formation—biomass) increases the macroparameters μ and k_S and, correspondingly, deforms the growth rate g

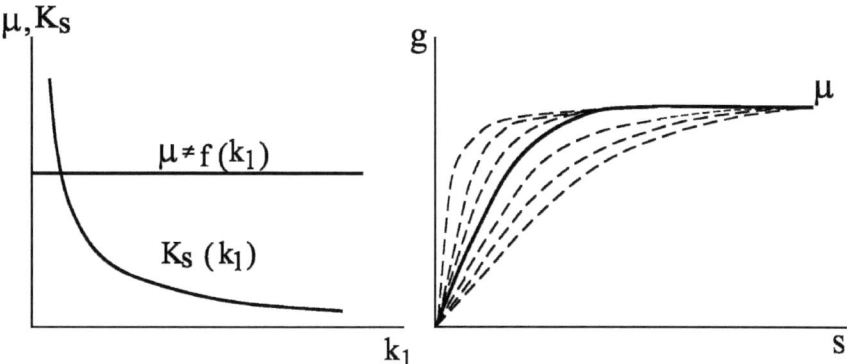

Fig. 9 Increasing k_1 does not change μ but decreases/increases k_S; correspondingly, the fan of growth rate curves g is formed with constant maximal specific rate μ

analysis will solve the set problem on the phenomenological level. This can be realized using the electronic databases on kinetic microorganism parameters from museum collections. The study of different "mechanistic" models of microparameter (or microprocesses) relations can be conducted in a parallel way.

Obviously, the cycle closure can be determined by the degree of dependence between population microparameters in the process of their evolutionary changes. Therefore, this question requires specific consideration in the context of the discussed problem.

6 General Degree of Biocycle Closure

Let us *generalize* the notion about the BMC closure degree on any biosystem with cycle, *including natural ones*.

The measure of *effective* (visible) *closure* of matter cycle in superorganism systems, which will integrate the biotic closure and thermodynamic *openness* in systems, will be introduced analogically to correlations (1). For systems open on the matter let us indicate through Cl_i^* the following value:

$$Cl_{i*} = Cl_i = \frac{\sum_k \dot{\Omega}_{ik}(Aia, Bib)}{\sum_k \dot{\Omega}_{ik}(Aia, Bib) + \sum_l \dot{U}_{il}(Aia, Bib)} \qquad (22)$$

where a are the channels transferring the matter into the system from outside; b the channels transferring the matter from the system, *except deadlock matters*; Aia the rate of ith biogenic element introduction into the system from outside through channel "a," e.g. migrations, inflow, artificial correction, etc.; Bib the rate of ith biogen removal (taking out) from the system through channel "b," e.g. yield with complete or partial removal from the system, emigration flows, current, etc.

In this case, ($\dot{\Omega}$) is the rate of any biogenic element flow from the processes, *in which there is production of biogenic elements in the ecosystem* and, in this or that way, *are intercepted by producers*, whereas in numerator (and correspondingly in the denominator) there is the sum of all biogen flows produced (constructed) within various processes in CC, and *not only related to heterotrophic processes* and directed to producers.

Further, let Ai be the rate of ith biogen introduction into the system from outside through all channels "a" ($A_i = \sum_a Aia$), Bi the rate of ith biogen taking out from the system through all channels "b" ($B_i = \sum_b Bib$). Then the degree (coefficient) of the biotic matter cycle closure (on ith element) for any system *in general case* will be the following:

$$\text{Cl}_i = \lim_{Ai, Bi \to 0} \text{Cl}_i^* = \lim_{Ai, Bi \to 0} \frac{\sum_k \dot{\Omega}_{ik}(Ai, Bi)}{\sum_k \dot{\Omega}_{ik}(Ai, Bi) + \sum_l \dot{U}_{il}(Ai, Bi)} \qquad (23)$$

All is analogically for Cl, coefficient of biotic cycle closure on matter on the whole:

$$\text{Cl} = \lim_{A, B \to 0} \text{Cl}^* \qquad (24)$$

where Cl^* is the coefficient of the *efficient system closure* on material on the whole; A is the rate of material introduction into the system on all "i" elements and through all "a" channels; B is the rate of the material elimination (removal) from the system through all "b" channels and on all "i" elements.

Hence, the degree of biotic cycle closure in superorganism systems (of ecosystem level and higher) for any biogenic element and matter on the whole is defined by correlation of the biogen flow rate *from all processes which involve production of biogenic elements in the biosystem and are directed (biogens) to producers (or, in this or that way, are intercepted by producers)* to the sum of rates of the same flow and matter coming to the deadlock.

Thus, the degree (coefficient) of biotic cycle closure for some chemical element in any superorganism system with the matter cycle is determined in general according to Eq. 23 taking into consideration Eq. 22, and for the matter on the whole according to Eq. 24, considering the formula (1b).

Let us emphasize the moments which allowed introduction of general measure of the biotic matter closure for any superorganism system with biotic matter closure: the presence of produced and reduced matter flows in the system (at that, the introduction of the reduced matter part is directed exactly to producers); obligatory naturally (biotically) formed deadlocks; marking of system boundaries; separation of thermodynamic and biotic closure terms.

Notably, the characteristic of closure in the present form is dynamic and allows estimating the closure degree on any element and matter on the whole, current, average and stationary, and comparing the systems according to these indices and making the respective decisions and actions, if necessary.

This measure helps to increase the biocycle closure degree on any biogenic element and matter on the whole for *any* superorganism system, including biosphere and BLSS (in practice).

7 Conclusions

Thus, by investigating the role of biotic turnover in the economy of nature, we discern two major characteristics—closedness and intensity. Closedness specifies the turnover mostly as a phenomenon, while the intensity as a process. The fact that the species composition of the ecosystems is not accidental and is defined in certain links-relations between the characteristics of species makes the system and forms the biocycle (*co*evolutionary characteristics and relations—analogue of stoichiometric correlations). Demonstration and registration of such relations—conditions of biotic cycle formation and maintenance and its complete closure conditions (CC)—is the most important and necessary moments of analysis at the mathematical description of superorganism systems. The record of these correlations in terms of species' kinetic characteristics allows answering the following questions: what organisms (with what characteristics) are able to form and sustain highly closed matter biocycle in natural and artificial biosystems and what are the ways of changing the biotic closure of these ecosystems? It allows avoiding "exotic" functioning regimes in *models* which break the law of matter conservation in the system.

The general measure of biotic matter cycle closure in superorganism systems of ecosystem (and higher) level for any ith biogenic element (Cl_i) and for the matter on the whole (Cl) was suggested. The *immanent* relation of the *suggested* measure of biocycle closure and closure conditions was mentioned, meaning that the closure coefficient on some element or matter on the whole can reach 1 if the corresponding CC are realized, and vice versa.

Several mechanisms of formation and sustenance of the matter biotic cycle in superorganism systems were suggested and studied. The ways related to increase of buffer volume of the system on total amount of matter within and on "horizontal" species diversity—formation of trophic nets—would be taken into consideration.

The evolutionary mechanism of sustaining high degree of matter cycle closure (at least global) based on the high buffer state of matter content in the system and realized practically according to trigger scheme by switching from one global limiting element to another with further introduction of the first into the cycle seems interesting. For example, this is the excretion of carbon from the global cycle in the form of coal, oil and gas with further return back to the cycle or interchange of redox properties, e.g. periodic land lowering with further sinking and vice versa. At least, such mechanics explains the paradox that the cycle closure degree (even high) is always subunit and the global system exists and prospers for a long time. We can say that in such mechanism the structural instability generates functional stability.

And, finally, we can suggest another mechanism illustrating the mechanics of cycle formation and retention in all superorganism systems from ecosystem to biosphere. It is noteworthy that natural ES are not 100% closed on biocycle and have the compensatory mechanism in the form of open exchange with other systems = with the environment. This mechanism works when closed on the biosphere level on the whole—biosphere is corrected by matter inflow as a result of volcanic activity. This is the mechanism which compensates non-complete closure of the biosphere and allows its long-term existence. Indeed, there are data showing that the volcanic activity, and underwater firstly, is substantial on our planet. Thus, the correlation of biogenic and abiogenic oxygen in the atmosphere, which is showed by the isotope composition of atmospheric oxygen, is approximately 50% to 50% on the mass (Bgatov, 1985). Degassing basalt magmatic rocks are the suppliers of free oxygen in the atmosphere. At that, the activity of underwater volcanos is very strong. For balancing the isotope composition of atmospheric oxygen, it would be sufficient if deep waters of the ocean supplied only 0.5% of diluted oxygen into the atmosphere per year. By mass, it is equal to the mass of oxygen formed in the atmosphere every year by photosynthesis. A similar situation occurs with other elements. Thus, solfatars of only one Ebeko volcano (Kamchatka) excrete the following into the atmosphere every year (in tons): 4,381,830 of H_2O, 1,078,790 of CO_2, 11,530 of SO_2, 5150 of HCl, 2880 of H_2S, 1830 of N_2, 330 of CH_4, 10 of H_2 (Bgatov, 1985). This requires special consideration but goes out of the scope of this work.

Therefore, we can state the existence of ecological and evolutional mechanisms of maintaining the highly closed matter biocycle in superorganism systems and compensatory mechanisms, which possess an exogenous character for ecosystems and endogenous for biosphere. Special periods of processes flowing in different biosystems are very important. Therefore, the hierarchical analysis of various types of cycle closure becomes important as well. The current closure is maintained by the "ecological" type of closure. The material for formation and maintenance of the biocycle high closure is provided by the evolutionary process and, correspondingly, the "evolutionary" type of closure occurs during the evolutionary periods. And, finally, the closure on a geological time-scale should be taken into account. Analysis of principles and regularities of these types of enclosures and, respectively, connecting (packing) their processes seem a prospective problem of theoretical biology and important contribution in understanding the formation of system relations and various, though, hierarchically packed, biosystems.

References

Abrosov, N.S., Gubanov, V.G. and Kovrov, B.G. (1976) A theoretical study of closed ecosystem functioning. Izv. SO AN SSSR 15. Ser. Biol. Vyp. 3, 3–9 (in Russ.).

Adamovich, V.A., Terskov, I.A. and Degermendzhy, A.G. (1987) Effect of autostabilization of growth controlling factors and interactions in community. Doklady AN 295, 1236–1239 (in Russ.).

Alekseyev, V.V. (1973) Dynamic stability of water biocenoses. Vodnye resursy 3, 156–166 (in Russ.).

Bartsev, S.I. (2002) Systematic approach to construction of closed ecological life support systems. In: J.I. Gitelson and N.S. Pechurkin (Eds), Ecological Biophysics. Logos, Moscow, vol. 3, pp. 190–226 (in Russ.).

Bartsev, S.I., Mezhevikin, V.V., Okhonin, V.A., Doll, S.C. and Rao, N.S. (1996) Life support system (LSS) designing: principle of optimal reliability. 26th ICES, SAE Paper 961365.

Beyers, R.J. (1964) The microcosm approach to ecosystem biology. Am. Biol. Teacher 26, 491–498.

Bgatov, V.I. (1985) The History of Earth Atmosphere Oxygen. Nedra, Moscow (in Russ.).

Bolsunovsky, A.Ya. (1999) Microalgae in culture and reservoirs: Ecological and biophysical mechanisms of predomination. Abstract of Ph.D. thesis in biology. Krasnoyarsk (in Russ.).

Bolsunovsky, A.Ya. and Kosinenko, S.V. (2000) Intracellular pool of cyanobacteria *Spirulina platensis* culture phosphorus. Microbiology 69(1), 135–137 (in Russ.).

Bolsunovsky, A.Ya., Kosinenko, S.V. and Khromechek, E.B (1997) Study of the oxygen regime of microalgae *Spirulina platensis* culture. Biotechnology 7–8, 60–68 (in Russ.).

Burdin, K.S. (1978) Application of man-made microecosystems to solve theoretical and applied problems of ecology. In: V.D. Fedorov (Ed.), Man and Biosphere. MGU, Moscow, pp. 124–144 (in Russ.).

Carpenter, S.R. (1996) Microcosm experiments have limited relevance for community and ecosystem ecology. Ecology 77, 677–680.

Degermendzhy, A.G., Pechurkin, N.S. and Shkidchenko, A.N. (1979) Autostabilization of Growth-Controlling Factors in Biological Systems. Nauka, Novosibirsk (in Russ.).

Finn, J.T. (1976) Measures of ecosystem structure and function derived from analysis of flows. J. Theor. Biol. 56, 363–380.

Finn, J.T. (1978) Cycling index: a general definition for cycling in compartment models, Environmental chemistry and cycling processes. DOE Symposium Series 45, CONF 760429, National Technical Service, Spring-field, Virginia, USA, pp. 138–164.

Folsome, C.E. and Hanson, I.A. (1986) The emergence of materially closed system ecology. Ecosystem Theory and Application. Wiley, New York, pp. 269–288.

Gitelson, J.I., Kovrov, B.G., Lisovsky, G.M., Okladnikov, Yu.N., Rerberg, M.C., Sidko, F.Ya. and Terskov, I.A. (1975) Experimental Ecological Manned Systems. Problems of Space Biology. Nauka, Moscow (in Russ.).

Gitelson, J.I., Lisovsky, G.M. and MacElroy, R. (2003) Manmade Closed Ecological Systems. Taylor & Francis, New York.

Gubanov, V.G. and Degermendzhy, A.G. (2002) Biophysical investigations of aquatic ecosystems. In: J.I. Gitelson and N.S. Pechurkin (Eds), Ecological Biophysics. Logos, Moscow, vol. 2, pp. 247–359 (in Russ.).

Gubanov, V.G. and Degermendzhy, A.G. (2003) Closure of biotic matter cycle in superorganism systems (methodological aspect). Degree of biotic closure. In: T.G. Volova (Ed.), Ecological Biophysics. SB RAS, Novosibirsk, pp. 318–333 (in Russ.).

Gubanov, V.G., Kovrov, B.G. and Fishtein, G.N. (1984) Closed microecosystems—a new test-object for biophysical and ecological research. In: I.A. Terskov (Ed.), Biophysical Methods of Ecosystem Research. Nauka, Novosibirsk, pp. 34–44 (in Russ.).

Kamshilov, M.M. (1966) Organic matter cycle the problem of life sense. ZhOB 27(3), 282–298 (in Russ.).
Kamshilov, M.M. (1979) Evolution of Biosphere. Nauka, Moscow (in Russ.).
Kovrov, B.G. and Fishtein, G.N. (1978) Experimental closed microecosystems containing unicellular organisms. Continuous Cultivation of Microorganisms, 7th International Symposium, Prague.
Lisovsky, G.M. and Tikhomirov, A.A. (2002) Subject and principles of constructing closed ecological systems. In: J.I. Gitelson and N.S. Pechurkin (Eds), Ecological Biophysics. Logos, Moscow, vol. 3, pp. 133–154 (in Russ.).
Morozov, G.I. (1977) Theoretical Fundamentals in Designing Life Support Systems. Problems of Space Biology. Nauka, Moscow (in Russ.).
Nelson, M., et al. (1993) Meaning of "Biosphere-2" in studying the ecosystem processes. Vestnik RAN 11, 1024–1034 (in Russ.).
Nixon, S.W. (1969) A synthetic microcosm. Limnol. Oceanogr. 14, 142–145.
Odum, Eu. (1975) Fundamentals of Ecology. Mir, Moscow (in Russ.).
Pechurkin, N.S. (1982) Energy Aspects of Super-organism Systems Development. Nauka, Novosibirsk (in Russ.).
Pechurkin, N.S. (1988) Energy and Life. Nauka, Novosibirsk (in Russ.).
Pinevich, V.V., Verzilin, N.N. and Mikhailov, A.A. (1970) An investigation into *Spirulina platensis*—a novel subject for a highly intensive cultivation. Plant Physiol. 17, 1037–1046 (in Russ.).
Sugiura, K. (1988) A materially-closed aquatic-ecosystem: a useful tool for determining changes of ecological processes in space. Biol. Sci. Space 15(2), 115–118.
Svirezhev, Yu.M. (1978) On the length of trophic chain. ZhOB 39(3), 373–379 (in Russ.).
Taub, F.B. (1969a) A biological model of freshwater community: a gnotobiotic ecosystem. Limnol. Oceanogr. 14, 136–142.
Taub, F.B. (1969b) Gnotobiotic models of freshwater communities. Verh. Int. Ver. Limnol. 17, 485–496.
Taub, F.B. (1974) Closed ecological systems. Annu. Rev. Ecol. Syst. 5, 139–160.
Tikhomirov, A.A., Ushakova, S.A., Manukovsky, N.S., Lisovsky, G.M., Kudenko, Yu.A., Kovalev, V.S., Gubanov, V.G., Barkhatov, Yu.V., Gribovskaya, I.V., Zolotukhin, I.G., Gros, J.B. and Lasseur, Ch. (2003) Mass exchange in an experimental new-generation LSS model based on biological regeneration of environment. Adv. Space Res. 31(7), 1711–1720.
Vernadsky, V.I. (1940) Biogeochemical Essays. AN SSSR, Moscow-Leningrad (in Russ.).
Vernadsky, V.I. (1967) Biosphere. Mysl, Moscow (in Russ.).
Voronin, G.I. and Polivoda, A.I. (1967) Life Support of Space Crafts Crew. Machine construction, Moscow (in Russ.).
Yablokov, A.V. and Yusufov, A.G. (1989) Evolution Teaching (Darwinism). Vysshaya shkola, Moscow (in Russ.).

Chromosomes and Continents

I. I. Kiknadze, L. I. Gunderina, M. G. Butler, W. F. Wuelker, and J. Martin

Abstract The high level of inversion polymorphism and, correspondingly, the abundance of inversion banding sequences (BSs) of polytene chromosomes in the banding sequence pool of *Chironomus* species permit scientists to reconstruct the cytogenetic evolution of the genus and to evaluate the role of structural rearrangements in the genome during population divergence and speciation. We performed a quantitative assessment of the important role of inversion polymorphism in the differentiation of natural populations and demonstrated the adaptive significance of different gene orders in populations of species occurring in different regions. For the first time, it has been shown that the BS pools of populations of the same species on different continents differed much in the sets and frequencies of gene inversion orders. BS pools of populations on each continent were found to contain continent-specific BSs in addition to sequences occurring on several continents. This intraspecies diversity of the linear organization of the genome is one of the major factors maintaining the evolutionary stability of species in dramatically different environments. In addition to endemic species-specific sequences, the BS pool of the genus *Chironomus* contains sequences common for different species, cytocomplexes, and continents. These sequences, termed basic sequences, are very important for reconstruction of genome divergence in the course of evolution. It is suggested that they are close to the initial primitive sequences existing on ancient supercontinents, whereas continent-specific BSs were formed after continent separation. Comparison of all currently known BSs in the sequence pool of the genus *Chironomus* showed that the genomes of the most distant species differed by more than 90 inversion breaks, causing changes of their linear structure. In such cases, conserved genome regions span about 10 bands.

I. I. Kiknadze
Institute of Cytology and Genetics SB, RAS, Novosibirsk, Russia
e-mail: kiknadze@bionet.nsc.ru

1 Introduction

As shown by cytogenetic analysis of chromosomal evolution, the divergence of animal karyotypes is mediated mainly by para- and peri-centric inversions, altering the gene order in linkage groups (Dobzhansky, 1970; Eichler and Sankoff, 2003; King, 1993; Navarro and Barton, 2003; White, 1977). Comparative genomics confirms this conclusion on the basis of molecular analysis of genomes and proteomes. The genomes of distant species, such as man, mouse, Drosophila, and Anopheles, differ mainly by the order of genes in linkage groups rather than the number and the set of genes (Ayala and Coluzzi, 2005; Marques-Bonet et al., 2004; Zdobnov et al., 2002).

Alteration of gene order in chromosomes during speciation can be traced visually in dipterans, which possess polytene chromosomes with distinct banding sequences (Beermann, 1972; Bridges, 1935; Painter, 1934; Zdobnov et al., 2002). Species of the genus *Chironomus* (Diptera, Chironomidae) are particularly appropriate for this purpose. They occur on all continents (except for Antarctica) with highly variable environments and include numerous groups of sibling species, whose members differ in the amount of cytological or genetic change that has accompanied speciation. By global analysis of banding sequences (BS) in the genus *Chironomus*, we have traced gene order change during species divergence and dispersal. We studied BSs in 46 Eurasian endemic species (the Palearctic biogeographic subzone), 14 North American species (the Nearctic subzone), five Holarctic species, whose populations inhabit two continents in two subzones, and 24 endemic species from Australia and New Zealand (Australasian biogeographic zone). In addition, we invoked data on BSs studied in four Central African species (Ethiopian biogeographic zone). The species studied belonged to two subgenera: *Chironomus* and *Camptochironomus*.

Before proceeding to the characteristics of BS divergence on different continents revealed by global analysis, let us consider the main parameters of the structures of karyotypes and BSs in species of the genus *Chironomus*.

2 Karyotype Structure in the Genus *Chironomus*

Species of the genus *Chironomus* possess seven conserved linkage groups, corresponding to seven chromosome arms (A, B, C, D, E, F, and G). The chromosome arms typically form four chromosomes in the *Chironomus* karyotype ($n = 4$). Different species possess different arm combinations: chromosomes AB CD EF G in the thummi cytocomplex, AE CD BF G in the pseudothummi cytocomplex, AB CF DE G in the camptochironomus cytocomplex, etc. In rare cases, the acrocentric chromosome is not arm G, e.g., the columbiensis group with AG, CD, BF and E. A few species of the genus have reduced chromosome numbers ($n = 3$, $n = 2$) owing to fusion of short arm G with another arm (Keyl, 1962; Kiknadze et al., 1989, 1991, 1996a; Martin, 1979; Michailova, 1989; Wuelker, 1980).

The seven linkage groups correspond to seven main BSs in the karyotype of each species. These main BSs are the most abundant in natural populations of a species. As the karyotypes of the overwhelming majority of *Chironomus* species are highly polymorphic due to various inversions, alternative BSs occur in each species in addition to the seven main ones. They can occur at relatively high frequencies, although not in all populations. There are also rare and unique BSs, found in the heterozygous state in occasional larvae and in few populations. Each arm sequence is designated with a unique identifier, which includes an abbreviation of the species name and the BS number. For example, arm A of *C. plumosus* has 12 BSs: pluA1, pluA2,..., pluA12; arm B has 8 BSs: pluB1,..., pluB8, etc. The set of BSs found in a particular species forms its banding sequence pool. The pool size depends on the level of chromosome polymorphism of the species. In the genus *Chironomus*, the pool size within a species varies within the range 7–60 BSs.

Speciation is accompanied by karyotype divergence mediated by the fixation of various inversion BSs. The role of chromosome variability in speciation is most clear in groups of sibling species. For example, karyotypes of virtually all of about a dozen species belonging to gr. *plumosus* have common BSs in some arms, but other arms differ by species-specific fixed inversion sequences. The number of arms with homologous BSs varies from one to six per genome. Correspondingly, the more homologous arms that are conserved between the karyotypes of species, the closer are these species, and vice versa. This conclusion is confirmed by Nei's cytogenetic distances. Species divergence can be enhanced by variability of frequencies of homologous BSs (Gunderina and Kiknadze, 2000; Gunderina et al., 1999).

Common BSs do not occur only in sibling species of the genus *Chironomus*. They can be found between more distant species, even those belonging to different cytocomplexes (Keyl, 1962; Wuelker, 1980). Such BSs found in different cytocomplexes are termed "basic" and considered to be primitive, existing before the formation of cytocomplexes. Moreover, some basic BSs are found not only in European species, to which earlier BS studies were dedicated, but also in species from other continents: North America (Wuelker, 1980; Wuelker and Martin, 1971, 1974; Wuelker et al., 1968), Australia (Martin, 1971, 1979), South America (Wuelker and Morath, 1989; Wuelker et al., 1989), and India (Saxena, 1995). However, the overall number and geographical distribution of basic BSs have not been known until recently. We attempt to recognize those BSs in the *Chironomus* banding sequences pool that can be assigned as "basic," trace their dispersal on various continents, and assess their contribution to the chromosomal evolution of the genus. Unfortunately, we have to confine our study to three of seven arms (A, E, and F), because BSs of Australian species are currently known only for these arms.

3 Divergence of Banding Sequences in the Genus *Chironomus* on Various Continents

As mentioned above, the genus *Chironomus* is virtually cosmopolitan. However, it is represented on each continent mainly by endemic species (Saether, 2000; Shobanov et al., 1996). In addition to endemic species, the genus includes few species with vast ranges, covering several continents and several biogeographic zones. In particular, Holarctic species occur on two continents belonging to the Holarctic biogeographic zones: Northern Eurasia (Palearctic) and North America (Nearctic). For this reason, the BS designations are preceded by symbols for the corresponding biogeographic zones or subzones, to specify the geographic occurrence of various basic BSs. Banding sequences confined to the Palearctic are designated with p' (e.g., p'pluA1); BSs confined to the Nearctic, with n' (e.g., n'plu A9); BSs found on both continents of the Holarctic, with h' (e.g., h'pluA2); Australasian BSs, with a' (e.g., a'oppA1); and Central African (Ethiopian biogeographic zone), with e' (e.g., e'allA1).

We can now consider BS features in species from various continents.

Of the *Holarctic species*, we have studied BS divergence in *C. plumosus*, *C. entis*, *C. pallidivittatus*, *C. anthracinus*, and *C. annularius*. Populations of these species occur in Northern Eurasia and North America (Butler et al., 1999; Gunderina et al., 1999; Kiknadze et al., 1996a,b, 2000). Generally, Holarctic *Chironomus* species are characterized by high levels of chromosomal polymorphism and, correspondingly, large banding sequence pools (Table 1). Unexpectedly, comparison of BSs between populations of each of these species on different continents shows that Eurasian and North American populations differ dramatically in BS sets. These differences are related mainly to the fact that different endemic BSs occur on either continent. Eurasian populations of each species have their own p'BS groups, and North American populations, n'BS groups (Table 1). Naturally, in addition to these continent-specific BS, the banding sequence pools of each species include Holarctic h'BSs, common for both continents. The numbers of the various BS categories in the sequence pools of Holarctic species are presented in Table 1. The pools of all Holarctic species are dominated by p'BSs (over 50%); n'BSs constitute ca. 35%; and h'BSs are much scarcer, ca. 15%. Thus, the presence of p'- and n'BSs, endemic for continents, is indicative of strong cytogenetic population divergence between the continents, whereas the presence of common h'BSs points to their common origin. It should be mentioned that intercontinental population divergence can be enhanced by differences in the prevalence of certain BS categories. First, if any endemic BS occurs as a homozygote (i.e., close to fixation), the degree of divergence increases. Second, frequencies of h'BSs strongly influence population divergence; that is, if any of them are scarce in the Palearctic but abundant in the Nearctic, or vice versa, the divergence of populations increases. Note also that the Nearctic populations have lower levels of chromosome polymorphism than Palearctic ones as reflected in the general

Table 1 Cytogenetic divergence of the species of the genus *Chironomus*

Species	Number of populations	Number of BS per species (BS pool)	Cytogenetic structure of the BS pool
Holarctic species			
C. plumosus	73	54	(28p + 8h + 18n)
C. entis	39	52	(26p + 6h + 20n)
C. annularius	10	25	(13p + 6h + 6n)
C. pallidivittatus	17	29	(19p + 5h + 5n)
C. anthracinus		17	(2p + 12h_3n)
Palearctic species			
C. acutiventris	2	27	(27p - -)
C. bernensis	3	16	(16p - -)
C. setivalva		8	(8p - -)
C. heterodentatus	6	24	(24 - -)
C. sokolovae	4	17	(17p - -)
C. agilis	3	8	(6p + 2h -)
C. balatonicus	17	60	(58p + 2h -)
C. borokensis	7	11	(9p + 2h -)
C. muratensis	4	17	(15p + 2h -)
C. nudiventris	4	19	(15p + 4h -)
C. tentans	25	45	(39p + 6h -)
C. sinicus		10	(9p + 1h -)
C. suwai		10	(8p + 2h -)
C. nuditarsis		16	(15p + 1h -)
Nearctic species			
C. crassicaudatus	1	11	(- - 11n)
C. utahensis	7	15	(- - 15n)
C. sp. B2	2	9	(- - 9n)
C. sp. B3	1	15	(- - 15n)
C. matures		13	(- - 13n)
C. atrella	2	16	(- 2h + 14n)
C. sp Is	3	9	(- 2h + 7n)
C. cuccini	2	9	(- 5h + 2n)
C. staegeri		17	(- 2h + 15n)

heterozygosity of populations and the mean number of heterozygous inversions per individual.

The differences between populations in their BS sets and frequencies allow quantitative assessment of cytogenetic distances between them within particular continents and between continents. The cytogenetic distance dendrogram shown in Fig. 1 (left) convincingly shows that intercontinental population divergence exceeds intracontinental divergence by an order of magnitude.

The available information indicates that the continental isolation of populations of Holarctic species caused their dramatic cytogenetic differentiation, resulting in the appearance of specific inversion BSs on each continent.

Nevertheless, Nearctic and Palearctic populations of each Holarctic species retain common h'BSs, pointing to the common origin of the populations. These h'BSs can be assigned to the basic BSs discussed earlier. Basic BSs common in the Palearctic are predominant on both continents; hence, Palearctic populations of Holarctic species are likely to be older. This conclusion is supported by the lower level of chromosome polymorphism in Nearctic populations.

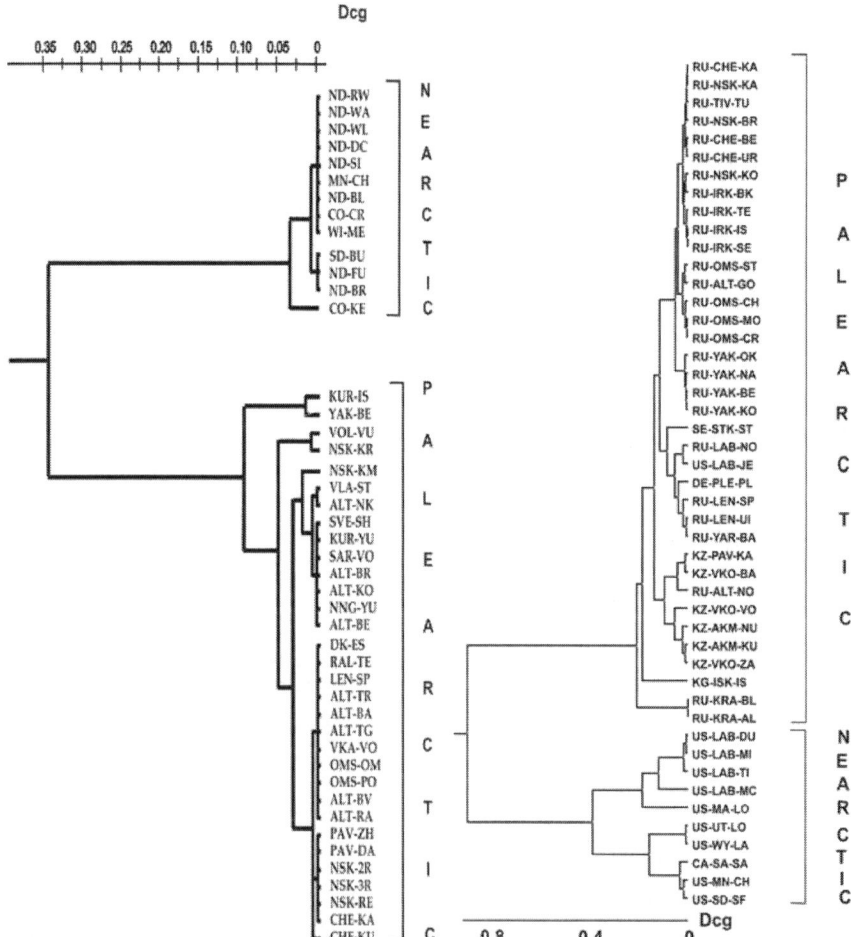

Fig. 1 Dendrogram of the divergence of Palearctic (Eurasia) and Nearctic (North America) *Chironomus plumosus* populations constructed from cytogenetic distances between populations (on the *left*). Dendrogram of the divergence of populations of sibling species, Palearctic *Chironomus tentans* and Nearctic *C. dilutus*, constructed from cytogenetic distances (*Dcg*) between populations (on the *right*)

Few Holarctic basic BSs (hb'BSs) have been found. They are listed in Table 2: four in arm A, four in arm E, and three in arm F. Some of the hb'BSs occur not only in Eurasia and North America but also on other continents. For example, one of the four arm A hb'BSs (Table 2) is common for four continents and, correspondingly, three biogeographic zones: Holarctic, Ethiopian, and Australian. We designate this sequence as heab'A. It is the most abundant in Eurasian species.

Three of four arm E hb'BSs are detected on other continents. One of them (hab'E) is widespread over Eurasian species, less common in North American species, and scarce in Australia. Another sequence, heab'E, occurs on all continents studied. It is most abundant in North Eurasia and Australia. It has been reported that heab'E also occurs in Brazil, South America (Wuelker and Morath, 1989). It is not inconceivable that further studies will show it to be cosmopolitan, although there is no data on cosmopolitan BSs in Chironomidae. A third sequence, heb'E, occurs in Eurasian, North American, and Central African species. It was also found in India (Saxena, 1995), although it could not be identified precisely because of poor image quality.

On arm F (Table 2), one of the three hb'F BSs was found on all continents studied and, correspondingly, designated as heab'F. Another sequence, pab'F, was found on two continents: Eurasia and Australia, being abundant in Australia. Sequence heab'F was also detected in Brazil, South America (Wuelker and Morath, 1989; Wuelker et al., 1989). Like heb'E, it may be cosmopolitan.

The banding sequence pools of many *North Eurasian* (*Palearctic*) endemic species contain only Eurasian p'BSs (Table 1), However, we found that the chromosome pools of some Palearctic endemics also contain h'BS, common for Northern Eurasia and North America. Typically, a banding sequence pool contains no more than two such h'BSs, but some species have four to six h'BSs. In particular, six hb'BSs have been found in the chromosome pool of *C. tentans*. This species shares these sequences, first of all, with its Nearctic sibling species *C. dilutus*, diverging relatively recently (Martin et al., 2002). It can be conjectured that species with few h'BSs are more divergent, whereas species with greater numbers of h'BSs are less divergent from Palearctic species. A total of 12 hb'BSs have been detected in arms A, E, and F of Eurasian species. They are identical to those described above for European populations of Holarctic species and have similar dispersal on continents.

The banding sequence pools of most *North American endemic* species studied contain only n'BSs, found only in *Nearctic* (Table 1). However, hb'BSs present also in Palearctic species were found in the pools of several Nearctic endemic species (Table 1). Generally, such sequences are few (two h'BSs), but some Nearctic endemics, e.g., *C. cucini*, have as many as five h'BSs. This fact may point to close relationship of this species to Palearctic ones. The Holarctic basic BSs present in banding sequence pools of North American species are the same as those found in Eurasian species (Table 2).

By now, BSs have been determined in banding sequence pools of four *Central African species* (*Ethiopian* biogeographic zone): *C. alluaudi, C. transvaalensis, C.*

Table 2 Geographic dispersal of the banding sequences (BS) of the species of the genus *Chironomus*

Arm	BS	BS symbol	Species having BS in Palearctic	Nearctic	Ethiopian	Australasian
A	01a-02c 10a-12c 03i-02d 09e-04a 13a-19fC	heab'A	p'hol, p'mls, p'pan, p'yos, p'sax, p'Al.1, p'rii, p'tnu, p'esa, h'lon	n'sta, n'cuc, n'maj, n'tar, h'lon	e'all, e'sp.3	a'feb, a'clo, a'inc
	01a-02c 10a-12a 13b-13a 04a-04c 02g-02d 09e-04d 02h-03i 12c-12b 13c-19f C	hb1'A	h'plu, h'ann, h'ant, p'bor, p'suw, p'ber, p'bon, p'com, p'sp.J	h'plu, h'ann, h'ant		
	01-02c 10a-12a 03f-02h 04d-09e 02d-02g 04c-04a 13a-13b 03g-03i 12c-12b 13c-19f C	hb2'A	p'cin, h'ant	h'ant		
	01a-19f C	hb3'A	p'pig, h'rip	h'rip		
E	01a-03e 05a-10b 04h-03f 10c-13gC	hab'E	p'abe, p'plu, p'agi, p'agi2, p'AL1, p'bel, p'beh, p'bon, p'bor, p'cin, p'cur, p'fra, p'fre, p'sp J, p'jon, p'lac, p'mel, p'nud, p'ocu, p'res, p'rii, p'sor, p'spTu3, p'use, h'lon, p'tnu, h'ann	n'cuc, n'sta, n'Ave, n'spIs, n'neo, n'tar, n'dec, h'ant, h'lon		a'gro
	01a-03e 10b-03f 10c-13g C	heab'E	h'plu, p'aci, p'apr, p'lur, p'uli, p'yos, p'spYa2, p'hpi, p'pil, p'wue	h'plu, n'atr		a'opp, a'tim, a'tep, a'sam, a'pse, a'mad, a'aus, a'aus, a'for, a'jac, a'sp5, a'sp6, a'sp8, a'sp9, a'sp7
	01a-13g C	heb' E	p'pig, h'rip, p'pst, p'hol, p'spYa4	h'rip	e'all	
	01a-02e 10g-10c 03f-04h 10b-05a 03e-03a 11a-13g C	hb3'E	p'mur, h'ent	h'ent		
F	01a-10d 17d-11a 18a-23f C	hb1'F	p'abe, p'alb, p'bal, p'bor, p'bel, p'fra, p'jon, p'rii, p'sor, p'tnu	n'cuc, n'maj, n'tar		
	01a-01d 06e-01e 07a-10d 17d-11a 18a-23f C	hb2'F	h'plu, h'ent, p'bon, p'spJ, p'mur, p'nud, p'use, h'ant	h'plu, h'ant		
	01a-23f C	heab'F	p'pig, h'rip, p'nig, p'res, h'lon, p'mel, p'spTul, h'ann, h'ant	n'Ave, h'rip, h,ant, h'ann, h'lon	e'all	a'gro

sp.3, and *C. formosipennis*. Some data on BSs in some of these species were formerly reported by Wuelker (1980) and Martin (1979). Analysis shows that the pool of only one of four species, *C. transvaalensis*, includes only BSs endemic for Central Africa. The other three species have basic BSs (heb'BSs). For example, three arms of *C. alluaudi*, A, E, and F, have heb'BSs (Table 2). The heb'BSs of arms A and F are common to four continents, and the basic sequence of arm E occurs on three continents. Other arms have endemic e'BSs. The banding sequence pool of *C. sp.3* has the basic heb'BS similar to *C. alluaudi* and species of Eurasia and North America. Other arms have e'BSs endemic for Africa. The banding sequence pool of *C. formosipennis* has one basic heab'BS characteristic of four continents. Other arms have endemic e'BSs (Table 2).

The most prominent feature of the banding sequence pools of *Australian and New Zealand endemic species* (*Australasian* biogeographic zone) is their smaller divergence from each other. Of 24 species, 17 have common BSs linking them to the central species *C. oppositus*. In addition, the BSs of Australian endemics are simpler than in Eurasian and North American ones; they differ from standard BSs by smaller numbers of inversions.

As in Eurasia, North America, and Central Africa, the banding sequence pools of all species in Australia are dominated by endemic Australian BSs (a'BSs). However, many species also have some of the basic BSs occurring in Eurasia, North America, and Central Africa. As is evident from Table 2, such common BSs are present in all the three arms studied: A, E, and F. Some of them occur on all continents, others on three, and still others on two. Five basic BSs were found in the Holarctic, Ethiopian, and Australasian biogeographic zones.

4 Basic Banding Sequences as Markers of Evolutionary Divergence of Species Genomes

Thus, we have analyzed 271 BSs in three arms (A, E, and F) of 87 *Chironomus* species and found 11 BSs occurring on several continents, which we regard as basic BSs. They include two heab'BSs, common for four continents (Eurasia, North America, Central Africa, and Australia); two hab'BSs, common for three continents (North Eurasia, North America, and Australia); one heb'BS, common for the other three continents (North Eurasia, North America, and Central Africa), and six sequences (hb'BSs and pab'BSs) common for two continents (Eurasia + North America or Eurasia + Australia, respectively).

It is conceivable that heab'BSs, common to four continents, are the oldest. They may have belonged to a common ancestor of the genus *Chironomus*, existing even on Pangaea supercontinent (Fig. 2). The appearance of *Chironomidae* is recorded in the range from the Triassic to Cretaceous, before the splitting of Pangaea into Laurasia (Northern Hemisphere: Eurasia, North America, and Greenland) and Gondwana (Southern Hemisphere: Australia, Africa, South America, and Antarctica).

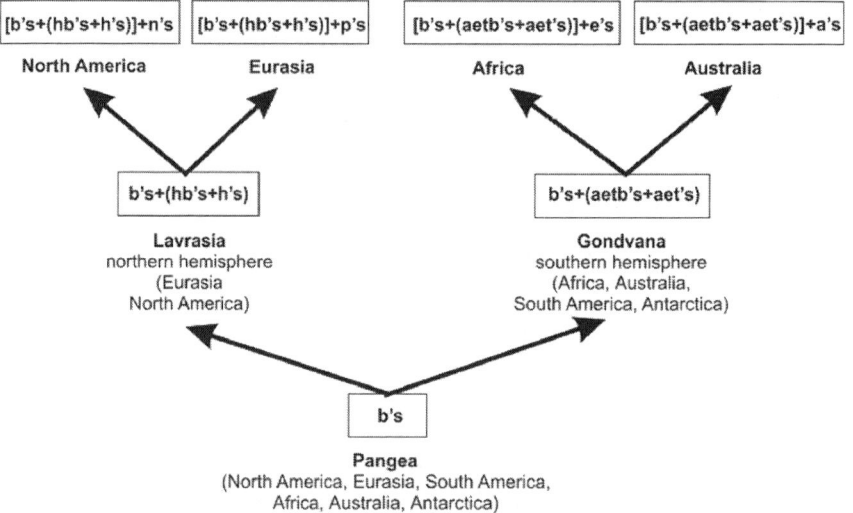

Fig. 2 The hypothetical pattern of divergence of banding sequence (BS) of polytene chromosomes in the genus *Chironomus* during formation of continents

Chironomidae are recorded in the late Permian (Paleozoic)–early Triassic (Mesozoic), when the supercontinent Pangaea existed, encompassing all continents. The BSs of these *Chironomidae* can be designated as original ones (o′s). They included basic sequences (b′s), which remained unchanged in the course of evolution. Later, the basic sequences existed in Laurasia, Gondwana, and on separate continents. After splitting of Pangaea into Laurasia and Gondwana, BS divergence was mediated by different inversions in the Northern and Southern Hemispheres because of different environmental conditions. Correspondingly, basic b′s could partly survive in Laurasia, and new inversion sequences common for Eurasia and North America (Holarctic h′s) could arise. Some of these h′s were shared by several species; that is, they were basic Holarctic, hb′s. Sequences common for future Australia (Australasian zone, a′), Africa (Ethiopian zone, e′), and South America (Neotropic zone, t′) formed in Gondwana. These sequences are designated as aet′s. They included basic sequences common for several species (aetb′s). In addition, part of b′s survived, reflecting the Pangaean origin of the BSs.

Later, during formation of each continent, continent-specific endemic sequences arose: Palearctic (p′s), Nearctic (n′s), Australian (a′), and Ethiopian (e′s). In addition, some BSs formed in Pangaea (b′s), Laurasia (h′s, hb′s), and Gondwana (ae′s, aeb′s) survived on continents. Sequences common for Eurasia and Australia, pab′BS, can also be assigned to that time. Later, hb′BSs found in Northern Eurasia and North America could arise in Laurasia. Common intracontinental p′, n′, e′, and a′BSs may reflect BS divergence in endemic species. These banding sequences may be considered the youngest.

As indicated above, basic BSs are the minority of the banding sequence pools of species. The most numerous BSs in chromosome pools are endemic. They are likely to have diverged independently on each continent. It is reasonable to suggest that mutagenic effects on chromosomes differed depending on environmental features on each continent (volcanism and glaciations), types of bottom sediments in water pools inhabited by chironomid larvae, etc. This suggestion is clearly confirmed by the fact that inversion breakpoints in endemic BSs are strictly continent-specific.

Study of the divergence of basic and endemic BSs allows the tracing of the history of gene orders during the evolution of the chironomid chromosome pool.

5 Scenario of BS Divergence

By now, BS divergence in the genus *Chironomus* has been studied for five of seven chromosome arms (Gunderina et al., 2005; Kiknadze et al., 2003, 2004). However, as said above, we had to confine ourselves to BSs of three arms, A, E, and F. Phylograms of BS divergence in these arms are shown in Figs. 3a–c.

Arm A (Fig. 3a) is one of the most variable arms in many species of the genus. However, the phylogram shows that there is a notably large group of species whose karyotypes contain the basic sequence heab'BS. This group includes Palearctic, Nearctic, and Holarctic species. The basic sequence heab'BS was productive in the evolution of arm A, giving rise to BSs present in the majority of Palearctic and Nearctic species. The left part of the phylogram indicates Palearctic species of the genus having various inversion BSs of arm A. In particular, inversion 4d-2g produced the basic Palearctic sequence pb'BS, occurring in eight species. In turn, various inversions in this sequence gave rise to a large group of hb'BSs and other BSs. Other BSs in arm A of Palearctic species were less productive, although their inversions also resulted in the formation of species-specific BSs.

The right part of the phylogram shows BSs characteristic of the banding sequence pools of Nearctic species. It is obvious that the formation of Nearctic BSs was mediated by quite different inversions with breakpoints located at other chromosomal sites than in the Palearctic. Moreover, the Nearctic species do not exhibit such a large number of basic BSs.

Arm E (Fig. 3b) is the least variable in the genus *Chironomus*. The presence of basic BSs in this arm in a great number of species is prominent in the phylogram of BS divergence. The most widespread is the Holarctic basic BS, hab'. It was found in at least 39 species, including p, n, and h species. Another basic BS, heab', is present in 11 species, also including p, n, and h species. The third BS, hb3, is less widespread than the first and second ones.

ARM A

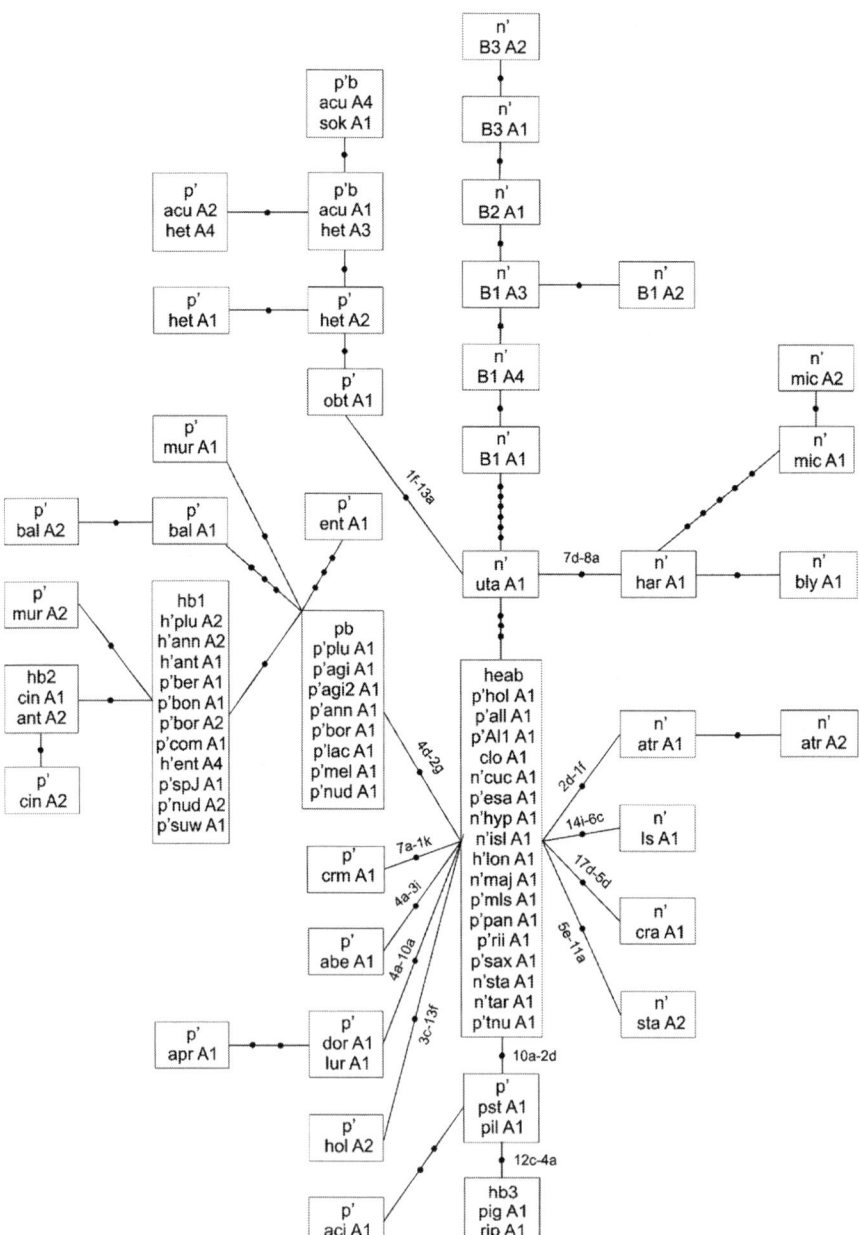

Fig. 3a Phylograms of the divergence of inversion banding sequences of the arm A in species of the genus *Chironomus*. *Black circles* designate numbers of inversions between sequences. Key inversions from basic sequences to Palearctic (*left*) and Nearctic (*right*) ones are indicated *above arrows*. Homologous basic sequences occurring in different species are *boxed*. For further explanations, see text

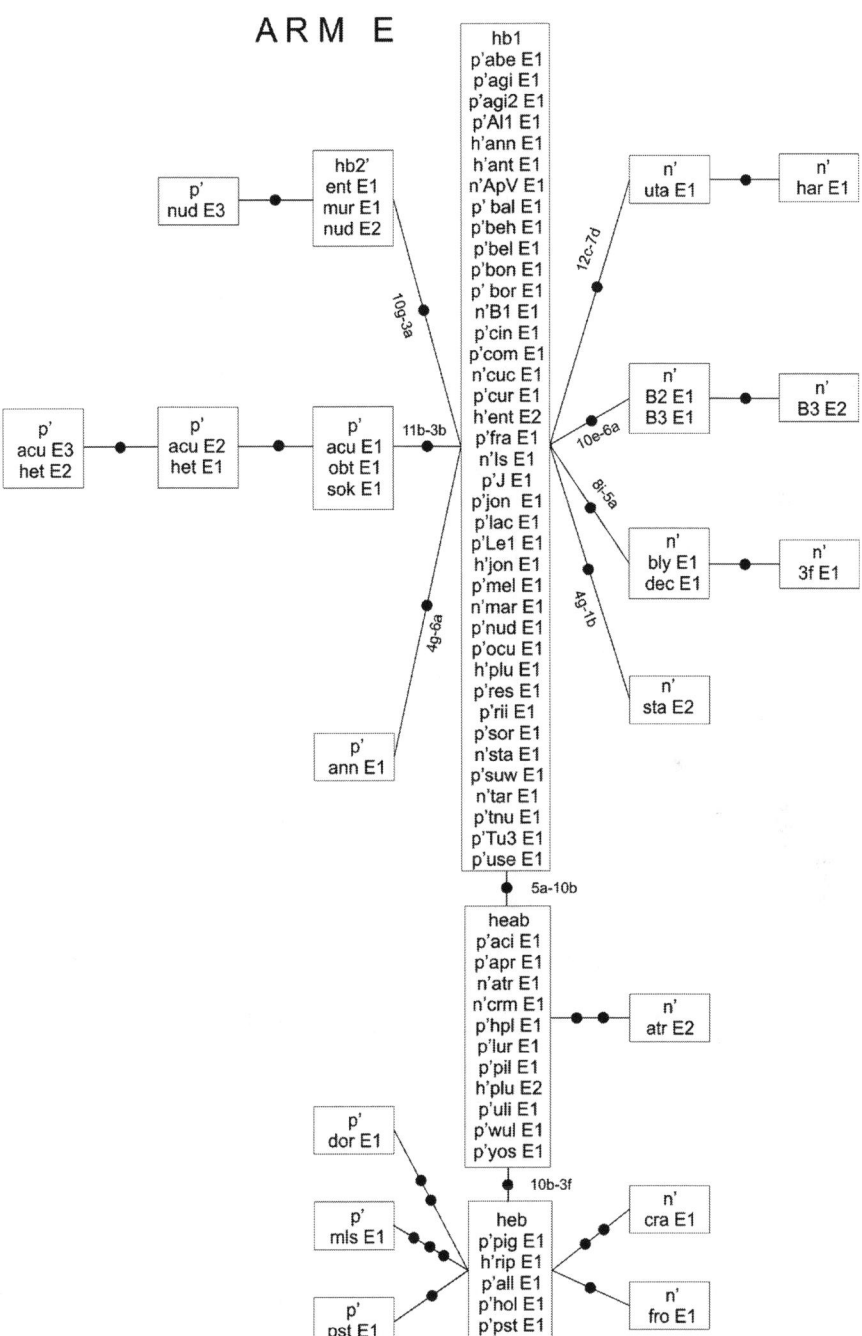

Fig. 3b Phylograms of the divergence of inversion banding sequences of the arm E in species of the genus *Chironomus*

ARM F

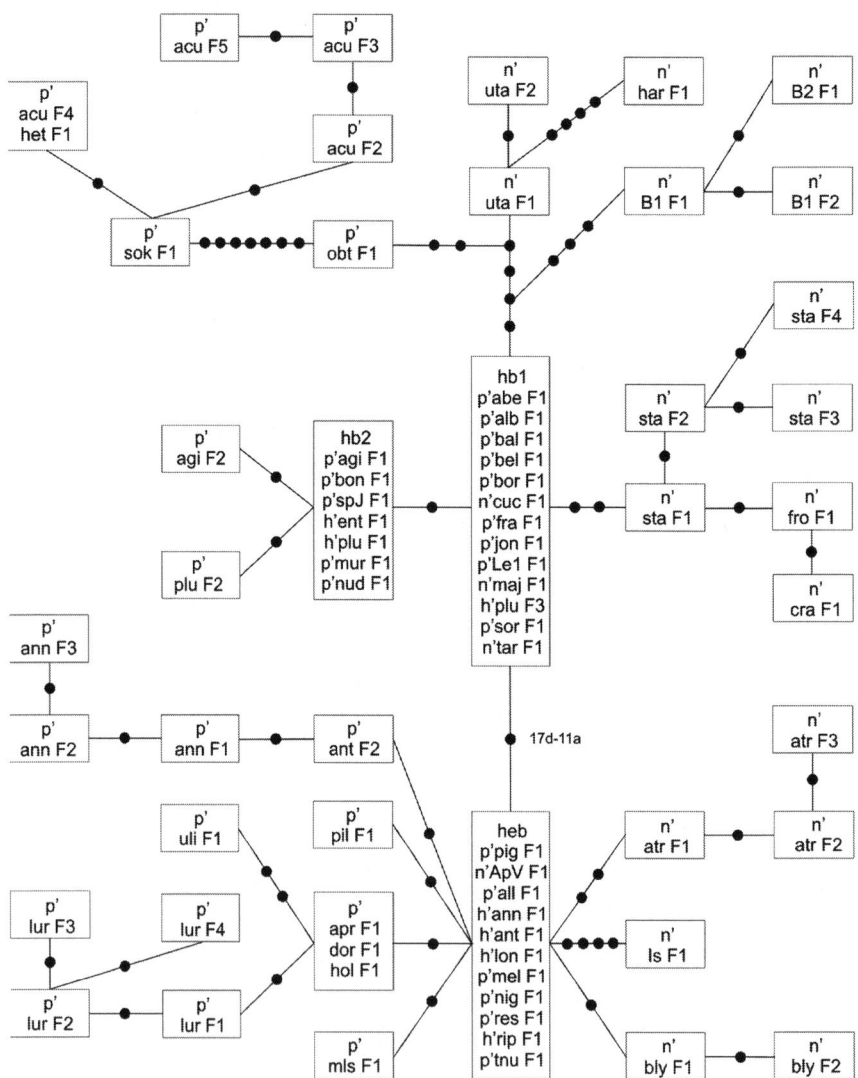

Fig. 3c Phylograms of the divergence of inversion banding sequences of the arm F in species of the genus *Chironomus*

Unlike arm A, arm E exhibits simpler BS divergence in both the Palearctic and Nearctic, mediated by fewer inversions. However, as in arm A, the divergence of BS in arm E involved different BSs on different continents.

Arm F (Fig. 3c) is intermediate between arms A and E with regard to inversion polymorphism. The phylogram clearly shows several species groups,

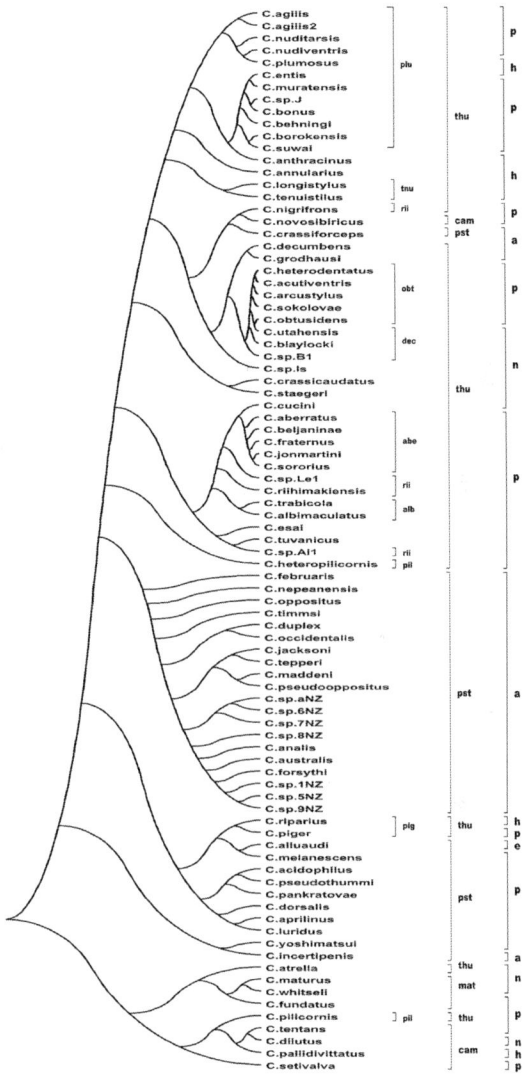

Fig. 4 The NJ-phylogenetic AEF tree constructed on the basis of analysis of pairwise similarity between banding sequences in three chromosome arms (A, E, and F) of polytene chromosomes in 84 *Chironomus* species. Groups of sibling species are designated as *plu, tnu, rii, obt, dec, abe, alb, pil,* and *pig*; cytocomplexes as thu, cam, pst, mat; and biogeographic zones as p, n, h, a, e. For further explanations, see text

each of which has a characteristic basic BS. Groups with hb1 and heab BSs include approximately equal numbers of species. Somewhat fewer species have the hb2 basic BS. The most productive sequences proved to be hb1 and heab. These BSs produced Palearctic and Nearctic sequences; however, BSs formed in the Palearctic and Nearctic were related to quite different inversions. This fact indicates that different gene orders are advantageous on different continents. It is noteworthy that, in addition to continent-specific BSs, there are groups of common basic BSs, indicative of the common origin of *Chironomus* species.

6 Dispersal of Species and Banding Sequences

Analysis of BS divergence allows tracing not only of their evolution but also of the putative pathways of species dispersal. As indicated above, the presence of basic BSs common for the four continents analyzed suggests that ancestors of *Chironomus* species existed even in Pangaea. Then, the primary BSs diverged mostly by means of inversions and, probably, gave birth to the endemic species of Eurasia, North America, Central Africa, and Australasia. The subsequent history of the Earth included multiple continent aggregations and breakups, which may have been accompanied by migrations of species between continents. The most appropriate group for studying such migrations is the Holarctic species, whose populations occur on several continents.

As early as 1970s, the variability of BSs in chromosomes of *C. tentans* was studied and it was suggested that this species had colonized North America from Europe through Siberia and the Bering land bridge (Acton and Scudder, 1971), because the BS pool of its Alaskan populations was intermediate between Europe and North America. By that time, nothing was known about the banding sequence pools of Siberian *C. tentans* populations, but Acton and Scudder suggested that the divergence of European BSs had begun in Siberia. We pioneered a parallel study of BS variability in Siberian, European, and North American *C. tentans* populations (Kiknadze et al., 1996b, 1998) and found that Siberian populations were similar to European ones in BS pools and different in the presence of a few unique Siberian BSs and frequencies of some main or alternative BSs. According to cytogenetic distances, Siberian and European populations belong to one Palearctic cluster (Fig. 1 right). In contrast, North American populations differ significantly from European and Siberian ones by the presence of endemic Nearctic BSs and frequencies of some Holarctic BSs. For example, h′tenF3 and h′tenA4, rare in Siberian populations, are predominant in North America. Owing to the drastic difference in BS sets and frequencies between Palearctic and Nearctic *C. tentans* populations, the cytogenetic distances between them is at the level of differences between sibling species (Fig. 4). On this basis, the formerly single Holarctic species *C. tentans* has been divided into two species: Palearctic *C. tentans* and Nearctic *C. dilutus* (Kiknadze et al., 1996b; Shobanov et al., 1999). A similar pattern of BS divergence between Palearctic and Nearctic populations is observed in other Holarctic species: *C. pallidivittatus, C. plumosus, C. entis, C. anthracinus, C. annularius*. However, the differentiated populations of these species still belong to one species (Butler et al., 1999; Kiknadze et al., 1998, 2000, 2004, 2005).

In all Holarctic species studied, Siberian populations are more polymorphic in BS pools, population heterozygosity, and mean numbers of heterozygous inversions than Nearctic ones. The intracontinental divergence of these populations is greater. These data show that Siberian populations are older than Nearctic ones. Moreover, there is one Siberian (Yakutian) population of *C. tentans* containing the rare h′tenF3 and h′tenA4 BSs, which became characteristic of North America. This leads us to the conclusion that North American populations descended from Siberian populations after their migration through the Bering land bridge. Obviously, migration of European populations via Greenland is not impossible, but this way is much more difficult.

For understanding the cause of migration of Siberian populations of Holarctic species to North America, it should be emphasized that the last Quaternary glaciation covered the bulk of North America and Europe for a long time, whereas Siberia never experienced complete glaciation. Separate glaciation zones in Siberia were confined mainly to mountainous regions. Correspondingly, the chironomid fauna of North America was much depauperated after completion of the solid glaciation, and its diversity may have been restored by migration of Siberian Chironomidae in the periods of existence of the Bering bridge.

One may suggest that profound changes of BSs after migration of species to North America resulted from colonization stress. A peculiar "chromosomal revolution" during the migration is likely to have occurred as follows: The intervention of Siberian populations presumably occurred in early Pleistocene just after the end of the last glaciation in North America (Kiknadze et al., 1998, 2000; Martin and Porter, 1973) or in the late Pliocene (Martin et al., 2002). It is likely that the dramatic change of environmental conditions resulted in increasing frequencies of chromosome rearrangements, because stress conditions favored variability, mediated by recombinations and mutations (Belyaev and Borodin, 1980). The low population size, related to either mass-scale death of individuals under extreme conditions or to colonization of new econiches according to the founder principle, favored rapid change of BS frequencies and fixation.

Ecogeographical analysis of a number of Holarctic animal and plant taxa shows that the degree of taxonomic difference between them correlates well with the duration of intercontinental isolation caused by the disappearance of the Bering bridge (Beringia in Kainozoi, 1976). It is suggested that animal populations isolated in the late Würm–Wisconsin epoch (ca. 30 kyr BP) belong to one species; species isolated in the late Riss–Illinoian epoch (ca. 75 kyr BP) are semispecies or sibling species; species isolated in the late Mindel–Kansan epoch or earlier (300–1000 kyr BP) are genuine allopatric species. The discovered correlation between the taxonomical state and time of continental isolation of populations of Holarctic species suggests that the ancestors of the species recognized by us (*C. dilutus* and *C. tentans*) came to North America from Eurasia in early Pleistocene or late Pliocene and evolved to the state of sibling species. Populations of other Holarctic species migrated later and diverged to the level of geographic populations. Therefore, another explanation for the different degrees of divergence between populations of different Holarctic species may be relevant to at least some of the species. The different divergence may reflect different abilities of the genomes to respond to colonization stress and environmental selection. The latter hypothesis is supported by molecular data on the divergence of two chironomid mitochondrial genes. They indicate that *C. tentans*, diverging to two sibling species, and *C. pallidivittatus*, diverging only to the level of differentiated populations in the Palearctic and Nearctic, synchronously migrated to North America (Martin et al., 2002).

7 Chromosome Evolution and Speciation

Comparison of BSs in species of the genus *Chironomus* allows assessment of evolutionary relationships between species and construction of phylogenetic trees. The number of breakpoints altering banding order in BSs under comparison was used as a measure of the degree of BS divergence. The degree of BS divergence was assessed in five chromosome arms (A, C, D, E, and F) in *Chironomus* species by a special-purpose method (Gunderina et al., 2005; Gusev et al., 2001; Kiknadze et al., 2003). The degree of divergence, i.e., number of breakpoints discriminating BS, increased with the rank of taxonomic difference between species (Gunderina et al., 2005). In this report, we analyze the tree constructed from data on breakpoint numbers in arms A, E, and F. The neighbor-joining tree constructed from breakpoint numbers in three chromosome arms in pairwise comparisons of BSs from 87 chironomid species (Saitou and Nei, 1987) illustrates a complex branched hierarchy (Fig. 4). It is apparent that BSs of species from different cytocomplexes and groups of closely related species generally form distinct clusters on the tree. The cluster first separated from the tree includes two distinct subclusters. One of them is formed by BSs from species of the camptochironomus-cytocomplex, the other by those from the maturus-cytocomplex. Then a succession of three clusters is separated. They are formed by BSs of species of the pseudothummi-cytocomplex. The rest of the tree includes BSs of species of the thummi-cytocomplex. This is an intricately branched set of clusters. Most of them correspond to separate groups of closely related species: *plumosus*, *albimaculatus*, *aberratus*, *pilicornis*, *tenuistylus*, and *piger*. Two groups, *decorus* and *obtusidens*, are bulked into one cluster. The resulting tree shows that BS evolution is in good agreement with the morphologically determined taxonomic positions of species in the genus *Chironomus*.

The tree clearly illustrates the correlation between the geographic distribution of BSs of chironomid species studied and their phylogenetic relationships. Our analysis shows that this correlation is not unambiguous. In most cases, the geographic distribution of BSs is clearly linked to the taxonomic positions of species. Such a linkage is obvious in the case of BSs of species of the pseudothummi-cytocomplex, grouped into three clusters. One cluster is formed by Palearctic BSs, the second one by Australasian ones. The third cluster includes only two BSs: one Australian and one Palearctic. Nevertheless, the BSs of the species of the camptochironomus-cytocomplex belong to a single cluster independent of the geographic region of their occurrence. The same regularity is observed with species of the maturus-cytocomplex. The BSs of species belonging to the thummi-cytocomplex occur on all continents, as clearly illustrated by the tree. It is apparent that in most cases BSs of species belonging to one group of closely related species occur in the same geographic regions. Thus, the geographic diversity of BSs in the thummi-cytocomplex is determined by different geographic occurrences of groups of closely related species present in this cytocomplex.

This regularity has some exceptions. For example, the cluster with Palearctic BSs of species belonging to the pseudothummi-cytocomplex also includes Holarctic BSs of *C. riparius*, a species of the thummi-cytocomplex. Palearctic BSs of *C. novosibiricus*, belonging to the camptochironomus-cytocomplex, and Australian BSs of *C. crassiforceps*, belonging to pseudothummi-cytocomplex, are clustered with Palearctic BSs of *C. nigrifrons* from the thummi-cytocomplex and belong to the system of BS clusters of the thummi-cytocomplex. This phenomenon may be related to secondary translocation of corresponding chromosome arms (Keyl, 1962), which occurred relatively recently in these species in the course of *Chironomus* karyotype evolution. The result is alteration of the arm combinations forming a chromosome and, correspondingly, transition of the species to another cytocomplex.

The diversity of geographic clustering of BSs in different cytocomplexes can be related to the differences in the time of divergence of these species and the strategy of colonization on different continents by these species. The BSs of species of the camptochironomus-cytocomplex belong to one cluster independent of their geographic occurrence. This may be related to the fact that the ancestor of the Palearctic species *C. tentans* and the Nearctic species *C. dilutus*, as well as *C. pallidivittatus*, was a Holarctic species. The complete but relatively recent geographic segregation of these species has not resulted in their complete cytogenetic divergence and formation of independent clusters on the tree. On the other hand, species of the pseudothummi-cytocomplex are considered the oldest in the genus *Chironomus*, and this has permitted the manifestation of both cytogenetic and geographic factors in determining their evolutionary divergence. As a result, BSs of species of this cytocomplex are combined into distinct clusters in agreement with their geographic occurrence. Thus, the AEF tree shows relationships between banding sequences and species and illustrates the geographic occurrence of *Chironomus* species in the past.

References

Acton, A.B. and Scudder, G. (1971) The zoogeography and races of *Chironomus tentans*. Limnologica 8, 83–92.
Ayala, F.J. and Coluzzi, M (2005) Chromosome speciation: human, *Drosophila*, and mosquitoes. PNAS 102, 6535–6542.

Beermann, W. (1972) Chromosomes and genes. *Results and Problems in Cell Differentiation* 4, 1–33.
Belyaev, D.K. and Borodin, P.M. (1980) The influence of stress on the frequency of crossingover in chromosome 2 of mouse. Rep. USSR Acad. Sci. 253, 727–729.
Beringia in Cenosoic (1976) Far Eastern Dept. Acad. Sci. USSR, Vladivostok.
Bridges, C.B. (1935) Salivary gland chromosome maps, with a key to the banding of chromosomes of *Drosophila melanogaster*. J. Hered. 26, 60–64.
Butler, M.G., Kiknadze, I.I., Golygina, V.V., Martin, J., Istomina, A.G., Wuelker, W., Sublette, J.E. and Sublette, M.F. (1999) Cytogenetic differentiation between Palearctic and Nearctic populations of *Chironomus plumosus* L. (Diptera, Chironomidae). Genome 42, 797–815.
Dobzhansky, Th. (1970) Genetics of Evolutionary Process. Columbia Univ. Press, New York.
Eichler, E.E. and Sankoff, D. (2003) Structural dynamics of eukaryotic chromosome evolution. Science 301, 793–797.
Gunderina, L.I. and Kiknadze, I.I. (2000) Divergence of karyofunds in sibling species of the plumosus group (Chironomidae, Diptera). Russ. J. Genet. 36, 265–272.
Gunderina, L.I., Kiknadze, I.I. and Golygina, V.V (1999) Intraspecific differentiation of the cytogenetic structure in natural populations of *Chironomus plumosus* L., the central species in the group of sibling species (Chironomidae: Diptera). Russ. J. Genet. 35, 142–150.
Gunderina, L.I., Kiknadze, I.I., Istomina, A.G., Gusev, V.D. and Miroshnichenko, L.A. (2005) Divergence of the polytene chromosomes banding sequences as a reflection of evolutionary rearrangements of the genome linear structure. Russ. J. Genet. 41, 130–137.
Gusev, V.D., Nemytikova, L.A. and Chuzhanova, N.A. (2001) Rapid method for identification of interconnections between functionally and/or evolutionarily related biological sequences. Mol. Biol. (Russ.) 35, 1015–1022.
Keyl, H-G. (1962) Chromosome evolution dei Chironomus. II. Chromosomenumbauten und phylogenetische Beziehungen der Arten. Chromosoma 13, 464–514.
Kiknadze, I.I., Blinov, A.G. and Kolesnokov, N.N. (1989) Molecular and cytological organization of chironomid genome. In: V. Shumny (Ed.), Structural and Functional Organization of Genome. Nauka SD, Novosibirsk, pp. 4–58.
Kiknadze, I.I., Shiliva, A.I., Kerkis, I.E, Shobanov, N.A., Zelentzov, N.I., Grebenjuk, L.P., Istomina, A.G., and Prasolov, V.A. (1991) Karyotypes and Morphology of Larvae in Tribe Chironomini. Nauka SD, Novosibirsk.
Kiknadze, I.I., Istomina, A.G., Gunderina, L.I., Salova, T.A., Aimanova, K.G. and Savvinov, D.D. (1996a) Banding Sequences Pool of Chironomids in Jakutian Criolitozone. Nauka SD, Novosibirsk.
Kiknadze, I.I., Butler, M.G., Aimanova, K.G., Gunderina, L.I. and Cooper, K. (1996b) Geographic variation in polytene chromosome banding pattern of the Holactic midge *Camptochironomus tentans* (Fabr.). Can. J. Zool. 74, 171–191.
Kiknadze, I.I., Butler, M.G., Aimanova, K.G., Andreeva, E.N., Martin, J. and Gunderina, L.I. (1998) Divergent cytogenetic evolution in Nearctic and Palearctic populations of sibling species in subgenus *Camptochironomus* Kieffer. Can. J. Zool. 76, 361–376.
Kiknadze, I.I., Butler, M.G., Golygina, V.V., Martin, J., Wuelker, W., Sublette, J.E. and Sublette, M.F. (2000) Intercontinental cytogenetic differentiation in *Chironomus entis* Shobanov, a holarctic species in plumosus-group (Diptera, Chironomidae). Genome 43, 857–873.
Kiknadze, I.I., Gunderina, L.I., Istomina, A.G., Gusev, V.D. and Nemytikova, L.A. (2003) Similarity analysis of inversion banding sequences of *Chironomus* species (breakpoint phylogeny). In: N. Kolchanov and R. Hofestaedt (Eds), Bioinformatics of Genome Regulation and Structure. Kluwer Acad. Press, Dordrecht, pp. 245–253.
Kiknadze, I.I., Gunderina, L.I., Istomina, A.G., Gusev, V.D. and Miroshnichenko, L.A. (2004) Reconstruction of chromosomal evolution in genus *Chironomus*. Euras. Entomol. J. 3, 265–273.

Kiknadze, I.I., Wuelker, W.G., Istomina, A.G. and Andreeva, E.N. (2005) Banding sequences pool in *Chironomus anthracinus* Zett. (Diptera, Chironomidae) in Palearctic and Nearctic. Euras. Entomol. J. 4, 13–27.

King, M. (1993) Species Evolution: The Role of Chromosome Change. Cambridge University Press. Cambridge New York.

Marques-Bonet, T., Caceres, M. and Bertranpetit, A. (2004) Chromosomal rearrangements and the genomic distribution of gene-expression divergence in humans and chimpanzees. Trends Genet. 20, 524–529.

Martin, J. (1971) A review of genus *Chironomus* (Diptera, Chironomidae). IY. The karyosystematics of australis group in Australia. Chromosoma 35, 418–430.

Martin, J. (1979) Chromosome as tools in taxonomy and phylogeny of Chironomidae (Diptera). Entomol. Scand. 10, 67–74.

Martin, J. and Porter, D. L (1973) The salivary gland chromosomes of *Glyptotendipes barbipes* (Staeger) (Diptera, Chironomidae): description of inversions and comparison of Nearctic and Palearctic karyotypes. Stud. Nat. Sci. (Portales, New Mexico) 1, 1–25.

Martin, J., Guriev, V. and Blinov, A. (2002) Population variability in *Chironomus (Camptochironomus)* species (Diptera, Nematocera) with Holarctic distribution: evidence of mitochondrial gene flow. Insect Mol. Biol. 11, 387–397.

Michailova, P.V. (1989) The polytene chromosomes and their significance to the systematics of the family Chironomidae, Diptera. Acta Zool. Fenn. 186, 107.

Navarro, A. and Barton, N.H. (2003) Accumulating postzygotic isolation genes in parapatry: a new twist on chromosomal speciation. Evolution 57, 447–459.

Painter, T.S. (1934) Salivary chromosomes and the attack on the gene. J. Hered. 25, 465–476.

Saether, O. (2000) Zoogeographical patterns in Chironomidae (Diptera). Verh. Int. Verein Limnol. 27, 290–302.

Saitou, N. and Nei, M. (1987) The neighbor-joining method: a new method for reconstructing phylogenetic trees. Mol. Biol. Evol. 4, 406–425.

Saxena, S. (1995) Basic patterns n the chromosome evolution of the genus *Chironomus* (Diptera): polytene chromosomes of the three Indian species *C. plumatisetigerus*, *C. calipterus*, and *C.* sp. In: P. Cranston (Ed.), Chironomids: From Gene to Ecosystems. CSIRO, Australia, pp. 39–48.

Shobanov, N.A., Shilova, A.I. and Belyanina, S.I. (1996) The size and the structure of genus *Chironomus* Meigen (Diptera, Chironomidae): the review of word fauna. In: N. Shobanov and T. Zinchenko (Eds), Ecology, Evolution and Taxonomy of Chironomidae. Tolijatti, Borok, pp. 44–96.

Shobanov, N.A., Kiknadze, I.I. and Butler, M.G. (1999) Palearctic and Nearctic *Chironomus (Camptochironomus) tentans* (Fabricius) are different species. Entomol. Scand. 30, 311–322.

White, M.J. (1977) Animal Cytology and Evolution. Cambridge Univ. Press, Melbourne.

Wuelker, W. (1980) Basic pattern in the chromosome evolution of the genus *Chironomus* (Diptera). Z. Zool. Syst. Evol. 18, 112–123.

Wuelker, W. and Martin, J. (1971) Karyosystematics of the *Chironomus staegeri* group. Stud. Nat. Sci. (Portales, New Mexico) 1, 22–34.

Wuelker, W. and Martin, J. (1974) A review of the genus *Chironomus* (Diptera, Chironomidae). YI. Cytology of the maturus-komplex. Stud. Nat. Sci. (Portales, New Mexico) 1, 1–24.

Wuelker, W. and Morath, E. (1989) South American *Chironomus* (Diptera). Karyotypes and their relations to North America. Acta Biol. Debr. Oecol. Hung. 2, 389–397.

Wuelker, W., Sublette, J.E. and Martin, J. (1968) Zur Cytotaxonomie nordamerikanischer *Chironomus*-Arten. Ann. Zool. Fenn. 5, 155–158.

Wuelker, W., Sublette, J.E., Morath, E. and Martin, J. (1989) *Chironomus columbiensis* n. sp. in South America and *C. anonymus* Williston in North America—closely related species . Stud. Neotr. Fauna Environ. 24, 121–136.

Zdobnov, E.M., von Mering, C. and Letunic, I. (2002) Comparative genome and proteome analysis of *Anopheles gambiae* and *Drosophila melanogaster*. Science 298, 149–159.

Part VI
Biosphere And Human Being

Genetic Landscape of the Central Asia and Volga–Ural Region

E. K. Khusnutdinova, M. A. Bermisheva, I. A. Kutuev, B. B. Yunusbayev, and R. Villems

Abstract The study of the Volga–Ural region and the Central Asia populations is carried out on the basis of the analysis of SNP and microsatellites of Y-chromosome, and also mtDNA hypervariable segment I and coding region. Principally new data on relationship, reciprocal location, degree of similarity and distinction of populations are received. Genetic relationships between populations of these regions are investigated.

1 Introduction

Studies of genetic variation within and between present day human populations provide unique information about the past human history. Phylogeographic reconstructions based on the analysis of uniparentaly inherited mitochondrial DNA (mtDNA) and Y-chromosome markers have become very popular in recent years owing to their ability to provide information about the place and the time of demographic events which can be associated with prehistoric events in human history. Because inferences about the past population history are based on the current phylogeographic pattern, it is essential to study present-day genetic diversity using as many as possible population samples in order to cover all possible geographic regions inhabited by human.

Numerous studies in the recent years have yielded considerable progress in understanding the phylogeography of Y-chromosome and mtDNA lineages mainly in the western part of Eurasia. In contrast eastern regions like the Central Asia and the Volga–Ural region have not been investigated systematically to the level, currently available for the western Eurasian populations. We conducted a number of field expeditions in order to obtain population samples from the

E. K. Khusnutdinova
Institute of Biochemistry and Genetics, Department of Genomics Ufa, Russia
e-mail: ekkh@anrb.ru

Volga–Ural region and Central Asia (Table 1) and here we provide a brief review of our earlier and currently obtained results.

It is common to treat the Central Asia and the Volga–Ural region as a crossroad area witnessed in the past numerous waves of population migrations. According to archeological and paleoanthropological data, southern Urals until Early Medieval Age was inhabited by an europeoid population which can be associated with the currently living representatives of different divisions of the Europeoid major race (Pontic type of the Indo-Mediterranean race and Suburalic type which prevail among Volga basin Finnic populations) (Akimova, 1974). Ancient population of the Volga basin could be linked to the present day Baltic and Suburalic types of the Europeoid major race (Akimova, 1974). Alexeev and Gokhman (1984) reviewed paleoanthropological studies and demonstrated that ancient population of the Central Asia region between Neolithic times and until Iron Age was predominantly europeoid (Alexeev and Gokhman, 1984). Turkic-speaking nomadic people from the south Siberia and later Tataro-Mongol invaders are considered to be relatively recent newcomers into Central Asia (Alexeev, 1974; Kuzeev, 1992).

2 Materials and Methods

A total of 1949 individuals were sampled from two regions during field expeditions in 1993–2003 (see details in Table 1). Four Turkic-speaking populations from the north Caucasus region: Karachays (106), Karanogays (102), Kuban Nogays (110) and Kumyks (107) were included into the analysis to provide data for comparison. Informed consent was obtained from all donors. The ethnic origin of sampled individuals was ascertained up to three generations, as determined through oral interviews. Genomic DNA was extracted from peripheral blood lymphocytes using a standard phenol–chloroform method (Mathew, 1984). We also included data on Yakuts, Altaics, Shors, Nganasans, Hants, Mansi, Selkups, Tuvinians, Dolgans, Turks, Azerbaijanians and Estonians for comparison (Khusnutdinova et al., 2002).

The fragment 16,024–16,400 of mtDNA hypervariable segment I (HVS I) was sequenced in a Perkin–Elmer ABI 377 DNA sequencer with a DYEnamic ET kit (Amersham Pharmacia Biotech). The nucleotide sequences obtained were compared with the Cambridge reference sequence (CRS) (Anderson et al., 1981; Andrews et al., 1999). Haplogroups were identified by restriction fragment length polymorphism analysis (RFLP analysis) of 39 polymorphic sites of the mtDNA coding region as described elsewhere (Macaulay et al., 1999; Richards et al., 1998, 2000; Torroni et al., 1996). Forty-two binary polymormphisms (M9, M89, YAP (M1), M174, M40, M35, M130, M48, 12f2, M267, M62, M172, M12, M201, M285, M342, P20, P15, P16, M286, M406, M287, M170, M253, P37, M223, M52, M231, Tat (M46), P43, M128, M175, M20, M70, 92R7, M207, M242, M173, SRY 1532, M73, M269 and M124) were genotyped using either RFLP analysis or direct sequencing, following the hierarchy of the Y-chromosome phylogeny (Jobling and Tyler-Smith, 2003; YCC, 2002). Principal component

Table 1 Geographic and linguistic description of studied populations

Population	Country, region and district	N	Linguistic affiliation
Volga-Ural region			
Bashkirs[a]	Northwestern, northeastern, southwestern and southeastern ethnogeographic groups of Bashkortostan	258	Turkic branch of Altaic Language family
Gaininsk Bashkirs[a]	Perm region	60	
Tatars[a]	Al'met'evskii and Elabuzhskii districts of Tatarstan	228	
Chuvashis[a]	Morgaushskii district of Chuvashia	55	
Komi-Permyaks[a]	Komi-Permyak autonomous district	74	Finno-Ugric subfamily of Uralic language family
Komi-Zyryans[a]	Sysol'skii district of the Komi Republic	62	
Mari[a]	Zvenigovskii district of Marii El	136	
Mordvinians[a]	Staro-Shaiginskii district of Mordovia	102	
Udmurts[a]	Malo-Purginskii district of Udmurtia and the Tatyshlinskii district of Bashkortostan	101	
Central Asia			
Uzbeks	Samarkandskaya, Khorezmskaya, Tashkentskaya provinces of Uzbekistan	103	Turkic branch of Altaic Language family
Kazakhs	Kazakhstan	331	
Uygurs	Kazakhstan	121	

[a]mtDNA data from Bermisheva et al. (2002)

analysis based on the haplogroup frequencies was performed using Statistica v. 5.5 (Statsoft, 1999).

3 mtDNA Variation

The analysis of mtDNA variation among the Volga–Ural populations (Tatars, Bashkirs, Chuvashis, Mari, Mordvins, Udmurts, Komi) revealed that the majority of maternal lineages are those distributed among western European and eastern European populations. Relatively high frequencies of east Asian (those specific to east Asian populations) G, D, C, Z and F haplogroups among both Turkic-speaking population (Bashkirs) and Finno-Ugric speaking populations (Udmurts and Komi-Permyaks) possibly indicate significant genetic contribution of Siberian and Central Asian populations. Comparable high frequencies of east Asian and European lineages are consistent with the intermediate position of the Volga–Ural region but when we take into account linguistic affiliation of individual populations some inconsistent features in distribution of frequencies of mtDNA lineages become apparent.

It is generally believed that the first Turkic-speaking people were native to a region spanning from Central Asia across throughout Siberia (Alexeev and Gokhman, 1984). We examined distribution of mtDNA haplogroups among 18 Turkic-speaking populations scattered throughout Eurasia by assuming that westward expansion of Turkic tribes involved spread of mainly east Asian maternal lineages. Frequencies of east Asian mtDNA lineages ranged from 1% in Gagauz to 99% in Yakuts, with an increasing cline from west to northeast (Color Plate 7, see p. 399). While the frequency cline is consistent with the westward direction of migration proportion of east Asian mtDNA lineages within individual populations is too small to suggest significant genetic input associated with south Siberian populations. For example, Turkic-speaking populations residing in the Volga–Ural region (Tatars and Chuvashis) are more closely related genetically to their geographic neighbors than to their linguistic neighbors elsewhere. It looks like as if indigenous populations of the region adopted Turkic languages without much genetic input as for example Azerbaijanians in Caucasus (Nasidze et al., 2003) and Turks in Anatolia (Cinnioglu et al., 2004). Limited presence of east Asian mtDNA lineages in these two Volga–Ural populations, most likely reflect consequences of language replacements that occurred via elite dominance. In general this observation is consistent with our previous suggestion that the geographic proximity rather than linguistic relatedness best explains genetic similarities between populations (Khusnutdinova et al., 2002).

It should be noted that spread of east Asian lineages from Turkic homeland imply that east Asian lineages were predominant in south Siberia and expectation of high frequencies of east Asian lineages among Turkic-speaking people does not take into account admixture process that occurred in the course of westward migration. More realistic explanation could be offered if we take into account paleoanthropological data from south Siberian and Central Asian area. Excavations at burial sites belonging to Turkic people of VII–VIII AD in Tuva (south Siberian region in Russia) demonstrated detectable europeoid traits in mongoloid human skeletal remains (Alexeev and Gokhman, 1984). Moreover, Alexeev and Gokhman (1984) based on their review of paleoanthropolical studies showed that westward expansion of Turkic people from south Siberia through Central Asia was accompanied by gradual admixture process with indigenous europeoid population (Alexeev and Gokhman, 1984). Thus, limited presence of east Asian mtDNA lineages in the Volga basin (Tatars, Chuvashis) as well as in Caucasus (Azerbaijanians) and Anatolia (Turks) (Cinnioglu et al., 2004; Nasidze et al., 2003) cannot be explained by limited genetic input. In the light of paleoanthropological data more realistic explanation could be the arrival of already admixed Turkic-speaking populations in these three regions which are remote from Turkic homeland. Finally, because of Turkic people were admixed and their migrations were accompanied by admixture process estimation of proportion of gene flow associated with their arrival turns out to be problematic or even impossible.

Geographic frequency distribution of Asian mtDNA lineages among 17 Finno-Ugric populations, with some exceptions, forms a similar west to east increasing cline.

Like in the Turkic-speaking groups Asian mtDNA lineages dominate in northeastern Siberia, reaching maximum frequency in Nganasans (80%) and decrease towards west: for example, in Estonians it is almost absent. A high proportion (60–70%) of typical west Eurasian mtDNA lineages among Hants, Mansi and Selkups living in the west Siberia does not fit to the opposite west east frequency gradient of European lineages. These populations are characterized by high frequency of U4 and low frequency of haplogroup W, which is typical of Finno-Ugric populations of Volga–Ural region. This may be the result of west to east migration from Volga–Ural region (Bermisheva et al., 2002, 2004; Tambets et al., 2003; Villems et al., 2002).

So far we analyzed the mtDNA haplogroup diversity by dividing mtDNA variants into two broad groupings (European and Asian). In order to test whether the frequency distribution of lineages correlates with our previous findings we summarized mtDNA haplogroup frequencies using the principal component analysis (PCA). Figure 1 demonstrates the projection of the Volga–Ural, Caucasus and Central Asian populations in the plot of two principal components of mtDNA variability. As illustrated by the PC plot distribution of populations approximates the previously demonstrated frequency cline. It is evident from the plot that the studied populations are clustered according to their geographic positions. Some deviations, for example, the close position of Nogays in the plot to the Central Asian populations can be explained by the high proportion of typical east Asian mtDNA lineages in this population.

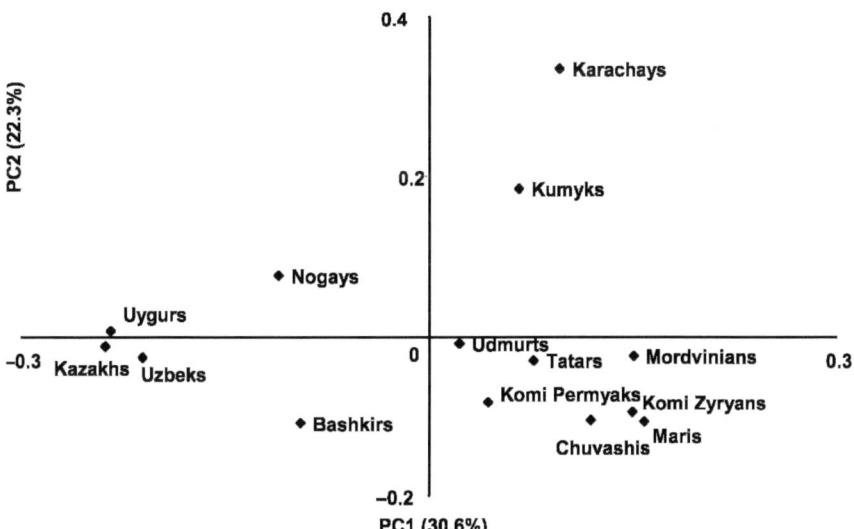

Fig. 1 PC plot based on mtDNA haplogroups frequencies for the 14 populations

Our preliminary summary of obtained results is that the predominantly europeoid population of Central Asia and the Volga–Ural region experienced significant gene flow from eastern mongoloid populations that reshaped eventually the genetic (mtDNA) landscape of the Volga–Ural region and Central Asia.

4 Y-Chromosome Variation

A total of 25 distinct haplogroups were defined in the combined Y-chromosome pool of the Volga–Ural and Central Asian populations. Most paternal lineages of the Volga–Ural and the Central Asian populations belong to the western Eurasian Y-chromosome haplogroups. Sub-clades of widely distributed R haplogroup together with northern Eurasian haplogroup N predominate in all studied populations, constituting about 25% of paternal lineages in Kazakhs and up to 91% in Bashkirs. Except Kazakhs, carrying C (25%) and J (18%) lineages in relatively high frequencies, all studied populations share considerable portion of eastern European lineages. Haplogroup N3, which is spread among northern Eurasian populations (Rootsi et al., 2007; Tambets et al., 2004; Zerjal et al., 2001) according to recent phylogeographic reconstruction has dispersed throughout north Eurasia either from northern China or southern Siberia (Rootsi et al., 2007). High incidence of this lineage among Udmurts (55%) and other Volga–Ural Finno-Ugric populations (Mari, Komi-Ziryans and Komi-Permyaks) is consistent with the scenario, according to which N3 bearing ancient populations witnessed numerous bottlenecks while migrating from Siberia to eastern Europe. Turkic-speaking populations in the Volga–Ural region (Chuvashis, Tatars and Bashkirs) also share this haplogroup but in lower frequencies. The high frequency of this lineage in the Volga–Ural region is likely to be associated with ancient Finno-Ugric tribes and somewhat lower frequency in neighboring Turkic-speaking groups is either a "signature" of Finno-Ugric genetic input or the legacy of common ancestral population for these two linguistically distinct groups. Indeed, taking into account that Karanogays from northeastern Caucasus (N3—3%), Kyrgyz and Dungans from Central Asia (Zerjal et al., 2002) as well as 11 Altaic-speaking south Siberian populations carry this haplogroups (frequency ranges from 2% in Shors to 25% in Tofalars) (Derenko et al., 2006), it is possible that N3 lineage was a common Siberian genetic background of Finno-Ugric and Turkic tribes in the past. Overall, haplogroup N subclades (N2, N3) regardless they were brought by the ancestors of Turkic or Finno-Ugric populations represent a significant amount of southern Siberian genetic legacy.

R subclades (R1b3, R1a1), which are widespread in Europe and West Asia, are distributed unevenly in the Central Asia and the Volga–Ural region, ranging from 1% in Kazakhs to 82% in Trans-Ural Bashkirs. The place and the time of origin of these two ancient lineages are not well established yet (Kivisild et al., 2003). There is an increasing evidence that Y-chromosome haplogroup R1b3 is not specific to west Europe but rather represent a genetic legacy of ancient widely spread population of Eurasia. Recently, Al-Zahery et al. (2003) and Cinnioglu et al. (2004) demonstrated that R1b3 Y-chromosomes in Middle Eastern populations were associated with the Taq I ht35 restriction profile (haplotype of the complex 49a,f RFLP polymorphic system) while west European populations were characterized by the ht15 haplotype. The latter is restricted mainly to Europe and has west east decreasing gradient. Al-Zahery et al. (2003) suggested that haplotype 35, which had uniform distribution across west Eurasia, settled west Eurasia earlier and that haplotype 15 could derived from it by a single mutation event. That mutation probably occurred within Europe and its high incidence in western Europe could be a result of the population bottleneck event after the Last Glacial Maximum (Al-Zahery et al., 2003). These findings are very important because they suggest subdivision of R1b3 lineage into yet undefined lineages, one of which (ht15) is mainly confined to Europe. In this regard decreasing frequency cline of haplogroup R1b3 from western Europe towards eastern Europe and unexpectedly high frequency of this lineage in the eastern most fringe, i.e. southern Urals (>82%) and in Central Asian highlands among Tajiks, Kyrgyz and Altai (>50%) (Kivisild et al., 2003) imply that R1b3 Y-chromosomes in these two eastern regions cannot be associated with the west east population expansion from Europe. In light of the earlier presence of europeoid population in both regions this proposal seems to be plausible. Further genetic evidence is necessary in order to test the above mentioned proposal.

Populations located in the Volga–Ural region exhibit a complex mtDNA and Y-chromosome lineage composition, consisting mainly of western Eurasian lineages, with somewhat lower but comparable contribution of eastern Asian lineages. Central Asians exhibit higher frequencies of east Asian lineages compared with their northern Volga–Ural neighbors. Overall, the pattern of genetic variation in these two regions point to a stronger genetic influence of Turkic-people on Central Asian populations compared with northern neighbors. Finally, genetic relationships between studied populations can be best explained by geographic proximity rather than linguistic affiliation.

References

Akimova, M.S. (1974) Anthropological investigations in Bashkiria. In: V.P. Alexeev (Ed.), Anthropology and Genogeography. Nauka, Moscow, pp. 77–97.

Alexeev, V.P. (1974) Geography of Human Races. Nauka, Moscow.

Alexeev, V.P. and Gokhman, I.I. (1984) Anthropology of the Asian Part of the USSR. Nauka, Moscow.

Al-Zahery, N., Semino, O., Benuzzi, G., Magri, C., Passarino, G., Torroni, A. and Santachiara-Benerecetti, A.S. (2003) Y-chromosome and mtDNA polymorphisms in Iraq, a crossroad of the early human dispersal and of post-Neolithic migrations. Mol. Phylogenet. Evol. 28, 458–472.

Anderson, S., Bankier, A.T., Barrell, B.G., de Bruijn, M.H.L., Coulson, A.R., Drouin, J. et al. (1981) Sequence and organization of the human mitochondrial genome. Nature 290, 457–465.
Andrews, R.M., Kubacka, I., Chinnery, P.F., Lightowlers, R.N., Turnbull, D.M. and Howell, N. (1999) Reanalysis and revision of the Cambridge reference sequence for human mitochondrial DNA. Nat. Genet. 23, 147.
Bermisheva, M.A., Tambets, K., Villems, R. and Khusnutdinova, E. K. (2002) Diversity of mitochondrial DNA haplogroups in ethnic populations of the Volga–Ural region. Mol. Biol. 36, 802–812.
Bermisheva, M.A., Kutuev, I.A., Korshunova, T. Yu., Dubova, N.A., Villems, R. and Khusnutdnova, E.K. (2004) Phylogeographic analysis of mitochondrial DNA in the Nogays: a strong mixture of maternal lineages from Eastern and Western Eurasia. Mol. Biol. 38, 516–523.
Cinnioglu, C., King, R., Kivisild, T., Kalfoglu, E., Atasoy, S., Cavalleri, G.L., Lillie, A.S., Roseman, C.C., Lin, A.A., Prince, K., Oefner, P.J., Shen, P., Semino, O., Cavalli-Sforza, L.L. and Underhill, P.A. (2004) Excavating Y-chromosome haplotype strata in Anatolia. Hum. Genet. 114, 127–148.
Derenko, M., Malyarchuk, B., Denisova, G.A., Wozniak, M., Dambueva, I., Dorzhu, C., Luzina, F., Miscicka-Sliwka, D. and Zakharov, I. (2006) Contrasting patterns of Y-chromosome variation in South Siberian populations from Baikal and Altai-Sayan regions. Hum. Genet. 118, 591–604.
Jobling, M.A. and Tyler-Smith, C. (2003) The human Y chromosome: an evolutionary marker comes of age. Nat. Rev. Genet. 4, 598–612.
Khusnutdinova, E., Bermisheva, M., Malyarchuk, M. et al. (2002) Towards a comprehensive understanding of the east European mtDNA heritage in its phylogeographic context. Meeting "Human Origins and Disease" Cold Spring Harbor, p. 90.
Kivisild, T., Rootsi, S., Metspalu, M., Mastana, S., Kaldma, K., Parik, J., Metspalu, E., Adojaan, M., Tolk, H.-V., Stepanov, V., Golge, M., Usanga, E., Papiha, S.S., Cinnioglu, C., King, R., Cavalli-Sforza, L., Underhill, P.A. and Villems, R. (2003) The genetic heritage of earliest settlers persist in both the Indian tribal and caste populations. Am. J. Hum. Genet. 72, 313–332.
Kuzeev, R.G. (1992) The People of the Central Volga Region and Southern Ural. Nauka, Moscow.
Macaulay, V.A., Richards, M.B., Hickey, E., Vega, E., Cruciani, F., Guida, V., Scozzari, R., Bonne-Tamir, B., Sykes, B. and Torroni, A. (1999) The emerging tree of West Eurasian mtDNAs: a synthesis of control-region sequences and RFLPs. Am. J. Hum. Genet. 64, 232–249.
Mathew, C.C. (1984) The isolation of high molecular weight eukaryotic DNA. In: J.M. Walker (Ed.), Methods in Molecular Biology. Haman Press, New York, pp. 31–34.
Nasidze, I., Sarkisian, T., Kerimov, A. and Stoneking, M. (2003) Testing hypotheses of language replacement in the Caucasus: evidence from the Y-chromosome. Hum. Genet. 112, 255–261.
Richards, M.B., Macaulay, V., Bandelt, H.J. and Sykes, B.C. (1998) Phylogeography of mitochondrial DNA in Western Europe. Ann. Hum. Genet. 62, 241–260.
Richards, M., Macaulay, V., Hickey, E., Vega, E., Sykes, B., Guida, V., Rengo, C., Sellitto, D., Cruciani, F., Kivisild, T., Villems, R., Thomas, M., Rychkov, S., Rychkov, O., Rychkov, Y., Golge, M., Dimitrov, D., Hill E. et al. (2000) Tracing European founder lineages in the Near Eastern mtDNA pool. Am. J. Hum. Genet. 67, 1251–1276.
Rootsi, S., Zhivotovsky, L.A., Baldovic, M., Kayser, M., Kutuev, I.A., Khusainova, R., Bermisheva, M.A., Gubina, M., Fedorova, S., Ilume, A.M., Khusnutdinova, E.K., Osipova, L.P., Stoneking, M., Ferak, V., Parik, J., Kivisild, T., Underhill, P. and Villems, R. (2007) A counter-clockwise northern route of the Y-chromosome haplogroups N from Southeast Asia towards Europe. Eur. J. Hum. Genet. 15, 204–211.

StatSoft, Inc. (1999) STATISTICA for Windows (Computer program manual). Tulsa, OK: StatSoft, Inc., Web: http://www.statsoft.com.

Tambets, K., Tolk, H., Kivisild, T., Metspalu, E., Parik, J., Voevoda, M., Damba, L., Golubenko, M., Stepanov, V., Puzerev, V., Bermisheva, M., Knushnudtinova, E., Usanga, E., Rudan, P. and Villems, R. (2003) Complex signals for population expansions in Europe and beyond. In: P. Bellwood and C. Renfrew (Eds.), Examining the Farming/Language Dispersal Hypothesis. McDonald Institute for Archaeological Research, Cambridge, pp. 449–458.

Tambets, K., Rootsi, S., Kivisild, T., Help, H., Serk, P., Loogvali, E.-L., Tolk, H.-V., Reidla, M., Metspalu, E., Pliss, L., Balanovsky, O., Pshenichnov, A., Balanovska, E., Gubina, M., Zhadanov, S., Osipova, L., Damba, L. et al. (2004) The Western and Eastern roots of the Saami—the story of genetic "outliers" told by mitochondrial DNA and Y chromosomes. Am. J. Hum. Genet. 74, 661–682.

Torroni, A., Huoponen, K., Francalacci, P., Petrozzi, M., Morelli, L., Scozzari, R., Obinu, D., Savontaus, M.L. and Wallace, D.C. (1996) Classification of European mtDNA from an analysis of three European populations. Genetics 144, 1835–1850.

Villems, R., Rootsi, S., Tambets, K., Adojaan, M., Orekhov, V., Khusnutdinova, E. and Yankovsky, N. (2002) Archaeogenetics of Finno-Ugric-speaking populations. In: K Julku (Ed.), The Roots of Peoples and Languages of the Northern Eurasia. Gummerus OY, Juvaskyla, pp. 271–284.

Y Chromosome Consortium (YCC) (2002) A nomenclature system for the tree of human Y-chromosomal binary haplogroups. Genome Res. 12, 339–348.

Zerjal, T., Beckman, L., Beckman, G., Mikelsaar, A.-V., Krumina, A., Kucinskas, V., Hurles, M.E. and Tyler-Smith, C. (2001) Geographical, linguistic, and cultural influences on genetic diversity: Chromosomal distribution in Northern European populations. Mol. Biol. Evol. 18, 1077–1087.

Zerjal, T., Wells, R.S., Yuldasheva, N., Ruzibakiev, R., and Tyler-Smith, C. (2002) A genetic landscape reshaped by recent events: Y-chromosomal insights into Central Asia. Am. J. Hum. Genet. 71, 466–482.

Problems of Reconstruction of Paleoenvironment and Conditions of the Habitability of the Ancient Man by the Example of Northwestern Altai

A. K. Agadjanian

Abstract Using the example of the northwestern Altai, the reconstruction of environment and conditions of habitats of the Paleolithic man are considered. It is shown that, in the Late Pleistocene of the Anui valley, the climate became more continental on the background of alternation of dry and humid climatic phases. Communities show a mosaic structure. Clear relationships between natural conditions of the Late Pleistocene, the structure of animal populations, and the state of the population and mode of life of the Paleolithic man are established.

1 Introduction

A topical problem in the study of the biosphere is the reconstruction of its development at the late stage of geological history, the position and role of the Paleolithic man in its structure. The resolution of these questions is important for a better understanding of the role of the modern man in the ecology of natural systems, an appraisal of the effect of economic activity on various components of the biosphere. Large regional ecological crises connected with human activity occurred at least 2000 years ago and even earlier. It is necessary to detect and study in detail these events and compare with the development of natural systems. This will provide the elaboration of prognostic models for the behavior of modern and future natural associations and the revelation of factors and parameters of human impact on biological resources and, hence, the development of measures for environmental protection.

A. K. Agadjanian
Paleontological Institute, Russian Academy of Sciences, Moscow, Russia
e-mail: aagadj@paleo.ru

2 Material

The area studied is located in the western part of the Altai Mountains, in the middle and upper reaches of the Anui River. This is a system of mountain ranges separated by deep valleys, with the altitudinal fluctuations from 500 to 2300 m above sea level. Altitudinal zonation of vegetation is well pronounced. The mountain slopes are mainly covered with taiga forests. River valleys mostly contain meadow–steppe associations. Communities of the nival type occur at elevations above 1800 m.

Both closed (Denisova, Kaminnaya, Ust'-Kanskaya caves) and open (Ust'-Karakol, Anui-2, Anui-3, etc.) Paleolithic sites were examined.

Open Paleolithic sites and later human habitats were studied in Russia for a long time. Classical works in this field were performed by Pidoplichko (1929, 1934, 1936), Gromov (1953, 1957, 1961) and Gromov and Fokanov (1980). Mammal remains from the Paleolithic beds of the Anui basin were collected and identified from the first stage of their study (Agadjanian, 1998, 1999; Agadjanian and Shun'kov, 1999, 2001; Agadjanian et al., 1999; Derevyanko and Markin, 1992; Derevyanko and Molodin, 1994; Derevyanko et al., 1998, 1999; Germonpré, 1993; Ivleva, 1990; Ovodov and Ivleva, 1986; Shun'kov and Agadjanian, 2000; Vasil'ev and Grebnev, 1994).

Apparently, an important factor of the accumulation of material in open sites was man. All Paleolithic sites of the Desna, Dnieper, and the lower reaches of the Don have yielded abundant marmot bones. Beavers are abundant in later settlements in the central part of the Russian Plain. Hares occur in almost all sites. Marmot bones from the Kamennaya Balka site in the southern Russian Plain were exposed to fire and occurred in the fire-spots (Agadjanian, 2006). Marmot bones from the Betovo Paleolithic site on the Desna River show clear traces of dissection of the skeleton by sharp cutting tools. Beaver bones from the Mesolithic settlement of Yazykovo in the Tver Region and the D'yakovo site in Kolomenskoe (Moscow) have traces of flaying, dissection of corpses, and separation of meat from the bones. The widespread use of marmot as food is corroborated by the material from Paleolithic sites of France (Patou, 1987).

3 Results

3.1 Contemporaneous Animals of the Anui Valley

The study performed shows the pattern of altitudinal zonation and the structure of the small mammal community. The following biotopes are recognized in the Anui drainage basin: floodplain–meadow, meadow–steppe, larch–birch, birch–pine, cedar pine forests, subalpine moss–shrub, mountain–steppe petrophilous, and nival sedge-grass. Each association is distinguished by its small mammal assemblage. It is shown that the vicinity of Denisova cave is dominated by the vole *Clethrionomys rutilus* Pallas, which comprised the majority of small mammals in almost all biotopes (39% of animals caught). On the forested slopes and cedar pine forests, this vole composes up to 80% of animals. Even on dry forestless slopes, in the river and meadow biotopes, the proportion of the root vole is 20–30%. The second most abundant group is composed of common voles of the genus *Microtus*, which occur in the biotopes with a low relief. *Microtus arvalis* Pallas is most abundant (18.5%). The floodplain is inhabited by *M. oeconomus* Pallas (4%). *Apodemus uralensis* Pallas and *A. peninsulae* Thomas are common on dry slopes covered with acacia (up to 30–40% of small mammals). In the lower part of the slopes, with a well-developed soil cover, the Altai mole *Asioscalops altaica* Nikolsky occurs frequently. The proportion of other small mammal species is less than 3%; however, this poorly corresponds with their role in communities. In particular, the chipmunk *Eutamias sibiricus* Laxmann frequently occurs at all altitudes from the floodplain of the Anui to the near-top mountain sites. The ground squirrel *Spermophilus undulatus* Pallas inhabits flat slopes with poor herbage in the Anui and Karakol valleys. The Baraba (or Dahurian) hamster *Cricetulus barabensis* Pallas sometimes occurs in the grassy-forb sites. In biotopes with dense forbs on the floodplain and flat slopes, two species of *Sicista*, *S. betulina* Pallas and *S. napaea* Hollister, are recorded. On high terraces and flat slopes, the zokor *Myospalax myospalax* Laxmann is rather common. It is safe to assume that at present the Anui valley is dominated by taiga mammal assemblages. In particular, they prevail in the vicinity of Denisova cave position. Meadow communities are less significant, while steppe associations are absent from this area. However, isolated representatives of these associations occur in agricultural and meadow communities.

3.2 Small Mammals from the Pleistocene of Denisova Cave

Deposits of Denisova cave yielded a total of 34,663 of identifiable bone specimens. Taphocenoses from different sites of the cave floor are rather similar; this makes the results obtained rather reliable. The list of small vertebrate taxa from the Pleistocene of Denisova cave includes more than 40 species.

The distribution of vertebrate bone remains over the sedimentary sequence of Denisova cave allows the reconstruction of environmental changes in the vicinity of the cave during the accumulation of the Pleistocene beds. Comparison of the modern fauna of the Anui canyon, its slopes, and adjacent sites of the valley with the fossil fauna (taphocenosis) from the Pleistocene beds of Denisova cave shows their significant differences. In particular, the modern fauna of the Anui basin lacks characteristic elements of steppe communities, such as marmot, Eversmann's hamster, mole-vole (*Ellobius*), lemming, and *Lagurus lagurus*. All these taxa occur here as fossils. As for the species present in both modern and fossil communities, they substantially differ in abundance. The modern fauna of the Anui basin is dominated by voles of the genus *Clethrionomys*, characteristic indicators of taiga conditions. The Pleistocene fauna of the cave is dominated by the voles *Stenocranius gregalis* and *Alticola strelzovi*, which inhabit dry steppes and tundras or highland steppes. In the modern fauna, Asian wood mice of the genus *Apodemus* (*A. uralensis* and *A. peninsulae*), representatives of the southern taiga subzone, play an important role. These species are extremely rare in the Pleistocene beds of the cave. In the modern fauna, the second most important group after red-backed voles is common voles (*Microtus*), specialized vegetarians living in herbage meadows. In the fossil fauna of Denisova cave, they compose not more than 5%. The pika *Ochotona* is currently not numerous and has a patchy distribution in the area studied, but it is relatively common in the fossil record. This indicates changes in the composition of the small mammal fauna at the Pleistocene–Holocene boundary. They were expressed in a reduction of nival and steppe elements of the biota and increase in the proportion of taiga species due to reorganization of ecosystems in the Anui valley and, probably, throughout the Altai Mountains at the Late Pleistocene–Holocene transition.

The turnover of ecosystems should not be overestimated. Both the fossil record and extant fauna constantly include regional endemic species, such as the Altai mole *A. altaica* and the zokor *M. myospalax*. Several characteristic East Siberian species occur in both the modern biota and the fossil record: long-tailed ground squirrel, chipmunk, Siberian ruddy vole, flat-headed vole, Baraba hamster, and other forms. This suggests that changes in the ecological composition of communities proceeded against a background of a generally stable zoogeographic situation. The community of the area studied mainly originated from the Altai Mountains and adjacent territories.

Nival (*Alticola*) and steppe (*Spermophilus*, *Stenocranius*, and *Lagurus*) taxa dominate along the studied sequence pointing to a more important role of communities of open landscapes than at present. The area of scree slopes occupied by mountain–steppe petrophyte associations was much larger. High terraces, gentle slopes, and near-top plateaus were occupied by light steppe vegetation rather than by high meadow herbage with dense turf, as today. The presence of marmots in the taphocenosis, though probably underestimated, confirms this. An indicator of dry steppes and semideserts is the mole-vole *Ellobius*, which sporadically occurs in the Pleistocene beds of Denisova cave. Today, its nearest populations occur in steppe Altai. The continuous presence of the Altai pika indicates extensive scree areas.

However, the domination of steppe and nival communities was not constant and complete. All Pleistocene beds of the section contain remains of the Siberian ruddy vole *C. rutilus*. This is evidence of a stable existence of forest vegetation in the site investigated in the Anui valley. The presence of chipmunk and, especially, red squirrel and flying squirrel, typical arboreal species, fully confirms this conclusion. The sites of larch–birch and birch–pine and possibly Siberian cedar pine forests, similar to modern forests, existed throughout the Late Pleistocene.

3.3 Small Mammals from the Ust'-Karakol Paleolithic Site

All specimens of the Ust'-Karakol site come from the natural sequence of soils, loesses, floodplain, and alluvial deposits. A larger part of the material belongs to species that lived directly in this part of the valley. The structure and texture of beds strongly suggest that vertebrate bones and mollusk shells were included in the strata without a distinct effect of dynamic or biogene processes. This gives a special significance to the material of the open site Ust'-Karakol. The total number of determined specimens is 1908. The list of small mammals of the taphocenosis includes about 30 species. Small mammal bones are unevenly distributed over the beds.

Ground squirrels and zokor were the major components of the small mammal community in the Anui valley during the accumulation of the Pleistocene strata. The group of codominants includes shrews, mole, red-backed vole, narrow-skulled vole, and, partly, common hamster and root voles. The group of rare species includes jerboa, mole-vole, Lagurus, lemmings of the tribe Lemmini, the gray voles *M. hyperboreus* and *M. agrestis*, the water vole Arvicola, and pika. The group of very rare species includes red squirrel, chipmunk, marmot, Baraba hamster, birch mouse, and the Asian highland vole Alticola. In all the features listed, the fossil fauna of Ust'-Karakol is essentially different from the taphocenosis of Denisova cave; this reflects local biotopic conditions. In contrast to Denisova cave, the fauna completely lacks bats and has very rare Alticola and very few hamsters.

The proportions of the major ecological groups of small mammals and their dynamics trace the general pattern of the change in faunal composition and environment in the Upper Pleistocene of the Anui valley near the mouth of the Karakol River. In general, from Bed 19B to Bed 11A, the assemblage was dominated by ground squirrels, while from Bed 11B, by zokor. Forest species, such as the red squirrel, chipmunk, red-backed voles mostly occur in the lower part of the section (Beds 18A to 11A). Most records of Lemmini also come from this interval (Beds 18B to 14). In the taphocenosis of the upper part of the section, starting from Bed 11A, an important role is played by the common hamster and narrow-skulled vole with the total domination of zokor. Thus, essential changes of paleogeographic conditions at the boundary of Beds 11B and 11A are assumable.

4 Mammal Fauna and Activity of the Paleolithic Man

The above data show that environmental and climatic changes in the Pleistocene of the Anui basin were directional. The Late Pleistocene is characterized by a general decrease in the heat supply and an increase in the continentality of the climate. The general direction of the process was complicated by periodic fluctuations caused by alternation of relatively dry and more humid climatic phases. Vertical zonation was more complex than in the present day. The mosaic structure of communities existed during the entire Middle and Late Pleistocene. It was determined by climatic features, slope aspect, orientation of mountain ridges and river valleys. Even little changes in average annual temperature and humidity resulted in increasing expansion of one or several zonal types of ecosystems, creating an overall unique mosaic of landscape conditions. These events influenced the history of the formation and development of the Paleolithic man.

Against the background of gradual transformations of the environment inferred from the material of Paleolithic sites of the Anui valley, the sharp and significant drop in the number of bats at the transition from Bed 22 to Bed 19 and higher beds of Denisova cave is very surprising. This phenomenon is not accidental because it is well expressed in different squares of the excavation. The thickness of Bed 22 is equal to that of the whole above-lying sequence of Pleistocene deposits. It essentially differs from it in lithology, structure, texture, and color. Even taken alone, this points to a change in sedimentation mode between Beds 21 and 19. This transition also shows a reduction in the abundance of small forest mammals and an increase in the proportion of steppe species. The composition of large mammals experienced drastic changes. The brown bear prevails among carnivores in Bed 22, whereas remains of the cave hyena dominate in the upper part of Pleistocene deposits. The proportion of bear bones becomes more than 11 times greater downward in the section. Remains of hyena demonstrate the reverse distribution, increasing almost twice upward in the section (Fig. 1). In the upper part of Pleistocene deposits, the proportion of ungulate bones increases (Fig. 2). Simultaneously, the amount of artifacts and tools grows drastically above Bed 22 (Derevyanko et al., 1998) (Fig. 3).

Judging from the small mammal fauna from the Pleistocene deposits of Denisova cave, the environment changed towards the expansion of open biotopes and a slight decline of forest vegetation during the accumulation of Beds 21 and 20. This inevitably resulted in increased areas and biomass of grass communities. The increase in the area and productivity of pastures led to an increase in population numbers and species diversity of ungulates, which, in turn, determined the growth of the population density of the Paleolithic man. As a consequence, humans visited the cave more frequently, which is reflected in the increased amounts of artifacts in cave sediments. Abundant charcoal and fire-spots in Beds 21 to 9 indicate that the presence of humans became long term and was accompanied by setting of fires. All these factors disturbed and negatively influenced bats and other inhabitants of the cave. This was the main reason for the sharp reduction of the bat colony.

The increased area of grasslands and density of ungulates resulted in the increase in the population of hyenas, as shown by the osteological material of Denisova cave. The quantitative ratio of bears and hyenas is also indicative. Bones of young animals prevail among bear remains in Bed 22. This suggests that bears used the cave in winter as hibernation shelter and birthplace of cubs. It is quite clear that the cave could not function simultaneously as a bear's lair and

Problems of Reconstruction of Paleoenvironment

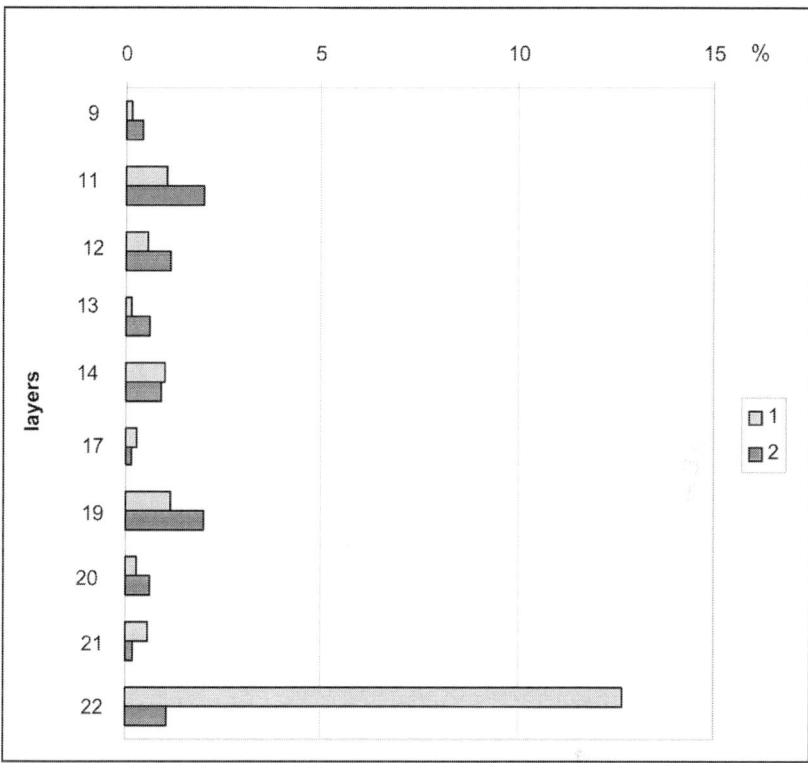

Fig. 1 Proportions of bear and hyena bones in the beds of Denisova cave. *1* abundance bones of bear; *2* abundance bones of hyena

a human dwelling. This suggests infrequent visits of humans to the cave during the accumulation of Bed 22, which is supported by the small amount of artifacts found. After the change of environmental conditions, an increase in the areas of meadow and steppe biotopes, growth in numbers of ungulates, hyenas, and humans, the bear was forced out of the cave. The facts listed above confirm this conclusion.

It is more difficult to explain how humans and hyenas could share the cave in the periods of their activity. It is known that hyenas need a protected shelter for giving birth to cubs only in spring and early summer. On the contrary, humans probably used the cave as a shelter primarily in fall and winter periods. Cave walls, especially combined with the use of fire, gave protection from the cold. It is the fire use that would have had an unfavorable effect on the bat colony which inhabited the cave at previous stages of its history. However, life in the cave also had its limitations for humans. Denisova cave is located in a narrow part of the Anui canyon, which restricts a wide outlook on the area and prevents hunting of moving herds of ungulates. Therefore, at the beginning of a warm season,

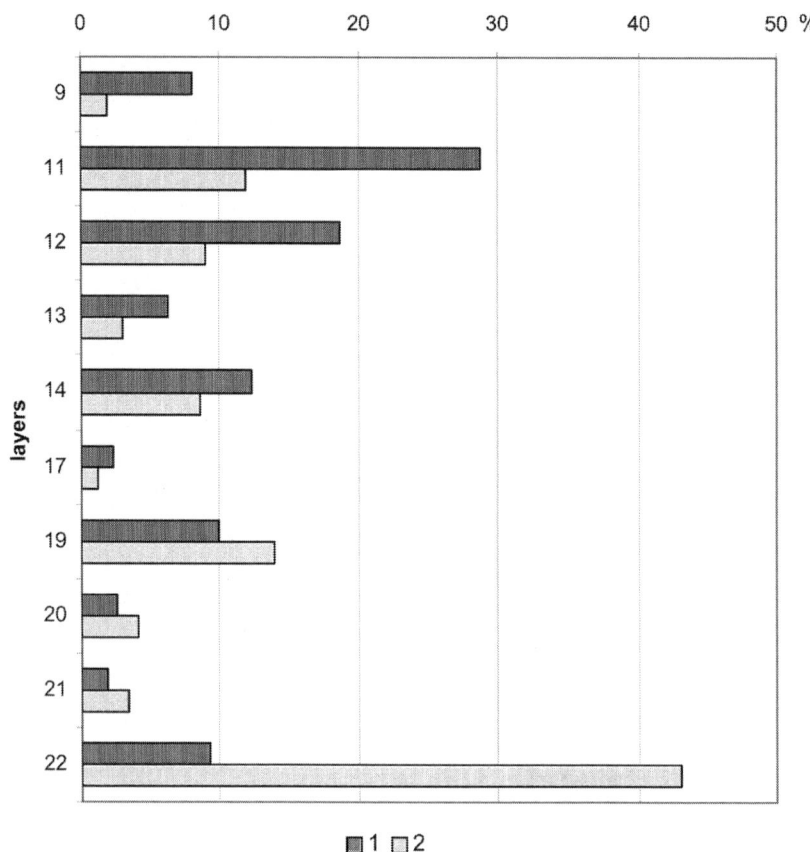

Fig. 2 Proportions of large mammal bones in the beds of Denisova cave. *1* abundance bones of large ungulate; *2* abundance bones of large predator

humans most likely moved to open-air sites, such as Ust'-Karakol, which provide an excellent view on the Anui and Karakol valleys and adjacent mountain slopes. This gave a visual control of an extensive space and, therefore, enabled a fast response by hunters to approaching ungulate herds.

There were probably other reasons why humans seasonally abandoned the cave. For example, during the long winter season, sleeping skins and the people themselves could have been increasingly infected with ectoparasites. In addition, during the spring and summer period, the air temperature inside the cave is much lower than outside. Measurements carried out in July to August of 2002 showed that, in the galleries of the cave, the day temperature keeps within the interval of 8–9 °C, and in the range of 12–14 °C in the central hall. On the same days, the daily fluctuations of the air temperature in the Anui valley ranged from 14 to 23 °C, the day temperature was at the level of 20–23 °C, sometimes reaching 31 °C. Therefore, additional efforts are required to keep the cave warm

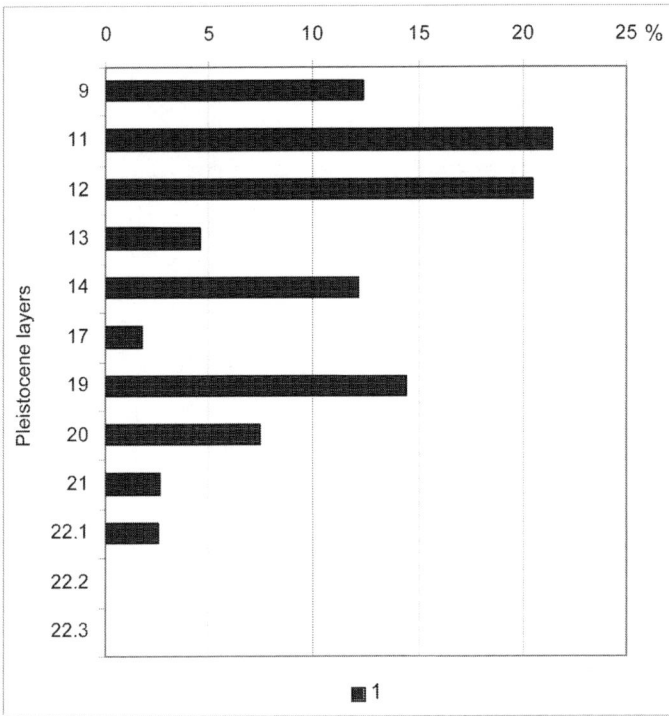

Fig. 3 Number of tools of the Paleolithic man from the beds of Denisova cave. *1* abundance of tools

in summer. It is also noteworthy that most of the dry brushwood in the cave's vicinity was consumed during the fall–winter period. In this situation, the move of its inhabitants to open air summer sites became almost inevitable. Thus, humans and hyenas could use the cave in different seasons of the year.

The data on bones of large mammals of Denisova cave reveals another regularity. As indicated by archeological materials, the presence of humans in the cave during the accumulation of Bed 22 was minimal, as well as their influence on the formation of the taphocenosis of this bed (Derevyanko et al. 2003). During this interval, bones of ungulates were brought into the cave only by carnivores. The quantitative ratio of bones of predators and ungulates in Bed 22 is 4.73:1, which corresponds to their natural ratio in taphocenoses formed without human influence. This ratio of predator to victim bones would have been retained during all the time of the formation of the cave's taphocenoses in the absence of humans. However, the actual data indicate a steady increase through the time in relative (and absolute) numbers of bones of herbivores.

The proportions reported allow the share of bones brought by predators and by humans to be calculated for every bed of Denisova cave. The results obtained are shown in Fig. 4.

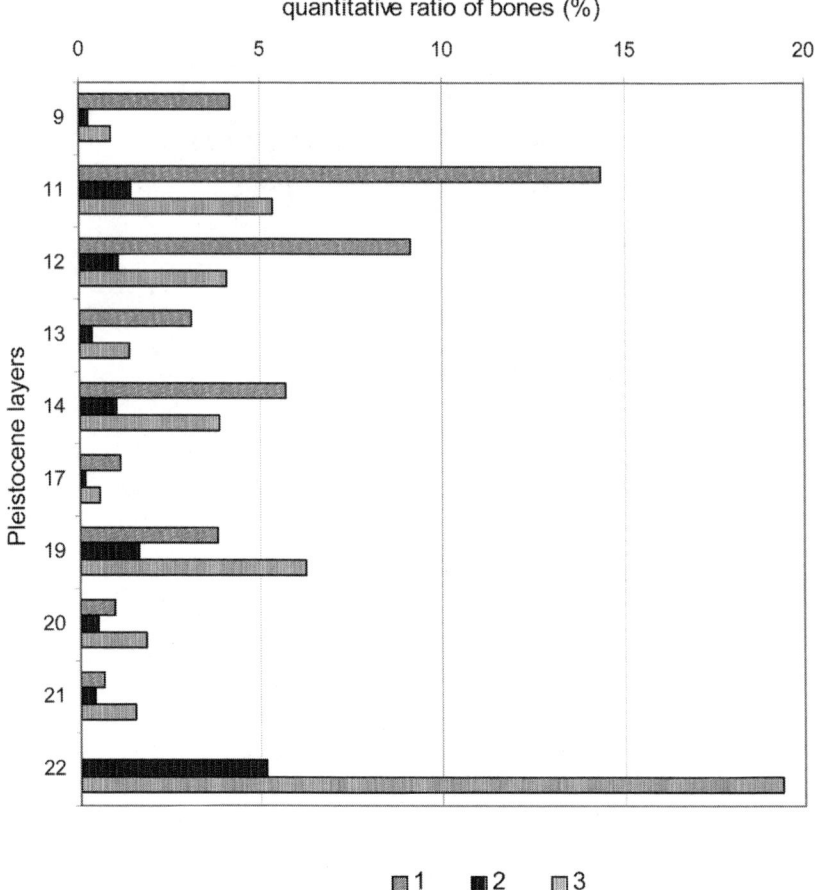

Fig. 4 Proportions of large herbivore bones brought to Denisova cave by large carnivores and humans during the Late Pleistocene. *1* bones abundance of large ungulate brought to Denisova cave by Paleolithic humans; *2* bones abundance of large ungulate brought to Denisova cave by large carnivores; *3* bones abundance bones large carnivores in Denisova cave

Figure 4 clearly shows a sharp increase during the Late Pleistocene in the amount of bones of herbivores brought by humans. This, in turn, reflects the increased human pressure on the populations of ungulates and the environment in general.

The data presented reveal the existence of intricate links between natural conditions of the Late Pleistocene, the composition of animal communities, the condition of populations, and the economy of prehistoric people. It is assumed that these links were not only unidirectional but had a feedback. The activity of Paleolithic people exerted a considerable influence on natural resources of the Anui valley throughout the last 100,000 years or even more.

Acknowledgment The study was supported by the programs of Presidium of the Russian Academy of Sciences "Origin and Evolution of the Biosphere," "Biological Resources of Russia: Fundamental Bases of Efficient Use," and the Russian Foundation for Basic Research, project no. 05-04-48493a.

References

Agadjanian, A.K. (1998) Small mammal fauna of Denisoava cave // Paleoecology and Paleolithic culture of century of Northern Asia and neighbouring territories. (The materials of international symposium), volume 1. Novosibirsk: Institute archeology and ethnography SB RAS. pp. 34–41

Agadjanian, A.K. (1999) Small mammals of Holocene beds of Denisova cave. Problems of Archaeology, Ethnography, and Anthropology of Siberia and Adjacent Territories. Inst. Arkheol. Etnogr. Sib. Otd. Ross. Akad. Nauk, Novosibirsk, vol. 5, pp. 226–231.

Agadjanian, A.K. (2006) Small mammals of the major layer of the Upper Paleolithic site of Kamennaya Balka II. Paleoecology of the Paleolithic Plain. Nauchnyi Mir, Moscow, Appendix no. 1, pp. 318–328.

Agadjanian, A.K. and Shun'kov, M.V. (1999) The rests of small mammal from sediments Paleolithic site Anuy-3. // the problem archeology, ethnography, anthropology of Siberia and neighbouring territories. Volume V. Institute archeology and ethnography SB RAS, Novosibirsk, pp. 6–10.

Agadjanian, A.K. and Shun'kov, M.V. (2001) The microtheriological characteristic of multilayer profil Paleolithic site Ust'-Karakol-1 // Problem archeology, ethnography, anthropology of Siberia and neighbouring territories. Novosibirsk: Institute archeology and ethnography SB RAS, Novosibirsk, Volume VII. pp. 37–42.

Agadjanian, A.K. Malaeva, L.M., Shun'kov M.V. (1999) Experience of reconstruction of a natural environment of the Paleolithic man of Denisova cave. // Ecology of ancient and modern communities. Tyumen: Institute of problems of development of North SB RAS, 1999, pp. 9–12.

Derevyanko A.P. and Markin, C.V. (1992) Musterian of Mountain Altai. Novosibirsk: Science. 225 p.

Derevyanko, A.P. and Molodin, V.I. (1994) Denisova Cave. Nauka, Novosibirsk.

Derevyanko, A.P., Agadjanian, A.K., Barysnikov, G.F., et al. (1998) Archaeology, Geology and Paleogeography of the Pleistocene and Holocene of the Altai Mountains. Inst. Arkheol. Etnogr. Sib. Otd. Ross. Akad. Nauk, Novosibirsk.

Derevyanko, A.P., Shun'kov, M.V., Agadjanian, A.K., et al. (1999) Multilayer paleolithic site Anui-3: results of work in 1999. Problems of Archaeology, Ethnography, and Anthropology of Siberia and Adjacent Territories. Inst. Arkheol. Etnogr. Sib. Otd. Ross. Akad. Nauk, Novosibirsk, vol. 5, pp. 111–116.

Derevyanko, A.P., Shun'kov, M.V., Agadjanian, A.K., et al. (2003) Natural Environment and Man in the Paleolithic of the Altai Mountains. Inst. Arkheol. Etnogr. Sib. Otd. Ross. Akad. Nauk, Novosibirsk.

Germonpré, M. (1993) Preliminary results of the taphonomic study of Denisova cave (based on the material of excavation in 1992). Altaica 2, 11–16.

Gromov, I.M. (1953) Vertebrate fauna from the Tardenauzskaya Murzak-Koba site in the Crimea. Mat. Issled. Arkheol. SSSR. 39, 459–462.

Gromov, I.M. (1957) Upper Quaternary rodents of the Samarskaya Luka river bend and conditions of the burial and accumulation of their remains. Tr. Zool. Inst. Akad. Nauk SSSR. 22, 112–150.

Gromov, I.M. (1961) Fossil Upper Quaternary rodents from the Submontane Crimea. Tr. Komiss. Izuch. Chetvertichn. Per. 17, 1–189.

Gromov, I.M. and Fokanov, V.A. (1980) On Late Quaternary rodent remains from Kudaro 1 cave. Kudaro Cave Site in Southern Ossetia. Nauka, Moscow, pp. 79–89.

Ivleva, N. G. (1990). Microtheriological materials from Okladnikov and Denisova cave on Altai // Complex researches of Paleolithic objects of Anuy basin. Novosibirsk: INF&F SB AN USSR. pp. 257–292.

Ovodov, N.D. and Ivleva, N.G. (1986) Pleistocene theriofauna of Denisova cave (Altai) based on material of excavation in 1982–1984. Quaternary Geology and Primeval Archaeology of Southern Siberia. Buryat. Fil. Sib. Otd. Akad. Nauk SSSR, Ulan-Ude, Part 1, pp. 88–90.

Patou, M. (1987) Les marmottes: animaux intrusifs ou gibiers des prehistoriques du Paleolithique. Archaeozoologia 1, 93–107.

Pidoplichko, I. (1929) Rodents and lagomorphs from a locality near the village of Zhuravtsy, Prilutsk Region. Antropologiya 3, 133–147.

Pidoplichko, I.G. (1934) Discovery of a mixed tundra-steppe fauna in the Quaternary beds of Novgorod-Seversk. Priroda 5, 80–82.

Pidoplichko, I.G. (1936) Fauna of the Gontsovo Paleolithic site. Priroda 1, 113–116.

Shun'kov, M.V. and Agadjanian, A.K. (2000) Paleogeography of the Paleolithic of Denisova cave. Arkheol. Etnogr. Antropol. Evraz. 2(2), 2–19.

Vasil'ev, S.K. and Grebnev, I.E. (1994) The Holocene mammal fauna from Denisova cave. In: A.P. Derevyanko and V.I. Molodin (Eds), Denisova Cave. Nauka, Novosibirsk, Part 1, pp. 167–180.

The Settling of the Ancient Man by the Example of North-Western Altai

A. P. Derevianko and M. V. Shunkov

Abstract Archaeological collections from the Pleistocene deposits of North-Western Altai are the best studied Quaternary materials from all of Northern and Central Asia. Stratified sites in that region have yielded the longest archaeological record, which includes Middle and Upper Pleistocene, covering all the stages of the Paleolithic, from the early to the late. Archaeological materials from these sites illustrate the gradual evolution of lithic industries and attest to the continuity of the technological traditions throughout the Paleolithic.

The Paleolithic complex in the Altai Mountains currently provides the most complete information about early human occupation in Northern and Central Asia. It is best represented in studies of multilayered Paleolithic sites in the Anui River valley of North-Western Altai. The general stratigraphic profile, which includes Middle and Upper Pleistocene deposits, has been defined based on data obtained from a number of sites in this region. Multidisciplinary research at the stratified Paleolithic sites in the Anui valley using archaeological, litho-stratigraphic and paleontological methods has made it possible to trace the origins and evolution of Paleolithic cultural traditions and to reconstruct the conditions of early human habitation throughout a long phase of the Pleistocene.

The occupation of the Altai by early humans is most likely connected with a northern migration wave of *Homo erectus* who expanded beyond the boundaries of the African continent and reached Asia. According to the dates that have been recently generated from the loess and soil from the Kuldara, Khonako II and Obi-Mazar-6 sites in Tajikistan, *Homo erectus* arrived in Central Asia in the chronological range of 600–900 ka (Ranov, 2001). The most archaic pebble tools that probably correspond to roughly that same

---

A. P. Derevianko
Institute of Archaeology and Ethnography, Siberian Branch, Russian Academy of Sciences
Novosibirsk, Russia
e-mail: derev@archaeology.nsc.ru

time have been reported from the northeastern piedmonts of Karatau in Kazakhstan (the Borykazgan, Tanirkazgan and Akkol sits) (Alpysbaev, 1979) and from the northern portion of the Valley of Lakes in Mongolia (Nariyn-Gol-17) (Derevianko et al., 2000c). These industries are characterized by irregular orthogonal cores, "citron" spalls, massive tools reminiscent of *racloirs* and large cutting tools resembling chopper/chopping tools.

The earliest stage of the human presence in the Altai Mountains is evidenced by archaic artifacts found in the red deposits at Karama site, dating from early Middle Pleistocene (Derevianko and Shunkov, 2005). Karama is located in the Anui valley 15 km downstream from the well-known archaeological site of Denisova Cave. The Karama site has yielded a cultural sequence of several human occupation horizons bearing distinct pebble tools attributable to the Lower Paleolithic in the time range of 600–800 ka BP.

The characteristic features of the lithic implements suggest their attribution to the Lower Paleolithic pebble industries. The assemblage of the products of primary reduction includes pebbles showing signs of core preparation with plain striking platforms and negative scars of parallel detachments and short non-faceted spalls. The collection of typologically distinct tools includes longitudinal and transverse *racloirs,* denticulate and notch-denticulate tools fashioned on short spalls and cutting tools of the chopper/chopping tool type with a convex flattened cutting edge and a trimmed massive back. Most pebble tools from Karama are characterized by archaic morphological features and comparatively advanced technology of secondary treatment.

The next stage in the development of the Altai Paleolithic is illustrated by the Early Mousterian industries from the basal sediments at Denisova Cave (strata 22 and 21) and from the alluvial sediments of stratum 19 at the lowermost portion of the Ust-Karakol site. Various dating methods suggest that the age of these lithological strata lies in the range of 133–282 ka, which corresponds to the second half of the Middle Pleistocene (Derevianko et al., 2003).

The most ancient industries of the Denisova Cave demonstrate the Levallois features in stone reduction and a preferable usage of flakes as blanks for tool manufacturing. Various types of *racloirs* and notch-denticulate tools prevail in the tool kit. Most spalls identified within the Ust-Karakol industry from stratum 19 show parallel edges on the dorsal face and a prepared platform. Such categories as *racloirs* with longitudinal and convergent edges, spur-like tools and notched tools with Clactonian and retouched *encoches* have been identified within the tool kit. A notable absence of tools made on complete pebbles and Acheulian bifaces, together with the features of parallel reduction and a set of typologically distinct implements made on standard blanks, all suggest a Middle Paleolithic attribution to the most ancient industries of the Denisova and Ust-Karakol.

The chronological attribution of the Altai early Mousterian industries to the Middle Pleistocene seems reasonable when comparing it to archaeological evidence from other Eurasian Paleolithic sites. Archaeological materials from Western and Central Europe have shown that pre- and early Mousterian industries with flake tools but without Acheulian bifaces appeared along with typical Acheulian technocomplexes as early as the initial Riss period (Bosinski, 1982; Tuffreau, 1982). It is known that *racloirs*, notches and denticulate tools were the most characteristic flake tools for certain early Mousterian industries (Laville, 1982). Recent geo-chronological estimations of the age of true Mousterian industries of the Tabun Cave in the Near East have suggested a period 250,000–270,000 years ago (Bar-Yosef, 1995; Mercier et al., 1995).

The Altai Middle Paleolithic industries continued their development in the Upper Pleistocene. Available Paleolithic evidence from the Altai testifies to the fact that the majority of Mousterian sites exhibit common features that evolved within a single Middle Paleolithic culture. However, various Altai technocomplexes reveal different proportions of the major technical–typological indices within this single cultural tradition. On the basis of these variations, two major types of industries have been established in the Altai Mousterian sites: industries with predominantly Mousterian technology and those with distinct Levallois tools.

The Mousterian group of industries includes collections recovered from the Denisova and Oklandikov cave sites as well as from the open-air Tiumechin-1 site. The primary reduction strategy is predominantly parallel and radial. Levallois reduction is apparent on only a few artifacts. In general, the impact of the Levallois technique on the technological process seems insignificant. The majority of tools were produced on medium-sized, short spalls. The collection of typologically distinct tools is dominated by Mousterian and notch-denticulate tools. Levallois implements are morphologically distinct but scarce. Various *racloirs* are most numerous. On the basis of the common technical–typological features noted within these materials, we propose categorizing these collections into a "Denisova variant of the Altai Mousterian" (Derevianko and Shunkov, 2002).

The Altai Middle Paleolithic industries included in the Levallois group possess the most distinct technical–typological features. This group includes the sites of Kara-Bom, Ust-Karakol, Anui-3, Ust-Kan Cave. These industries are characterized by the predominance of Levallois reduction, a developed technique of blade detachment, comparatively large numbers of tools fashioned on blades and Levallois spalls, a rather small variety of tool types where blades and Levallois points are most numerous. According to these specific characteristics of the Altai Levallois-Mousterian industries, they are designated as the "Kara-Bom variant of the Altai Middle Paleolithic" (Derevianko et al., 2000b).

The Middle Paleolithic industries from the multilayered sites of Anui-3 and Ust-Karakol demonstrate a well-developed Levallois technology of tool production and bifacial working. Within the Kara-Bom technical variant, materials included in these Levallois-Mousterian collections form a specific industrial type with distinct foliate bifaces (Derevianko and Shunkov, 2002).

The evidence available has not yet provided reliable grounds for associating the technological variants of the Altai Middle Paleolithic with distinct prehistoric human populations bearing independent cultural traditions. There is also currently insufficient evidence for considering the noted industrial variability of the Altai Middle Paleolithic as a purely chronological phenomenon. The Altai Paleolithic chronostratigraphy testifies to the long-term parallel development of two major industrial variants throughout the so-called Mousterian Würm chron. The initial stage of development of true Mousterian industries (e.g., Denisova Cave, stratum 22) is estimated as falling within the Middle Pleistocene, while its final stages (e.g., Okladnikov Cave) are associated with an absolute date of 33–44 ka. The age estimates for Ust-Karakol (stratum 18) and Kara-Bom (Mousterian horizon 1) suggest that Levallois-Mousterian industries occurred within the chronometric range of 100 to 44 ka. The current state of our knowledge allows us to hypothesize that differentiation of lithic industries took place within a single Middle Paleolithic culture as the result of various adaptation strategies to different environmental, seasonal, economic, productive and raw material factors, among others.

The specificity of the productive and economic activities of ancient populations at long-term and seasonal occupation sites may be regarded as one of the reasons for the variability noted in these archaeological assemblages. The pattern of lithic artifact distribution by stratum at the sites of Ust-Karakol and Anui-3 suggests regular, though relatively short-term, occupation by human ancestors. On the other hand, the diverse composition of the tool kit does not allow us to regard these sites as merely short-term encampments. Practically all occupation horizons at these sites yielded lithic collections illustrating the entire technological sequence of raw material utilization. Thus, these collections include instruments for primary stone working, principal products of stone reduction and a typologically diverse tool kit. The noted specificity of the tool kits, correlated with the structure of enclosing sediments, suggests the classification of these multilayered sites as sequences of episodic, seasonal occupation sites. This hypothesis is well supported by the topography of Ust-Karakol and Anui-3. Both sites are located in areas of the river valley, which are most beneficial for establishing seasonal hunting camps. The available evidence suggests that these Middle Paleolithic industries were primarily aimed at producing hunting equipment, such as Levallois points and foliate bifaces. Refitting analyses of the lithic artifacts from Ust-Karakol and petrographic analysis of knapped stone from Anui-3 indicate that the tools were produced there, rather than having been brought in from elsewhere.

The Mousterian collections associated with long-term occupation cave sites also include bifacially worked tools and classic Levallois implements. Generally, bifacial and Levallois traditions in stone reduction are less pronounced in the collections from presumed long-term occupation sites. This may be explained by the fact that solitary, typologically distinct products are not as apparent in the non-homogeneous concentration of waste accumulated at long-term sites. Most likely, distinct technical and typological characteristics dissolved in the homogeneous industrial context of long-term occupation sites.

Early Paleolithic pebble-tool technocomplexes, which have been recently discovered in the Altai, may hardly be considered as a basis for the development of Middle Paleolithic industries. Sources for the development of such industries are likely to be discovered in contiguous regions of North and Central Asia. Acheulian industries, which are characterized by tools produced on strategically planned, shaped spalls detached from well-prepared nuclei, i.e., technocomplexes exhibiting parallel (proto-prismatic) reduction strategies and Levallois flaking, may be considered as candidates for the Early Paleolithic genesis of their development. These cultural traditions may have originated in Acheulian industries of Western Asia: the Caucasus, the Levant and southern Arabia (Amirhanov, 1991; Bar-Yosef, 1994; Hours, 1975; Liubin, 1998).

Kazakhstan, adjacent to the Altai, has produced the most distinctive Acheulian-like technocomplexes from sites located in the northwestern piedmont of the Mugodjari Mountains (Derevianko et al., 2001). The Early Paleolithic sites of Mugodjari-3 to 6 represent concentrations of heavily to moderately abraded artifacts occurring on the surfaces of diluvial benches and on the crests of hills in the vicinity of quartz sandstone outcrops exploited as sources of raw material for tool production. These assemblages contain distinct foliate, ovoid and cordiform bifaces of the Acheulian type as well as nuclei exhibiting morphological features of Levallois reduction and various *racloirs* and notch-denticulate tools.

In Mongolia, Acheulian-like bifaces were first reported within the surface collections at such open-air sites as Bottom-of-the-Gobi and in the vicinity of Mount Yarkh (Okladnikov, 1986). Recent investigations in the southeastern Gobi Altai have provided new information supporting a model of the dissemination of Acheulian elements over Mongolia in the Paleolithic. Bifacially worked tools associated with Levallois products have been identified within the series of artifacts exhibiting heavy surficial aeolian abrasion at Tsakhiurtyn Hondii or "Flint Valley" (Derevianko et al., 1996), as well as from collections associated with the lower stratigraphic levels in Tsagaan Agui Cave (Derevianko et al., 2000a). Age estimates for the Tsagaan Agui sediments suggest a Lower Pleistocene origin for the local Levallois-Acheulian traditions.

The majority of the Mongolian Lower Paleolithic industries bears features characteristic of pebble tool traditions (Derevianko et al., 1990, 2000c). Numerous sites located have yielded rich collections of aeolian abraded artifacts including large polyhedral cores, Levallois and parallel nuclei with one flaking surface, "citron" spalls, various types of *racloirs*, notch-deniculate tools and choppers. Nearly all the early industries of Mongolia exhibit Levallois technical methods in stone reduction.

The Torgalyk A Early Paleolithic site located in Tuva has yielded Acheulian artifacts (Astahov, 1998). Its geo-morphological setting and the heavily aeolized state of the artifact surfaces allow age estimates of the Middle Pleistocene. Among other artifacts, cores exhibiting elements of Levalloisian reduction, longitudinal *racloirs*, massive points and *grattoirs*, notched and denticulate tools have also been identified. The collection also includes archaic bifaces in a variety of forms including *limandes* and proto-*limandes* as well as amygdaloidal and ovoid bifaces.

Bifacial stone reduction strategy has also been recorded at the Early Paleolithic sites of the southern Angara region (Medvedev, 1983). Judging by the relative stratigraphic position of these artifacts and the state of aeolian abrasion apparent on their surfaces, they cannot be younger than the OIS 6. All southern Angara lithic collections may be subdivided into one of two traditions: Tarakhaiski and Olonski. The Tarakhaiski group comprises industries with well-developed pebble tool technology. Primary reduction is based on the "citron" flaking strategy and chopper tools constitute a considerable proportion of the tool kit. The Olonski assemblages are more similar to Acheulian-type industries. This variant is characterized by bifacially, radially flaked core-like implements and specific quartzite micro-bifaces.

Eastwards, bifaces reminiscent of western Acheulian specimens have been reported from the territory limited within North China. Solitary Middle Pleistocene bifacially worked tools classified as handaxes (e.g., the Gongwangling locality at Lantian, Shanxi, and Kehe, Shanxi) and a cleaver (Zhoukoudian, Locality 13) were recovered from the loess plateau and Huanghe Basin (Jia and Huang, 1991). However, reliable evidence illustrating Levallois technology in Paleolithic industries has not yet been reported from China (Gao, 2000). The notable absence of developed, standard Levallois technologies in East Asian industries serves as a major argument supporting the hypothetical western origin of the technical and typological basis of the Lower Paleolithic in Central Asia.

In summary, this brief review of known Lower Paleolithic technocomplexes reported from those regions geographically contiguous with the Altai has shown that the majority of industries producing Acheulian-like bifaces are characterized by developed methods of parallel and Levallois reduction and by the production of tools on intentional blanks of standard size. The features noted support our hypothesis regarding the original development of the Altai Middle Paleolithic on the basis of local Lower Paleolithic cultural traditions bearing the Levallois reduction strategy.

The original development of the majority of the Altai Middle Paleolithic industries does not exclude close relationships with contiguous territories. In particular, such a supposition is supported by similarities noted in the characteristics of the Mousterian industries of the Altai and Central Asia. Most Central Asian Mousterian technocomplexes as well as the Altai Middle Paleolithic industries can be subdivided into two major technical variants: the true Mousterian (Montane Mousterian) and the Levallois-Mousterian (Ranov and Nesmeianov, 1973).

The major technical features of the Central Asian Levallois-Mousterian industries, such as those from Obi-Rakhmat, Khodjakent and Khudji, include the parallel reduction strategy for cores with prepared platforms and large numbers of laminar blanks and tools fashioned on large blades. Similar features are also characteristic of the Kara-Bom variant of the Altai Middle Paleolithic. Predominantly radial and parallel cores, a small number of blades and a typologically diverse series of *racloirs* typify the industries associated with long-term occupation cave sites like Teshik-Tash and Ogzi-Kichik. Such technical features are also noted within those industries included in the Denisova variant of the Altai Mousterian.

Previously identified variants of the Altai Middle Paleolithic also show features analogous with Paleolithic industries recorded in the Eastern Mediterranean. For instance, Levallois-Mousterian archaeological materials of the Kara-Bom type resemble the Tabun D Early Mousterian assemblages from the Levant (Bar-Yosef and Meignen, 1992; Marks, 1992). The high level of development of Levallois technology focused on parallel flaking in order to produce elongated spalls, including large blades and Levallois points, is the major characteristic feature shared by these industries. The resemblance of the Mousterian materials recovered from Okladnikov Cave in the Altai to the Yabrudian complexes recovered from Tabun Cave, Yabrud I rockshelter and other Paleolithic sites in the Levant is also noteworthy (Jelinek, 1982; Rust, 1950). These industries include many similar tool types, especially the numerous *déjeté* scrapers.

The analogous technical–typological features noted in the Middle Paleolithic technocomplexes of the Altai, Central Asia and Near East suggest their attribution to a single cultural domain. In this respect, western Central Asian sites seem to provide links between Middle Eastern and Central Asian industries.

Mousterian traditions dispersed over Asia to include the territory of Mongolia. The available materials provide evidence for similar tendencies in the formation and development of Middle Paleolithic industries. Thus, the Mousterian industry identified in the important cave site of Tsagaan Agui (Derevianko et al., 2000a) is characterized by parallel reduction of Levallois, proto-prismatic and narrow-face nuclei. Massive *racloirs*, notch-denticulate and spurred tools represent major tool categories within tool kit. In Gorny Altai, the industries of the Denisova variant of the Middle Paleolithic represent the closest analog with these Mongolian materials.

The alternative variant of the Mongolian Middle Paleolithic is represented by Levallois industries. The Barlagiin-Gol-1 site located in the southeastern Mongolian Altai (Derevianko and Petrin, 1987) and the Orkhon-1 stratified site located in the southern Khangai Mountains yielded archaeological collections which demonstrated high indices of Levallois technique utilization in tool production. Levalloisian tools, including cores, points, blades and flakes, represent the most numerous categories in these tool kits. The Orkhon-1 archaeological materials indicate the contemporaneous existence of Levalloisian and Mousterian industries in the Mongolian Paleolithic. The Levallois technocomplexes of Mongolia particularly resemble the Kara-Bom variant of the Altai Middle Paleolithic. It has been noted that both technical traditions evolved within the Mongolian Paleolithic. Both developmental trends are illustrated in the available archaeological materials attributable to the transitional period from the Middle to the Upper Paleolithic. The true Mousterian variant is illustrated by the archaeological collection associated with deposits from the third sedimentation cycle in Tsagaan Agui Cave, while the Levallois-Mousterian variant has been identified in materials recovered from the Orok-Nuur-1 and 2 sites (Derevianko et al., 2000c).

The cultural continuity apparent in the development of the Mongolian Paleolithic suggests the formation of Mousterian traits on the basis of a local Lower Paleolithic tradition with Levalloisian technology. The analogous features of the major Mousterian variants allow us to include Mongolia, primarily the Mongolian and Gobi Altai regions, and Altai Mountains in a single geographical unit representing the development of a distinctive Middle Paleolithic culture.

In this respect, the anthropological identification of the Middle Paleolithic population of the Altai presents particular interest. Formerly, it was believed that odontological remains of hominid fossils recovered from the Mousterian layers at Denisova and Okladnikov Caves represent clear European Neanderthal features (Turner, 1990). However, Alekseev argued that scanty anthropological materials did not allow unambiguous Neanderthal identification of the fossils. The noted morphological features are interpreted as belonging to physically modern humans (Alekseev, 1998). Additional analyses of the fossil collection from the Altai caves have shown that despite certain archaic features, these remains most likely belong to representatives of physically modern humans—early *Homo sapiens sapiens* (Shpakova, 2001).

Around 50,000–40,000 years ago, Initial Upper Paleolithic industries were formed as a result of the continuous transformation of Middle Paleolithic industries in the Altai. The Altai Initial Upper Paleolithic technocomplexes exhibit certain features in common. However, each assemblage possesses its own specific characteristics, on the basis of which the whole body of Altai Initial Upper Paleolithic industries may be subdivided into two major groups reflecting a particular developmental trend: the Ust-Karakol and Kara-Bom trajectories.

Such Altai technocomplexes as Ust-Karakol, Denisova Cave, Anui-3, Tiumechin-4 are included in the Ust-Karakol variant. This variant is characterized by the parallel reduction of Levallois and single platform cores as well as by the addition of new methods aimed at repetitive detachment of elongated blanks from prismatic, conical and narrow-face cores, including wedge-shaped varieties. As a result of the application of this progressive technology, the technique of microblade flaking developed, aimed at the production of microblades themselves and at fashioning specific Upper Paleolithic tool forms. The tool kits identified within these industries still include numerous *racloirs* and notch-denticulate tools. The Upper Paleolithic tools constitute several new types that have not been noted in earlier collections. Most interesting are the so-called Aurignacian forms including end-scrapers on blades, carinated scrapers fashioned with micro-laminar removals, dihedral burins, large blades retouched throughout their whole perimeters and backed microblades. Bifacially worked tools, especially classical foliate bifaces, constitute an important characteristic feature of these assemblages. Another important feature is the occurrence of various bone implements recovered in Denisova Cave stratum 11.

The bone tool collection includes eyed needles, piercers, cylindrical beads with annular incisions, bead blanks, a ring fragment made of mammoth tusk and pendants made of various animal teeth (Color Plate 8, see p. 400). This bone implement collection from Denisova Cave is the earliest thus far recorded in the Paleolithic of North and Central Asia.

The lithic industries from Kara-Bom, Kara-Tenesh and possibly Maloyalomansky Cave represent the Kara-Bom variant of the Initial Upper Paleolithic. The laminar technique is most pronounced in the technological processes of these industries. Most cores exhibit parallel flaking patterns suggesting the production of elongated spalls. The Levallois technique is still apparent; however, certain new technical methods began to be employed, in particular microlaminar flaking of cores including narrow-face varieties. Large blades were the principal intended product, as more than one half of the collection of tools is fashioned on such blanks. Notch-denticulate tools are still numerous in the collection, although Upper Paleolithic tools fashioned mostly on large blade blanks dominate the tool kit. The following Upper Paleolithic categories have been identified: end-scrapers, dihedral burins, knives with retouched backs, long points with flattened ventral faces and blades showing retouch along their longitudinal margins. Certain so-called Aurignacian elements were also noted within these collections, as well as scarce bifaces and objects of adornment made from animal teeth. However, such artifacts are scarce and do not form a discernible stable technical–typological series. Repetitive production of large blades and blade-based tools represents the principal technological feature of the Kara-Bom tradition.

Two technological trends in the development of the Altai Upper Paleolithic have been identified on the basis of excavated technocomplexes attributable to the Initial Upper Paleolithic. These trends may be extrapolated over a broader area of North and East Asia because of the crucial geographical and chronological position of the Altai. The Kara-Bom tradition was responsible for the dissemination of blade-based industries over this territory. The Ust-Karakol technical variant stimulated the development of industries based on a narrow-face reduction strategy, micro-flaking technology and the production of foliate bifaces.

The middle stage of the Altai Upper Paleolithic is best illustrated by the artifacts recovered in association with the lowermost culture-bearing horizons of the Anui-2 site. A series of radiocarbon dates in the range 27–23 ka has been generated for this site. Most cores show the parallel reduction strategy including the prismatic reduction and detachment of flakes from the narrow face of the cores. Microblade cores of the narrow-face, wedge-shaped and prismatic varieties are especially noteworthy. A considerable share of elongated blanks including microblades attests to a well-developed laminar technique. The relevant tool kit is predominated by the Upper Paleolithic tools including *grattoirs*, burins, chisel-like tools and piercers. The major typological feature specific for the industry is the series of microtools comprising Gravettian points, miniature *grattoirs*, piercers and microblades showing abrupt retouch on the margins.

The final stage of the Altai Upper Paleolithic is represented in the archaeological materials of cave sites. The Upper Paleolithic industries of Denisova and Kaminnaya caves demonstrate the further development of the laminar technique. Comparing to the assemblages associated with previous stages, these final Upper Paleolithic industries comprise a greater amount of elongated spalls, in which the share of microblades also increases. Blades were used as blanks for manufacturing longitudinal varieties of *racloirs*, end-scrapers, burins, chisel-like tools, denticulate and notched tools. Backed microblades and foliate bifaces represent the most typologically perfect tool types. A set of bone tools and adornments includes eyed needles, piercers, pendants made of deer teeth cylindrical beads and ring-shaped beads made of ostrich egg shell.

In general, lithic industries of the final stage Upper Paleolithic demonstrate a continuous development. This continuity is reflected in a combination of certain archaic (radial and Lavallois cores, Mousterian *racloirs* and notch-denticulate tools) and clear Upper Paleolithic (prismatic and narrow-face nuclei, end-scrapers, dihedral burins, backed microblades, composite tools) elements both in primary reduction strategy and in typology of lithic and bone implements. Additionally, these archaeological materials demonstrate a continuous development of Paleolithic traditions apparent in a wider application of laminar reduction technique, in microblade production in particular.

References

Alekseev, V. (1998) The physical specificities of Paleolithic hominids in Siberia. In: The Paleolithic of Siberia. Univ. of Illinois Press, Urbana, IL, pp. 329–335.

Alpysbaev, H.A. (1979) Pamiatniki nizhnego paleolita Yuzhnogo Kazahstana. Nauka Kaz. SSR, Alma-Ata.

Amirhanov, H.A. (1991) Paleolit Yuga Aravii. Nauka, Moscow.

Astahov, S.N. (1998) Paleolit Tuvy. In: Drevnie kultury Tsentralnoi Azii i Sankt-Peterburg. Kult-inform-press, St Petersburg, pp. 109–114.

Bar-Yosef, O. (1994) The Lower Palaeolithic of the Near East. J. World Prehist. 8(3), 211–265.

Bar-Yosef, O. (1995) The Lower and Middle Paleolithic in the Mediterranean Levant: chronology and cultural entities. In: Man and Environment in the Paleolithic. Univ. de Liege, Liege, pp. 247–263. (ERAUL, 62).

Bar-Yosef, O. and Meignen, L. (1992) Insights into Levantine Middle Paleolithic cultural variability. In: The Middle Paleolithic: Adaptation, Behavior, and Variability. Univ. Museum, Univ. of Pennsylvania, Philadelphia, PA, pp. 163–182. (Univ. Museum Monograph; 72. Univ. Mus. Symp. Ser.; 4).

Bosinski, G. (1982) The transition Lower/Middle Paleolithic in North-Western Germany. In: The Transition from Lower to Middle Paleolithic and the Origin of Modern Man. British Archaeological Reports, Oxford, pp. 165–177. (BAR. International Ser.; 151).

Derevianko, A.P. and Petrin V.T. (1987) Kompleks kamennoi industrii s yuzhnogo fasa Mongolskogo Altaia. In: Arkheologia, etnografia i antropologia Mongolii. Nauka, Novosibirsk, pp. 5–27.

Derevianko, A.P. and Shunkov, M.V. (2002) Middle Paleolithic industries with foliate bifaces in Gorny Altai. Archaeol. Ethnol. Anthropol. Eurasia 1, 16–42.

Derevianko, A.P. and Shunkov, M.V. (2005) The Karama Lower Paleolithic site in the Altai: initial results. Archaeol. Ethnol. Anthropol. Eurasia 3, 52–69.

Derevianko, A.P., Dorj, D., Vasilevskii, R.S., Larichev, V.E., Petrin, V.T., Deviatkin, E.V. and Malaeva, E.M. (1990) Kamennyi vek Mongolii: Paleolit i neolit Mongolskogo Altaia. Nauka, Novosibirsk.

Derevianko, A.P., Olsen, J.W., Tseveendorj, D., Petrin, V.T., Zenin, A.N., Krivoshapkin, A.I., Reeves, R.W., Deviatkin, E.V. and Mylnikov, V.P. (1996) Archaeological Studies Carried Out by the Joint Russian–Mongolian–American Expedition in Mongolia in 1995. IAET SO RAN, Novosibirsk.

Derevianko, A.P., Olsen, J.W., Tseveendorj, D., Krivoshapkin, A.I., Petrin. V.T. and Brantingham P.J. (2000a) The Stratified Cave site of Tsagaan Agui in the Gobi Aita i (Mongolia). Archaeol. Ethnol. Anthropol. Eurasia 1, 23–36.

Derevianko, A.P., Petrin, V.T. and Rybin, E.P. (2000b) The Kara-Bom Site and characteristics of the Middle-Upper Paleolithic Transition in the Altai. Archaeol. Ethnol. Anthropol. Eurasia 2, 33–51.

Derevianko, A.P., Petrin, V.T., Tseveendorj, D., Deviatkin, E.V., Larichev, V.E., Vasilevskii, R.S., Zenin, A.N. and Gladyshev, S.A. (2000c) Kamennyi vek Mongolii: Paleolit i neolit severnogo poberezhia Doliny Ozer. IAET SO RAN, Novosibirsk.

Derevianko, A.P., Petrin, V.T., Gladyshev, S.A., Zenin, A.N. and Taimagambetov, Z.K. (2001) Acheulian complexes from the Mugodjari Mountains (North-Western Asia). Archaeol. Ethnol. Anthropol. Eurasia 2, 20–36.

Derevianko, A.P., Shunkov, M.V., Agadjanian, A.K., Baryshnikov, G.F., Malaeva, E.M., Ulianov, V.A., Kulik, N.A., Postnov, A.V. and Anoikin, A.A. (2003) Prirodnaya sreda I chelovek v paleolite Gornogo Altaya. IAET SO RAN, Novosibirsk.

Gao, X. (2000) Core reduction at Zhoukoudian locality 15. Archaeol. Ethnol. Anthropol. Eurasia 3, 2–12.

Hours, F. (1975) The Lower Paleolithic of Lebanon and Syria. In: Problems in Prehistory: North Africa and the Levant. Southern Methodist Univ. Press, Dallas, pp. 249–271.

Jelinek, A.J. (1982) The Tabun Cave and Palaeolithic man in the Levant. Science 216, 1369–1375.

Jia, L. and Huang, W. (1991) The Paleolithic cultures of China. Quatern. Sci. Rev. 10, 519–521.

Laville, H. (1982) On the transition from "Lower" to "Middle" Paleolithic in South-West France. In: The transition from Lower to Middle Paleolithic and the origin of Modern Man. British Archaeological Reports, Oxford, pp. 131–137. (BAR. International Ser.; 151).

Liubin, V.P. (1998) Ashelskaia epoha na Kavkaze. Peterburgskoe Vostokovedenie, St Petersburg.

Marks, A. (1992) Typological variability in the Levantine Middle Paleolithic. In: The Middle Paleolithic: Adaptation, Behavior, and Variability. Univ. Museum Univ. of Pennsylvania, Philadelphia, PA, pp. 127–142.

Medvedev, G.I. (1983) Paleoliticheskie obitateli iuga Sibirskogo ploskogoria i drevnie kultury Severnoi Ameriki. In: Pozdnepleistotsenovye i rannegolotsenovye kulturnye sviazi Azii i Ameriki. Nauka, Novosibirsk, pp. 36–41.

Mercier, N., Valladas, H. and Valladas, G. (1995) Flint thermoluminescence dates from the CFR Laboratory at GIF: contributions to the study of the chronology of the Middle Palaeolithic. Quatern. Sci. Rev. 14, 351–364.

Okladnikov, A.P. (1986) Paleolit Mongolii. Nauka, Novosibirsk.

Ranov, V.A. (2001) Loess-paleosoil formation of Southern Tajikistan and the loess Palaeolithic. Praehistoria 2, 7–27.

Ranov, V.A. and Nesmeianov, S.A. (1973) Paleolit i stratigrafia antropogena Srednei Azii. Donish, Dushanbe.

Rust, A. (1950) Die Hohlenfunde von Jabrud (Syrien). Karl Wachholtz, Neumunster.

Shpakova, E.G. (2001) Paleolithic human dental remains from Siberia. Archaeol. Ethnol. Anthropol. Eurasia 4, 64–76.

Tuffreau, A. (1982) The transition Lower/Middle Paleolithic in Northern France. In: The transition from Lower to Middle Paleolithic and the origin of Modern Man. British Archaeological Reports, Oxford, pp. 137–151. (BAR. International Ser.; 151).

Turner, C.G. (1990) Paleolithic Siberian dentition from Denisova and Okladnikov Caves, Altaiskiy Kray, USSR. Curr. Res. Pleistocene 7, 65–66.

Evolutionary History of Wheats—the Main Cereal of Mankind

N. P. Goncharov, K. A. Golovnina, B. Kilian, S. Glushkov,
A. Blinov and V. K. Shumny

Abstract An attempt in integrating the results of different comparative-genetic analyses of wheats and their molecular taxonomy has been made; the correspondence of earlier evolutionary specifications to the phylogeny within the genus *Triticum* species has been estimated. The relationships have been established based on chloroplast and nuclear DNA sequence data. One phylogenetic tree has been constructed based on the chloroplast sequences, and several phylogenetic groups have been found within the genera *Triticum* and *Aegilops*. It has been shown that *Aegilops speltoides* was a donor of the plasmon for all polyploid wheat species, whereas the chloroplast genomes of the diploid *Triticum* species are close to other *Aegilops* species. Nuclear *Acc-1* and *Pgk-1* genes have been used as molecular markers for the A and B genomes of the *Triticum* species. No variability has been found in these genes within polyploid wheats. In contrast, three variants of these genes have been detected in diploid A genome *Triticum*. The detailed analysis showed that one of these variants was a progenitor for all A genomes of all polyploid *Triticum* species; the second variant is close to the B genomes of *Ae. speltoides*; and the third one is unique for wild diploid wheats. The inheritance of two domesticated and taxonomically important characters was studied in the ancient hexaploid wheat *Triticum antiquorum*. It was shown that the recessive gene controlling spherical grain was allelic to the *s* gene determining the same character in the endemic Indian species *T. sphaerococcum*. The dominant genes of *T. antiquorum* and *T. sphaerococcum* controlling compact ears were proved to be non-allelic to the corresponding *T. compactum* gene. Results of molecular analysis indicated the close relationship of all hexaploid wheat species.

Annual, self-pollinated coarse-grained plants spread over vast territories in the Mediterranean climate. Their grains are "convenient" for intensive gathering and, what is not less important, for long-term preservation. The transition

---

N. P. Goncharov
Institute of Cytology and Genetics, Siberian Branch of the Russian Academy of Sciences, Novosibirsk, Russia
e-mail: gonch@bionet.nsc.ru

from gathering wild cereals towards modern plant breeding is complex and still a matter of discussion.

The history of cultivated plants is closely interwoven with that of mankind. Many investigations based on genetic, comparative-genetic, molecular, archeological and geographobotanical analysis have succeeded in identifying the progenitors of cultivated species, their phylogeny and place of domestication (Goncharov et al., 2007). Independent domestications of four main cereals wheat, barley, rice, and maize produced similar results (Harlan, 1992). The earliest signs of domestication appear in Pre-pottery Neolithic B (Nesbitt, 2001). Archeological data provided evidence that barley (*Hordeum* ssp.) and wheat (*Triticum* ssp.) were among the first domesticated plants. Their cultivation was commenced on the threshold of late Stone Age (Nesbitt, 2001).

In the course of time, wheat became the main cultivated crop covering the largest area among all cultivated plants. Curiously, diploid cultivated einkorn wheat *Triticum monococcum* L. was the basic crop of the Shumer culture, whereas tetraploid emmer wheat *T. dicoccum* was cultivated in ancient Egypt.

De Candole (1885) already considered the problem on the origin of wheat species separately having no data for resolving this point on the origin of cultivated wheat species in general. Since then it is still a matter of discussion.

Timid attempts in the complex consideration of wheat species origin faltered due to the absence of large and representative *Triticum* collections. Only the world wheat collection of the N.I. Vavilov Institute of Plant Industry (St. Petersburg, Russia) and the Kyoto University (Kyoto, Japan), have been scrupulously collected and studied during 100 years. These comprehensive collections are a unique possibility for researchers to look deeper into the *Triticum* phylogeny.

Several wheat domestication schemes suggested in the past do not present our knowledge today. Some of them are misleading or contrary.

On the other hand, modern comparative-genetic and molecular methods might allow us to get deeper insights into phylogenetic relationships within *Triticum* and the related species.

The aim of our research was to provide new data to reconstruct the wheat evolution based on chloroplast and nuclear gene loci.

1 Material and Methods

1.1 Plant Materials

Plants used for chloroplast sequences are indicated in Golovnina et al. (2007).

Ten *T. urartu* Thum. ex Gandil. accessions (K-33869, PI 428217, PI 428297, PI 427328, PI 428197, PI 538736, Ig-44829, Ig-45296, Ig-116196, Ig-116198), four *T. boeoticum* Boiss. accessions (K-14384, K-20741, K-25811, K-28300) and 14 *T. monococcum* accessions (K-20970, K-20400, K-18105, KT3-5, G-1777,

PI 355517, PI 277137, PI 427927, PI 428175, PI 362610, PI 355523, PI 349049, PI 326317, PI 94743) were used for nuclear sequence analysis. Additional sequence data were obtained from Kilian et al. (2007).

Two *T. antiquorum* Heer ex Udach. accessions K-56397 and K-56398 from Tajikistan and two *T. sphaerococcum* Perciv. accessions K-23790 from India and K-23824 from Pakistan were used for comparative-genetic analyses. The genes determining the compact spike were tested for allelism using the Finnish cultivar Vakka, Buryatian (Russia) accession WAG 8226, and the American accession CI 3090 of *T. compactum* Host.

1.2 Total DNA Isolation and PCR Amplification

Total DNA was isolated from 50–170 mg of fresh leaves using the standard CTAB method (Rogers and Bendich, 1985). The primer combinations to amplify the chloroplast *trnT–trnL* intergenic spacer, the *trnL* intron and the *trnK* intron are those described in Golovnina et al. (2007).

In order to amplify A and B genome specific fragments of nuclear DNA, two multiple alignments were developed using sequence data of *Acc-1* and *Pgk-1* genes (GenBank accession numbers AF343496–AF343536 and AF343474–AF343495, respectively, and those from Kilian et al. (2007):

DQ290259–DQ290360; DQ290363–DQ290375; DQ364823–DQ364846;
DQ290658–DQ290757; DQ290760–DQ290771; DQ364891–DQ364912
for different genomes available among *Triticum* and *Aegilops* L. representatives. Based on these alignments, one pair of B genome specific primers (Acc 3T sense 5′-GCTCATATGGTATATTATGTTCC-3′, Acc 3T antisense 5′-TTTAGGCA-CAGAAATAACAT-3′) and six different primers for genome A (AccT1s 5′-GGACTTAGTTTTTTGTCGTCAGTT-3′, AccT1a 5′-GAAAAAAACGCA-GCCCAATT-3′, AccT1a new 5′-CTTCCAAACGTAAGGACCAATACA-3′, PgkT4s 5′-GCTTGGCTCCCCTTGTGCCCCG-3′, PgkT1s new 5′-GGCAT-TGAGGTATTCTTTTGTTCCACTTCCAC-3′, PgkT1a 5′-CACACTTCTCC-AGCAGGGATTCGA-3′) were designed.

All PCR reactions were performed in a 20 µl volume containing 65 mM Tris-HCl (pH 8.9), 16 mM $(NH_4)_2SO_4$, 1.5 mM $MgCl_2$, 200 µM of each dNTP, 0.5 µM of each primer, 20–50 ng genomic DNA template, and 1 U of Taq DNA polymerase. The PCR program had an initial strand separation step at 94°C for 3 min followed by 30 cycles of denaturation at 94°C for 30 s, annealing at 42°C for 42 s for chloroplast sequences and at 52°C for 42 s for nuclear sequence, and elongation at 72°C for 1 min.

The PCR products were analyzed in agarose electrophoresis and extracted from gel with a Qiaquick Gel Extraction Kit (Qiagene; according to the manufacturer's protocol).

1.3 DNA Sequencing and Phylogenetic Analysis

200 ng of the PCR product was used in a 10 μl cycle sequencing reaction with the ABI BigDye Terminator Kit on an ABI 377 DNA sequencer. The nucleotide sequences were aligned using ClustalX software (Thompson et al., 1997) edited using the GenDoc Version 2.6.002 (Nicholas et al., 1997). The phylogenetic tree was generated by the Neighbor-Joining method using MEGA 3.1 program (Kumar et al., 2004). Statistical support for the tree was evaluated by bootstrapping (1000 replications) (Felsenstein, 1985).

2 Results and Discussion

The origin of polyploid wheat species is almost like a "detective story," not all parts are understood so far.

Besides, wild diploid (*T. boeoticum* A^bA^b genome–*T. urartu* A^uA^u genome) as well as wild tetraploid wheat species *T. dicoccoides* (Körn. ex Aschers. et Graebn.) Schweinf. (BBAA genome) and *T. araraticum* Jakubz. (GGAA genome) are morphologically not differentiated from one another and not distinguishable in archeological excavations (Nesbitt, 2001).

It is further unknown whether the first cultivated naked wheat species in Europe was tetraploid or hexaploid.

At the same time, the possibility of their identification is connected with a search for possible wild species involved in cultivated wheat origin and their domestication.

2.1 Pile-Dwelling Wheat

Heer (1865) was the first who described *T. antiquorum* (BBAADD genome) on the basis of grain remains in archeological excavations in Switzerland. It is possible that *T. antiquorum* was among the first cultivated hexaploid wheat species in Europe.

T. antiquorum could have played an important role during the early cultivation of hexaploid wheats and it is possible that Asia occupied a central place in this process (Udachin, 1982). The drawback of wheat phylogenetic schemes currently proposed consists in the absence of data on the character (type) of genetic control of morphological taxonomically important traits that are identical in their phenotypic manifestation. Since living *T. antiquorum* has been found in Tajikistan (Udachin, 1982), it is maybe possible to solve some phylogenetic-related problems for the origin of cultivated hexaploid wheat in Europe. The inheritance of taxonomically important characters was therefore studied in pile-dwelling wheat *T. antiquorum* and two additional contemporary hexaploid wheat species.

These external morphological traits are very often arbitrary. Two basic traits—spherical grains and compact spike—make this wheat distinct from all the other ones.

2.1.1 Spherical Grains

Table 1 summarizes the data on the segregation with respect to the grain shape in F_2 hybrids of *T. sphaerococcum* and *T. antiquorum* accessions and the results of checking the hypothesis that spherical grain genes of these species are allelic. Likewise, no segregation by grain shape was detected in F_2 plants for crosses of K-56397 *T. antiquorum* × K-23790 *T. sphaerococcum*. Monogenic segregation was observed in F_2 hybrids with common wheat (Table 1). This character is controlled monogenically by a recessive gene in K-23790 of *T. sphaerococcum*. Since the character is determined by recessive alleles in both species, F_1 hybrids would express the wild-type (normal) phenotype in the case of non-allelic genes; i.e., the grains would be non-spherical. All 15 grains were spherical in the F_1 hybrid of K-56398 *T. antiquorum* and K-23790 *T. sphaerococcum*. Therefore, the recessive gene determining spherical grain in K-56398 of *T. antiquorum* is allelic to that of K-23790 of *T. sphaerococcum*.

2.1.2 Compact Spike

The segregations observed in the F_2 hybrids of *T. compactum* cultivar Vakka with accession K-56397 of *T. antiquorum* and accession K-23790 of *T. sphaerococcum* with CI 3090 of *T. compactum* are shown in Table 2. The results suggest that the dominant genes, controlling compact spike for *T. compactum* are non-allelic to the genes determining compact spike in *T. antiquorum* and *T. sphaerococcum*. Thus, the relevant *T. antiquorum* and *T. sphaerococcum* genes are different from the dominant *C* gene, which is responsible for compact ear in *T. compactum*.

We designated it as *C2*. The origin of the dominant genes *C* and *C2* in hexaploid wheats is unknown because compact spike accessions of

Table 1 Inheritance of grain shape in F_2 hybrids of hexaploid wheats with *T. antiquorum* and the test for allelic genes controlling spherical grains in *T. antiquorum* and *T. sphaerococcum*, respectively

Cross combination	Number of F_2 hybrid plants with grain		x^2	
	spherical	normal	1:3	1:15
Triple Dirk D × K-23790 *T. sphaerococcum*	14	44	0.02	31.67
Vrn8 × K-56398 *T. antiquorum*	18	82	2.61	23.56
K-56397 *T. antiquorum* × K-23790 *T. sphaerococcum*	238	0	–	–

Table 2 Inheritance of compact spike in *T. compactum*, *T. sphaerococcum*, and *T. antiquorum*

Cross combination	Number of F$_2$ hybrid plants		x^2	
	Compact	Normal	3:1	15:1
Vakka *T. compactum* × K-20900 *T. aestivum*	47	19	0.51	57.22
Vakka *T. compactum* × K-56397 *T. antiquorum*	79	3	19.92	0.94
K-23790 *T. sphaerococcum* × CI 3090 *T. compactum*	105	17	7.97	1.86

Ae. squarrosa, the D genome donor of hexaploid wheats, have not been discovered (Goncharov, 2002). However, the presence of different non-allelic dominant genes in *T. antiquorum* and *T. compactum* does not indicate a single occurrence of this taxonomically important mutation in hexaploid wheats. Furthermore, based on the data obtained for non-allelism of genes controlling compact spike in studied hexaploid wheat species, we cannot use the earlier suggested schemes of hexaploid wheat species origin, as non-allelism of genes implies their independent origin. Hence, making up new phylogenetic schemes (for example, see Udachin (1982) among others) of wheat origin and new methods are necessary. For this purpose molecular markers for all wheat genomes would be useful in order to detect and to understand their relationships. Here we provide new sequence data for two chloroplast and two nuclear gene loci.

2.2 Chloroplast Evidence of Wheat Evolution

The analysis using all known wheat species including also *Aegilops* species was based on chloroplast *matK* gene comparison along with *trnL* (tRNA-Leu) intron sequences of some species (Fig. 1). Based on the neighbor-joining tree, all analyzed wheat and *Aegilops* species are subdivided into four related groups (Fig.1). Polyploid wheat species are divided only into two groups-Emmer I (*T. dicoccoides* and other BBAA *Triticum* species, not shown) and Timopheevii II (*T. araraticum*, *T. timopheevii*, other G genome wheats, and *Ae. speltoides*) dividing B and G genome wheat species. This result corroborates with the previous suggestion of a diphyletic origin of polyploid wheats based on earlier hybridological, cytological and molecular analyses (Kilian et al., 2007; Lilienfeld and Kihara, 1934; Mori et al., 1995). Group III comprises the diploid AA genome wheats (*T. boeoticum*, *T. monococcum*, *T. urartu*). *Aegilops* section *Sitopsis* and *Vertebrata* members (not shown) and artificial *Aegilotricum* and *T. palmovae* are within group IV. Each group I–III includes both wild and cultivated wheat species.

Various *Triticum* and *Aegilops* species were implicated as donors of genomes of these polyploid wheats (Kerby and Kuspira, 1986). Among all the species analyzed for the *Sitopsis* section of *Aegilops* in this study, *Aegilops speltoides*

Tausch (included in group II) is most closely related to the polyploid B or G genome. Based on the data presented, both *trnL* and *trnK* intron sequences of *Ae. speltoides* are more variable than the corresponding sequences from all other *Aegilops* and diploid *Triticum* species. Sequences were previously obtained (Golovnina et al., 2007) and submitted in GenBank.

This observation strongly coincides with the previous results based on nucleotide variations of the four other chloroplast non-coding regions and microsatellite repeat motifs (Yamane and Kawahare, 2005) and the *ndhF* gene (Kilian et al., 2007).The topology of the trees in both studies clearly demonstrates that the *Ae. speltoides* ancestor branched out before a separation of wild diploid *Triticum* and *Aegilops* species (Fig. 1).

Based on our results (Golovnina et al., 2007), it is proposed that one *Ae. speltoides* ancestor was involved the first polyploidization event of wheat species. It is likely that there were two ancestor forms of *Ae. speltoides* involved in a two-step hybridization, i.e. independent events (Emmer and Timopheevii groups). The high degree of intraspecific variation observed among *Ae. speltoides* accessions and differentiation into B and G genome of polyploid wheats support this hypothesis. The G genome and plasmon of the section *Timopheevii* species (clade II) appears evolutionarily younger and is closely related to the

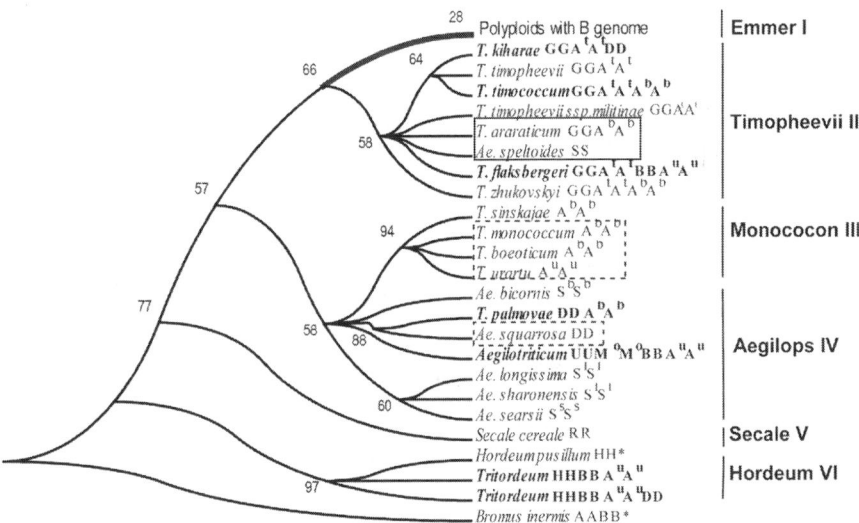

Fig. 1 Neighbor-Joining phylogenetic tree based on the comparison of *matK* sequences. Four observed clusters are shown by solid lines on the right. The genome composition for each species is indicated. Synthetic wheats are represented in *bold letters*. For all species belonging to the Emmer group (24 representatives, see Table 1 in Golovnina et al., (2007)) are indicated as "Polyploids with B genome." Based on the indel event in the *trnL* intron sequence of some analyzed species, representatives with observed insertions are marked by *solid boxes* and the rest ones by *dotted boxes*. Asterisks denote species from which the *matK* sequence was obtained from the GenBank. Bootstrap values are shown

contemporary *Ae. speltoides*, whereas the polyploid *Triticum* species (clade I) with the B genome occurred as a result of one more ancient hybridization event with the *Ae. speltoides* ancestor.

2.3 Nuclear Loci

The presence of four different wheat genomes—A, B, D, and G whose various combinations form three groups of *Triticum* species on their ploidy (di-, tetra- and hexaploids)—are well known. The origin of wheat genomes was a matter of discussion since more than seven decades. The A genome is found only in *Triticum* species and is subdivided into two genomes—A^u and A^b, according to the sources of their origin, i.e. two wild diploid wheat species—*T. urartu* and *T. boeoticum*. *T. urartu* was the A genome donor. *Ae. speltoides* was the donor of both B and G genomes, and *Ae. squarrosa* L. (syn. = *Ae. tauschii* Coss.) was that of D genome.

In the present study we have focused our research on the genome A and genome B. We selected two nuclear gene loci *Acc-1* and *Pgk-1*, because comprehensive datasets were available in gene banks and we provided new data from so far not investigated species.

Total DNA from different *Triticum* and *Aegilops* species has been amplified with A and B genome specific primer combinations. These primers were designed based on unique indels and nucleotide substitutions. The sequencing procedure was conducted with primers complementary to the flanking regions of specificity. The results of PCR analysis of both *Aegilops* and polyploid *Triticum* species completely confirmed the correct choice of primers. The PCR fragments of the expected size have been obtained for homologous genes tested in the samples where the corresponding genomes were present.

In contrast to the polyploid *Triticum* species, some samples of the diploid A genome wheat species (*T. urartu*, *T. boeoticum*, and *T. monococcum*, showed unexpected results. PCR amplification with non-A genome specific primers amplified the A genome fragments also successfully, vice versa, PCR amplification with A genome specific primers appeared to be negative. Such results have been obtained for both *Acc-1* and *Pgk-1* genes. The results of PCR amplification with B genome specific *Acc-1* primer combinations are shown in Fig. 2.

In the next step, we took a more detailed analysis and focused first on different geographical variants of diploid wheat species of *T. urartu*, *T. boeoticum*, and *T. monococcum*. At the same time, Kilian et al. (2007) published data on *Acc-1* and *Pgk-1* gene sequences obtained from different wheat species, which were also used for our analysis. The results of these comparative analysis for *Acc-1* and *Pgk-1* gene sequences are summarized in Figs. 3a and b.

All analyzed *Acc-1* sequences can be divided into two groups due to a 46 bp deletion from position 1067 to position 1151, and some nucleotide substitutions specific for each group (Fig. 3a). The first group with the deletion comprises the

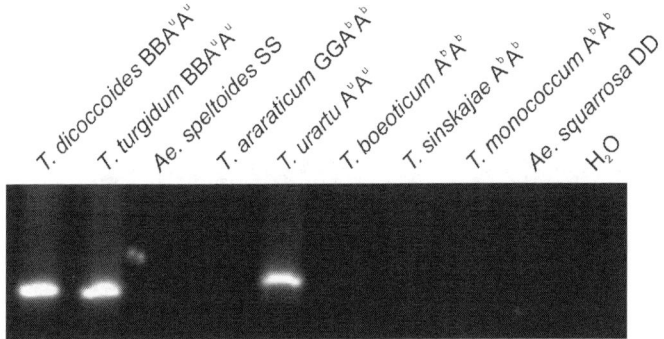

Fig. 2 PCR amplification with primers Acc 3T sense/Acc 3T antisense which were initially considered to be specific for genome B

sequences from polyploid wheat A genomes and also from all three diploid A genome wheat species. The second group integrates the sequences of *Acc-1* genes from *Ae. speltoides*, those of polyploid wheat B, G, and D genomes and also the sequences of all three diploid A genome *Triticum* species.

Thus, diploid A genome wheat sequences are found in both groups. To explain these results, it is necessary to postulate that the wheat genome A originated after the separation of the genus *Triticum* progenitor from the common progenitor with *Aegilops*. Later, both A- and B-like A genomes spread among the three diploid wheat species—*T. urartu*, *T. boeoticum*, and *T. monococcum* (Fig. 4).

The analysis of the *Pgk-1* gene sequences is even more phylogenetically informative, and it allows us to divide all the obtained sequences into several groups. First, four groups are definitely outlined in *Pgk-1* sequence comparisons according to their position to one of the polyploid wheat genomes A, B, G, or D by the presence of specific indels and specific nucleotide substitutions (Fig. 3b). Among the diploid *Triticum* species we have found three different sequences of the *Pgk-1* gene fragment. One of these sequences was found in *T. urartu*, which is identical to those from the A genomes of the polyploid wheats. This fact supports that *T. urartu* was the donor of genome A in all polyploid wheats. Two other variants of the *Pgk-1* gene were determined in (1) *T. boeoticum* and *T. monococcum* and (2) *T. urartu* and *T. boeoticum*, respectively. These sequences are unique and different from the others found in both *Triticum* and *Aegilops* genomes, although the first group (1) is closer to the *Pgk-1* gene from *Ae. speltoides*.

Finally, based on the analyses of *Acc-1* and *Pgk-1* sequences, we can conclude that three diploid *Triticum* species may contain several different genomes in contrast to the polyploid species in which no heterogeneity has been found. The real significance of these results may be confirmed by genetic experiments for crossing diploid species with different genomes selected in our analysis.

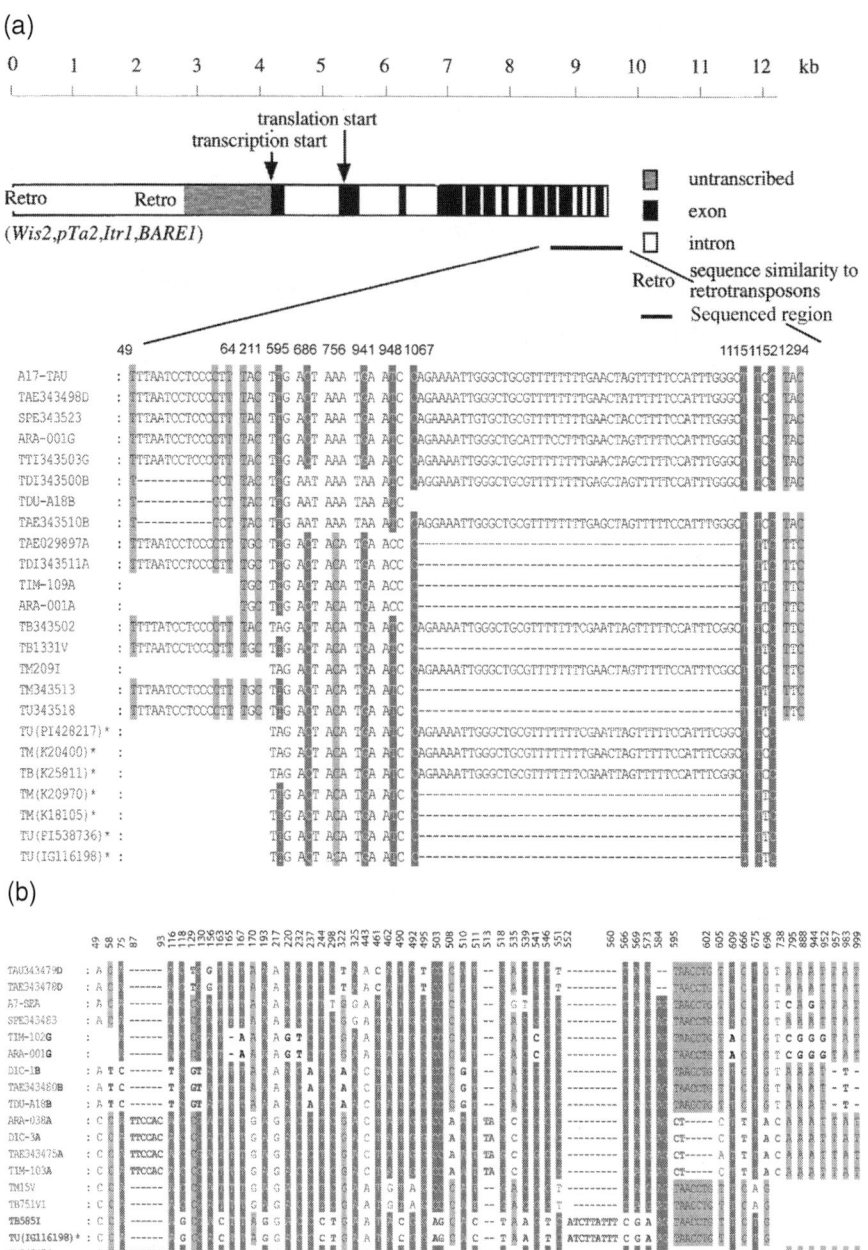

Fig. 3a Structure of the plastid *Acc-1* gene together with the alignment of the region investigated. Numbers at the top on each column indicate positions in the whole alignment. *Asterisks* denote sequences obtained in the present study, others were included from Kilian et al. (2007). TAU–*Ae. tauchii*, TAE–*T. aestivum*, SEA–*Ae. searsii*, SPE–*Ae. speloides*, TIM–*T. timopheevii*, ARA–*T. araraticum*, DIC–*T. diccocoides*, TDU–*T. durum*, TM–*T. monococcum*, TB–*T. boeoticum*, TU–*T. urartu*. Letters at the end of the sequence name indicate the genome

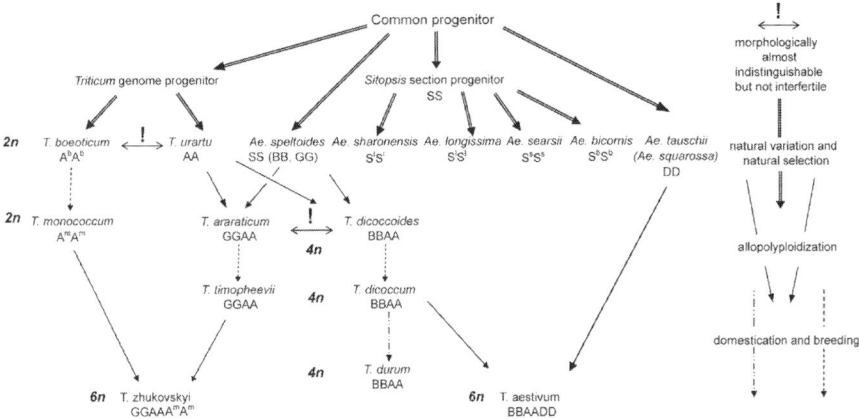

Fig. 4 Phylogenetic scheme of *Triticum* and *Aegilops* evolution (revised from Kilian et al. 2007)

2.4 Evolutionary Scenario of Genus Triticum

Based on the comparative and phylogenetic analysis of the chloroplast and nuclear sequences from different *Triticum* and *Aegilops* species obtained in the present study and including published data, we can propose the following evolutionary scenario for genus *Triticum*:

1. According to the chloroplast data, we can conclude that *Ae. speltoides* was the donor of the plasmon during polyploid wheat evolution.
2. The analysis of nuclear *Acc-1* and *Pgk-1* gene sequences carried out in this research allows us to hypothesize that a minimum of three different A genome donor lines existed. In contrast, *Ae. speltoides* had only two, which were designated as wheat B and G genomes.
3. Interrelations of the three diploid wheat species *T. urartu*, *T. boeoticum*, and *T. monococcum* are not well studied, yet. However, it is possible to make some preliminary conclusions. First, it is obvious that wild *T. urartu* was the donor of genome A of all polyploid wheat species, as only in this species the *Pgk-1* gene sequence is identical to that of this gene sequence of polyploid wheats. Second, a wild einkorn *T. boeoticum Pgk-1* haplotype occurs also in cultivated einkorn

Fig. 3b Alignment of the *Pgk-1* gene sequences. Only significant invariable sites are shown. Numbers at the top on each column indicate positions in the whole alignment. Genome specific sites or sites specific for diploid species are indicated in bold letters: green–A genome specific, blue–B genome specific, red–D genome specific, black–G genome, purple–specific for two diploid species. Asterisk denotes sequence obtained in the present study, others were used from Kilian et al. (2007). TAU–*Ae. tauchii*, TAE–*T. aestivum*, SEA–*Ae. searsii*, SPE–*Ae. speloides*, TIM–*T. timopheevii*, ARA–*T. araraticum*, DIC–*T. diccocoides*, TDU–*T. durum*, TM–*T. monococcum*, TB–*T. boeoticum*, TU–*T. urartu*. Letters at the end of the sequence name indicate the genome

T. monococcum. This supports former findings that *T. monococcum* originated from *T. boeoticum* (Beijerinck, 1884; Salamini et al., 2002).

Further experiments are still necessary in the future to get deeper insights into the relationships of A genome wheats.

Acknowledgment We are grateful to Prof. N.A. Kolchanov for rapt attention to this investigation and critical reading of the manuscript; Dr. H. Bockelman (the National Small Grains Collection, Aberdeen, USA), Dr. T. Kawahara (Graduate School of Agriculture of Kyoto University, Kyoto, Japan), Drs. R.A. Udachin, O.P. Mitrofanova and N.A. Anfilova (N.I. Vavilov All-Russian Institute of Plant Industry, St.-Petersburg, Russia) and Dr. J. Valkoun (International Centre for Agriculture Research in the Dry Areas, Aleppo, Syria) for supplying the seeds of wheat species. The research was partially financed on the Subprogram II of Program basic research N25 of the Russian Academy of Sciences "The Origin and Evolution of Biosphere."

References

Beijerinck, M.W. (1884) Über die Dastarde ruischen *Triticum monococcum* x *Triticum dicoccum*. Nederl. Krint. Arch. II. Serv., 189–255.
De Candole, A. (1885) Origin of cultivated plants. K. Rikker, St.-Petersburg, 490 pp. (in Russian).
Felsenstein, J. (1985) Confidence limits on phylogenies: an approach using the bootstrap. Evolution 39, 783–791.
Golovnina, K.A., Glushkov, S.A., Blinov, A.G., Mayorov, V.I., Adkison, L.R. and Goncharov N.P. (2007) Molecular phylogeny of the genus *Triticum* L. Plant. Syst. Evol. 264, 195–216.
Goncharov, N.P. (2002) Comparative genetics of wheats and their related species. Siberian University Press, Novosibirsk, 252 pp. (in Russian with English Summary).
Goncharov, N.P. Glushkov, S.A. and Shumny, V.K. (2007) Domestication of cereal crops in Old World: in search of a new approach to solving old problem. Zhournal Obschei Biologii 68, 125–147.
Harlan, J.R. (1992) Crops and Man, 2nd ed. Am. Soc. Agronomy, CSSA, Madison, Wisconsin, 284 p.
Heer, O. (1865) Die Pflanzen der Pfahlbauten. Neujahrsblatt der Naturforschenden Gesellschaft (Zrich fr das Jahr 1866). 68, 1–54.
Kerby, K. and Kuspira, J. (1986) The phylogeny of polyploid wheats *Triticum aestivum* (bread wheat) and *Triticum turgidum* (macaroni wheat). Genome 29, 722–737.
Kilian, B., Ozkan, H., Deusch, O., Effgen, S., Brandolini, A., Kohl, J., Martin, W. and Salamini, F. (2007). Independent wheat B and G genome origins in outcrossing *Aegilops* progenitor haplotypes. Mol. Biol. Evol. 24, 217–227.
Kumar, S., Tamura, K. and Nei, M. (2004) MEGA3: Integrated software for molecular evolutionary genetics analysis and sequence alignment. Briefings Bioinfomat. 5, 150–163.
Lilienfeld, F. and Kihara, H. (1934) Genomanalyse bei *Triticum* und *Aegilops*. V. *Triticum timopheevi* Zhuk. Cytologia 6, 87–122.
Mori, N., Liu, Y.-G. and Tsunewaki, T. (1995) Wheat phylogeny determined by RFLP analysis of nuclear DNA. 2. Wild tetraploid wheats. Theor. Appl. Genet. 90, 129–134.
Nesbitt, M. (2001) Wheat evolution: integrating archaeological and biological evidence. The Linnean. 3, 37–59. Special issue.

Nicholas, K.B., Nicholas, H.B., Jr. and DeerWeld II D.W. (1997) GeneDoc: analysis and visualization of genetic variation. EMBnet News 4, 14.

Rogers, S.O. and Bendich, A.J. (1985) Extraction of DNA from milligram amounts of fresh, herbarium and mummified plant tissues. Plant Mol. Biol. 5, 69–76.

Salamini, F., Özkan, H., Brandolini, A., Schäfer-Pregl, R. and Martin, W. (2002) Genetics and geography of wild cereal domestication in the Near East. Nat. Rev. Genetics 3, 429–441.

Thompson, J.D., Gibson, T.J., Plewniak, F., Jeanmougin, F. and Higgins, D.G. (1997) The CLUSTAL_X windows interface: flexible strategies for multiple sequence alignment aided by quality analysis tools. Nucleic Acids Res. 15, 4876–4882.

Udachin, R.A. (1982) On the possible current existence of *Triticum antiquorum* Heer, Nauch.-Tekhn. Byul. VNII Rastenievod 119, 72–73.

Yamane, K. and Kawahara, T. (2005) Intra- and interspecific phylogenetic relationships among diploid *Triticum-Aegilops* species (*Poaceae*) based on base-pair substitutions, indels, and microsatellites in chloroplast non-coding sequences. Am. J. Bot. 92, 1887–1898.

Subject Index

aaRS, *see* Aminoacyl-tRNA synthetases
AAS, *see* Atom absorption spectroscopy
Acc-1 genes, 403, 412
Aegilops species, 408
Aldohexose-2,4,6-triphosphates, 105
Aldopentoses, 103
Aldotetrose- and aldopentose-2, 4-diphophates, 105
Allose-2,4,6-triphosphate, 105
Alticola strelzovi, 382
γ-alumina, 105
Amidotriphosphate, 105
Aminoacyl-tRNA synthetases, 254–255, 257, 259
Ammonia and carbohydrates, heterocycles synthesis and, 108
Ammonium chloride (NH_4Cl), 333
Anaerobic lithotrophic eubacteria, *see Carboxydothermus*
Anhysteretic remanence (ARM), 230
Anui drainage basin, 380–381
Apatite $Ca_5(OH, F, Cl)(PO_4)_3$, 106
Apodemus uralensis, 381
Aquifex, 28
Arabinocytidine-3'-phosphate, 111
Archaean concretions, isotopic composition of, 169–173
Archaeoglobus, 28
Archean atmosphere, origin, 26
Archetype, 5
Asioscalops altaica, 381
Astrocatalysis hypothesis
 and planetology, 45–46
 "RNA world" and life origin, 46–50
 universe evolution and life origin, 43–45
Atom absorption spectroscopy, 229
Autocatalytic systems evolution, in flow reactor, 115
 auto-oligomerase reaction, 122
 computer model and conditions of real experiment, 120–122
 computer simulation, results, 122–123
 phase-separated systems (PSS), 118–122
 presuppositions for problem elimination, 117

Baikal rift zone, microbial mats in lakes and hydrotherms of, 185
 hot spring phototrophic, 187–192
 non-phototrophic biofilms, 192–193
 of saline and soda lakes, 193–194
 terminal destruction processes in, 194
 types of, 186
Banded iron formations, 29, 31, 168
 associated carbonates, 179–180
Banding sequences, in genus *Chironomus*, 346–347
 dispersal of species and, 360–362
 divergence of, 348–353, 355–360
 as markers of evolutionary divergence of species genomes, 353–355
Barlagiin-Gol-1 site, 397
Barley (*Hordeum* ssp.), 404
Batrachuperus, 301
BIFs, *see* Banded iron formations
Binary hammerhead ribozymes (binRz), 144
 rate constants, for stages of catalytic cycles of, 147–148
 RNA cleavage by, 145–146
Biological cycle closure
 closure coefficients, 332–334, 340
 ecological principle of, 335
 "evolutionary" type of closure, 335–337
 general degree of, 339–341
 microevolutionary population parameters, interaction of, 337–339

Biological life support systems, 326, 334
Biological realm, hierarchical scale-free representation, 65
   referents
      architecture of complexity, two directions in, 73–75
      evolutionary equation, 77–79
      hierarchy, growing of, 79–81
      identification, 67–70
      from linnaean hierarchy to hierarchy of complexity, 71–73
      signs, space of, 76–77
      transition thresholds and irreversibility of transition, 75–76
      synthetic theory of evolution (STE) and, 66, 72, 75
Biomineralization, 228
   composition, structure, and functions, evolution, 211–213
   modern understanding and types of, 209–211
   silicon, *see* Silicon biomineralization
Biotic matter cycle, 325–327
   establishment, conditions for, 329–332
Bistate system, as prototype of living organism, 158–161
BLSS, *see* Biological life support systems
Blue-green algae, *see* Cyanobacteria
BMC, *see* Biotic matter cycle
BSs, *see* Banding sequences
Butlerov reaction, *see* Formose reaction

Caenorhabditis elegans, 17, 275
CAGE analysis, 278
Calcium phosphate $Ca_3(PO_4)_2$, 106
Calderobacterium, 28
Calothrix sp., 216
Cambridge reference sequence, 370
Cannizzaro reaction, 101
Carbohydrates, 99
   C_2-C_3, prebiotic synthesis of, 102–104
   cytidine ribonucleotides synthesis, on sugar phosphate, 110–111
   formose reaction, 100–102, 104
   heterocycles synthesis from, 108
   ribose, 100
   selective prebiotic synthesis of, 104
      of carbohydrates phosphates, 105–106
      catalyzed by natural minerals, 105
      co-condensation of lower carbohydrates and formaldehyde, 106–107
      putative prebiotic synthesis of, 107–108
   "sugar model" by A.L. Weber, 108–110
Carbonaceous meteorites and comets, biosphere origin and, 53–54
   filamentous cyanobacteria and sulfur bacteria, 55
   Orgueil and Murchison, cyanobacteria microfossils in, 58–61
   origin and distribution of life, 56–58
Carboxydothermus, 28
$\delta^{13}C_{carb}$, of Early Precambrian carbonates, 167–168
Central Antarctic Ice Sheet, 54
Central Asia and Volga–Ural region, genetic landscape of, 369
   materials and methods, 370–371
   mtDNA variation, 371–374
   Y-chromosome variation, 374–375
Chaetoceros muelleri, 221
Chetverikov's principle, 300
Chironomus, genus
   basic BSs, as markers of evolutionary divergence of species genomes, 353–355
   BSs divergence, 348–353, 355–360
   chromosome evolution and speciation, 362–363
   karyotype structure, 346–347
   species dispersal and BSs, 360–362
Chloroflexus aurantiacus, 187–188
Chromosomes
   and continents, *see Chironomus*, genus
   evolution
      cytogenetic analysis of, 346
      and speciation, 362–363
   and speciation, 311
      S. murinus (house musk shrew), 314–316
      Sorex araneus (common shrew), 316–319
      Thrichomys, South American caviomorph rodent, 312–314
Chroococcalean cyanobacteria, 59
Clethrionomys rutilus, 381
Closure coefficients, 332–334
Co-condensation, of lower carbohydrates and formaldehyde, 106–107
CO_2- dominated atmosphere, 27
Coevolution, 207
   biomineralization
      composition, structure, and functions, evolution, 211–213

Subject Index

modern understanding and types of, 209–211
CO^2–H_2 bacterial filter, 201–202
"Coherent ecosystems", 242
Coherent evolution, 18
Comet P/Halley, 57
Comets and carbonaceous meteorites, biosphere origin and, 53–54
  filamentous cyanobacteria and sulfur bacteria, 55
  Orgueil and Murchison, cyanobacteria microfossils in, 58–61
  origin and distribution of life, 56–58
Common shrew, chromosome races of, 316–319
C_{org} recycling
  after c. 2000 Ma, 179
  prior to c. 2000 Ma, 178–179
Cricetulus barabensis, 381
CRS, see Cambridge reference sequence
CTAB method, 405
Cyanobacteria
  changes in argillaceous minerals and, 227, 232–236
    biological activity of, in presence of clays, 230–232
    Microcoleus chthonoplastes, for experimental study, 228–230
  communities, tropical structures, 35–38
  filamentous, biosphere origin and, 55
  microfossils of, in Orgueil (CI1) and Murchison (CM2), 58–61
  silicification, 215–217
"Cyanobacterial mats", see Microbial mats, in lakes and hydrotherms of Baikal rift zone
2′,3′–cyclic phosphates, 129–130, 133–135, 142
Cylindrotheca fusiformis, 221
Cytidine-2′,3′-cyclophosphate, 111
Cytidine ribonucleotides synthesis, on sugar phosphate, 110–111
Cytochrome b gene, 301
Cytogenetic divergence, of genus Chironomus, 349

Darwin's selection, 119
DDC model, see Duplication-degeneration-complementation model
DDGF equations, 337
Denisova cave, 381–382, 392
Desulfurococcus-type organotrophs, 28

Diatom valves, biosilification, 219
  early stages morphogenesis of, 220–221
  silicic acid transporters discovery, 221–223
"Dirty snowball" model, 56
Ditylum brightwellii, 220
Drosophila melanogaster, 17, 275
Duplication–degeneration–complementation model, 261–262
  See also Gene duplications, evolution by
DYEnamic ET kit, 370

Early Precambrian carbonates, $\delta^{13}C_{carb}$ of, 167–168
EC model and G-value paradox, 269–270
  See also Gene duplications, evolution by
ElectroScan environmental scanning electron microscope (ESEM), 58
Energy dispersive X-ray spectroscopy (EDS), 53, 58
Entophysallis, 54
Ephydatia fluviatilis, 218
Epigenetic complementation (EC) model, 262
  See also Gene duplications, evolution by
eRF3, eukaryotic release factor (RF), 273–274
  evolution of genes encoding for, 278–279
  structural organization of, 275–278
Eukaryotes, translation termination system evolution in
  eRF3, termination factor
    evolution of genes encoding for, 278–279
    structural organization of, 275–278
  GSPT2, as new phylogenetic marker, 279–282
  termination factors, evolutionary origin of, 273–275
Eutamias sibiricus, 381

Field emission scanning electron microscopes (FESEM), 58
Fischer–Tropsch synthesis, 46
Formose reaction, 93–94, 100–102, 104, 128

Gallus gallus, 275
Garga spring, microfossils, 191–192
Gene duplications, evolution by
  gene copies, epigenetic regulation
    duplicate genes survival, repositioning and, 263–264
    duplication–degeneration – complementation model, 261–262

Gene duplications, evolution by (*cont.*)
  EC model and G-value paradox, 269–270
  epigenetic complementation model, 262
  mutational asymmetry, repositioning and, 265–266
  young duplicates, adaptive evolution of, 266–269
  genetic code origin, duplication in tRNA
    tRNA aminoacylation, simultaneous sense–antisense coding and, 257–260
    tRNAs domains and paradox of two codes, 253–256
    tRNAs points to ancient duplication, dual complementarity in, 256–257
Genetic code, origin, *see* tRNA, duplication
Geosphere and biosphere evolution, stages, 3–20, 15–16
  archetype, canalization of, 5
  cacosphere, 20
  climate evolution, stages of, 9
  coherent evolution, 18
  cyanobacteria, emergence of, 12
  DNA/RNA/protein-based life and, 4
  energy-producing enzyme systems and, 15
  K/Na ratio and, 12
  mantle, evolution of, 6
  moon-like stage of Earth, 4–5
  "plume dropper" formation, 12
  Rayleigh number (Ra), 10
  SELEX experiments, 7
  World Ocean, 10, 11, 14
Gli/Ci genes expression, 289
Global paleogeographical reconst-ructions, 303
Glycolaldehyde phosphate synthesis, 105–106
GSPT1 protein, 275
GSPT2 genes, 278–282
GSPT2, as new phylogenetic marker, 275, 279–282

Hamersley basin, 169
Hbs1 protein, 274
Hedgehog (Hh) signaling cascade system, 285–286
  genes, evolutionary mode of, 292
  modeling of, 287–289
  molecular evolution of, 290–292
Hexose-2,4,6-triphosphates, 105
Holarctic basic BSs (hb′BSs), 351
Homo erectus, 391

Homo sapiens, 19, 87
Homo sapiens sapiens, 398
Hot spring phototrophic microbial mats, 187–192
  See also Microbial mats, in lakes and hydrotherms of Baikal rift zone
House musk shrew, chromosome races of, 314–316
ht15 haplotype, 375
Hydrocarbon-oxidizing bacterial filter, 200–201
Hydrogenobacter, 28
Hydrogenotrophic chemosynthetic microbes, 27–28
2-Hydroxymethyl glycerol, 103
Hynobius, 301

"Incoherent ecosystems", 242

Kaufmann's theory, 88
Kimura's neutral evolution theory, 286
K-23790 *T. sphaerococcum*, 407

Levallois technique, 393
Lobry de Bruyn Alberda van Ekenstein reaction, 101
Lomagundi-Jatuli isotopic event, OM recycling and, 180
Lubomirskia baicalensis, 217–218

Magnetite, 28
Mammalian phylogeny, divergence events, 302–304
Marmot bones, from Betovo Paleolithic site, 380
Mastigocladus laminosus, 29
matK gene, 408–409
Metanosaeta thermoacetophila, 29
Michaelis–Menten constant, 337
Microbial biosphere, 23
  actualistic principle, limits, 26–27
  post-prokaryotic evolution, 25
  prokaryotic evolution, 24
  relict microbial communities
    acidophilic, 29
    anaerobic lithotrophic eubacteria, 28
    cyano-bacterial community, trophic structure of, 35–38
    hydrogenotrophic chemosynthetic microbes, 27–28
    landscape formation, in prokaryotic biosphere, 30–35

Subject Index

thermophilic cyanobacterial community, 30
Microbial mats, in lakes and hydrotherms of Baikal rift zone, 185
  hot spring phototrophic, 187–192
  non-phototrophic biofilms, 192–193
  of saline and soda lakes, 193–194
  terminal destruction processes in, 194
  types of, 186
Microfossils, 31, 57
  of cyanobacteria, in Orgueil (CI1) and Murchison (CM2), 58–61
Microtus arvalis, 381
Miller's theory, 88
Mitochondrial DNA (mtDNA), 369
  variation, among Volga–Ural populations, 371–374
Murchison (CM2) and Orgueil (CI1), cyanobacterial microfossils in, 58–61
Mutational asymmetry, of duplicate genes, 265–266
Mycoplasma genitalium, 17, 69
Myospalax myospalax, 381
Myotis lucifugus, 279

Nanobacteria, 210
Naphthidiobiosis, zone, 202–203
"Natural selection", 88, 90–91, 95–96, 300
Natural minerals, carbohydrates synthesis and, 105
Navicula salinarum, 220
NJ-phylogenetic AEF tree, 359
Non-phototrophic biofilms, 192–193
North-western Altai
  ancient man settling and, 391–400
  Paleolithic man habitats reconstruction and, 379
    Anui valley, contemporaneous animals of, 380–381
    Denisova cave, small mammals from, 381–382
    mammal fauna and activity of Paleolithic man, 383–389
    Ust'-Karakol site, small mammals from, 383
Nostoc sp., 60
Notch-denticulate tools, 399
Novosibirsk–Tomsk F1 hybrids, 317–318

Ochromonas ovalis, 222
OM recycling, Lomagundi-Jatuli isotopic event and, 180

Oparin–Holdane–Bernal theory, 88
Ordovician period, 241–242
  biota of, evolution, 245–248
  ecological quilds of, 243
  global geological events, 244
Orgueil (CI1) and Murchison (CM2), cyanobacterial microfossils in, 58
  cyanobacteria morphotypes in, 59–61
Origin of life, prebiotic phase of
  autocatalytic system, 89, 95
  capacity for self-replication phenomenon, 87
  DNA and RNA, 87
  Engels theory, 85–86
  formosa reaction, 93–94
  Kaufmann's theory, 88
  Miller's theory, 88
  natural selection, 88, 90–92, 95–96
  Oparin–Holdane–Bernal theory, 88
  physicochemical definition for, 96
Orkhon-1 stratified site, 397

Palaeoproterozoic concretions, isotopic composition, 173–177
Paleolithic man habitats reconstruction, Northwestern Altai and, 379
  Anui valley, contemporaneous animals of, 380–381
  Denisova cave, small mammals from, 381–382
  mammal fauna and activity of, 383–389
  Ust'-Karakol site, small mammals from, 383
Particulated organic matter, 36
PCA, *see* Principal component analysis
PDMPO, fluorescent, 220
Peach latent mosaic viroid (PLMVd), 133
PFD, *see* Prion forming domain
Pgk-1 genes, 403
Phormidium laminosum, 29, 33
Phylogenetic reconstruction, 299
  Chetverikov's principle, 300
  mammalian faunas, basic events in, 304–306
  mammalian phylogeny, divergence events in, 302–304
  zoogeographical dating and, 306–308
Phylogeographic reconstructions, 369
Planetology, astrocatalysis hypothesis and, 45–46
"Plume dropper" formation, 12
POC, *see* Particulated organic matter
Podospora anserina, 276

Prebiotic organic microsystem, 153
  bistate system, as prototype of living organism, 158–161
  fundamental properties, 154–155
  stabilized bifurcation, as starting point of life, 155
    chemical system, properties, 156–157
    living cell, characteristics of, 157–158
Principal component analysis, 373
Prion forming domain, 276
Prokaryotic biosphere, landscape formation, 30
  amphibial landscapes, 31, 33
  halophilic community and, 34
  Precambrian soils and, 32
  stromatolites formation and, 33–34
Pseudohynobius, 301
"Purple and green mats", *see* Microbial mats, in lakes and hydrotherms of Baikal rift zone

R1b3 Y-chromosomes, 375
Rayleigh number (Ra), 10
RCPs, *see* Repeat-containing proteins
Real enzymatic "bottleneck" reaction, 337–338
Recombination reaction, RNA molecules formation and, 132
  intramolecular transesterification reaction, 133
  ligation reaction, 133–134
  Watson–Crick base pairing, 135–136
  See also RNA world, evolution
Referent concept, hierarchical scale-free representation of biological realm and
  architecture of complexity, two directions in, 73–75
  evolutionary equation, 77–79
  hierarchy, growing of, 79–81
  identification, 67–70
  from linnaean hierarchy to hierarchy of complexity, 71–73
  signs, space of, 76–77
  transition thresholds and irreversibility of transition, 75–76
  two-dimensional schema of, 74
Repeat-containing proteins, 277
Restriction fragment length polymorphism analysis (RFLP analysis), 370, 375
Rhodamine 123 (R 123), 220
Ribose-2,4-diphosphate, 93–94, 105, 128
Ribozymes, *trans* hammerhead, 139–140
  binary hammerhead ribozymes and, *see* Binary hammerhead ribozymes (binRz)
  RNA ligation by, 141–144
  variants of, 142
RNA ligation, by *trans* hammerhead ribozymes, 141–144
RNA world, evolution, 44, 100
  astrocatalysis and life origin, 46–50
  catalytic RNAs emergence, 131–132
  first RNA monomers, 127–129
  recombination reaction, 132–136
  RNA oligomers, prebiotic synthesis of, 129–131
  trans hammerhead ribozymes and, *see* Trans hammerhead ribozymes
  Watson–Crick base pairing, 129, 131, 135–136
RS-space, evolutionary equation on, 77–79

Saline and soda lakes, microbial mats, 193–194
  See also Microbial mats, in lakes and hydrotherms of Baikal rift zone
Satellite tobacco ringspot virus, 141–142
S. cerevisiae, 274, 276
Shizosaccharomyces pombe, 276
Silica deposition vesicles (SDVs), 216, 220–221
Silicic acid transporter protein, 221–223
Silicon biomineralization
  cyanobacteria, silicification, 215–217
  diatoms, 219
    early stages morphogenesis, of diatom valves, 220–221
    silicic acid transporters discovery, 221–223
  sponges, 217–219
Simple homogenous closed ecosystem, on matter supply
  base model description, 327–329
  BMC establishment, conditions for, 329–332
SIT, *see* Silicic acid transporter protein
sit genes, 222–223
Slmb genes, 290, 294
smo gene, 290
SOPM method, 277
Sorex araneus, *see* Common shrew, chromosome races of
Speciation, chromosomes and, 311
  S. murinus (house musk shrew), 314–316
  Sorex araneus (common shrew), 316–319

Subject Index

Thrichomys, South American caviomorph rodent, 312–314
"SPECTROSCAN MAKC-GV" XRF crystal diffraction scanning spectrometer, 229
Spermophilus undulatus, 381
Spirulina platensis, 332–333
Sponges, silicon biomineralization and, 217–219
SRB, *see* Sulphate-reducing bacteria
STE, *see* Synthetic theory of evolution
Stenocranius gregalis, 382
sTRSV, *see* Satellite tobacco ringspot virus
Subsurface microbiology, 199
  CO_2–H_2 bacterial filter zone, 201–202
  evaluation of, 203
  hydrocarbon-oxidizing bacterial filter, zone of, 200–201
  naphthidiobiosis zone, 202–203
  structural organization of, 200
Su(fu) and *fu* genes, 290
"Sugar model" by A.L. Weber, 108–110
"Sulfuric mats", *see* Microbial mats, in lakes and hydrotherms of Baikal rift zone
Sulphate-reducing bacteria, 36, 167
Suncus murinus, *see* House musk shrew, chromosome races of
Synthetic hydroxylapatite $Ca_5(OH)(PO_4)_3$, 106
Synthetic theory of evolution, 66, 72, 75
Synura petersenii, 222

Template-directed synthesis model, 130
  See also RNA world, evolution
Tethya aurantia, 218
Thalassiosira weissflogii, 220
Thermodesulfobacterium, 29
Thrichomys, South American rodent, 312–314
Trans hammerhead ribozymes, 139–140
  binary hammerhead ribozymes and,*see* Binary hammerhead ribozymes (binRz)
  RNA ligation by, 141–144
  variants of, 142
Translation termination system evolution, in eukaryotes
  eRF3, termination factor
    evolution of genes encoding for, 278–279
    structural organization of, 275–278
    GSPT2, as new phylogenetic marker, 279–282

termination factors, evolutionary origin of, 273–275
Triticum ssp., *see* Wheat, evolutionary history
tRNA, duplication
  tRNA aminoacylation, simultaneous sense–antisense coding and, 257–260
  tRNAs domains and paradox of two codes, 253–256
  tRNAs points to ancient duplication, dual complementarity in, 256–257
  See also Gene duplications, evolution by
trnT–trnL intergenic spacer, 405

Uracil, 128
Ust'-Karakol site, 383, 392

Vega Dust Mass Spectrometer, 57
Vibrating sample magnetometer (VSM), 230
Vivianite $Fe_3(PO_4)_2$, 106
Volga–Ural region and Central Asia, genetic landscape of, 369
  materials and methods, 370–371
  mtDNA variation, 371–374
  Y-chromosome variation, 374–375

Watson–Crick base pairing, 129, 131, 135–136
Wheat, evolutionary history, 403
  chloroplast evidence of, 408–410
  DNA sequencing and phylogenetic analysis, 406
  genus *Triticum*, 413–414
  nuclear loci, 410–413
  pile-dwelling wheat, 406–408
  plant materials, 404–405
  total DNA isolation and PCR amplification, 405
Wollastonit equilibrium, 27

X-ray diffraction (XRD), 229

Y-chromosome
  markers, 369
  variation, in Volga–Ural and Central Asian populations, 374–375
Young duplicates, adaptive evolution, 266–269
  See also Gene duplications, evolution by

Zoogeographical dating, phylogenetic reconstruction and, 306–308

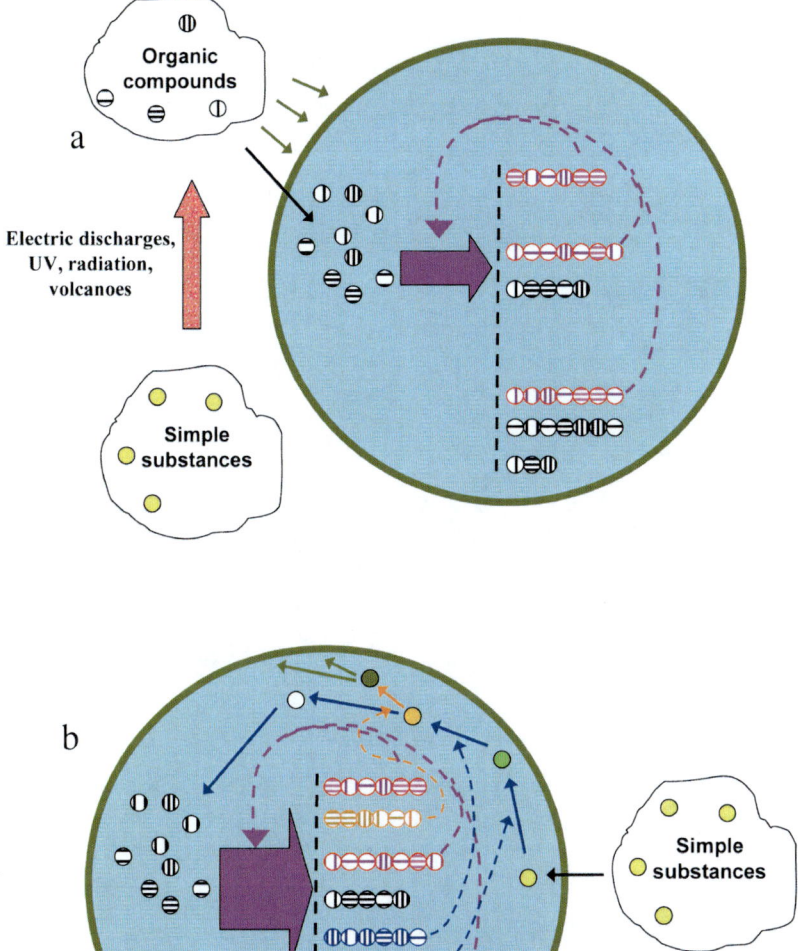

Color Plate 1. Scheme of the origin of pre-biotic metabolism on the base of multivariate oligomer auto-catalyst (energy flow is not shown): **a** initial stage, when auto-oligomerase appears and begins to catalyze its own production; **b** final stage, when PSS can support its own existence only by energy flow and simple chemical substances

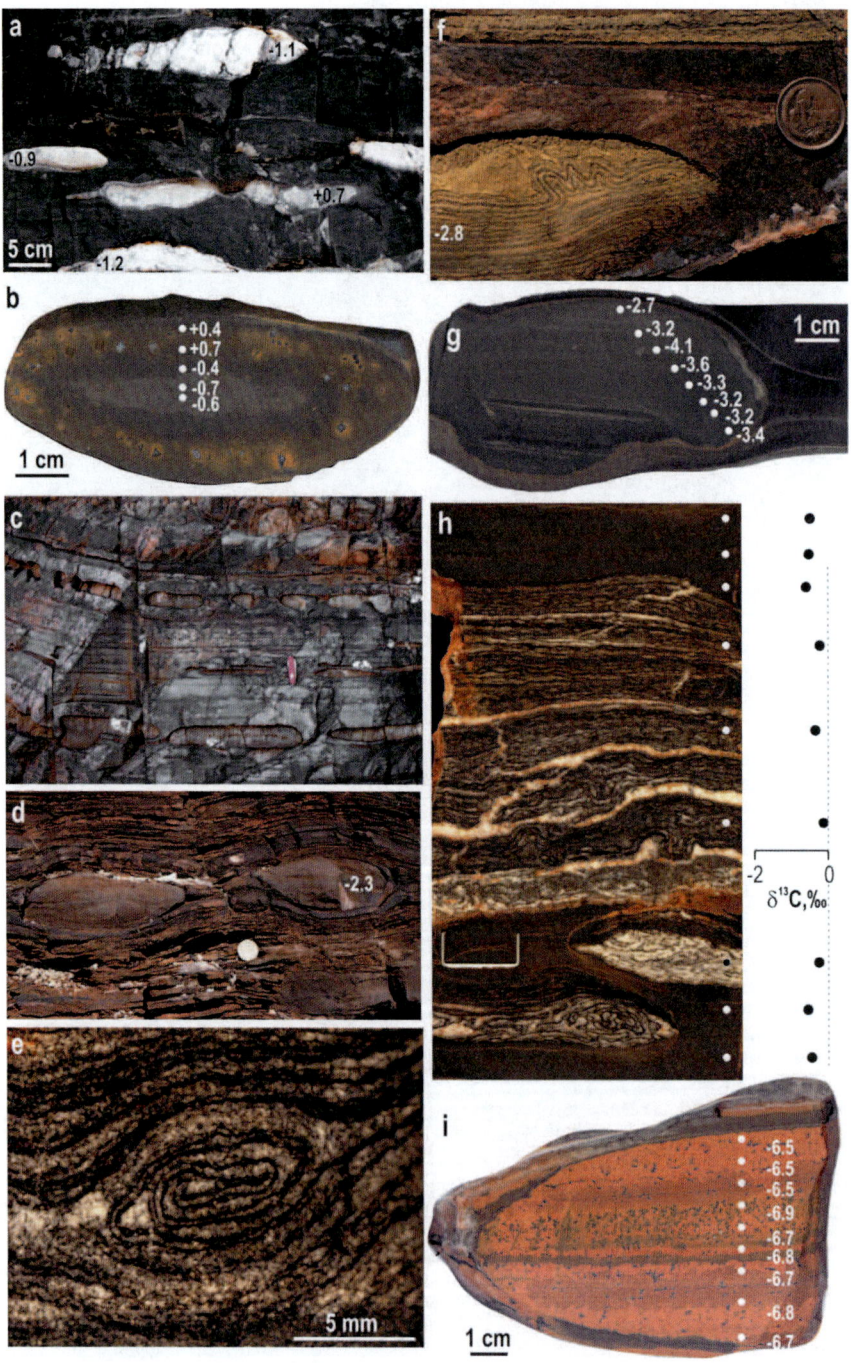

Color Plate 2. Archaean carbonate concretions (*numbers* on concertions denote $\delta^{13}C_{carb}$ obtained from whole-rock samples). *Boomplaas Formation, Kaapvaal Craton, South Africa*:

a lensoidal calcite concretions (*white to pale grey*) in "black schist" (C_{org} = 0.3%) at the base of the Campbellrand-Malmani carbonate platform. $\delta^{13}C_{carb}$ obtained from several concretions ranges between –1.2 and +0.7‰ ($n = 19$); **b** isotopic profile through concretion showing a limited range between –0.7 and +0.7‰. *Wittenoom Formation, Pilbara Craton, Western Australia*: **c** recessive-weathering carbonate concretions form nice "trains" with consistent shapes along specific bedding planes below the Main Tuff Interval (Hassler, 1993); knife is 9 cm long; **d** two large carbonate concretions at same stratigraphic level with same internal stratigraphy (roll-up structures in lower halves but none in upper halves); note differential compaction of less well-cemented strata; coin is 2.1 cm in diameter; sample no. 92026 (Table 3 in "The Ancient Anoxic Biosphere Was Not as We Know It") was collected nearby from same outcrop; **e** roll-up structure in otherwise lumpy-laminated dolostone; **f** shale-hosted carbonate concretion with internal folds ending abruptly at upper edge of concretion whereas laminae in lower part pass continuously into shale; coin is 1.7 cm in diameter; sample no. 92489-A (Table 3 in "The Ancient Anoxic Biosphere Was Not as We Know It") was collected nearby from same outcrop; **g** isotopic profile through shale-hosted, reworked, calcite concretion (sample no. 96255, Tables 1 and 3 in "The Ancient Anoxic Biosphere Was Not as We Know It") located beneath Main Tuff Interval; note significant differential compaction of shale around concretion; **h** isotopic profile through marl (massive, *dark grey*) and dolomite with "lumpy lamination" that in places deformed into roll-up structures and folds; ovoid masses with roll-up structures at bottom are probably concretionary in origin; the *white zones* are sparry dolomite veinlets; sample no. 92011 (Tables 1 in "The Ancient Anoxic Biosphere Was Not as We Know It"); staple for scale—9 mm tip to tip; **i** isotopic profile through weathered to *pink colour* ferroan dolomite concretion encased in weathered shale adjacent to the Jimblebar iron mine; sample no. 88273 (Tables 1 and 3 in "The Ancient Anoxic Biosphere Was Not as We Know It")

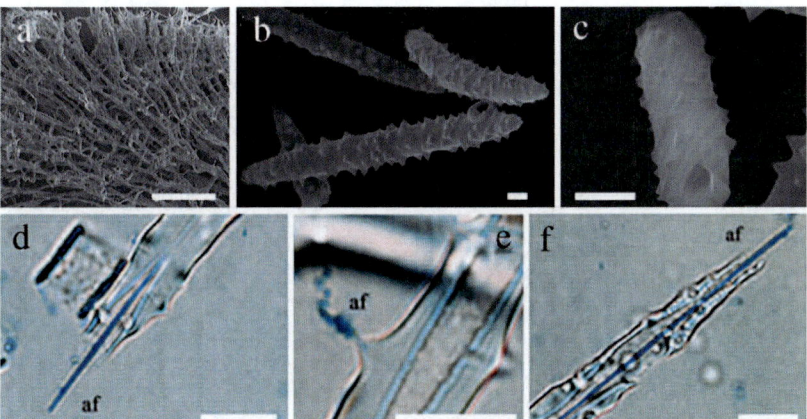

Color Plate 3. Spicule proteins of *Lubomirskia baicalensis*. **a** Sponge skeleton consisting of spicules; **b** distinct spicules; **c** the central channel, SEM. **d–f** Stages of spicule dissolution in 6 M NH$_4$F, finding of axial filaments (*af*) under LM. Scale bars: **a** 1000 μm; **b–f** 10 μm

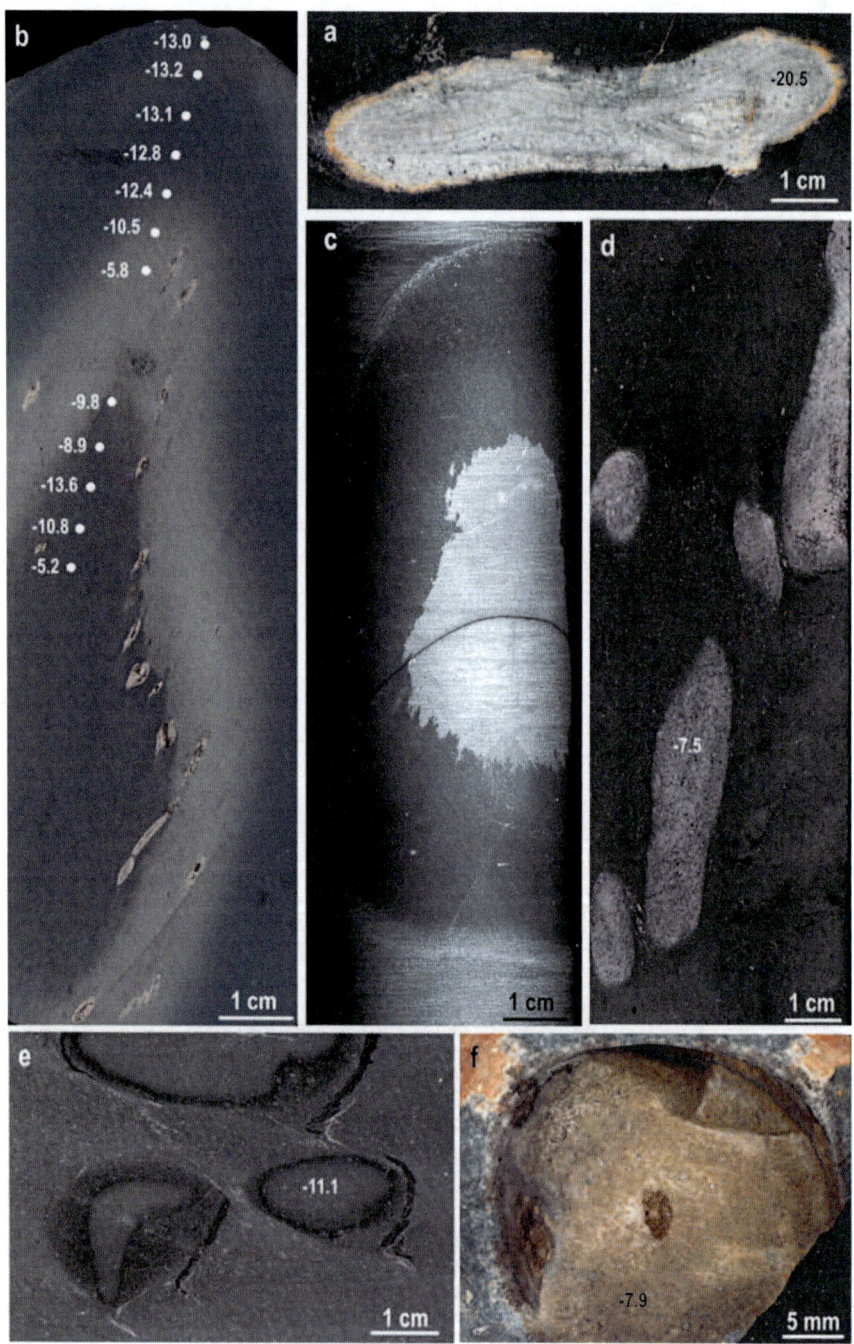

Color Plate 4. 2050 to 1970 Ma diagenetic carbonate concretions from the Fennoscandian Shield (*numbers* on concretions denote $\delta^{13}C_{carb}$ obtained from whole-rock samples). *C. 2050*

Ma Il'mozero Formation, Imandra/Varzuga Greenstone Belt: **a** greywacke-hosted (C_{org} = 0.14%), reworked, calcite concretion retaining small-scale trough cross-bedding of host sediment. *2004 Ma Pilgujärvi Sedimentary Formation, Pechenga Greenstone Belt:* **b** isotopic profile through multizoned calcite concretion hosted by turbiditic, sulphidic (S = 3.7%) "black-schist" (C_{org} = 0.25%); middle zone contains numerous pyrrhotite concretions (*pale yellow*). **c** Multizoned, calcite concretion in turbiditic, sulphidic (S = 5.5%) "black-schist" (C_{org} = 1.5%). **d** Massive lensoidal and spherical calcite concretions hosted by ferropicritic tuff. **e** Irregular zoned calcite concretions in vocaniclastic greywacke. *Pilgujärvi Volcanic Formation, Pechenga Greenstone Belt:* **f** brown weathering, zoned, calcite concretion with pyrrhotite core (*dark brown*) hosted by 1970 Ma felsic tuff

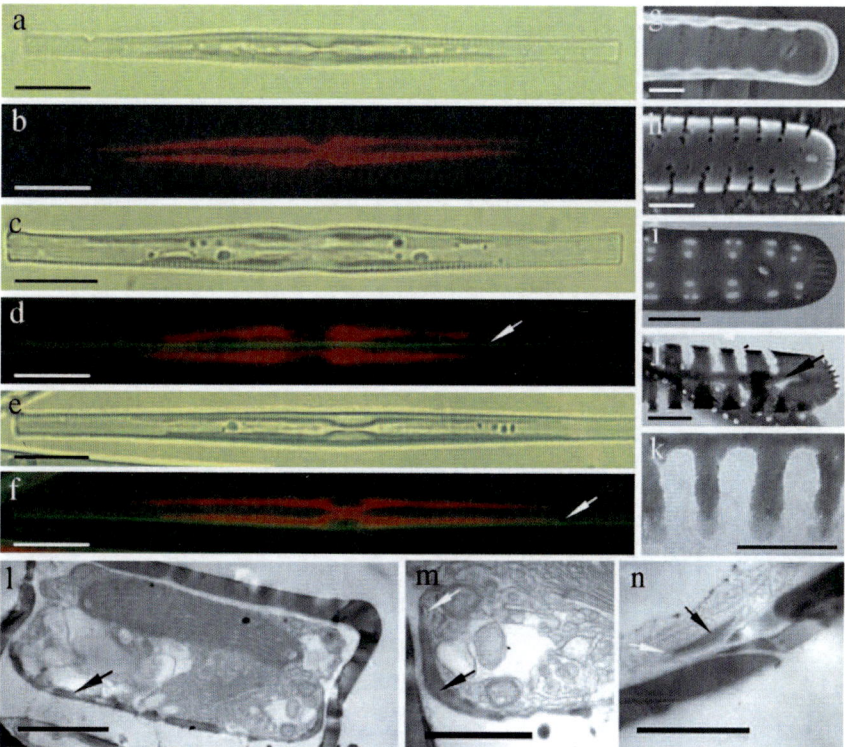

Color Plate 5. Morphogenesis of *Synedra acus*. **a, c, e** LM; **b, d, f** fluorescence microscopy; **g, h** SEM; **i–n** TEM (L–N cross-sections). **a, b** Without staining (*red* chloroplasts), **c–f** vital staining with rhodamine 6G (*green* new valves, *arrows*). **c, d** 24 h of the incubation: divided cells still remain together. **e, f** 48 h of the incubation: fluorescing of new valve only. **g, i** Mature valves, **h** formation of areolae, **j** development of a rimoportula from a silica loop (*arrow*), **k** growing virgae. **l–n** New silica structures (*black arrows*) formed inside SDV (silicalemma *white arrows*): **l** valve, **m** mantle of valve, **n** girdle band. Scale bars: **a–f** 10 μm; **j–m** 1 μm; **n** 0.5 μm

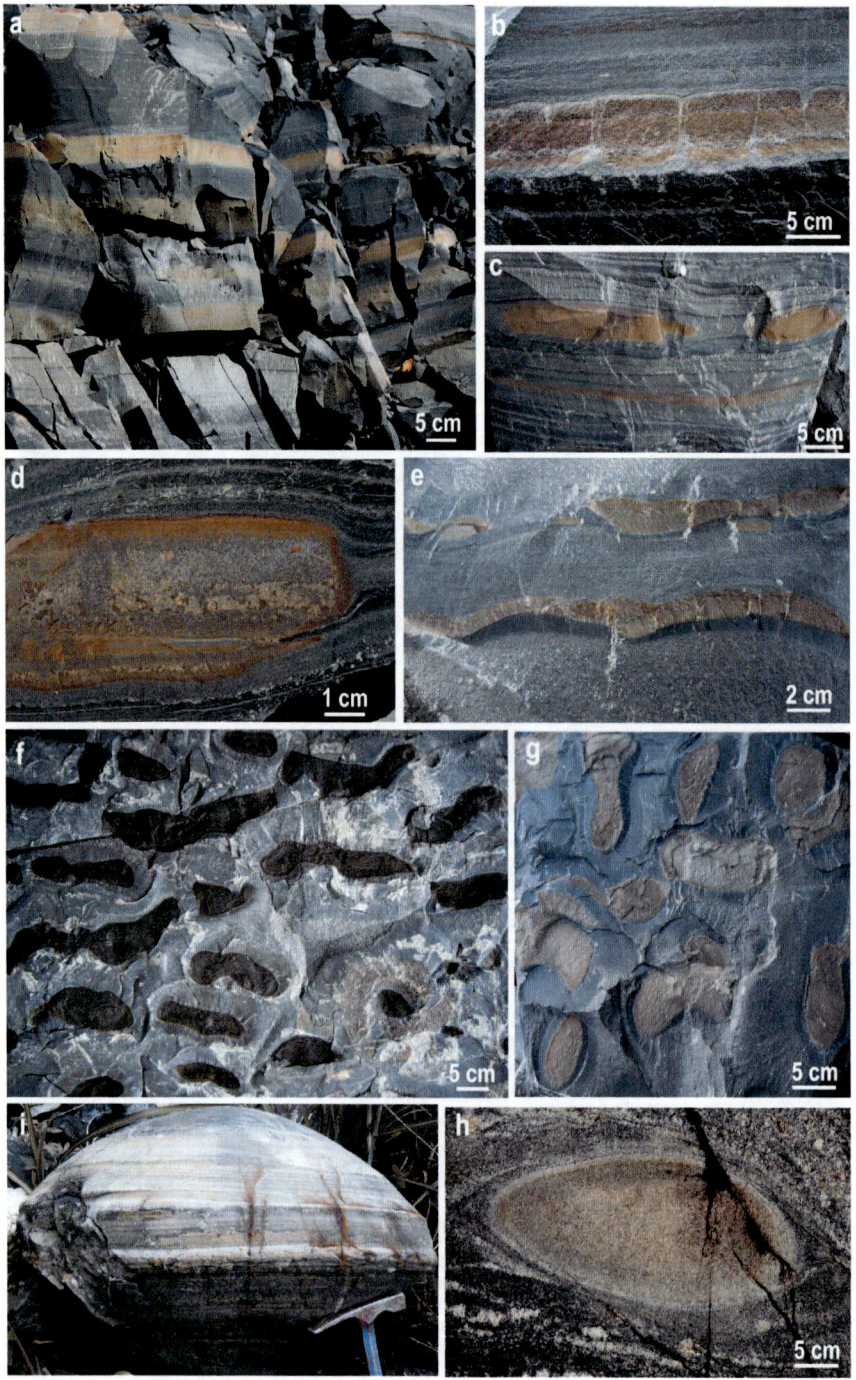

Color Plate 6. 1900 Ma and younger Proterozoic carbonate concretions. *c. 1900 Ma Kondopoga*

Formation, Onega Basin, eastern Fennoscandian Shield: **a** numerous, brown weathering, ferroan calcite, concretionary beds in lacustrine, C_{org}-rich, volcaniclastic, turbiditic rhythmites; **b** brown weathering, ferroan calcite, concretionary bed affected by sedimentary boudinage; **c** brown weathering, ferroan calcite, concretionary bed and lensoidal concretions with consistent shape along specific bedding plane in lacustrine, rhythmically bedded greywacke; **d** brown weathering, ferroan calcite, concretion around which there has been significant differential compaction of less-well cemented strata; **e** sandy ripples and channels cemented by ferroan calcite; **f** and **g** bedding surface showing sub-parallel and sub-perpendicular sandy ripples in turbiditic siltstone cemented by ferroan calcite. *C. 1870 Ma Ladoga Series, Ladoga Basin, south-eastern Fennoscandian Shield*: **h** paragneiss-hosted, calcite–diopside–forsterite (primary calcitic) concretion surviving a high-temperature amphibolite facies metamorphism accompanied by partial melting. *Neoproterozoic Doushantuo Formation, Yangtze Gorge area, China*: **i** large carbonate concretion fallen out of C_{org}-rich shale

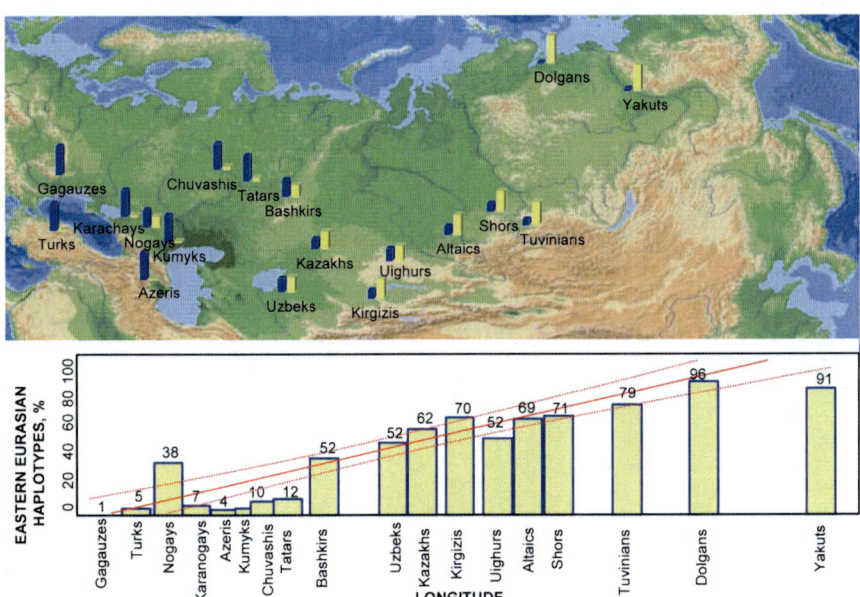

Color Plate 7. West to east frequency gradient of East Asian mtDNA lineages among 18 Turkic-speaking populations of Eurasia

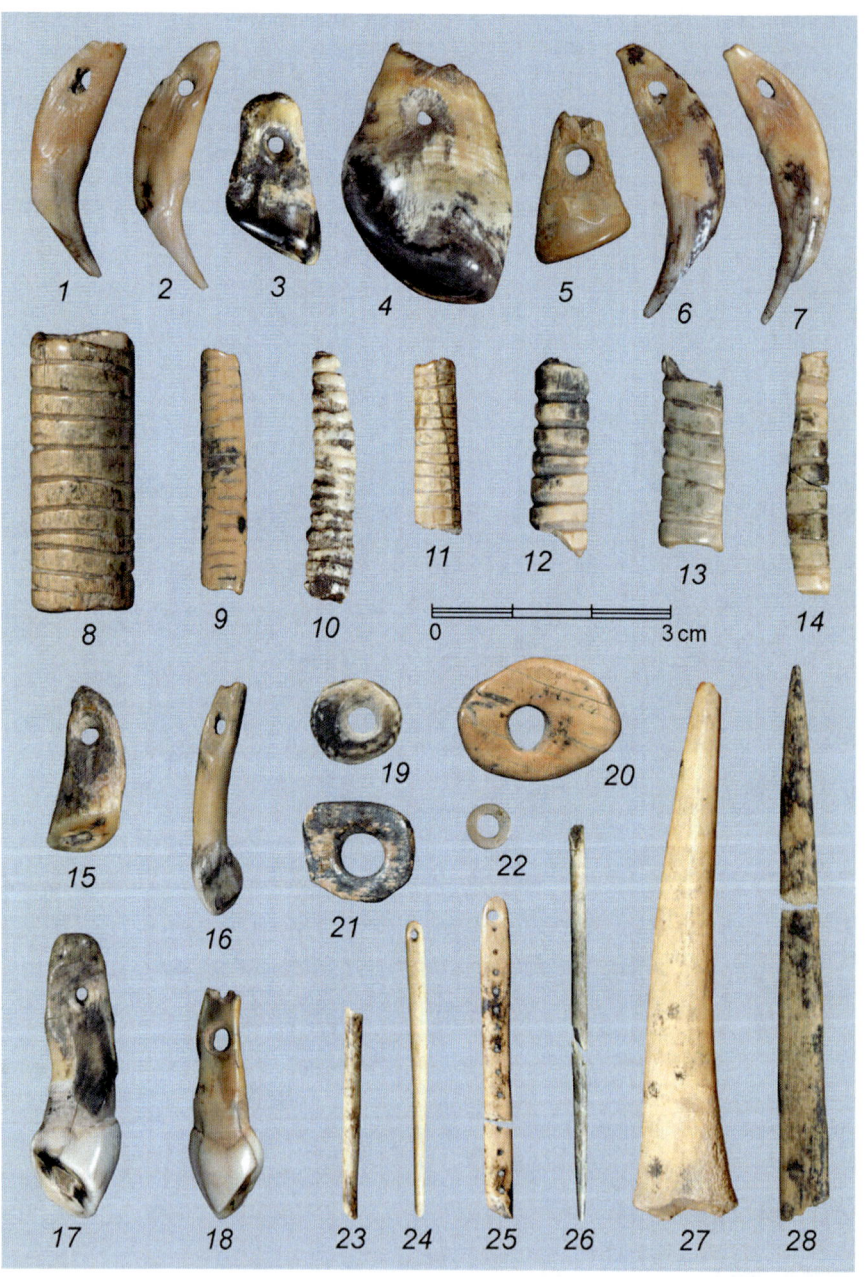

Color Plate 8. Early Upper Paleolithic ornaments and tools made of bone (*8–14, 20, 21, 23–28*), animal teeth (*1–7, 15–18*), mammoth ivory (*19*), and egg shell (*22*) from the Denisova Cave